CONTEMPORARY MATHEMATICS

704

Advances in Ultrametric Analysis

14th International Conference
p-adic Functional Analysis
June 30–July 4, 2016
Université d'Auverge, Aurillac, France

Alain Escassut
Cristina Perez-Garcia
Khodr Shamseddine
Editors

AMS AMERICAN MATHEMATICAL SOCIETY

2010 *Mathematics Subject Classification.* Primary 06F99, 11S82, 12J10, 12J25, 30D35, 37P05, 40A30, 46A35, 46S10, 47A53.

Library of Congress Cataloging-in-Publication Data

Names: International Conference on p-adic Functional Analysis (14th : 2016 : Aurillac, France) | Escassut, Alain, editor. | Perez-Garcia, C., 1956– editor. | Shamseddine, Khodr, 1966– editor.

Title: Advances in ultrametric analysis : 14th International Conference on p-adic Functional Analysis, June 30–July 4, 2016, Université d'Auvergne, Aurillac, France / Alain Escassut, Cristina Perez-Garcia, Khodr Shamseddine editors.

Description: Providence, Rhode Island : American Mathematical Society, [2018] | Series: Contemporary mathematics ; volume 704 | Includes bibliographical references.

Identifiers: LCCN 2017042943 | ISBN 9781470434915 (alk. paper)

Subjects: LCSH: Functional analysis–Congresses. | p-adic analysis–Congresses. | AMS: Order, lattices, ordered algebraic structures–Ordered structures–None of the above, but in this section. msc | Number theory–Algebraic number theory: local and p-adic fields–Non-Archimedean dynamical systems. msc | Field theory and polynomials–Topological fields–Valued fields. msc | Field theory and polynomials–Topological fields–Non-Archimedean valued fields. msc | Functions of a complex variable–Entire and meromorphic functions, and related topics–Distribution of values, Nevanlinna theory. msc | Dynamical systems and ergodic theory–Arithmetic and non-Archimedean dynamical systems–Polynomial and rational maps. msc | Sequences, series, summability–Convergence and divergence of infinite limiting processes–Convergence and divergence of series and sequences of functions. msc | Functional analysis–Topological linear spaces and related structures–Summability and bases. msc | Functional analysis–Other (nonclassical) types of functional analysis– Functional analysis over fields other than **R** or **C**. msc | Operator theory–General theory of linear operators–(Semi-) Fredholm operators; index theories. msc

Classification: LCC QA319 .I58 2018 | DDC 515/.7|dc23 LC record available at https://lccn.loc.gov/2017042943

Contemporary Mathematics ISSN: 0271-4132 (print); ISSN: 1098-3627 (online)

DOI: `http://dx.doi.org/10.1090/conm/704`

∞ The paper used in this book is acid-free and falls within the guidelines
established to ensure permanence and durability.
Visit the AMS home page at `http://www.ams.org/`

10 9 8 7 6 5 4 3 2 1 23 22 21 20 19 18

Contents

Preface

This volume contains papers based on lectures given at the 14th International Conference on p-adic Functional Analysis which was held in Aurillac (France), at the Institut Universitaire de Technologie de Clermont-Ferrand (Université d'Auvergne) from June 30th to July 4th, 2016. The conference followed a tradition of thirteen earlier meetings which took place at Laredo (Spain, 1990), Santiago (Chile, 1992), Clermont-Ferrand (France, 1994), Nijmegen (The Netherlands, 1996), Poznan (Poland, 1998), Ioanina (Greece, 2000), Nijmegen (The Netherlands, 2002), Clermont-Ferrand (France, 2004), Concepción (Chile, 2006), East-Lansing (Michigan, USA, 2008), Clermont-Ferrand (France, 2010), Winnipeg (Manitoba, Canada, 2012), and Paderborn (Germany, 2014).

The 15 papers included in this book show recent developments in various areas of non-Archimedean analysis such as non-Archimedean Banach spaces, functions on Levi-Civita fields, spectral properties of ultrametric Banach rings and Banach algebras of differential functions, ultrametric Calkin algebras, new applications of the p-adic Nevanlinna theory, problems linked to dynamical systems on $\mathbb{Q}_p$, representation of p-adic functions with sub-coordinates, and techniques of summation for p-adic series. A survey paper recalls how ultrametric analysis appeared in the mathematical world and became an important domain of research; and it gives a nice classification of non-Archimedean fields.

The three editors are very grateful to the participants in the conference and to other colleagues who helped in the refereeing process, particularly José Aguayo, Jesus Araujo, Bertin Diarra, Helge Glöckner, Anatoly Kochubei, Albert Kubzdela, Herminia Ochsenius, and Elena Olivos. We are also very grateful to Nicolas Mainetti, director of the Institut Universitaire de Technologie de Clermont-Ferrand and to the personnel of that institute (including students) for organizing the conference in Aurillac, a nice old city in Southern Auvergne.

Finally, we thank all the authors and referees as well as the American Mathematical Society, particularly Chistine Thivierge, for their role in producing this book of the proceedings of the conference, which is an important contribution to the knowledge of the latest results in ultrametric analysis.

The conference was dedicated to the memory of Lucien van Hamme, a professor at the Vrije Universiteit Brussel who worked in p-adic analysis until he retired in 2001. With his colleague Nicole de Grande de Kimpe he organized a conference on p-adic analysis at Hengelhoef in 1986. From 1990 until 2002, he took part in all conferences of the new series begun at Laredo. He passed away on October 27, 2014. He was the thesis advisor of several students, particularly Ann Verdoodt and Stany Desmedt. With Nicole de Grande de Kimpe, he did great work in p-adic analysis in Brussels. Our 14th Conference gave us the opportunity to honor his memory.

The next conference in the series is scheduled to be held in Poznan (Poland) in July, 2018.

Alain Escassut, Christina Perez-Garcia, Khodr Shamseddine

Contemporary Mathematics
Volume **704**, 2018
http://dx.doi.org/10.1090/conm/704/14158

Summary on non-Archimedean valued fields

Angel Barría Comicheo and Khodr Shamseddine

ABSTRACT. This article summarizes the main properties of ultrametric spaces, valued fields, ordered fields and fields with valuations of higher rank, highlighting their similarities and differences. The most used non-Archimedean valued fields are reviewed, like a completion in the case of the p-adic numbers fields and the Levi-Civita fields, or like an algebraic closure as is sometimes the case of the Puiseux series fields. Also a study of spherically complete valued fields is presented and their characterization as maximally complete fields is given where the Hahn fields and the Mal'cev-Neumann fields (their p-adic analogues) play an important role as "spherical completions". Finally several of the metric, topological, algebraic and order properties of the most used non-Archimedean valued fields are collected in a catalog where we can appreciate the analogy between fields that have the same characteristic as their residue class fields and fields that do not satisfy this property.

Introduction.

When a graduate student or any mathematician in general begin their journey in non-Archimedean Analysis, the freedom of choice for the base valued field presents different cases to consider, most of which do not occur when we work with the fields of real or complex numbers. The existence of discrete valuations, or the possible non-existence of vectors of norm 1 sometimes lead to the use of new procedures for proofs of results in Functional Analysis. The need and type of these procedures depend greatly on the properties of the base valued field. This article is prepared as a summary and will serve as a tool and source of references for the mathematician who wants to get familiar with the most commonly used non-Archimedean valued fields and/or needs to consider different valued fields in order to understand the potential extension of results that are valid for a particular valued field. Also this article helps one realize that any non-Archimedean valued field can be embedded in a Hahn field or a p-adic Mal'cev-Neumann field, thus the structure of a non-Archimedean valued field is not as diverse as one initially may think. The unique and remarkable features of this article are the catalog of valued fields (section 8) that collects all the fields and properties discussed in this paper, and the classification of valued fields diagram (section 9).

2010 *Mathematics Subject Classification.* Primary 12J10, 12J20, 12J25.
This research was supported by a doctoral scholarship for the first author from the Advanced Human Capital Program of the Chilean National Commission for Scientific and Technological Research (CONICYT-PFCHA/Doctorado en el Extranjero/72130551).

1. Ultrametric Spaces.

A valued field is a mathematical entity with a topological and an algebraic structures that will be defined and studied in section 2. In this section we will introduce the notion of an ultrametric space which is a metric space used to study the metric and topological properties of a non-Archimedean valued field without worrying about its algebraic structure.

1.1. The Strong Triangle Inequality and its Consequences.

DEFINITION 1.1. A ***metric*** on a set X is a function $d : X \times X \to \mathbb{R}$ satisfying the following properties for all $x, y, z \in X$:

(1) $d(x, y) \geq 0$, and $d(x, y) = 0$ if and only if $x = y$,
(2) $d(x, y) = d(y, x)$,
(3) $d(x, y) \leq d(x, z) + d(z, y)$ (triangle inequality).

The pair (X, d) is called a ***metric space***.

NOTATION 1.2. Let (X, d) be a metric space and let $a \in X$ and $r > 0$. The sets $B(a, r) := \{x \in X : d(x, a) < r\}$ and $B[a, r] := \{x \in X : d(x, a) \leq r\}$ are called the ***open and closed balls of center a and radius r***, respectively. The family of open balls forms a base of neighbourhoods for a uniquely determined Hausdorff topology on X. This topology is called the ***topology induced by d on X***. With respect to this topology the open balls are open sets and the closed balls are closed sets in X.

The ***diameter*** of a non-empty set $Y \subset X$ is $diam(Y) := \sup\{d(x, y) : x, y \in Y\}$ and the ***distance*** between two non-empty sets $Y, Z \subset X$ is $dist(Y, Z) := \inf\{d(y, z) : y \in Y, z \in Z\}$.

The set of values of a metric $d : X \times X \to \mathbb{R}$ is denoted and defined by $d(X \times X) := \{d(x, y) : x, y \in X\}$.

DEFINITION 1.3. A metric $d : X \times X \to \mathbb{R}$ is called an ***ultrametric*** when it satisfies the so-called ***strong triangle inequality*** $d(x, y) \leq \max\{d(x, z), d(z, y)\}$ for all $x, y, z \in X$. In that case, the pair (X, d) is called an ***ultrametric space***.

EXAMPLES 1.4. (1) Let X be a set and $d : X \times X \to \mathbb{R}$ be the discrete metric, i.e. $d(x, y) = 1$ if $x \neq y$ and $d(x, x) = 0$ for all $x, y \in X$. Then (X, d) is an ultrametric space. In this case, for $x \in X$, we have that $B[x, 1] = B[x, r]$ for all $r \geq 1$. Also, for each y in $B(x, r)$, we have that $B(x, r) = B(y, r)$. As we will see, it is not rare to find a ball of an ultrametric space with infinitely many radii, and where each point of a ball is the center of such ball.

(2) Let p be a fixed prime. The p-adic metric on $\mathbb{Z}$ is defined by $d(n, m) = 0$ if $n = m$, and for $n \neq m$, $d(n, m) = p^{-r}$ where r is the largest non-negative integer such that p^r divides $m - n$. The pair $(\mathbb{Z}, d)$ is an ultrametric space.

(3) Let $\mathbb{R}[x]$ be the ring of all polynomials with real coefficients. For each nonzero polynomial $p(x) = a_0 + a_1 x + \cdots + a_n x^n$ in $\mathbb{R}[x]$, put $\lambda(p) = \min\{i : a_i \neq 0\}$. Thus the map $d : \mathbb{R}[x] \times \mathbb{R}[x] \to \mathbb{R}$ defined by

$$d(p, q) := \begin{cases} e^{-\lambda(p-q)} & , \text{ if } p \neq q \\ 0 & , \text{ if } p = q \end{cases}$$

is an ultrametric on $\mathbb{R}[x]$.

(4) Let $\mathbb{N}$ be the set of positive integers and $d : \mathbb{N}\times\mathbb{N} \to \mathbb{R}$ be the map defined by

$$d(m,n) := \begin{cases} \max\{1+\frac{1}{m}, 1+\frac{1}{n}\} & , \text{ if } m \neq n \\ 0 & , \text{ if } m = n. \end{cases}$$

Then $(\mathbb{N}, d)$ is an ultrametric space.

(5) Any subset of a non-Archimedean valued field $(K, |\ |)$ with the map $(x, y) \mapsto |x-y|$ constitutes an ultrametric space (non-Archimedean valued fields will be introduced in the next section). Notice that with this example we have listed all ultrametric spaces, since W. Schikhof proved in [**39**] that any ultrametric space can isometrically be embedded into a non-Archimedean valued field.

Let (X, d) be a metric space. The following condition is called the ***Isosceles triangle principle***: If $d(x, z) \neq d(z, y)$ then $d(x, y) = \max\{d(x, z), d(z, y)\}$, i.e. every triangle with vertices in X is isosceles.

THEOREM 1.5 ([**32**, p. 3], [**49**, 2.A]). *Let (X, d) be a metric space. The metric d is an ultrametric if and only if it satisfies the Isosceles triangle principle.*

The following theorem collects the most remarkable results of an ultrametric space, all direct consequences of the strong triangle inequality.

THEOREM 1.6. *Let (X, d) be an ultrametric space. Then the following properties are satisfied:*

(1) *Each point of a ball is a center of the ball.*
(2) *Each ball in X is both closed and open ("**clopen**") in the topology induced by the ultrametric.*
(3) *Each ball has an empty boundary.*
(4) *Two balls are either disjoint, or one is contained in the other.*
(5) *Let $a \in Y \subset X$. Then $diam(Y) = \sup\{d(x, a) : x \in Y\}$.*
(6) *The radii of a ball B form the set $\{r \in \mathbb{R} : r_1 \leq r \leq r_2\}$, where $r_1 = diam(B)$, $r_2 = dist(B, X\backslash B)$ ($r_2 = \infty$ if $B = X$). It may happen that $r_1 < r_2$, so that a ball may have infinitely many radii.*
(7) *If two balls B_1, B_2 are disjoint, then $dist(B_1, B_2) = d(x, y)$ for all $x \in B_1, y \in B_2$.*
(8) *Let $U \neq \emptyset$ be an open subset of X. Given a sequence $(r_n)_n$ in $(0, \infty)$, strictly decreasing and convergent to 0, there exists a partition of U formed by balls of the form $B[a, r_n]$, with $a \in U$ and $n \in \mathbb{N}$.*
(9) *Let $\varepsilon \in \mathbb{R}^+$. For $x, y \in X$ the relation $d(x, y) < \varepsilon$ is an equivalence relation and induces a partition of X into open balls of radius ε. Analogously for $d(x, y) \leq \varepsilon$ and closed balls.*
(10) *Let $Y \subset X$, B a ball in X, $B \cap Y \neq \emptyset$. Then, $B \cap Y$ is a ball in Y.*
(11) *Let $(x_n)_n$ be a sequence in X converging to $x \in X$, then for each $a \in X \setminus \{x\}$, there exists $N \in \mathbb{N}$ such that $d(x_n, a) = d(x, a)$ for all $n \geq N$.*
(12) *There are not new values of an ultrametric after completion, i.e. if $(X^\wedge, d^\wedge)$ is the completion of (X, d), then $d(X \times X) = d^\wedge(X^\wedge \times X^\wedge)$.*
(13) *A sequence $(x_n)_n$ on X is Cauchy if and only if $\lim_{n\to\infty} d(x_n, x_{n+1}) = 0$.*

PROOF. The property (8) can be found in [**41**, Theorem 18.6], while (5) in [**40**, 1.D]. The property (3) follows directly from (2) and the remaining proofs can be found in [**32**, pp. 3–4]. □

1.2. Compactness and Separability.

THEOREM 1.7 ([**41**, 19.2], [**40**, 1.E]). *Let (X,d) be a compact ultrametric space. Then the following statements are satisfied:*

(1) *$d(X\times X)$ is countable, it has a maximum, and it has 0 as the only possible accumulation point. Thus if $d(X\times X)$ is infinite, then $d(X\times X)=\{0\}\cup\{r_n : n\in\mathbb{N}\}$, where $(r_n)_n$ is a strictly decreasing sequence of $(0,\infty)$ such that $\lim_{n\to\infty} r_n = 0$.*
(2) *X is complete and separable.*
(3) *Every open covering of X has a finite refinement formed by disjoint balls.*
(4) *Every partition of X by clopen sets is finite.*
(5) *If $Y\subset X$ is clopen, $\emptyset\neq Y\neq X$, then $dist(Y, X\setminus Y)$ is positive.*

THEOREM 1.8 ([**32**, 1.1.3]). *If X is a locally compact ultrametric space then there exists a partition of X consisting of compact balls.*

THEOREM 1.9 ([**41**, 19.3]). *An ultrametric space is separable if and only if the set of all its balls is countable.*

1.3. Ultrametrizability.

DEFINITION 1.10. A topology on a set X is called ***ultrametrizable*** if it is the topology induced by an ultrametric. A topological space is called ultrametrizable if its topology is ultrametrizable.

If X is a topological space, then a subset of X is a ***clopen*** set if it is closed and open simultaneously.

A topological space is ***zero-dimensional*** if it has a base of clopen sets.

From 1.6 (2) we can deduce the following:

THEOREM 1.11. *Every ultrametrizable space is zero-dimensional, and hence totally disconnected.*

The converse of the previous theorem is false, since P. Roy [**38**] has constructed a zero-dimensional complete metric space that is not ultrametrizable. But we have a partial converse, which is a generalization of 1.9:

THEOREM 1.12 ([**12**, 2]). *A space is zero-dimensional, Hausdorff and has a countable base if and only if it is ultrametrizable and separable.*

The following result gathers different characterizations of ultrametrizable spaces.

THEOREM 1.13. *Let X be a topological space. The following statements are equivalent:*

(1) *X is ultrametrizable.*
(2) *X is homeomorphic to a subspace of a countable product of discrete spaces.*
(3) *X is homeomorphic to a subspace of a non-Archimedean valued field.*
(4) *X is metrizable and has large inductive dimension 0 (for every two closed, mutually disjoint subsets A and B of X, there exists a clopen $U\subset X$ with $A\subset U$ and $B\subset X\setminus U$. In the literature this is denoted by $\operatorname{Ind} X = 0$).*
(5) *X is metrizable and has covering dimension 0 (every finite open covering of X has a refinement that is a partition formed by clopen sets. In the literature this is denoted by $\dim X = 0$).*

(6) *X is metrizable and its topology is non-Archimedean (it is Hausdorff and it has a base $\mathcal{B}$ such that if $A, B \in \mathcal{B}$ then either $A \subset B$ or $B \subset A$ or $A \cap B = \emptyset$).*

PROOF. The equivalence (1)⇔(2) is proved in [**49**, 2.1] and the equivalence (1)⇔(3) is proved in [**39**], while the equivalence (1)⇔(4) is proved in [**12**, Theorem II]. The equivalence (4)⇔(5) is due to the equivalence $\operatorname{Ind} X = 0 \Leftrightarrow \dim X = 0$ (see [**31**, 2.3]). The implication (1)⇒(6) follows from 1.6(d) and the implication (6)⇒(4) follows from [**30**, p. 121] □

THEOREM 1.14 ([**49**, 2.1]). *A countable product of ultrametrizable spaces is ultrametrizable.*

1.4. Spherical Completeness. Recall that a metric space is said to be Cauchy complete if every Cauchy sequence is convergent, or equivalently, if each nested sequence of closed balls whose radius approaches to 0, has a non-empty intersection. This motivates the following:

DEFINITION 1.15. An ultrametric space is called ***spherically complete*** if each nested sequence of balls has a non-empty intersection.

REMARK 1.16. The concept of spherical completeness plays a key role as a necessary and sufficient condition for the validity of the Hahn-Banach theorem in the non-Archimedean context (see [**49**, 4.10, 4.15]). At the end of this subsection we will discuss other important features of spherically complete spaces: a fixed point theorem and best approximations.

THEOREM 1.17 ([**49**, 2.3]). *Let (X, d) be an ultrametric space. The following statements are equivalent:*

(1) *(X, d) is spherically complete.*
(2) *For any collection $(B_i)_{i\in I}$ of balls in X such that $B_i \cap B_j \neq \emptyset$ for any $i, j \in I$, we then have $\bigcap_{i\in I} B_i \neq \emptyset$.*
(3) *Every sequence of balls $B[a_1, \varepsilon_1] \supset B[a_2, \varepsilon_2] \supset \cdots$ for which $\varepsilon_1 > \varepsilon_2 > \cdots$ has nonempty intersection.*

It is clear that a spherically complete ultrametric space is Cauchy complete, but the converse is not always true. For instance, the space of the example 1.4(4) is a complete ultrametric space that is not spherically complete. However there is a partial converse:

LEMMA 1.18 ([**43**, Lemma 1.7]). *Suppose that (X, d) is a Cauchy complete ultrametric space. If 0 is the only accumulation point of the set $d(X \times X)$, then X is spherically complete.*

LEMMA 1.19. *If an ultrametric space is compact then it is spherically complete.*

PROOF. It is enough to recall that if X is compact, then every collection of closed subsets of X, with the finite intersection property, has non-empty intersection. □

The following three results show methods to construct or identify ultrametric spaces that are not spherically complete. For that we need the following:

DEFINITION 1.20. A metric on X is ***dense*** if for each ball B in X the set $d(B \times B)$ is dense in the interval $[0, diam(B)]$.

THEOREM 1.21 ([**41**, 20.4]). *If (X, d) is a complete ultrametric space and d is a dense metric, then there exists a subspace of X that is complete but not spherically complete.*

THEOREM 1.22 ([**41**, 20.5]). *If (X, d) is a separable ultrametric space and d is a dense metric, then X is not spherically complete.*

THEOREM 1.23 ([**43**, p. 5]). *Let (X, d) be a complete ultrametric space and let $B_1 \supset B_2 \supset \cdots$ be a decreasing sequence of balls in X such that $diam(B_1) > diam(B_2) > \cdots$ and $\inf\{diam(B_n) : n \in \mathbb{N}\} > 0$. Then the subspace $X \setminus (\bigcap_n B_n)$ is complete but not spherically complete.*

The concept of spherical completeness is geometrical rather than topological.

THEOREM 1.24 ([**49**, 2.F]). *Let (X, d) be a complete ultrametric space. Then the formula*

$$\sigma(x, y) := \inf\{2^{-n} : n \in \mathbb{Z}, d(x, y) \leq 2^{-n}\}$$

defines an ultrametric σ such that $d \leq \sigma \leq 2d$, and (X, σ) is spherically complete.

A fixed point theorem. One of the attributes of spherically complete ultrametric spaces is that they satisfy a stronger version of the fixed point theorem for complete metric spaces.

DEFINITION 1.25. Let (X, d) be a metric space. A function $f : X \to X$ is called a ***shrinking map*** when $d(f(x), f(y)) < d(x, y)$ for all $x, y \in X$, $x \neq y$. If there exists $k \in (0, 1)$ such that $d(f(x), f(y)) < kd(x, y)$ for all $x, y \in X$, $x \neq y$, then f is called a ***contraction***.

The fixed point theorem for complete metric spaces is the following:

THEOREM 1.26 ([**48**, 3.7.4]). *Every contraction of a complete metric space has a unique fixed point.*

Notice that the map $f : [1, \infty) \to [1, \infty)$, $f(x) = x + \dfrac{1}{x}$ is a shrinking map defined on a complete space but it is not a contraction and has no fixed point. In fact, $x = f(x)$ does not have a solution. Thus by 1.26, f cannot be a contraction. However:

$$|f(x) - f(y)| = |x - y|\left(1 - \frac{1}{xy}\right) < |x - y|.$$

For spherically complete ultrametric spaces we have the following:

THEOREM 1.27 ([**35**, 2.3]). *Every shrinking map of a spherically complete ultrametric space has a unique fixed point.*

Best Approximations.

DEFINITION 1.28. Let Y be a subset of an ultrametric space (X, d). Let $a \in X$ and $b \in Y$. Then b is a ***best approximation*** of a in Y if $d(a, b) = dist(a, Y)$.

Another attribute of a spherically complete ultrametric space is the existence of best approximations as it is stated in the following result.

THEOREM 1.29 ([**41**, 21.2]). *Let $Y \neq \emptyset$ be a spherically complete ultrametric space embedded in an ultrametric space X. Then each $x \in X$ has a best approximation in Y, i.e.* $\min\{d(y, x) : y \in Y\}$ *exists.*

In general, best approximations are not unique.

THEOREM 1.30 ([**41**, 21.1]). *Let $Y \neq \emptyset$ be a subset of an ultrametric space X. Suppose that Y has no isolated points. If an element $a \in X \setminus Y$ has a best approximation in Y then it has infinitely many.*

2. Valued fields.

2.1. Archimedean and non-Archimedean valuations.

DEFINITION 2.1. Let K be a field. A ***valuation*** on K is a map $|\ |: K \to \mathbb{R}$ satisfying the following axioms for all $x, y \in K$:

(1) $|x| \geq 0$, and $|x| = 0$ if and only if $x = 0$,
(2) $|xy| = |x||y|$,
(3) $|x + y| \leq |x| + |y|$.

The pair $(K, |\ |)$ is called a valued field.

It is not hard to see that $|1_K| = 1$, $|-x| = |x|$ and $|x^{-1}| = |x|^{-1}$ for $x \neq 0$. In the rest of the article we will denote the set $K \setminus \{0\}$ by K^*.

DEFINITION 2.2. A valuation $|\ |$ on K is called ***non-Archimedean*** if it satisfies the ***strong triangle inequality***: $|x + y| \leq \max\{|x|, |y|\}$ for all $x, y \in K$. Otherwise it is called ***Archimedean***.

THEOREM 2.3 ([**49**, 1.1], [**41**, lemma 8.2]). *Let $(K, |\ |)$ be a valued field. The following conditions are equivalent:*

(1) *$|\ |$ is non-Archimedean.*
(2) *If $a, b \in K$ and $|a| < |b|$, then $|b - a| = |b|$ (Isosceles triangle principle).*
(3) *The set $\{|n1_K| : n \in \mathbb{N}\}$ is bounded.*
(4) *$|n1_K| \leq 1$ for every $n \in \mathbb{N}$.*
(5) *$|2 \cdot 1_K| \leq 1$.*

EXAMPLES 2.4. (1) The usual absolute values in $\mathbb{R}$ and $\mathbb{C}$ are valuations, and since the set $\{|n| : n \in \mathbb{N}\}$ is unbounded, they are Archimedean.

(2) Let K be a field. The map defined by $|x| = 1$ for $x \neq 0$ and $|0| = 0$ is called ***trivial valuation*** and is a non-Archimedean valuation.
(3) If K is a finite field, then the trivial valuation is the only valuation on K, for if there exists $x \in K^*$, $|x| \neq 1$, then the set $\{|x^n| : n \in \mathbb{Z}\}$ is infinite.
(4) If $char(K) \neq 0$ then any valuation on K is non-Archimedean. Indeed, in this case the prime subfield of K (the subfield of K generated by 1_K) is finite. Thus the set $\{|n1_k| : n \in \mathbb{N}\}$ is bounded.
(5) Let p be a prime number. The ***p-adic valuation*** $|\ |_p$ on $\mathbb{Q}$ is defined by
$$|0|_p = 0 \quad \text{and} \quad \left|p^v \frac{m}{n}\right| = \frac{1}{e^v}$$
where e is the base of the natural logarithms, $v \in \mathbb{Z}$, and $n, m \in \mathbb{Z} \setminus \{0\}$ are not divisible by p. Since the set $\{|n1|_p : n \in \mathbb{Z}\} = \{e^{-v} : v \in \mathbb{N}\} \cup \{0\}$ is bounded in $\mathbb{R}$, the p-adic valuation on $\mathbb{Q}$ is non-Archimedean.
(6) Let p be an irreducible polynomial in $K[x]$, where K is any field. The p-adic valuation $|\ |_p$ on the rational function field $K(x)$ (the quotient field of the ring of polynomials $K[x]$) is defined by $|0|_p = 0$ and
$$\left|p^v \frac{f}{g}\right| = \frac{1}{e^v}$$

where $v \in \mathbb{Z}$, and $f, g \in K[x] \setminus \{0\}$ are not divisible by p. This valuation is non-Archimedean since $\{|n1|_p : n \in \mathbb{Z}\} = \{0, 1\}$. For a description of all valuations on $K(x)$ see [**36**, 3.1.K] and [**2**].

A valuation $|\ |$ on a field K defines the metric $d(x, y) := |x - y|$ for $x, y \in K$. In particular, any valuation on K induces a metrizable topology on K. If the valuation $|\ |$ is non-Archimedean, then the induced metric d is an ultrametric.

DEFINITION 2.5. If two valuations induce the same topology on K, then they are called ***dependent***. If this is not the case then they are called ***independent***.

THEOREM 2.6 ([**14**, 1.1.2]). *Let $|\ |_1$ and $|\ |_2$ be two non-trivial valuations on a field K. The following conditions are equivalent:*

(1) *$|\ |_1$ and $|\ |_2$ are dependent.*
(2) *For all $x \in K$, $|x|_1 < 1$ implies $|x|_2 < 1$.*
(3) *There exists $\lambda > 0$, such that for all $x \in K$, $|x|_1 = (|x|_2)^\lambda$.*

NOTATION 2.7. The symbol $|\ |_0$ will denote the usual absolute value on any subfield of the field of the complex numbers $\mathbb{C}$.

THEOREM 2.8. *Let $(K, |\ |)$ be a valued field and let $|\ |_1 := (|\ |)^\lambda$ for $\lambda \in \mathbb{R}^+$. The following statements are satisfied:*

(1) *if $|\ |$ is non-Archimedean, then $|\ |_1$ and $|\ |$ are dependent valuations.*
(2) *if $|\ |$ is Archimedean, then $R := \{\rho \in \mathbb{R}^+ : |\ |^\rho$ is a valuation$\} = (0, r]$ for some $r \geq 1$. Moreover, if $\lambda \in R$, then $|\ |_1$ and $|\ |$ are dependent.*

PROOF. (1): the verification of the fact that $|\ |_1$ is a non-Archimedean valuation is straight forward. Since $|x| < 1 \Leftrightarrow |x|_1 < 1$, by 2.5 it follows that $|\ |_1$ and $|\ |$ are dependent valuations on K. (2): the proof of this statement can be found in [**36**, 1.2.D and 1.2.E]. □

The next result is an approximation theorem for pairwise-independent valuations on a field.

THEOREM 2.9 ([**14**, 1.1.3]). *Let K be a field and $|\ |_1, |\ |_2, \ldots, |\ |_n$ non-trivial pairwise-independent valuations on K. If $x_1, x_2, \ldots, x_n \in K$ and $\varepsilon \in \mathbb{R}^+$, then there exists $x \in K$ such that $|x - x_i|_i < \varepsilon$, for all $i \in \{1, \ldots, n\}$.*

The following theorem implies that in the examples 1.4 we have listed all the possible valuations on $\mathbb{Q}$ up to dependency.

THEOREM 2.10 ([**49**, 1.2](Ostrowski's Theorem)). *Let $|\ |$ be a non-trivial valuation on $\mathbb{Q}$. If $|\ |$ is Archimedean, then there exists $\lambda \in (0, 1]$, such that, $|\ | = (|\ |_0)^\lambda$. Otherwise, there are a prime p and $\lambda > 0$ such that $|\ | = (|\ |_p)^\lambda$.*

2.2. Completion of valued fields.

THEOREM 2.11 ([**14**, 1.1.4]). *Let $(K, |\ |)$ be a valued field. There exists a Cauchy complete valued field $(\widehat{K}, \widehat{|\ |})$, and an embedding $i : K \to \widehat{K}$, such that $|x| = \widehat{|i(x)|}$ for all $x \in K$, and the image $i(K)$ is dense in $\widehat{K}$. If $(\widehat{K}', \widehat{|\ |}', i')$ is another such trio, then there exists a unique isomorphism $\varphi : \widehat{K} \to \widehat{K}'$ satisfying $\widehat{|\varphi(x)|}' = \widehat{|x|}$ for all $x \in \widehat{K}$ and making the following diagram commutative:*

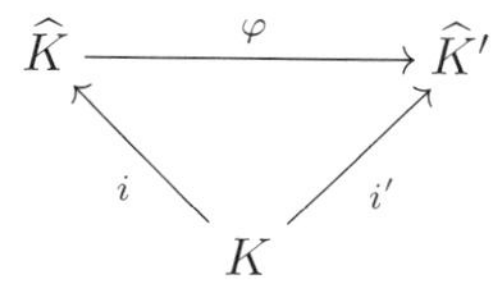

DEFINITION 2.12. A pair $(\widehat{K}, \widehat{|\ \ |})$ as in 2.11 is called a ***completion*** of the valued field $(K, |\ \ |)$.

REMARK 2.13. Let $(K, |\ \ |)$ be a valued field and $(\widehat{K}, \widehat{|\ \ |})$ its completion with embedding $i : K \to \widehat{K}$ such that $|x| = \widehat{|i(x)|}$ for all $x \in K$. Then

$$\{|n1_K| : n \in \mathbb{N}\} = \{\widehat{|i(n1_K)|} : n \in \mathbb{N}\} = \{\widehat{|n1_{\widehat{K}}|} : n \in \mathbb{N}\}.$$

Therefore the completion of an Archimedean valued field is an Archimedean valued field and the completion of a non-Archimedean valued field is a non-Archimedean valued field. In the latter case, we have that $|K| := \{|x| : x \in K\} = \{\widehat{|x|} : x \in \widehat{K}\} =: \widehat{|\widehat{K}|}$ by 1.6(12).

2.3. Completion of Archimedean valued fields.

THEOREM 2.14 ([**47**, 15.2.1](Gelfand-Tornheim's Theorem)). *Let $(K, \|\ \ \|)$ be a normed algebra over $(\mathbb{R}, (|\ \ |_0)^\lambda)$ for some $\lambda \in (0,1]$. If K is a field, then either $K = \mathbb{R}\cdot 1_K$ or $K = (\mathbb{R}\cdot 1_K)(i)$, where, for the second possibility, $i \in K$ and $i^2 = -1$.*

As a consequence of the Gelfand-Tornheim's Theorem, we obtain a full description of all the Archimedean valued fields and their completions.

THEOREM 2.15 ([**47**, 15.2.2]). *If $(K, |\ \ |)$ is an Archimedean valued field, then there exist $\lambda \in (0,1]$ and a field monomorphism $\sigma : K \to \mathbb{C}$, such that $|x| = (|\sigma(x)|_0)^\lambda$. Furthermore, if $(K, |\ \ |)$ is Cauchy complete, then either $\sigma(K) = \mathbb{R}$ or $\sigma(K) = \mathbb{C}$.*

2.4. Completion of non-Archimedean valued fields. In this subsection we will give a brief construction of our first non-trivial examples of Cauchy complete, non-Archimedean valued fields. Notice that in a field K with the trivial valuation, every Cauchy sequence is eventually constant, and thus K is complete. Regarding the fields already presented, we anticipate that $\mathbb{Q}$ is not Cauchy complete with respect to the usual absolute value $|\ \ |_0$ nor with respect to any p-adic valuation since $\mathbb{Q}$ is not a Baire space (for details see 2.5). Similarly, the field of rational functions $K(x)$ is not Cauchy complete with respect to the x-adic valuation (this is not trivial and it is proved in 2.5).

As we will see, in contrast with the Archimedean case, in the non-Archimedean case there are different families of Cauchy complete non-Archimedean valued fields.

Let $(K, |\cdot|)$ be a non-Archimedean valued field. Then $K^* = \{\lambda \in K : \lambda \neq 0\}$ is a multiplicative group and so is the ***value group*** of $(K, |\ \ |)$ defined by $|K^*| = \{|\lambda| : \lambda \in K^*\}$. The value group can either be dense in $(0, \infty)$ or discrete, where, for the second possibility, if the valuation is not trivial, then the group $|K^*|$ is cyclic with generator $\rho := \max\{r \in |K^*| : r < 1\}$ ([**47**, B.5.2]). Any element $\pi \in K$ for which $|\pi| = \rho$ is called a ***uniformizer*** for $|\ \ |$. Furthermore, $B[0,1]$, the "closed" unit disk in K, is a ring and $B(0,1)$, the "open" unit disk in K, is a maximal ideal of $B[0,1]$ ([**49**, p. 4]).

DEFINITION 2.16. Let $(K, |\ |)$ be a non-Archimedean valued field. The quotient field $B[0,1]/B(0,1)$ is called the ***residue class field*** of K. Moreover, if $|K^*|$ is discrete in $(0,\infty)$, then the valuation $|\ |$ is said to be ***discrete*** and if $|K^*|$ is dense in $(0,\infty)$, then the valuation is said to be ***dense***.

EXAMPLES 2.17. Fields with discrete valuation.

(1) Consider $(\mathbb{Q}, |\cdot|_p)$ for a prime number p. In this case,
$$|\mathbb{Q}^*| = \{e^{-n} : n \in \mathbb{Z}\} = \langle e^{-1}\rangle = \langle |p|_p\rangle$$
and therefore p is a uniformizer for $|\ |_p$.
$$B[0,1] = \left\{\frac{m}{n} : m, n \in \mathbb{Z},\ n \text{ is not divisible by } p\right\}$$
and
$$B(0,1) = \left\{\frac{pm}{n} : m, n \in \mathbb{Z},\ n \text{ is not divisible by } p\right\}.$$
Notice that $B[0,1] = (\mathbb{Z}\setminus(p))^{-1}\mathbb{Z}$, i.e. it is the localization of the ring $\mathbb{Z}$ at the prime ideal $(p) = p\mathbb{Z}$, and $B(0,1) = pB[0,1]$ is the maximal ideal of $B[0,1]$. Using the first isomorphism theorem for rings, we conclude that $B[0,1]/B(0,1) \simeq \mathbb{Z}/p\mathbb{Z}$. In other words, the residue class field of $(\mathbb{Q}, |\cdot|_p)$ is isomorphic to $\mathbb{F}_p$. For a simpler but larger proof see [**36**, p. 62].

(2) Consider $(K(x), |\ |_p)$ for an irreducible polynomial $p \in K[x]$, where K is a field. In this case, p is a uniformizer for $|\ |_p$,
$$B[0,1] = \left\{\frac{f}{g} : f, g \in K[x],\ g \text{ is not divisible by } p\right\}$$
and
$$B(0,1) = \left\{\frac{pf}{g} : f, g \in K[x],\ g \text{ is not divisible by } p\right\}$$
Analogously to the previous example, $B[0,1]$ is the localization of the ring $K[x]$ at the prime ideal (p), and $B(0,1) = pB[0,1]$ is the maximal ideal of $B[0,1]$. Thus the residue class field of $(K(x), |\ |_p)$ is isomorphic to $K[x]/(p)$. In particular, if $p(x) = x$, then the residue class field of $(K(x), |\ |_p)$ is isomorphic to K. For another proof see [**36**, p. 88].

THEOREM 2.18 ([**14**, 1.3.4]). *Let $(K, |\ |)$ be a non-Archimedean valued field and $(\widehat{K}, \widehat{|\ |})$ its completion. If k and $\widehat{k}$ are their respective residue class fields, then $k \simeq \widehat{k}$ and $|K^*| = \widehat{|\widehat{K}^*|}$.*

The following result is going to bring us an explicit description of the completions of $(\mathbb{Q}, |\ |_p)$ and $(K(x), |\ |_p)$.

THEOREM 2.19 ([**14**, 1.3.5]). *Let $|\ |$ be a discrete valuation on the field K, with uniformizer π and residue class field k. Then every element $x \in K^*$ can be written uniquely as a convergent series*
$$x = r_v\pi^v + r_{v+1}\pi^{v+1} + r_{v+2}\pi^{v+2} + \cdots = \lim_{n\to\infty}\sum_{i=v}^{n} r_i\pi^i$$
where $v = \log_{|\pi|}|x|$, $r_v \neq 0$, and the coefficients r_i are taken from a set $R \subset B[0,1]$ of representatives of the residue classes in k (i.e., the canonical map $B[0,1] \to k$ induces a bijection of R onto k).

REMARK 2.20. If π is a uniformizer then $|K^*| = \{|\pi|^n : n \in \mathbb{Z}\}$. Thus, for every $x \in K^*$, there is $v \in \mathbb{Z}$ such that $|x| = |\pi|^v$. Hence $\log_{|\pi|} |x| \in \mathbb{Z}$.

DEFINITION 2.21. Let p be a prime number. The completion of the field $(\mathbb{Q}, |\ |_p)$ is called the ***field of p-adic numbers*** and is denoted by $\mathbb{Q}_p$.

We checked in 2.17 that p is a uniformizer for $(\mathbb{Q}, |\ |_p)$ and that $\mathbb{F}_p$ is its residue class field. By 2.18, $\mathbb{Q}_p$ has the same residue class field and p as a uniformizer. According to 2.19, by taking $R = \{0, 1, \ldots, p-1\}$, this field has the following description

$$\mathbb{Q}_p = \left\{ \sum_{i=v}^{\infty} r_i p^i : v \in \mathbb{Z}, r_i \in R,\ r_v \neq 0 \right\} \cup \{0\},$$

where the valuation on $\mathbb{Q}_p$ is defined as $|0| = 0$ and $\left| \sum_{i=v}^{\infty} r_i p^i \right| = e^{-v}$ when $r_v \neq 0$.

The closed disk in $\mathbb{Q}_p$ of center 0 and radius 1, is the ***ring of p-adic integers***:

$$B[0,1] = \mathbb{Z}_p := \left\{ \sum_{i=0}^{\infty} r_i p^i : r_i = 0, 1, \ldots, p-1 \right\}.$$

Be aware of the fact that addition of two 'series' of the form $\sum_{i=v}^{\infty} r_i p^i$ is not coefficientwise, as the set $\{0, 1, \ldots, p-1\}$ is not closed under addition. As a simple example observe that choosing $p = 7$,

$$5p^i + 4p^i = 5p^i + 2p^i + 2p^i = 7p^i + 2p^i = p^{i+1} + 2p^i.$$

DEFINITION 2.22. Let K be a field and $p = x \in K[x]$. The completion of $(K(x), |\ |_p)$ is called the ***field of formal Laurent series*** and denoted by $K((x))$.

In 2.17, we saw that x is a uniformizer for $(K(x), |\ |_x)$ and that K is the residue class field of this valued field. By 2.18, $K((x))$ has x as a uniformizer and K as the residue class field as well. As a consequence of the previous theorem, this field has the following description

$$K((x)) = \left\{ \sum_{i=v}^{\infty} r_i x^i : v \in \mathbb{Z}, r_i \in K,\ r_v \neq 0 \right\} \cup \{0\},$$

where the valuation on $K((x))$ is defined as $|0| = 0$ and $\left| \sum_{i=v}^{\infty} r_i x^i \right| = e^{-v}$ when $r_v \neq 0$. The closed disk in $K((x))$ of center 0 and radius 1, is the ***ring of formal power series***:

$$B[0,1] = K[[x]] := \left\{ \sum_{i=0}^{\infty} r_i x^i : r_i \in K \right\}.$$

Notice that $K((x))$ is the quotient field of $K[[x]]$ ([**36**, 3.1.L]). Here the addition of two such series is defined coefficientwise and the multiplication of two nonzero series is defined as follows:

$$\left(\sum_{j=v_1}^{\infty} s_j x^j \right) \cdot \left(\sum_{k=v_2}^{\infty} t_k x^k \right) = \sum_{i=v}^{\infty} r_i x^i,$$

with $v = v_1 + v_2$, $r_i = \sum_{j+k=i} s_j t_k = \sum_{j=v_1}^{\infty} s_j t_{i-j} = \sum_{k=v_2}^{\infty} s_{i-k} t_k$, where $s_j = 0$ for $j < v_1$ and $t_k = 0$ for $k < v_2$.

Now, the following result is a direct consequence of 1.18.

THEOREM 2.23. *The valued fields $\mathbb{Q}_p$ and $K((x))$ are spherically complete.*

The following theorem shows how different a Cauchy complete non-Archimedean valued field is from their Archimedean analogs $\mathbb{R}$ and $\mathbb{C}$.

THEOREM 2.24 ([**4**, II.1.1]). *Let $(K, |\ |)$ be a Cauchy complete non-Archimedean valued field. If $(x_n)_n$ is a sequence of elements of K, then*

$$\sum_{n=1}^{\infty} x_n \text{ is convergent in } K \Leftrightarrow \lim_{n\to\infty} x_n = 0.$$

2.5. Incompleteness of $\mathbb{Q}$ and $K(x)$. In the following we will discuss the incompleteness of $\mathbb{Q}$ and $K(x)$ with respect to their non-Archimedean valuations. Since $\mathbb{Q}_p$ is the completion of $(\mathbb{Q}, |\ |_p)$, we can identify the elements of $\mathbb{Q}$ as elements of $\mathbb{Q}_p$. For example,

$$\frac{1}{1-p} = \sum_{i=0}^{\infty} p^i \qquad \text{and} \qquad -1 = \sum_{i=0}^{\infty} (p-1)p^i.$$

The algorithm to compute the coefficients of the p-adic expansion of an arbitrary rational number can be found in [**37**] in the proof of the following result.

THEOREM 2.25 ([**37**, 5.3]). *Let $x = \sum_{i=v}^{\infty} r_i p^i \in \mathbb{Q}_p$ $(v \in \mathbb{Z},\ 0 \le r_i \le p-1)$. Then x is a rational number if and only if the sequence $(r_i)_i$ of digits of x is eventually periodic, i.e. there exists $n \in \mathbb{N}$ such that the subsequence $(r_i)_{i\ge n}$ is periodic.*

We can use 2.25 to prove the incompleteness of $(\mathbb{Q}, |\ |_p)$. Consider $x = \sum_{i=0}^{\infty} p^{i^2} \in \mathbb{Q}_p$. Note that x is the limit of a convergent sequence of rational numbers, say $a_n = \sum_{i=0}^{n} p^{i^2}$. In fact, $|x - a_n|_p = e^{-(n+1)}$ for every $n \in \mathbb{N}$. Thus $(a_n)_n$ is a Cauchy sequence in $(\mathbb{Q}, |\ |_p)$ but 2.25 implies that $x \notin \mathbb{Q}$, i.e. $(a_n)_n$ is not convergent in $(\mathbb{Q}, |\ |_p)$. Another argument to prove the incompleteness of $(\mathbb{Q}, |\ |_p)$ is the following: if $(\mathbb{Q}, |\ |_p)$ is Cauchy complete, then it must be isomorphic to its completion. But $\mathbb{Q}$ is countable and $\mathbb{Q}_p$ is not, leading to a contradiction. Among the proofs of the incompleteness of $(\mathbb{Q}, |\ |_p)$ that don't invoke $\mathbb{Q}_p$, we highlight the following. By the Baire category theorem ([**28**, 48.2]), the space $(\mathbb{Q}, |\ |_p)$ is not Cauchy complete. In fact, for every $q \in \mathbb{Q}$, the set $\mathbb{Q} \setminus \{q\}$ is open and dense in $(\mathbb{Q}, |\ |_p)$, since $(q + p^{N+n})_n$ is a sequence of elements of $\mathbb{Q} \setminus \{q\}$ that converges to q, with $N = -\ln(|q|_p)$. Nevertheless $\bigcap_{q\in\mathbb{Q}} (\mathbb{Q} \setminus \{q\}) = \emptyset$, therefore, the space $(\mathbb{Q}, |\ |_p)$ is not Baire, hence it is not complete.

Let K be any field. Since $K((x))$ is the completion of $(K(x), |\ |_x)$, we can identify the elements of $K(x)$ as elements of $K((x))$. For example,

$$\frac{1}{1-x} = \sum_{i=0}^{\infty} x^i \qquad \text{and} \qquad \frac{1}{1-x-x^2} = 1 + \sum_{i=1}^{\infty} F_i x^i,$$

where $F_1 = F_2 = 1$ and $F_n = F_{n-1} + F_{n-2}$ $(n > 1)$ are the Fibonacci numbers.

The algorithm to compute the coefficients of the x-adic expansion of an arbitrary rational function is analogous to the algorithm used to compute the coefficients of the p-adic expansion of rational numbers.

For the identification of rational functions in $K((x))$, the periodicity of the coefficients of the x-adic expansion is not adequate for such description and a weaker condition is needed: linear recurrence.

DEFINITION 2.26. Let K be a field. A sequence $(r_n)_n$ of elements of K is ***linearly recurrent*** if there exist constants $c_1, c_2, \ldots, c_k \in K$ such that

$$r_{n+k} = c_1 r_n + c_2 r_{n+1} + \cdots + c_k r_{n+k-1} \tag{2.1}$$

for all $n \in \mathbb{N}$.

REMARK 2.27. Linearly recurrent sequences are generalizations of periodic sequences. A periodic sequence with least period k is a linearly recurrent sequence $(r_n)_n$ with coefficients $c_1 = 1$ and $c_2 = c_3 = \cdots = c_k = 0$ satisfying the relation 2.1.

THEOREM 2.28 ([**23**, V.5. Lemma 5], [**36**, 3.1.N]). *Let $f(x) = \sum_{i=0}^{\infty} r_i x^i \in K[[x]]$, where K is any field. For $m, s \geq 0$, let $A_{s,m}$ be the matrix $\{r_{s+i+j}\}_{0 \leq i,j \leq m}$:*

$$\begin{pmatrix} r_s & r_{s+1} & r_{s+2} & \cdots & r_{s+m} \\ r_{s+1} & r_{s+2} & r_{s+3} & \cdots & r_{s+m+1} \\ r_{s+2} & r_{s+3} & r_{s+4} & \cdots & r_{s+m+2} \\ \vdots & \vdots & \vdots & & \vdots \\ r_{s+m} & r_{s+m+1} & r_{s+m+2} & \cdots & r_{s+2m} \end{pmatrix}$$

and let $N_{s,m} := \det(A_{s,m})$. Then $f(x) \in K(x)$ if and only if there exist integers $m \geq 0$ and S such that $N_{s,m} = 0$ whenever $s \geq S$, or equivalently, the sequence $(r_n)_n$ is eventually linearly recurrent, i.e. there exists $t \in \mathbb{N}$ such that the subsequence $(r_n)_{n \geq t}$ is linearly recurrent.

Now we are in position to prove the incompleteness of $(K(x), |\ |_x)$. Let $f(x) = \sum_{i=0}^{\infty} x^{i^2} \in K((x))$. Notice that $f(x)$ is the limit of a sequence of rational functions, say $a_n(x) = \sum_{i=0}^{n} x^{i^2} \in K[x]$. Thus $(a_n(x))_n$ is a Cauchy sequence in $K(x)$. We will show that $(a_n(x))_n$ is not convergent in $K(x)$ by showing that $f(x) \notin K(x)$.

Given m and S non-negative integers, choose a large enough $n \in \mathbb{N}$ such that $S < (n-1)^2 < n^2 - m < n^2 < n^2 + m < (n+1)^2$. By putting $\ell := n^2 - m$ we have

$$S < (n-1)^2 < \ell < \ell + m = n^2 < \ell + 2m < (n+1)^2.$$

Since all the members of an anti-diagonal of $A_{\ell,m}$ are equal, it follows that,

$$A_{\ell,m} = \begin{pmatrix} 0 & 0 & 0 & \cdots & 0 & 0 & 1 \\ 0 & 0 & 0 & \cdots & 0 & 1 & 0 \\ 0 & 0 & 0 & \cdots & 1 & 0 & 0 \\ \vdots & \vdots & \vdots & \iddots & \vdots & \vdots & \vdots \\ 0 & 0 & 1 & \cdots & 0 & 0 & 0 \\ 0 & 1 & 0 & \cdots & 0 & 0 & 0 \\ 1 & 0 & 0 & \cdots & 0 & 0 & 0 \end{pmatrix}$$

and $N_{\ell,m} \neq 0$ with $\ell > S$. Then by 2.28, $f(x) \notin K(x)$.

2.6. Compact and Locally compact valued fields.

THEOREM 2.29 ([**36**, 1.7 Lemma 6]). *Every compact valued field is finite.*

THEOREM 2.30 ([**36**, 1.7 Theorem 5]). *Let $(K, |\ |)$ be a locally compact valued field. Then $(K, |\ |)$ is Cauchy complete.*

The next theorem classifies the locally compact Archimedean valued fields.

THEOREM 2.31 ([**36**, 1.7 Theorem 6]). *If $(K, |\ |)$ is a locally compact Archimedean valued field, then there exist $\lambda \in (0, 1]$ and a field monomorphism $\sigma : K \to \mathbb{C}$ such that $|x| = (|\sigma(x)|_0)^\lambda$ for all $x \in K$, satisfying either $\sigma(K) = \mathbb{R}$ or $\sigma(K) = \mathbb{C}$.*

PROOF. It follows from 2.30 and 2.15. □

The next two results describe every locally compact non-Archimedean valued field implicitly and explicitly respectively.

THEOREM 2.32 ([**41**, 12.2], [**49**, 1.B]). *A non-Archimedean valued field is locally compact if and only if it is Cauchy complete, its value group is discrete and its residue class field is finite.*

THEOREM 2.33 ([**49**, 3.U], [**7**, VI.9.3.1]). *Let $(K, |\ |)$ be a locally compact non-Archimedean valued field with residue class field k. The following statements are satisfied:*

(1) *If $char(K) = 0$, then K is a finite algebraic extension of K_0, where K_0 is the closure of the prime subfield of K. Furthermore, there exists an isomorphism $T : K_0 \to \mathbb{Q}_p$ satisfying $|x| = (|T(x)|_p)^\tau$ for all $x \in K_0$, where $p = char(k)$ and $\tau > 0$ is such that $|p1_K| = p^{-\tau}$.*
(2) *If $char(K) = p$, then $char(k) = p$ and K is isomorphic to $k((x))$. In particular, there exists a field monomorphism $\theta : k \to K$ such that the map $\sigma : k((x)) \to K$, $\sigma(\sum_{i=v}^{\infty} r_i x^i) = \sum_{i=v}^{\infty} \theta(r_i)\pi^i$ is an isometric isomorphism where π is a uniformizer for $|\ |$ and the valuation on $k((x))$ is defined by $\| \sum_{i=v}^{\infty} r_i x^i \| := |\pi|^v$ (for $r_v \neq 0$).*

COROLLARY 2.34. *If $(K, |\ |)$ is a locally compact non-Archimedean valued field, then it is spherically complete.*

PROOF. It follows immediately from 2.32 and 1.18. □

3. Ordered fields.

In this section we will present some examples of ordered fields and will discuss the concept of Archimedean extension of a field created by Hans Hahn in 1907 [**16**].

3.1. Formally real fields. By a ring, we will mean a commutative ring with unit $1 \neq 0$. Let A be a ring that is an ordered set such that its additive group $(A, +)$ is an ordered group (it has a total ordering which is compatible with the addition). The ring A is ordered if for all $x, y \in A$, $x > 0$ and $y > 0$ implies $xy > 0$. Note that an ordered ring is necessarily an integral domain. A field that is an ordered ring will be called an ***ordered field***.

DEFINITION 3.1. A field K is ***formally real*** if satisfies the following condition: given $a_1, \ldots, a_n \in K$, such that, $\sum_{i=1}^{n} a_i^2 = 0$, then $a_1 = \cdots = a_n = 0$.

The following result characterizes the formally real fields as the fields that can be ordered.

THEOREM 3.2 ([**3**, 1.70(5) and 1.71(6)]). *Let K be a field. The following conditions are equivalent:*

(1) *K is formally real,*
(2) *-1 is not a sum of squares in K,*
(3) *There exists an order $\leq$ on K such that $(K, \leq)$ is an ordered field.*

EXAMPLES 3.3. (1) If $char(K) \neq 0$ then, $0 = \sum_{i=1}^{char(K)} 1^2$. Hence K is not formally real. Thus if K is formally real, then $char(K) = 0$.
(2) The field of complex numbers $\mathbb{C}$ cannot be an ordered field, since $-1 = i^2$ and therefore it is not formally real.
(3) If K is an ordered field then we can define an order in $K((x))$, which is compatible with the addition and multiplication. Thus $K((x))$ can be ordered, and therefore it is formally real. Such order is defined as follows: for every $z \in K((x))$ there are $r_i \in K$ such that $z = \sum_{i=v}^{\infty} r_i x^i$. We say that $z > 0$ if $z \neq 0$ and $r_v > 0$. Then $z_1 > z_2$ if $z_1 - z_2 > 0$.
(4) $\mathbb{Q}_p$ is not formally real because if $p = 2$, then -7 is a square and if $p > 2$ then $1 - p$ is a square ([**36**, p. 144]). Recall that in a formally real field the squares are non-negative elements. Since $\mathbb{Q} \subset \mathbb{Q}_p$, $char(\mathbb{Q}_p) = 0$.

3.2. General Hahn fields and the Embedding theorem. Let's review the concept of Archimedean extension of a field and the general Hahn fields.

DEFINITION 3.4. Let S be an ordered group. Two elements $x, y \in S^*$ are ***comparable*** if there exist $n, m \in \mathbb{N}$ such that $|x|_0 < n|y|_0$ and $|y|_0 < m|x|_0$, where

$$|a|_0 := \begin{cases} a & , a \geq 0 \\ -a & , a < 0 \,. \end{cases}$$

Let K be an ordered field. The relation of being comparable is an equivalence relation on K^* and to denote 'x and y are comparable' we write $x \sim y$. This relation defines a partition of K^* into equivalence classes, which are called the ***Archimedean classes of*** K. The equivalence class of $x \in K$ is denoted by $[x]$. Let's denote the class of all the Archimedean classes by G_K.

THEOREM 3.5. *Let K be an ordered field. The class G_K is an ordered abelian group under the order $\prec$ and addition $+$ defined as follows: for every $x, y \in K^*$*

(1) $[x] \prec [y] \iff \forall n \in \mathbb{N}, n|y|_0 < |x|_0 \iff y \not\sim x$ *and* $|y|_0 < |x|_0$.
(2) $[x] + [y] := [xy]$

In this group, the neutral element is $[1_K]$, and $-[x] = [x^{-1}]$.

DEFINITION 3.6. An ordered field K is ***Archimedean*** if $G_K = \{[1_K]\}$, i.e. when every two elements in K^* are comparable.

THEOREM 3.7. *An ordered field K is Archimedean if and only if it satisfies the Archimedean property, i.e. for every $x \in K$, there exist $n \in \mathbb{N}$ such that $|x|_0 < n1_K$.*

PROOF. If K is Archimedean, then every $x \in K^*$ is comparable with 1_K. Hence K satisfies the Archimedean property. If K is not Archimedean then there exist $x \in K^*$ such that $[x] \prec [1]$. Thus $n1_K < |x|_0$ for every $n \in \mathbb{N}$. □

Thus for every ordered field K, the group G_K measures the 'Archimedicity' or the 'non-Archimedicity' of K. The field $\mathbb{R}$ of real numbers (the only ordered, Dedekind complete field up to isomorphism) is characterized by the fact that each Archimedean ordered field can be embedded in $\mathbb{R}$ ([**17**, 3.5]). Hans Hahn in [**16**] (1907) generalized this property (see 3.13) and by doing so he ended up with ordered fields that extend all the ordered fields with a given "level of Archimedicity".

DEFINITION 3.8. Let E/K be an extension of ordered fields, where the order on E restricted to K coincides with that of K. The field E is an ***Archimedean extension of*** K if every $x \in E$ is comparable to some $y \in K$. In that case, G_E and G_K are isomorphic ordered groups. An ordered field K is called ***Archimedean complete*** if it has no proper archimedean extension fields.

DEFINITION 3.9. Let K be an ordered field. If G is an ordered abelian group isomorphic to G_K, then we say that ***K is of type G*** and G is called an ***Archimedean group of K***.

THEOREM 3.10. *The field $\mathbb{R}$ is (up to isomorphism) the only Archimedean complete, ordered field of type $\{0\}$.*

PROOF. An ordered field K is of type $\{0\}$ and Archimedean complete if and only if K is Hilbert complete (K is Archimedean and has no proper ordered Archimedean field extensions), which is in turn equivalent to being Archimedean and Dedekind complete ([**17**, 3.11]). □

THEOREM 3.11 ([**3**, 6.20, 6.21, 7.32], [**11**, 2.15], [**16**]). *Let K be a field (not necessarily ordered) and G an ordered abelian group. The set*

$$K((G)) := \{f : G \to K : supp(f) \text{ is well-ordered}\},$$

where $supp(f) := \{x \in G : f(x) \neq 0\}$, is a field under the addition and multiplication defined as follows: for every $f, g \in K((G))$ and $x \in G$,

(1) $(f+g)(x) := f(x) + g(x)$,

(2) $fg(x) := \sum_{a+b=x} f(a)g(b)$

Fields of the form $K((G))$ are called ***general Hahn fields***.

When K is an ordered field we can define an order on $K((G))$ generalizing the definition of the order in $K((x))$ (3.3).

DEFINITION 3.12 (Ordered general Hahn fields). Let K be an ordered field and consider $\lambda : K((G))^* \to G$, $\lambda(f) = \min\{supp(f)\}$. For $f, g \in K((G))$ we define:

$$f < g \Leftrightarrow f \neq g \text{ and } (g-f)(\lambda(g-f)) > 0.$$

Then $(K((G)), \leq)$ is an ordered field.

Theorem 3.13 is crucial on ordered structures, while 3.14 generalizes 3.10.

THEOREM 3.13 ([**10**], [**18**, 3.1], [**3**, 1.64], [**11**, 1.35], [**16**] (Hahn's Embedding Theorem)). *If K is an ordered field, then for every archimedean group G of K, there exists an order-preserving field monomorphism σ from K into $\mathbb{R}((G))$ such that $\mathbb{R}((G))$ is an Archimedean extension of $\sigma(K)$.*

THEOREM 3.14 ([**10**, pp. 862–863], [**18**, 3.2], [**16**] (Hahn's Completeness Theorem)). *If G is an ordered abelian group, then the field $\mathbb{R}((G))$ is (up to isomorphism) the only Archimedean complete, ordered field of type G.*

3.3. Hahn Fields and Levi-Civita fields. In this subsection we will define a non-Archimedean valuation in some general Hahn fields and the family of the Levi-Civita fields will be introduced.

DEFINITION 3.15. A ***Hahn field*** is a general Hahn field $K((G))$ for which G is a subgroup of $(\mathbb{R}, +)$ and K is any field.

The distinctive characteristic of a Hahn field is that we can define in a natural way a non-Archimedean valuation on them.

THEOREM 3.16 ([**41**, A.9 pp. 288-292], [**42**, II.6 corollary, p. 51]). *Let G be a subgroup of $(\mathbb{R}, +)$ and K any field. If the map $|\ |: K((G)) \to \mathbb{R}$ is defined by*

$$|f| := \begin{cases} e^{-\min\{supp(f)\}} & , f \neq 0 \\ 0 & , f = 0, \end{cases}$$

then $(K((G)), |\ |)$ is a Cauchy complete non-Archimedean valued field with residue class field isomorphic to K and value group $|K((G))^*| = \{e^g \in \mathbb{R} : g \in G\}$. Moreover it is spherically complete.

When a field has a discrete valuation, then every nonzero element can be written in a unique way as a limit of a convergent power series (2.19). The following result shows that in some Hahn fields this also is possible when the valuation is dense.

THEOREM 3.17. *Let K be a field and let $d : \mathbb{Q} \to K$ be the function defined by*

$$d(x) := \begin{cases} 1 & , x = 1 \\ 0 & , x \neq 1. \end{cases}$$

Then d is an element of the field $K((\mathbb{Q}))$, and for any $r \in \mathbb{Q}$, we have that

$$d^r(x) = \begin{cases} 1 & , x = r \\ 0 & , x \neq r. \end{cases}$$

The value group of $(K((\mathbb{Q})), |\ |)$ is $\{e^{-r} = |d^r| = |d|^r : r \in \mathbb{Q}\}$. Furthermore, every nonzero element f in $K((\mathbb{Q}))$ is the sum of a convergent generalized power series with respect to the valuation on $K((\mathbb{Q}))$, specifically:

$$f = \sum_{r \in \mathbb{Q}} f(r) d^r = \sum_{r \in supp(f)} f(r) d^r.$$

Additionally, every generalized power series of the form $\sum_{r \in \mathbb{Q}} a_r d^r$ for which $\{r \in \mathbb{Q} : a_r \neq 0\}$ forms a well ordered subset of $\mathbb{Q}$, is convergent in $K((\mathbb{Q}))$, and if two series of such form differ in at least one coefficient then their sum are different.

THEOREM 3.18 ([**49**, 1.3]). *Let K be any field and let G be a subgroup of $(\mathbb{R}, +)$.*

$$L[G, K] := \{f : G \to K \mid supp(f) \cap (-\infty, n] \text{ is finite for every } n \in \mathbb{Z}\}$$

*is a subfield of $K((G))$. When we restrict the valuation of $K((G))$ to $L[G, K]$, the latter becomes a Cauchy complete, non-Archimedean valued field with residue class field isomorphic to K and value group $|L[G, K]^*| = \{e^g : g \in G\}$. Fields of the form $L[G, K]$ are called **Levi-Civita fields**.*

THEOREM 3.19. *Let K be any field and G be a subgroup of $(\mathbb{R}, +)$. Then:*

(1) *The fields $K((G))$ and $L[G, K]$ coincide if and only if G is discrete.*
(2) *The field $L[G, K]$ is spherically complete if and only if G is discrete.*

(3) *If K is an ordered field, then $K((G))$ is an Archimedean extension of $L[G,K]$ with respect to the order defined in* 3.12. *If in addition K is Archimedean, then both $K((G))$ and $L[G,K]$ are of type G (see* 3.9*).*

PROOF. For (1), if G is not discrete, then it is dense in $\mathbb{R}$ by [**47**, B.5.2]. Thus there exists a sequence $(\alpha_n)_n$ of elements of G which is strictly increasing and converging to 0. Hence the function $f: G \to K$ defined by

$$f(g) = \begin{cases} 1_K & , g \in \{\alpha_n : n \in \mathbb{N}\} \\ 0 & , \text{else} \end{cases}$$

has a well-ordered support such that $supp(f) \cap (-\infty, 0]$ is infinite. Therefore $f \in K((G)) \setminus L[G,K]$. Conversely, if G is discrete then there exists $r > 0$ such that $G = r\mathbb{Z}$ ([**47**, B.5.2]). Let S be a well-ordered subset of G. Since $S \cap (-\infty, n]$ is finite for all $n \in \mathbb{Z}$, it follows that $K((G)) = L[G,K]$. For the proof of the statement (2) we will use terminology and a result from the section 6. By 3.18 and 3.16 it follows that $K((G))$ is an immediate extension field of $L[G,K]$. Therefore $L[G,K]$ is maximally complete if and only if $L[G,K] = K((G))$, since $K((G))$ is always maximally complete ([**42**, corollary, p. 51], [**24**, p. 193]). By (1), $L[G,K]$ is maximally complete if and only if G is discrete. Thus (2) follows from 6.12. □

EXAMPLE 3.20. If $G = (\mathbb{Z}, +)$ and K is any field, then $L[\mathbb{Z}, K] = K((\mathbb{Z}))$ is isomorphic to $K((x))$ by the isomorphism $\varphi : K((x)) \to L[\mathbb{Z}, K]$ defined by

$$\varphi\left(\sum_{i=m}^{\infty} a_i x^i\right) = \sum_{i=m}^{\infty} a_i d^i,$$

with $|\varphi(q)| = |q|_x$ for each $q \in K((x))$, where $|\ |_x$ is the x-adic valuation in $K((x))$.

EXAMPLE 3.21. The fields $\mathbb{F}_p((\mathbb{Z})) = L[\mathbb{Z}, \mathbb{F}_p]$ and $\mathbb{Q}_p$ are Cauchy complete with respect to their valuations, both have the same value group $\{e^n : n \in \mathbb{Z}\}$, and their residue class fields are isomorphic to $\mathbb{F}_p$. However, these fields are not isomorphic since $L[\mathbb{Z}, \mathbb{F}_p]$ has characteristic p while $\mathbb{Q}_p$ has characteristic 0.

3.4. Real-closed field extensions of $\mathbb{R}$. In this section we will study real closed field extensions of $\mathbb{R}$ with non-Archimedean valuations, in particular, we will see that under some conditions, the smallest among such (proper) extensions is a Levi-Civita field. Recall that a field K is ***algebraically closed*** if every polynomial in $K[x]$ has a root in K. If L/K is a field extension then $a \in L$ is ***algebraic*** over K if it is the root of a polynomial in $K[x]$. If every element of L is algebraic over K, then L is an ***algebraic extension*** of K. Also, K is ***real-closed*** if K is formally real and does not admit a proper algebraic extension that is formally real.

THEOREM 3.22 ([**26**, Chapter XI], [**3**, 1.71(21),1.71(22)], [**8**, 5.4.4], [**6**, Chapter 5, Section 4, Lemma 4.1]). *Let K be a field. The following conditions are equivalent:*

(1) *K is real-closed,*
(2) *x^2+1 is irreducible in K and $K(i)$ is algebraically closed ($i^2 = -1$),*
(3) *K is an ordered field, each positive element of K has a square root and every $p \in K[x]$ of odd degree has a root in K,*
(4) *any sentence in the first-order language of fields is true in K if and only if it is true in $\mathbb{R}$,*
(5) *K is an ordered field and the intermediate value theorem holds for all polynomials over K.*

Therefore, in order to develop a theory of Calculus over ordered fields for which the intermediate value theorem holds, then our base field has to be real-closed.

THEOREM 3.23 ([**3**, 6.23 (1)–(2)]). *Let K be a field and G be an ordered abelian group. Then $K((G))$ is real-closed if and only if K is real-closed and G is divisible.*

THEOREM 3.24 ([**46**, p. 218]). *The Levi-Civita field $L[\mathbb{Q}, \mathbb{R}]$ is real-closed under the order defined in* 3.12.

DEFINITION 3.25. Let (x_n) be a sequence of elements in an ordered field K. Then (x_n) is ***Cauchy*** if for every 0-neighborhood U with respect to the order topology in K, there exists $N \in \mathbb{N}$ such that $x_m - x_n \in U$ for all $m, n \geq N$. The sequence (x_n) is ***convergent*** to $x \in K$ if for every 0-neighborhood U with respect to the order topology in K, there exists $N \in \mathbb{N}$ such that $x_n - x \in U$ for all $n \geq N$. An ordered field K is said to be ***Cauchy complete*** if every Cauchy sequence of K is convergent in the order topology.

THEOREM 3.26 ([**45**, 3.11]). *Let $K/\mathbb{R}$ be a field extension where K is a Cauchy complete ordered field such that: (1) the order on K extends the one in $\mathbb{R}$, (2) there exists $\delta \in K$ such that $0 < \delta < r$ for every $r \in \mathbb{R}^+$, and (δ^n) converges to 0 in the order topology, (3) for all $x \in K, x > 0$, for every $n \in \mathbb{N}$, there exists $y \in K, y > 0$ such that $x = y^n$. If d is the function defined in* 3.17, *then there exists an order-preserving field monomorphism $\sigma : L[\mathbb{Q}, \mathbb{R}] \to K$ defined by*

$$\sigma(f) = \sigma\left(\sum_{q \in supp(f)} f(q) d^q \right) = \sum_{q \in supp(f)} f(q) \delta^q.$$

In particular, we have the following result:

COROLLARY 3.27. *Let $K/\mathbb{R}$ be a field extension where K is a Cauchy complete ordered field such that: (1) the order on K extends the one in $\mathbb{R}$, (2) there exists $\delta \in K$ such that $0 < \delta < r$ for every $r \in \mathbb{R}^+$, and (δ^n) converges to 0 in the order topology, (3) it is real-closed. If d is the function defined in* 3.17, *then there exists an order-preserving field monomorphism $\sigma : L[\mathbb{Q}, \mathbb{R}] \to K$ defined by*

$$\sigma(f) = \sigma\left(\sum_{q \in supp(f)} f(q) d^q \right) = \sum_{q \in supp(f)} f(q) \delta^q.$$

REMARK 3.28. The embedding mentioned in 3.26 and 3.27 may exist even when the field K does not satisfy the condition (2). For example, the real closed field $\mathbb{R}((\mathbb{Q}[x]))$ extends $L[\mathbb{Q}, \mathbb{R}]$, but it does not satisfy the condition (2). In fact, we assert the following:

Claim: If K is an ordered field such that G_K has an infinite subset of pairwise non-comparable elements that is cofinal in G_K (3.4), then for each $x \in K^*$, the sequence $(x^n)_n$ does not converge to 0 in the order topology.

For example, the field $\mathbb{R}((\mathbb{Z}[x]))$ satisfies the hypothesis of the claim but $\mathbb{R}((\mathbb{Z}^3))$ does not. ***Proof:*** By the Hahn's embedding theorem 3.13, the field K can be embedded in $\mathbb{R}((G_K))$ so we can consider the Hahn valuation on K (5.7, (3)). Let $x \in K^*$ and $\lambda(x) = \min\{supp(x)\} \in G_K$. Then $\lambda(x^n) = n\lambda(x) \in G_K$ for all $n \in \mathbb{N}$. By hypothesis we can choose $\varepsilon \in K$, $\varepsilon > 0$ such that $\lambda(\varepsilon) > n\lambda(x)$ for all $n \in \mathbb{N}$. If $|x|_0 = \max\{x, -x\}$, then $\lambda(x^n) = \lambda((|x|_0)^n) = \lambda((|x|_0)^n - \varepsilon)$ for all $n \in \mathbb{N}$. Thus $((|x|_0)^n - \varepsilon)(\lambda((|x|_0)^n - \varepsilon)) = ((|x|_0)^n - \varepsilon)(\lambda((|x|_0)^n)) = (|x|_0)^n(\lambda((|x|_0)^n)) > 0$.

Therefore $(|x|_0)^n > \varepsilon$ for all $n \in \mathbb{N}$. In other words, $x^n \notin (-\varepsilon, \varepsilon)$ for all $n \in \mathbb{N}$. Hence, $(x^n)_n$ does not converge to 0 in the order topology.

The proof of 3.26 ([**45**, 3.11]) can be adapted to prove the following:

COROLLARY 3.29. *Let $K/\mathbb{R}$ be a field extension where K is real-closed and Cauchy complete with respect to a non-trivial valuation $\widehat{|\ |}$ such that $\widehat{|x|} = 1$ for all $x \in \mathbb{R}^*$. Then, for every $\delta \in K^*$ such that $\widehat{|\delta|} < 1$, there exists a field monomorphism $\sigma : L[\mathbb{Q}, \mathbb{R}] \to K$ defined by*

$$\sigma(f) = \sigma\left(\sum_{q \in supp(f)} f(q) d^q \right) = \sum_{q \in supp(f)} f(q) \delta^q,$$

satisfying $\widehat{|\sigma(f)|} = |f|^\tau$ for all $f \in L[\mathbb{Q}, \mathbb{R}]$, where $\tau > 0$ is such that $\widehat{|\delta|} = |d|^\tau$.

4. Algebraic closure of valued fields and their completions.

In this section we will present non-Archimedean valued fields that are algebraic closure of other valued fields, like certain Puiseux series fields or the algebraic closure of $\mathbb{Q}_p$. Also we will review the completion of algebraic closures like the field of p-adic complex numbers $\mathbb{C}_p$, or the completion of a Puiseux series field.

Recall that an ***algebraic closure*** of a field K is an algebraically closed, algebraic extension of K. Each field has an algebraic closure and any two algebraic closures of a field K are isomorphic by means of an isomorphism leaving K pointwise fixed ([**27**, I.8.25], [**20**, 66], [**26**, V 2.5, 2.9]). The algebraic closure of K will be denoted by K^a.

4.1. General Results.

THEOREM 4.1 ([**41**, 14.1] (Krull's Existence Theorem)). *Let K be a subfield of a field L and let $|\ |$ be a non-Archimedean valuation on K. Then there exists a non-Archimedean valuation on L that extends $|\ |$.*

THEOREM 4.2 ([**41**, 14.2] (Krull's Uniqueness Theorem)). *Let L/K be an algebraic field extension. If $|\ |$ is a non-Archimedean valuation on K such that $(K, |\ |)$ is Cauchy complete, then there exists a unique valuation on L that extends $|\ |$. This extension is also non-Archimedean.*

As a consequence of Krull's Theorems we have the following:

COROLLARY 4.3 ([**15**, 6.3]). *Let $(K, |\ |)$ be a Cauchy complete, non-Archimedean valued field. The valuation $|\ |$ can be uniquely extended to a valuation on K^a.*

THEOREM 4.4 ([**41**, 16.2, 16.3], [**15**, 6.4, 6.5, 6.6, 6.8]). *Let $(K, |\ |)$ be a Cauchy complete, non-Archimedean valued field with residue class field k. The following statements are true:*

(1) *the field k^a is the residue class field of K^a,*
(2) *the field k^a is infinite,*
(3) *$|(K^a)^*| = \{r \in (0, \infty) : r^n \in |K^*| \text{ for some } n \in \mathbb{N}\}$,*
(4) *the only valuation on K^a that extends the one in K is dense.*

THEOREM 4.5 ([**41**, 17.1], [**15**, 6.10]). *Let $(K, |\ |)$ be non-Archimedean valued field and let $|\ |'$ be the valuation on K^a that extends $|\ |$. Then the completion of $(K^a, |\ |')$ is algebraically closed.*

4.2. Examples.

THEOREM 4.6. *Consider the field $\mathbb{Q}_p^a$ with the only valuation that extends $|\ |_p$ on $\mathbb{Q}_p$, also denoted by $|\ |_p$. The following statements are satisfied:*

1. *$\mathbb{Q}_p^a$ is a proper extension of $\mathbb{Q}_p$,*
2. *the residue class field of $\mathbb{Q}_p^a$ is $\mathbb{F}_p^a$,*
3. *$|(\mathbb{Q}_p^a)^*| = \{e^r : r \in \mathbb{Q}\}$,*
4. *$(\mathbb{Q}_p^a, |\ |_p)$ is not locally compact,*
5. *$(\mathbb{Q}_p^a, |\ |_p)$ is not Cauchy complete,*
6. *$\mathbb{Q}_p^a$ is an infinite dimensional vector space over $\mathbb{Q}_p$.*

PROOF. To prove (1) it is enough to show that $\mathbb{Q}_p$ is not algebraically closed. In fact, the polynomial $x^2 - p$ is irreducible on $\mathbb{Q}_p$. Otherwise there would be an x in $\mathbb{Q}_p$ such that $|x^2|_p = |p|_p = e^{-1}$. Hence $|x| = \sqrt{e^{-1}} \in |\mathbb{Q}_p|$ which is not possible. The statements (2) and (3) follow from 4.4. The statement (4) follows from (3) and 2.32. The statements (5) and (6) are proved in [**41**, 16.6 and 16.7] respectively. □

DEFINITION 4.7. The completion of the field $(\mathbb{Q}_p^a, |\ |_p)$ is called the ***field of p-adic complex numbers*** and it is denoted by $\mathbb{C}_p$. The valuation on $\mathbb{C}_p$ which extends $|\ |_p$ will also be denoted by $|\ |_p$.

THEOREM 4.8. *The field of p-adic complex numbers satisfies:*

1. *the residue class field of $\mathbb{C}_p$ is $\mathbb{F}_p^a$,*
2. *$|\mathbb{C}_p^*| = \{e^r : r \in \mathbb{Q}\}$,*
3. *$(\mathbb{C}_p, |\ |_p)$ is not locally compact,*
4. *$\mathbb{C}_p$ is algebraically closed,*
5. *$\mathbb{C}_p$ is an infinite dimensional vector space over $\mathbb{Q}_p$,*
6. *$\mathbb{C}_p$ is separable,*
7. *$\mathbb{C}_p$ and $\mathbb{C}$ are isomorphic as fields.*
8. *$\mathbb{C}_p$ is not spherically complete.*

PROOF. The statements (1), (2) and (5) follow from 2.18 and 4.6, while (3) follows from (2) and 2.32. The statement (4) follows from 4.5 whereas (6) can be found in [**41**, 17.2], (7) can be found in [**49**, page 83] and (8) in [**41**, 20.6]. □

DEFINITION 4.9. Let K be a field. The set

$$\begin{aligned} K\langle\langle x\rangle\rangle &:= \bigcup_{n=1}^{\infty} K((x^{\frac{1}{n}})) \\ &= \left\{ \sum_{i=v}^{\infty} r_i x^{\frac{i}{n}} : v \in \mathbb{Z}, r_i \in K,\ r_v \neq 0, \text{ for some } n \in \mathbb{N} \right\} \cup \{0\}, \end{aligned}$$

is a field when we adopt the convention $x^{\frac{\ell p}{\ell q}} = x^{\frac{p}{q}}$ for all $\ell \in \mathbb{Z}$. In fact, if $x \in K((x^{\frac{1}{n}}))$ and $y \in K((x^{\frac{1}{m}}))$, then both x and y are elements of $K((x^{\frac{1}{nm}}))$, and therefore $x + y$ and xy are well-defined in $K((x^{\frac{1}{nm}}))$ and hence in $K\langle\langle x\rangle\rangle$.

The mapping $\theta : K\langle\langle x\rangle\rangle \to \bigcup_{n=1}^{\infty} K\big(\big(\frac{1}{n}\mathbb{Z}\big)\big)$ defined by

$$\theta\left(\sum_{i=v}^{\infty} r_i x^{\frac{i}{n}}\right) = \sum_{i=v}^{\infty} r_i d^{\frac{i}{n}},$$

is an isomorphism such that $|\theta(f)| = |f|$ for all $f \in K\langle\langle x\rangle\rangle$. Therefore $K\langle\langle x\rangle\rangle$ can be considered as a subfield of $K((\mathbb{Q}))$. When we restrict the valuation of $K((\mathbb{Q}))$ (see below of 2.22) to $K\langle\langle x\rangle\rangle$, the latter becomes a non-Archimedean valued field with residue class field isomorphic to K and value group equal to $\{e^r : r \in \mathbb{Q}\}$. Fields of the form $K\langle\langle x\rangle\rangle$ are called ***Puiseux series fields***.

With this field we obtain the following chain of field extensions:

$$K \subsetneqq K(x) \subsetneqq K((x)) \subsetneqq K\langle\langle x\rangle\rangle \subsetneqq L[\mathbb{Q}, K] \subsetneqq K((\mathbb{Q})).$$

THEOREM 4.10. *If K is any field, then $L[\mathbb{Q}, K]$ is the completion of $(K\langle\langle x\rangle\rangle, |\ |)$.*

PROOF. Consider the identification $K\langle\langle x\rangle\rangle = \bigcup_{n=1}^{\infty} K\left(\left(\frac{1}{n}\mathbb{Z}\right)\right)$. By 3.17, every element of the field $L[\mathbb{Q}, K]$ has the form $f = \sum_{i=v}^{\infty} r_i d^{\alpha_i}$ where $v \in \mathbb{Z}$, $r_i \in K$ and $(\alpha_i)_i$ is a strictly increasing sequence in $\mathbb{Q}$ such that $\{\alpha_i : i = v, v+1, \dots\} \cap (-\infty, n]$ is finite for every $n \in \mathbb{Z}$. Notice that the partial sum sequence $(\sum_{i=v}^{n} r_i d^{\alpha_i})_n$ is Cauchy in $\bigcup_{n=1}^{\infty} K\left(\left(\frac{1}{n}\mathbb{Z}\right)\right)$ with limit f. Hence the theorem holds by 2.11, since $\bigcup_{n=1}^{\infty} K\left(\left(\frac{1}{n}\mathbb{Z}\right)\right)$ is dense in $L[\mathbb{Q}, K]$ and the latter is Cauchy complete (3.18). □

The next result shows the interesting analogy of the trios $(\mathbb{Q}_p, \mathbb{Q}_p^a, \mathbb{C}_p)$ and $(K((x)), K\langle\langle x\rangle\rangle, L[\mathbb{Q}, K])$ when K satisfies the following conditions:

THEOREM 4.11. *If K is an algebraically closed field of characteristic 0, then:*

1. *$K((x))^a = K\langle\langle x\rangle\rangle$.*
2. *$L[\mathbb{Q}, K]$ is the completion of the field $(K\langle\langle x\rangle\rangle, |\ |)$.*
3. *$L[\mathbb{Q}, K]$ is algebraically closed.*

PROOF. The first statement is proved in [**13**, 13.15], the second statement is satisfied for any field K (4.10), and the last statement follows from 4.5. □

Now we are able to determine when the Intermediate Value Theorem is valid for polynomials over $K\langle\langle x\rangle\rangle$ and $L[\mathbb{Q}, K]$.

COROLLARY 4.12. *The following statements are equivalent:*

1. *K is real-closed.*
2. *$K\langle\langle x\rangle\rangle$ is real-closed.*
3. *$L[\mathbb{Q}, K]$ is real-closed.*

PROOF. (1) $\Longrightarrow$ (2) follows from [**5**, 2.6 Theorem 2.91] while (2) $\Longrightarrow$ (1) and (3) $\Longrightarrow$ (1) are proved in [**3**, 6.23 (1)]. Let's prove (1) $\Longrightarrow$ (3): if K is real-closed, then $K(i)$ is an algebraically closed filed of characteristic 0, where i is a root for $x^2 + 1 = 0$ (3.22). By 4.11 the field $K(i)\langle\langle x\rangle\rangle$ is algebraically closed and hence $L[\mathbb{Q}, K](i) = L[\mathbb{Q}, K(i)]$ is algebraically closed as well (4.5, 4.10). Finally, $L[\mathbb{Q}, K]$ is real-closed by 3.22. □

Now we will see that the analogy of the trios breaks down when the characteristic of K is positive.

THEOREM 4.13 ([**1**, p. 904], [**9**, pp. 64–65]). *If K is algebraically closed and $char(K) = p$, then the fields $K\langle\langle x\rangle\rangle$ and $L[\mathbb{Q}, K]$ are not algebraically closed, since the polynomial $p(z) = z^p - z - x^{-1} \in K\langle\langle x\rangle\rangle[z]$ has the following factorization in $K((\mathbb{Q}))$:*

$$p(z) = \prod_{i=0}^{p-1} \left(z + i - \sum_{i=1}^{\infty} x^{\frac{-1}{p^i}} \right).$$

In order to describe the algebraic closure of $K((x))$ when K is algebraically closed of positive characteristic, we need a fair amount of terminology on Regular Languages that will not be presented here but can be found with full details in [**22**]. Such a description reads as follows.

THEOREM 4.14 ([**22**, Theorem 10.6]). *Let K be an algebraically closed field of characteristic p and $x = \sum_i r_i x^i \in K((\mathbb{Q}))$. Then $x \in K((x))^a$ if and only if for all $n \in \mathbb{Z}$, y_n is p-quasi-automatic and $\mathrm{span}\{y_n : n \in \mathbb{Z}\}$ is a finite dimensional vector space over K, where*

$$y_n := \sum_{i \in [0,1) \cap \mathbb{Q}} r_{n+i} x^i .$$

COROLLARY 4.15. *The following statements are equivalent:*

(1) *K is algebraically closed and has characteristic* 0.
(2) *$K\langle\langle x \rangle\rangle$ is algebraically closed.*
(3) *$L[\mathbb{Q}, K]$ is algebraically closed.*

PROOF. (1) $\Longrightarrow$ (2) and (1) $\Longrightarrow$ (3) follow from 4.11. The implications (2) $\Longrightarrow$ (1) and (3) $\Longrightarrow$ (1) follow form [**3**, 6.23 (0)] and 4.13. □

5. General valuations.

Note that in 3.16, we have restricted our attention to those Hahn fields of Archimedean group embedded in $(\mathbb{R}, +)$ in order to define a valuation with values in $\mathbb{R}$. However it is possible to define a valuation with values in any ordered abelian group. If $|\ \ | : K \to \mathbb{R}$ is a valuation on a field K and we consider the function $v : K \to \mathbb{R} \cup \{\infty\}$ defined by $v(x) = -\ln |x|$, then we are shifting our attention to the additive structure of $\mathbb{R}$ rather than the multiplicative one of $(0, \infty)$. In particular, if $|\ \ |$ is a non-Archimedean valuation, then the strong triangle inequality now has the form: $v(x + y) \geq \min\{v(x), v(y)\}$ for all $x, y \in K$. In this section we will use this approach to define a generalization of a non-Archimedean valuation.

DEFINITION 5.1. Let K be a field and let $(G, +)$ be an ordered abelian group. A map $v : K \to G \cup \{\infty\}$ is a ***general valuation (or Krull valuation)*** on K if it satisfies:

(1) v is onto,
(2) $v(x) = \infty$ if and only if $x = 0$,
(3) $v(xy) = v(x) + v(y)$,
(4) $v(x + y) \geq \min\{v(x), v(y)\}$,

where ∞ is a symbol that satisfies, for all $g \in G$, the following axioms:

$$g < \infty \text{ and } \infty = \infty + \infty = g + \infty = \infty + g.$$

The group $G = v(K^*)$ is called the ***value group*** of (K, v) and the quotient $\{x \in K : v(x) \geq 0\}/\{x \in K : v(x) > 0\}$ is the residue class field of (K, v). If G is order-isomorphic to a subgroup of $(\mathbb{R}, +)$, then we say that v has ***rank* 1**. Otherwise v is of ***higher rank***. The general valuation v is called ***discrete*** if G is cyclic. In that case, if G is not trivial then it is isomorphic to $(\mathbb{Z}, +)$ and hence it has rank 1.

THEOREM 5.2 ([**4**, Chapter III, section 3]). *An ordered group $(G, +)$ is order-isomorphic to a subgroup of $(\mathbb{R}, +)$ if and only if it is Archimedean, i.e. for every $a, b \in G$, $b > 0$, there exists $n \in \mathbb{N}$ such that $a < nb$. In particular, every Archimedean ordered group is abelian.*

DEFINITION 5.3. Let K be a field with general valuation $v : K \to G \cup \{\infty\}$. The sets of the form $U_v[a,g] := U[a,g] := \{x \in K : v(x-a) \geq g\}$ for $a \in K, g \in G$, constitute a base for a topology on K called the ***valuation topology*** induced by v. When G is not trivial, the sets of the form $U_v(a,g) := U(a,g) := \{x \in K : v(x-a) > g\}$ also constitute a base for the valuation topology induced by v. The field K with this topology becomes a Hausdorff topological field ([**3**, 7.64]).

DEFINITION 5.4. Let K be an ordered field and let G_K be the ordered group of the Archimedean classes of K (3.4, 3.5). The map $\mu : K \to G_K \cup \{\infty\}$

$$\mu(x) := \begin{cases} [x] & , x \neq 0 \\ \infty & , x = 0, \end{cases}$$

is a general valuation on K and it is called ***the order valuation*** of K ([**3**, 1.61]).

THEOREM 5.5. *Let K be an Archimedean ordered field and let μ be the order valuation on K. Then the valuation topology induced by μ is the discrete topology.*

PROOF. Since $G_k = \{[1]\}$, then μ is the trivial valuation on K. □

THEOREM 5.6 ([**3**, 7.63, 7.64]). *Let K be a non-Archimedean ordered field and let μ be the order valuation on K. Then the order topology on K coincides with the valuation topology induced by μ.*

EXAMPLES 5.7.
(1) If $|\ | : K \to \mathbb{R}$ is a non-Archimedean valuation, then the function $v : K \to \mathbb{R} \cup \{\infty\}$ defined by $v(x) = -\ln|x|$ is a general valuation (after a suitable restriction of the codomain). The topology induced by $|\ |$ coincides with the valuation topology induced by v, since $B(x, e^{-r}) = U(x,r)$ and $B[x, e^{-r}] = U[x,r]$ for every $x \in K$ and $r \in \mathbb{R}$.
(2) If $v : K \to G \cup \{\infty\}$ is a general valuation of rank 1 on K and $i : G \to (\mathbb{R}, +)$ is an order-preserving homomorphism, then the map $|\ | : K \to \mathbb{R}$ given by

$$|x| := \begin{cases} e^{-i(v(x))} & , x \neq 0 \\ 0 & , x = 0, \end{cases}$$

is a non-Archimedean valuation on K. The topology induced by $|\ |$ coincides with the valuation topology induced by v, since $B(x,|y|) = U(x, v(y))$ and $B[x,|y|] = U[x, v(y)]$ for every $x, y \in K$.
(3) Let K be a field and G be any ordered abelian group. The map $\lambda : K((G)) \to G \cup \{\infty\}$ defined by

$$\lambda(f) := \begin{cases} \min\{supp(f)\} & , f \neq 0 \\ \infty & , f = 0, \end{cases}$$

is a general valuation on the general Hahn field $K((G))$ called ***the Hahn valuation*** on $K((G))$ (compare with the valuation in 3.16).

THEOREM 5.8. *Let K be an Archimedean ordered field, G a nontrivial ordered abelian group, μ the order valuation on $K((G))$ and λ the Hahn valuation on $K((G))$. The following statements are true:*
(1) *$\mu(f) = [f] = \{g \in K((G)) : \lambda(g) = \lambda(f)\}$ for every $f \neq 0$,*
(2) *the valuation topologies induced by μ and λ coincide with the order topology on $K((G))$ (defined in 3.12),*

(3) *If G is a subgroup of $(\mathbb{R},+)$, then the valuation $|\ |$ of the Hahn field $K((G))$ defined in* 3.16, *induces the order topology on $K((G))$.*

PROOF. To prove the first statement it is enough to note that f and g are comparable if and only if $\lambda(f)=\lambda(g)$. The second statement follows from 5.6 and the equalities $U_\mu(f,\mu(g))=U_\lambda(f,\lambda(g))$ and $U_\mu[f,\mu(g)]=U_\lambda[f,\lambda(g)]$ for all $f,g\in K((G))$. The last statement follows from part (2) and from example 5.7(1). $\square$

LEMMA 5.9. *Let F/K be a field extension and let $v: F\to G\cup\{\infty\}$ be a general valuation on F. The restriction $v|_K$ of v to K is a general valuation on K. Moreover, if τ_v ($\tau_{v|_K}$) denotes the valuation topology on F (on K) induced by v (by $v|_K$ respectively), and $\tau_v\cap K$ denotes the subspace topology on K induced by (F,τ_v), then $\tau_{v|_K}=\tau_v\cap K$.*

PROOF. After a suitable restriction of the codomain the map $v|_K$ is a general valuation on K. The second statement follows from the following equalities valid for all $x,y\in K$: $U_{v|_K}(x,v(y))=\{z\in K: v(x-z)>v(y)\}=U_v(x,v(y))\cap K$. $\square$

THEOREM 5.10. *Let K be a non-Archimeden ordered field. By* 3.13 *K can be embedded in $\mathbb{R}((G_K))$ where G_K is the group of Archimedean classes of K. Let:*

- *τ_K be the order topology on K,*
- *τ_μ be the valuation topology on K induced by the order valuation μ of K,*
- *τ_0 be the order topology on $\mathbb{R}((G_K))$,*
- *τ_η be the valuation topology on $\mathbb{R}((G_K))$ induced by the order valuation η of $\mathbb{R}((G_K))$,*
- *τ_λ be the valuation topology on $\mathbb{R}((G_K))$ induced by the Hahn valuation λ of $\mathbb{R}((G_K))$.*

Then $\quad\tau_K=\tau_\mu=\tau_{\lambda|_K}=\tau_{\eta|_K}=\tau_\lambda\cap K=\tau_0\cap K=\tau_\eta\cap K.$

PROOF. The equality $\tau_K=\tau_\mu$ follows from 5.6 while the equalities $\tau_\lambda\cap K=\tau_0\cap K=\tau_\eta\cap K$ follow from 5.8. The equalities $\tau_{\lambda|_K}=\tau_\lambda\cap K$ and $\tau_{\eta|_K}=\tau_\eta\cap K$ follow from 5.9. Finally we will show that $\tau_\mu=\tau_\eta\cap K$. For every $f\in K$, $\mu(f)=\{g\in K: g \text{ is comparable to } f\}$. Since K is embedded in $\mathbb{R}((G_K))$ we have that $K\subset\mathbb{R}((G_K))$. Hence $\mu(f)\subset\eta(f)=\{g\in\mathbb{R}((G_K)):\lambda(g)=\lambda(f)\}$. It follows that $\mu(f)=\mu(g)$ if and only if $\eta(f)=\eta(g)$ for all $f,g\in K$. Additionally, $\mu(f)<\mu(g)$ if and only if $\eta(f)<\eta(g)$. Hence $U_\eta(f,\eta(g))\cap K=U_\mu(f,\mu(g))$ for all $f,g\in K$. $\square$

REMARK 5.11. Let $(A,<)$ be an ordered set with order topology τ_A and let B be a subset of A. The order in A induces an order in B which induces a topology $\tau_<$ on B. In general the subspace topology $\tau_A\cap B$ on B may be different than the topology $\tau_<$ ([**28**, p. 90]). Also, it is well-known that if B is a convex subset of A, then $\tau_A\cap B=\tau_<$ ([**28**, 16.4]). The surprising thing about the previous result is that for every non-Archimedean ordered field K its order topology τ_K (which is the topology induced by the order induced by the order on $\mathbb{R}((G_K))$) always coincides with the the subspace topology $\tau_0\cap K$, even when K is not convex in $\mathbb{R}((G_K))$. For example, $L[\mathbb{Q},\mathbb{R}]$ is not a convex subfield of $\mathbb{R}((\mathbb{Q}))$ and by 5.10, the order topology on $L[\mathbb{Q},\mathbb{R}]$ coincides with the subspace topology inherited from $\mathbb{R}((\mathbb{Q}))$.

More details about general valuations can be found in [**36**, 13.1] and [**33**].

6. Maximal, Pseudo and Spherical completeness.

In this section we will present briefly the concepts of Maximal completeness, Pseudo-completeness and Spherical completeness together with their equivalences in fields with general valuations or with non-Archimedean valuations.

6.1. Pseudo-completeness. When we study the Cauchy completeness of a space, usually an equivalent condition using sequences is employed. For spherical completeness of non-Archimedean valued fields we also have an equivalent condition using sequences: the notion of pseudo-completeness.

DEFINITION 6.1. Let K be a field, $|\ |: K \to \mathbb{R}$ be a non-Archimedean valuation and $v : K \to G \cup \{\infty\}$ be a general valuation where G is an abelian ordered group. A well-ordered set $\{x_\rho\}$ of elements of K, without a maximal element, is ***pseudo-Cauchy*** (with respect to v) if $v(x_\sigma - x_\rho) < v(x_\tau - x_\sigma)$ for all $\tau > \sigma > \rho$. A sequence $(x_n)_n$ of elements of K is ***pseudo-Cauchy*** (with respect to $|\ |$) if there exists $n_0 \in \mathbb{N}$, such that $|x_n - x_m| < |x_m - x_k|$ for all $n > m > k \geq n_0$.

Notice that a sequence $(x_n)_n$ is pseudo-Cauchy with respect to $|\ |$ if and only if it is eventually pseudo-Cauchy with respect to $v(x) = -\ln|x|$.

LEMMA 6.2 ([**42**, II.4. Lemma 7], [**29**, 2.1. Lemma 4]). *Let K be a field, $|\ |$ be a non-Archimedean valuation on K and $v : K \to G \cup \{\infty\}$ be a general valuation.*

(1) *If the well-ordered set $\{x_\rho\}_\rho$ is pseudo-Cauchy, then for all $\sigma > \rho$, $v(x_\sigma - x_\rho) = v(x_{\rho+1} - x_\rho) =: \gamma_\rho$. Hence the net $(\gamma_\rho)_\rho$ is strictly increasing in G.*
(2) *If the sequence $(x_n)_n$ is pseudo-Cauchy, then for some $n_0 \in \mathbb{N}$ and all $k > n \geq n_0$, $|x_k - x_n| = |x_{n+1} - x_n| =: \mu_n$. Hence the sequence $(\mu_n)_n$ is eventually strictly decreasing in $(0, \infty)$.*

DEFINITION 6.3 ([**29**, 2.1. Lemma 5]). Let K be a field, $|\ | : K \to \mathbb{R}$ be a non-Archimedean valuation and $v : K \to G \cup \{\infty\}$ be a general valuation. An element $a \in K$ is a ***pseudo-limit*** of the pseudo-Cauchy set $\{x_\rho\}_\rho$ if $v(a - x_\rho) = \gamma_\rho$ for all ρ, or equivalently, if $v(a - x_\rho) < v(a - x_\sigma)$ for all $\sigma > \rho$. An element $a \in K$ is a ***pseudo-limit*** of the pseudo-Cauchy sequence $(x_n)_n$ if $|a - x_n| = \mu_n$ for all $n \geq n_0$, for some $n_0 \in \mathbb{N}$, or equivalently, if $|a - x_n| > |a - x_m|$ for all $m > n \geq n_0$, for some $n_0 \in \mathbb{N}$.

DEFINITION 6.4. Let K be a field, $|\ | : K \to \mathbb{R}$ be a non-Archimedean valuation and $v : K \to G \cup \{\infty\}$ be a general valuation. If every pseudo-Cauchy set of K has a pseudo-limit in K (w.r.t. v), then (K, v) is said to be ***pseudo-complete***. If every pseudo-Cauchy sequence of K has a pseudo-limit in K (w.r.t. $|\ |$), then $(K, |\ |)$ is said to be ***pseudo-complete***.

The following result shows the equivalence of pseudo-completeness and spherical completeness.

THEOREM 6.5 ([**29**, 2.1. Theorem 2]). *Let $(K, |\ |)$ be a non-Archimedean valued field. Then $(K, |\ |)$ is spherically complete if and only if it is pseudo-complete.*

6.2. Spherical completeness. In view of the fact that any ultrametric space can be embedded in a non-Archimedean valued field ([**39**]), the following definition is a generalization of the definition given in 1.15. Let K be a field with general valuation v and value group G. A set of the form $U[a, g] := \{x \in K : v(x - a) \geq g\}$ for $a \in K, g \in G$ is called a ***ball*** in K. Recall that the collection of all balls in K constitute a basis for the valuation topology on K (5.3).

DEFINITION 6.6. A field K with general valuation v is ***spherically complete*** if every nested sequence of balls $U[a_1, g_1] \supset U[a_2, g_2] \supset \cdots$ for which $g_1 < g_2 < \cdots$ has nonempty intersection.

THEOREM 6.7. *A field K with general valuation v is spherically complete if and only if it is pseudo-complete.*

PROOF. The proof of [**29**, 2.1. Theorem 2] can be adapted to our case. □

6.3. Maximal completeness. Just as a metric space has a completion, a non-Archimedean valued field has a "spherical completion". In this subsection we will formalize this notion, and study its existence, uniqueness and structure.

DEFINITION 6.8. A field F with general valuation v' is an ***immediate extension*** of a field K with general valuation v if

(1) F/K is a field extension,
(2) v' is an extension of v,
(3) $v'(F^*) = v(K^*)$,
(4) the residue class field of (F, v') is equal to the residue class field of (K, v).

If (K, v) admits no proper immediate extension, then it is ***maximally complete***.

In particular, if the general valuations have rank 1, then the previous definition can be rephrased as follows.

DEFINITION 6.9. A non-Archimedean valued field $(F, |\ |')$ is an ***immediate extension*** of a non-Archimedean valued field $(K, |\ |)$ if

(1) F/K is a field extension,
(2) $|\ |'$ is an extension of $|\ |$,
(3) $|F^*|' = |K^*|$,
(4) the residue class fields of $(F, |\ |')$ and $(K, |\ |)$ are equal.

If $(K, |\ |)$ admits no proper immediate extension, then it is ***maximally complete***.

EXAMPLE 6.10. The completion of a non-Archimedean valued field $(K, |\ |)$ is an immediate extension of $(K, |\ |)$ by 2.18.

THEOREM 6.11 ([**21**, Theorem 4], [**42**, II.6. Theorem 8]). *A field K with general valuation v is maximally complete if and only if it is is pseudo-complete, or equivalently, spherically complete.*

In particular, for non-Archimedean valued fields we have the following:

THEOREM 6.12 ([**29**, 2.3. Theorem 2, 2.1. Theorem 5]). *Let $(K, |\ |)$ be a non-Archimedean valued field. Then $(K, |\ |)$ is maximally complete if and only if it is pseudo-complete if and only if it is spherically complete.*

The next result guarantees the existence of a maximally complete immediate extension of any valued field.

THEOREM 6.13 ([**42**, II.3. Theorem 5], [**29**, 2.4. Theorem 1], [**24**, Theorem 24]). *For each field K with general valuation v, there exists at least one maximally complete immediate extension. This also holds for non-Archimedean valued fields.*

The uniqueness of a maximally complete immediate extension of a valued field is not always possible. However there are some conditions for which we have uniqueness. First, let's clarify the type of uniqueness we are referring to.

DEFINITION 6.14. Two immediate extensions (F_1, v_1) and (F_2, v_2) of a field K with general valuation v are ***analytically equivalent*** (over (K, v)) if there exists a field isomorphism $\sigma : F_1 \to F_2$ such that $v_1(x) = v_2(\sigma(x))$ for all $x \in F_1$ and $\sigma(y) = y$ for all $y \in K$. Similarly, two immediate extensions $(F_1, | \ |_1)$ and $(F_2, | \ |_2)$ of a non-Archimedean valued field $(K, | \ |)$ are ***analytically equivalent*** (over $(K, | \ |)$) if there exists a field isomorphism $\sigma : F_1 \to F_2$ such that $|x|_1 = |\sigma(x)|_2$ for all $x \in F_1$ and $\sigma(y) = y$ for all $y \in K$.

DEFINITION 6.15. Let the field K have a general valuation with value group G and residue class field k. If $char(k) = p > 0$, then the ***hypothesis A*** is the following pair of conditions:

(1) Any equation of the form $x^{p^n} + a_1 x^{p^{n-1}} + \cdots + a_{n-1} x^p + a_n x + a_{n+1} = 0$, with coefficients in k, has a root in k.
(2) $G = pG$.

Analogously for non-Archimedean valued fields we have the following:

DEFINITION 6.16. Let the field K have a non-Archimedean valuation $| \ |$ with value group $|K^*|$ and residue class field k. If $char(k) = p > 0$, then the ***hypothesis A*** is the following pair of conditions:

(1) Any equation of the form $x^{p^n} + a_1 x^{p^{n-1}} + \cdots + a_{n-1} x^p + a_n x + a_{n+1} = 0$, with coefficients in k, has a root in k.
(2) $|K^*| = (|K^*|)^p$.

THEOREM 6.17 ([**21**, Theorem 5], [**42**, VII.5. Theorem 4]). *Let K be a field with a general or a non-Archimedean valuation and let k be its residue class field.*

(1) *If $char(k) = 0$, then $char(K) = 0$ and the maximally complete immediate extension of K is uniquely determined up to analytical equivalence.*
(2) *If $char(k) = p > 0$, and k and the value group satisfy the hypothesis A, then the maximally complete immediate extension of K is uniquely determined up to analytical equivalence over K.*

EXAMPLE 6.18. If K is any field and G is any ordered abelian group, then the general Hahn field $K((G))$ with the Hahn valuation (5.7) is maximally complete (see [**42**, corollary, page 51], [**24**, page 193]). Hence any maximally complete immediate extension of the Levi-Civita field $L[G, K]$ is analytically equivalent to $K((G))$ when K and G satisfy the conditions of 6.17. Recall that the value group of $L[G, K]$ is G and its residue class field is isomorphic to K.

Note that if a valued field K has a divisible value group G and its residue class field k is algebraically closed, then G and k satisfy the hypothesis A. Consequently:

COROLLARY 6.19. *Let K be a field with a general valuation. If the value group of K is divisible and the residue class field of K is algebraically closed, then the maximally complete immediate extension of K is uniquely determined up to analytical equivalence over K.*

The uniqueness obtained in 6.19 is related to the uniqueness of the algebraic closure of a field in the the following sense:

THEOREM 6.20 ([**34**, Corollary 4], [**3**, 6.23. (0)]). *Let F be a maximally complete field with a value group G and a residue class field K. The field F is algebraically closed if and only if G is divisible and K is algebraically closed.*

REMARK 6.21. In [**21**, pp. 318–320] Irving Kaplansky presents a field that does not satisfy the Hypothesis A and has two maximally complete immediate extensions that are not analytically equivalent and not even isomorphic as fields. For a deeper understanding of the Hypothesis A, and generalizations of the uniqueness theorems of a maximally complete field see [**25**] where the Hypothesis A is put into a Galois theoretic setting.

7. Structures of Maximally Complete fields.

In this section we will describe the structures of maximally complete fields, which can be split into two cases: when the characteristics of the valued field and of its residue class field are equal (same characteristic case), and when they are different (mixed characteristic case).

7.1. Equal characteristic case.

DEFINITION 7.1 ([**42**, I.6, p. 23]). Let K be a field and G be an ordered abelian group. Let $f : G \times G \to K^*$ be a function such that for all $\alpha, \beta, \gamma \in G$:

$$f(\alpha,\beta) = f(\beta,\alpha)$$
$$f(0,0) = f(\alpha,0) = f(0,\beta)$$
$$f(\alpha,\beta+\gamma)f(\beta,\gamma) = f(\alpha+\beta,\gamma)f(\alpha,\beta).$$

The set $\{f(\alpha,\beta) : \alpha,\beta \in G\}$ is called a ***factor set*** for G in K. The field $K((G, f_{\alpha,\beta}))$ is defined as the group $(K((G)), +)$ with the multiplication induced by the relations $t^\alpha t^\beta = f(\alpha,\beta)t^{\alpha+\beta}$ for $\alpha,\beta \in G$. The Hahn valuation $\lambda : K((G, f_{\alpha,\beta})) \to G$, $\lambda(x) := \min\{\mathrm{supp}(x)\}$ is a general valuation on $K((G, f_{\alpha,\beta}))$.

THEOREM 7.2 ([**21**, Theorem 6], [**42**, VII.6 Theorem 6]). *Let F be a maximally complete field with value group G, residue class field k and $char(F) = char(k)$.*

1. *If $char(k) = 0$, then F is analytically equivalent to $(k((G, f_{\alpha,\beta})), \lambda)$ for some factor set $\{f(\alpha,\beta) : \alpha,\beta \in G\}$.*
2. *If $char(k) = p > 0$, and G and k satisfy the hypothesis A, then F is analytically equivalent to $(k((G, f_{\alpha,\beta})), \lambda)$ for some factor set $\{f(\alpha,\beta) : \alpha,\beta \in G\}$.*

With an extra hypothesis this result can be simplified as follows:

THEOREM 7.3 ([**21**, Theorem 8], [**42**, VII.6 Corollary]). *Let F be a maximally complete field with value group G and residue class field k such that $char(F) = char(k)$. Suppose that every element of k has an n-th root for all $n \in \mathbb{N}$.*

1. *If $char(k) = 0$, then F is analytically equivalent to $(k((G)), \lambda)$.*
2. *If $char(k) = p > 0$, and G and k satisfy the hypothesis A, then F is analytically equivalent to $(k((G)), \lambda)$.*

COROLLARY 7.4. *Let K be a field with a general valuation, a value group G and a residue class field k such that $char(K) = char(k)$. Suppose that every element of k has an n-th root for all $n \in \mathbb{N}$.*

1. *If $char(k) = 0$, then $(k((G)), \lambda)$ is the only maximally complete immediate extension of K up to analytical equivalence.*
2. *If $char(k) = p > 0$, and G and k satisfy the hypothesis A, then $(k((G)), \lambda)$ is the only maximally complete immediate extension of K up to analytical equivalence.*

PROOF. It follows immediately from 6.13, 6.17 and 7.3. □

COROLLARY 7.5 ([**21**, Corollary], [**42**, VII.6 Corollary], [**34**, Corollary 6]). *An arbitrary field K with general valuation v and residue class field k such that $char(K) = char(k)$, is analytically isomorphic to a subfield of $k^a((G))$ for some divisible ordered abelian group G. Furthermore, there exists a maximally complete immediate extension of K analytically isomorphic to a subfield of $k^a((G))$.*

The corollaries 7.4 and 7.5 are generalizations of the Hahn's Embedding Theorem (3.13) in the sense that K does not need to be an ordered field to be embedded in a general Hahn field. Nevertheless the value of the Hahn's Embedding Theorem lies in the order-preserving nature of the embedding when K is an ordered field.

DEFINITION 7.6. An ordered abelian group G is ***discrete*** if its isolated subgroups $\{G_\rho\}$ are well-ordered with respect to inclusion, and if each $G_{\rho+1}/G_\rho$ is isomorphic to $(\mathbb{Z}, +)$.

Another strong result about the structure of a maximally complete field is the following which does not invoke the hypothesis A.

THEOREM 7.7 ([**19**, Theorem]). *Let F be a maximally complete field with value group G and residue class field k such that $char(F) = char(k)$. If G is discrete, then F is analytically equivalent to $(k((G)), \lambda)$.*

COROLLARY 7.8. *Let K be a field with a general valuation, a discrete value group G, a residue class field k and $char(K) = char(k)$. Then $(k((G)), \lambda)$ is the only maximally complete immediate extension of K up to analytical equivalence.*

PROOF. It follows from 6.13 and 7.7. □

7.2. Mixed characteristic case.

THEOREM 7.9 ([**44**, II.5. Theorem 3]). *If R is a perfect field of characteristic $p > 0$, then there exists a unique ring $W(R)$ (up to isomorphism) of characteristic 0, Cauchy complete with respect to a discrete valuation v, $v(p) = 1 \in \mathbb{Z}$, and residue class field R. Also, the field of fractions $F(W(R))$ of $W(R)$ is the only field (up to isomorphism) of characteristic 0, Cauchy complete with respect to a discrete valuation v, $v(p) = 1 \in \mathbb{Z}$ and with residue class field R.*

In this context, $W(R) = \{x \in F(W(R)) : v(x) \geq 0\}$ is called the ring of Witt vectors with coefficients in R (see [**44**, II.5]). For example, if $R = \mathbb{F}_p$, then $W(R) = \mathbb{Z}_p$ and $F(W(R)) = \mathbb{Q}_p$.

In [**34**], Bjorn Poonen describes a maximally complete field of characteristic 0 with prescribed value group and perfect residue class field of positive characteristic. In other words, he describes an analog of a general Hahn field for the mixed characteristic case. The construction is as follows:

DEFINITION 7.10. Let R be a perfect field of characteristic $p > 0$, let G be an ordered abelian group and let $W(R)$ and $F(W(R))$ be as in 7.9. Define $W(R)((G))$ as the subring of the general Hahn field $F(W(R))((G))$ formed by all the generalized power series $\sum_g \alpha_g t^g$ where $g \in G$, $\alpha_g \in W(R)$, and $\{g : \alpha_g \neq 0\}$ is well-ordered. A series $\sum_g \alpha_g t^g \in W(R)((G))$ is ***null*** if for all $g \in G$, $\sum_{n\in\mathbb{Z}} \alpha_{g+n} p^n = 0$ in $F(W(R))$. The set N of all null series is an ideal of $W(R)((G))$ that contains the polynomial $t - p$ ([**34**, Proposition 3]).

THEOREM 7.11 ([**34**, Corollary 3]). *Let R be a perfect field of characteristic $p > 0$ and G be an ordered abelian group. The quotient*

$$L := W(R)((G))/N$$

is a field of characteristic 0, with a general valuation, a valuation group G and residue class field R. These fields are called ***p-adic Mal'cev-Neumann fields****.*

THEOREM 7.12 ([**34**, Proposition 4]). *Let R be a perfect field of characteristic $p > 0$ and let G be an ordered abelian group. If $S \subset W(R)$ is a set of representatives of the classes of R, then each series $\sum_g \alpha_g t^g$ in $W(R)((G))$ is null equivalent to a unique series of the form $\sum_g \beta_g t^g$ for $\beta_g \in S$. In other words, we have the following expansion for the elements of L:*

$$L = \left\{ \sum_{g\in G} \beta_g p^g : \beta_g \in S, \{g : \beta_g \neq 0\} \text{ is well-ordered} \right\}.$$

REMARK 7.13. The field $\mathbb{Q}_p$ is an example of a p-adic Mal'cev-Neumann field:

$$\mathbb{Q}_p = \mathbb{Z}_p((\mathbb{Z}))/N = W(\mathbb{F}_p)((\mathbb{Z}))/N = \left\{ \sum_{n=m}^{\infty} \beta_n p^n : m \in \mathbb{Z}, \beta_n = 0, 1, \ldots, p-1 \right\}.$$

The next result strengthens the analogy between general Hahn fields and p-adic Mal'cev-Neumann fields.

THEOREM 7.14 ([**34**, Theorem 1]). *Let G be an ordered abelian group and K be a perfect field of characteristic $p > 0$. The field $L = W(R)((G))/N$ is maximally complete under the Hahn valuation.*

REMARK 7.15. By 6.19, 7.11 and 7.14 the field $W(\mathbb{F}_p^a)((\mathbb{Q}))/N$ is the only maximally complete immediate extension of $\mathbb{Q}_p^a$ and of $\mathbb{C}_p$ up to analytic equivalence.

By 7.12 and 7.14, we can formulate the following corollary of 7.9:

COROLLARY 7.16. *If R is a perfect field of characteristic $p > 0$, then the p-adic Mal'cev-Neumann field $W(R)((\mathbb{Z}))/N$ is the only field (up to isomorphism) of characteristic 0, with a discrete valuation v such that the residue class field is R, $v(p) = 1 \in \mathbb{Z}$, and $W(R)((\mathbb{Z}))/N$ is Cauchy complete with respect to v. Furthermore $F(W(R)) = W(R)((\mathbb{Z}))/N$ is maximally complete.*

The next result is analogous to 7.5.

COROLLARY 7.17 ([**34**, Corollary 5, Corollary 6]). *An arbitrary field K of characteristic 0 with a general valuation and a residue class field k such that $char(k) = p > 0$, is analytically isomorphic to a subfield of $W(k^a)((G))/N$ for some divisible ordered abelian group G. Furthermore, there exists a maximally complete immediate extension of K analytically isomorphic to a subfield of $W(k^a)((G))/N$.*

8. Catalog of fields.

In the next table we summarize the fields and their properties presented so far. Let G be an ordered abelian group, let K be any field and let p be a positive prime integer. Assume that each of the listed fields has the usual non-Archimedean valuation (or general valuation) defined in this article and it is equipped with the respective induced topology. Denote the cardinalities of K and G by c and g respectively. In case that K is a perfect field of characteristic p, denote the field

$W(K)((G))/N$ by $\mathbb{K}_p^{sph}$ (see 7.10 and 7.11). Notice that the first 5 fields of the table form a chain of field extensions in the mixed characteristic case, i.e. each of the fields has a characteristic different than the characteristic of its residue class field:

$$\mathbb{Q} \subset \mathbb{Q}_p \subset \mathbb{Q}_p^a \subset \mathbb{C}_p \subset \mathbb{K}_p^{sph}.$$

The last 5 fields of the table form a chain of field extensions in the equal characteristic case, i.e. each of the fields has a characteristic equal to the characteristic of its residue class field:

$$K(x) \subset K((x)) \subset K\langle\langle x \rangle\rangle \subset L[\mathbb{Q}, K] \subset K((\mathbb{Q})).$$

Note that these chains are similar in several ways, for example the construction of the fields in the i-th position from the fields in the $(i-1)$-th position is identical (under certain conditions when it is necessary). Other metric, topological or algebraic similarities are easily seen from the table. If the m-th property is satisfied by the n-th field of the table, then the symbol ✓ will appear in the entry (m, n). Otherwise the symbol ✗ will take place. When a number (n) appears instead, the property is satisfied under certain conditions specified below the table.

	$\mathbb{Q}$	$\mathbb{Q}_p$	$\mathbb{Q}_p^a$	$\mathbb{C}_p$	$\mathbb{K}_p^{sph}$	$K(x)$	$K((x))$	$K\langle\langle x\rangle\rangle$	$L[\mathbb{Q}, K]$	$K((G))$
Totally disconnected	✓	✓	✓	✓	✓	✓	✓	✓	✓	✓
Separable	✓	✓	✓	✓	(1)	(2)	(2)	(2)	(2)	(2)
Cauchy complete	✗	✓	✗	✓	✓	✗	✓	✗	✓	✓
Spherically complete	✗	✓	✗	✗	✓	✗	✓	✗	✗	✓
Locally compact	✗	✓	✗	✗	(3)	✗	(4)	✗	✗	(4)
Algebraically Closed	✗	✗	✓	✓	(5)	✗	✗	(6)	(6)	(5)
Formally real	✓	✗	✗	✗	✗	(7)	(7)	(7)	(7)	(7)
Real-closed	✗	✗	✗	✗	✗	✗	(8)	(9)	(9)	(8)
Archimedean complete	✗	✗	✗	✗	✗	✗	(10)	✗	✗	(10)
Cardinality	$\aleph_0$	$\aleph_1$	$\aleph_1$	$\aleph_1$	c^g	$c^{\aleph_0}$	$c^{\aleph_0}$	$c^{\aleph_0}$	$c^{\aleph_0}$	c^g
Residue class field	$\mathbb{F}_p$	$\mathbb{F}_p$	$\mathbb{F}_p^a$	$\mathbb{F}_p^a$	K	K	K	K	K	K

(1) Under the assumption that G is a subgroup of $(\mathbb{R},+)$, the p-adic Mal'cev-Neumann field $\mathbb{K}_p^{sph}$ is separable if and only if K and G are countable (1.9). In particular, if $K=\mathbb{F}_p^a$ and $G=\mathbb{Q}$, then $\mathbb{K}_p^{sph}$ is separable and it is a field extension of $\mathbb{C}_p$, $\mathbb{Q}_p^a$, $\mathbb{Q}_p$ and $\mathbb{Q}$. Thus these subfields are separable.

(2) Under the assumption that G is a subgroup of $(\mathbb{R},+)$, the field $K((G))$ is separable if and only if K and G are countable (1.9). Since $K\subset K(x)\subset K((x))\subset K\langle\langle x\rangle\rangle\subset L[\mathbb{Q},K]\subset K((\mathbb{Q}))$, each of these fields is separable if and only if K is separable.

(3) Under the assumption that G is a subgroup of $(\mathbb{R},+)$, the p-adic Mal'cev-Neumann field $\mathbb{K}_p^{sph}$ is locally compact if and only if K is finite and G is cyclic (2.32, 7.11).

(4) Under the assumption that G is a subgroup of $(\mathbb{R},+)$, the field $K((G))$ is locally compact if and only if K is finite and G is cyclic (2.32, 3.16). In particular, $K((x))$ is locally compact if and only if K is finite.

(5) The field $K((G))$ is is algebraically closed if and only if K is algebraically closed and G is divisible (6.20, 6.18). In the case when K is a perfect field of characteristic p, the p-adic Mal'cev-Neumann field $\mathbb{K}_p^{sph}$ is algebraically closed if and only if K is algebraically closed and G is divisible (6.20, 7.14).

(6) Each of the fields $K\langle\langle x\rangle\rangle$ and $L[\mathbb{Q},K]$ is algebraically closed if and only if K is algebraically closed of characteristic 0 (4.15).

(7) The field $K((G))$ is formally real if and only if K is formally real (3.12). In particular, since $K\subset K(x)\subset K((x))\subset K\langle\langle x\rangle\rangle\subset L[\mathbb{Q},K]\subset K((\mathbb{Q}))$, each of these fields is formally real if and only if K is formally real.

(8) The field $K((G))$ is real-closed if and only if K is real-closed and G is divisible ([**3**, 6.23 (1)–(2)]). In particular, $K((x))$ is real-closed for no field K.

(9) Each of the fields $K\langle\langle x\rangle\rangle$ and $L[\mathbb{Q},K]$ is real-closed if and only if K is real-closed (4.12).

(10) $K((G))$ is Archimedean complete of type G if and only if K is isomorphic to $\mathbb{R}$ as ordered field (3.14).

9. Classification of fields.

The following diagram classifies the fields that admit a valuation or general valuation (non-trivial valuation when the field is infinite) in 6 non-overlapping classes. With a suitable choice of some of these classes it is possible to obtain partitions for: the class of ordered fields, the class of Archimedean valued fields, the class of non-Archimedean valued fields, and the class of fields with a general valuation. Below each of the 6 classes, there are some examples of their members.

Archimedean valued fields

non-Archimedean valued fields

Non-formally real Archimedean valued fields	Formally real Archimedean valued fields	Formally real fields where $G_K \neq \{0\}$ is non-Archimedean	Formally real non-Archimedean valued fields	Non-formally real non-Archimedean valued fields	non-formally real fields with general valuation of higher rank
$\mathbb{C}$, $\mathbb{Q}(i)$, $\mathbb{Q}(\sqrt{2}, i)$	$\mathbb{R}$, $\mathbb{Q}$, $\mathbb{Q}(\sqrt{3})$, $\mathbb{Q}(\pi)$	$\mathbb{R}((\mathbb{Z}^3))$, $L[\mathbb{R}^2, \mathbb{Q}]$, $\mathbb{Q}((x, y))$, $\mathbb{R}((L[\mathbb{Z}, \mathbb{Q}]))$	$\mathbb{R}(x)$, $\mathbb{R}((x))$, $\mathbb{Q}(e)\langle\langle x\rangle\rangle$, $L[\mathbb{Q}, \mathbb{R}]$, $\mathbb{Q}((\mathbb{Q}))$	$\mathbb{Q}_p$, $\mathbb{F}_q$, $\mathbb{C}_p$, $\mathbb{C}(x)$, $\mathbb{F}_q((x))$, $\mathbb{Q}^a\langle\langle x\rangle\rangle$, $\mathbb{Q}_p((\mathbb{Q}))$, $L[\mathbb{Q}, \mathbb{C}]$	$\mathbb{Q}_p((\mathbb{Z}^5))$, $\mathbb{F}_q((x, y))$, $\mathbb{C}((\mathbb{R} \oplus \mathbb{Q}))$, $L[\mathbb{Z}^2, \mathbb{Q}_p]$, $\mathbb{K}_p^{sph}$ with $K = \mathbb{F}_p$ and $G = \mathbb{Q}^3$

Ordered fields

Fields with general valuations

In the following, the 6 classes are ordered from left to right.

1. By 2.15 the second class can be described as the collection of all the fields that are isomorphic to a subfield of $\mathbb{R}$ or equivalently as the collection of all the Archimedean ordered fields.
2. By 5.2 the third class can be described as the collection of all the non-Archimedean ordered fields K for which G_K cannot be embedded in $(\mathbb{R}, +)$ ($G_K \neq \{0\}$ is not an Archimedean ordered group).
3. By 5.2 the fourth class can be described as the collection of all the non-Archimedean ordered fields K for which G_K can be embedded in $(\mathbb{R}, +)$ ($G_K \neq \{0\}$ is an Archimedean ordered group).

References

[1] Shreeram Abhyankar, *Two notes on formal power series*, Proc. Amer. Math. Soc. **7** (1956), 903–905, DOI 10.2307/2033558. MR0080647

[2] V. Alexandru, N. Popescu, and A. Zaharescu, *All valuations on $K(X)$*, J. Math. Kyoto Univ. **30** (1990), no. 2, 281–296, DOI 10.1215/kjm/1250520072. MR1068792

[3] Norman L. Alling, *Foundations of analysis over surreal number fields*, North-Holland Mathematics Studies, vol. 141, North-Holland Publishing Co., Amsterdam, 1987. Notas de Matemática [Mathematical Notes], 117. MR886475

[4] George Bachman, *Introduction to p-adic numbers and valuation theory*, Academic Press, New York-London, 1964. MR0169847

[5] Saugata Basu, Richard Pollack, and Marie-Françoise Roy, *Algorithms in real algebraic geometry*, 2nd ed., Algorithms and Computation in Mathematics, vol. 10, Springer-Verlag, Berlin, 2006. MR2248869

[6] J. L. Bell and M. Machover, *A course in mathematical logic*, North-Holland Publishing Co., Amsterdam-New York-Oxford, 1977. MR0472455

[7] N. Bourbaki, *Éléments de mathématique* (French), Masson, Paris, 1998. Algèbre commutative. Chapitre 10. [Commutative algebra. Chapter 10]; Profondeur, régularité, dualité. [Depth, regularity, duality]. MR2272929

[8] C. C. Chang and H. J. Keisler, *Model theory*, 3rd ed., Studies in Logic and the Foundations of Mathematics, vol. 73, North-Holland Publishing Co., Amsterdam, 1990. MR1059055

[9] Claude Chevalley, *Introduction to the Theory of Algebraic Functions of One Variable*, Mathematical Surveys, No. VI, American Mathematical Society, New York, N. Y., 1951. MR0042164

[10] A. H. Clifford, *Note on Hahn's theorem on ordered abelian groups*, Proc. Amer. Math. Soc. **5** (1954), 860–863, DOI 10.2307/2032549. MR0067882

[11] H. Garth Dales and W. Hugh Woodin, *Super-real fields*, London Mathematical Society Monographs. New Series, vol. 14, The Clarendon Press, Oxford University Press, New York, 1996. Totally ordered fields with additional structure; Oxford Science Publications. MR1420859

[12] J. de Groot, *Non-archimedean metrics in topology*, Proc. Amer. Math. Soc. **7** (1956), 948–953, DOI 10.2307/2033568. MR0080905

[13] David Eisenbud, *Commutative algebra. With a view toward algebraic geometry*, Graduate Texts in Mathematics, vol. 150, Springer-Verlag, New York, 1995. MR1322960

[14] Antonio J. Engler and Alexander Prestel, *Valued fields*, Springer Monographs in Mathematics, Springer-Verlag, Berlin, 2005. MR2183496

[15] Alain Escassut, *Analytic elements in p-adic analysis*, World Scientific Publishing Co., Inc., River Edge, NJ, 1995. MR1370442

[16] Hans Hahn, *Über die nichtarchimedischen größensysteme*, Hans Hahn Gesammelte Abhandlungen Band 1/Hans Hahn Collected Works Volume 1, Springer, 1995, pp. 445–499.

[17] James Forsythe Hall, *Completeness of ordered fields*, arXiv preprint arXiv:1101.5652 (2011).

[18] M. Hausner and J. G. Wendel, *Ordered vector spaces*, Proc. Amer. Math. Soc. **3** (1952), 977–982, DOI 10.2307/2031746. MR0052045

[19] Irving Kaplansky, *Maximal fields with valuations. II*, Duke Math. J. **12** (1945), 243–248. MR0012276

[20] Irving Kaplansky, *Fields and rings*, Chicago Lectures in Mathematics, University of Chicago Press, 1972.

[21] Irving Kaplansky, *Maximal fields with valuations*, Selected Papers and Other Writings, Springer Science & Business Media, 1995, pp. 1–22.

[22] Kiran S Kedlaya, *On the algebraicity of generalized power series*, arXiv preprint arXiv:1508.01836 (2015).

[23] Neal Koblitz, *p-adic numbers, p-adic analysis, and zeta-functions*, 2nd ed., Graduate Texts in Mathematics, vol. 58, Springer-Verlag, New York, 1984. MR754003

[24] Wolfgang Krull, *Allgemeine Bewertungstheorie* (German), J. Reine Angew. Math. **167** (1932), 160–196, DOI 10.1515/crll.1932.167.160. MR1581334

[25] Franz-Viktor Kuhlmann, Matthias Pank, and Peter Roquette, *Immediate and purely wild extensions of valued fields*, Manuscripta Math. **55** (1986), no. 1, 39–67, DOI 10.1007/BF01168612. MR828410

[26] Serge Lang, *Algebra*, 3rd ed., Graduate Texts in Mathematics, vol. 211, Springer-Verlag, New York, 2002. MR1878556

[27] Paul J. McCarthy, *Algebraic extensions of fields*, 2nd ed., Dover Publications, Inc., New York, 1991. MR1105534

[28] James R Munkres, *Topology*, Prentice Hall, 2000.

[29] Lawrence Narici, Edward Beckenstein, and George Bachman, *Functional analysis and valuation theory*, Marcel Dekker, Inc., New York, 1971. Pure and Applied Mathematics, Vol. 5. MR0361697

[30] P. Nyikos and H. C. Reichel, *On the structure of zerodimensional spaces*, Nederl. Akad. Wetensch. Proc. Ser. A **78** Indag. Math. **37** (1975), 120–136. MR0365527

[31] A. R. Pears, *Dimension theory of general spaces*, Cambridge University Press, Cambridge, England-New York-Melbourne, 1975. MR0394604

[32] C. Perez-Garcia and W. H. Schikhof, *Locally convex spaces over non-Archimedean valued fields*, Cambridge Studies in Advanced Mathematics, vol. 119, Cambridge University Press, Cambridge, 2010. MR2598517

[33] Yvette Perrin, *Fonctions analytiques dans les corps valués de rang supérieur à un* (French), Compositio Math. **49** (1983), no. 1, 51–74. MR699859

[34] Bjorn Poonen, *Maximally complete fields*, Enseign. Math. (2) **39** (1993), no. 1-2, 87–106. MR1225257

[35] Paulo Ribenboim, *The new theory of ultrametric spaces*, Period. Math. Hungar. **32** (1996), no. 1-2, 103–111, DOI 10.1007/BF01879736. MR1407913

[36] Paulo Ribenboim, *The theory of classical valuations*, Springer Monographs in Mathematics, Springer-Verlag, New York, 1999. MR1677964

[37] Alain M. Robert, *A course in p-adic analysis*, Graduate Texts in Mathematics, vol. 198, Springer-Verlag, New York, 2000. MR1760253

[38] Prabir Roy, *Nonequality of dimensions for metric spaces*, Trans. Amer. Math. Soc. **134** (1968), 117–132, DOI 10.2307/1994832. MR0227960

[39] W. H. Schikhof, *Isometrical embeddings of ultrametric spaces into non-Archimedean valued fields*, Nederl. Akad. Wetensch. Indag. Math. **46** (1984), no. 1, 51–53. MR748978

[40] W. H. Schikhof, *A crash course in p-adic analysis*, 2003.

[41] W. H. Schikhof, *Ultrametric calculus: An introduction to p-adic analysis*, Cambridge Studies in Advanced Mathematics, vol. 4, Cambridge University Press, Cambridge, 2006. Reprint of the 1984 original [MR0791759]. MR2444734

[42] O. F. G. Schilling, *The Theory of Valuations*, Mathematical Surveys, No. 4, American Mathematical Society, New York, N. Y., 1950. MR0043776

[43] Peter Schneider, *p-adic Lie groups*, Grundlehren der Mathematischen Wissenschaften [Fundamental Principles of Mathematical Sciences], vol. 344, Springer, Heidelberg, 2011. MR2810332

[44] Jean-Pierre Serre, *Local fields*, Graduate Texts in Mathematics, vol. 67, Springer-Verlag, New York-Berlin, 1979. Translated from the French by Marvin Jay Greenberg. MR554237

[45] Khodr Mahmoud Shamseddine, *New elements of analysis on the Levi-Civita field*, ProQuest LLC, Ann Arbor, MI, 1999. Thesis (Ph.D.)–Michigan State University. MR2700407

[46] Khodr Shamseddine and Martin Berz, *Analysis on the Levi-Civita field, a brief overview*, Advances in p-adic and non-Archimedean analysis, Contemp. Math., vol. 508, Amer. Math. Soc., Providence, RI, 2010, pp. 215–237, DOI 10.1090/conm/508/10002. MR2597696

[47] Niel Shell, *Topological fields and near valuations*, Monographs and Textbooks in Pure and Applied Mathematics, vol. 135, Marcel Dekker, Inc., New York, 1990. MR1075419

[48] Satish Shirali and Harkrishan L. Vasudeva, *Metric spaces*, Springer-Verlag London, Ltd., London, 2006. MR2161427

[49] A. C. M. van Rooij, *Non-Archimedean functional analysis*, Monographs and Textbooks in Pure and Applied Math., vol. 51, Marcel Dekker, Inc., New York, 1978. MR512894

Department of Mathematics, University of Manitoba, Winnipeg, Manitoba R3T 2N2, Canada

Email address: angelbarriac@gmail.com

Department of Physics and Astronomy, University of Manitoba, Winnipeg, Manitoba R3T 2N2, Canada

Email address: Khodr.Shamseddine@umanitoba.ca

Contemporary Mathematics
Volume **704**, 2018
http://dx.doi.org/10.1090/conm/704/14159

Non-compact Trace class operators and Schatten-class operators in p-adic Hilbert spaces

S. Basu

Dedicated to Professor T.S.S.R.K. Rao and Professor Ajit Iqbal Singh

ABSTRACT. We introduce Schatten-class operators in p-adic Hilbert spaces and study their properties and give several examples. We also show that the Trace class operators in p-adic Hilbert spaces strictly contains the class of completely continuous operators. We give examples of Trace class operators which are not completely continuous.

1. Introduction

Our primary goal in this paper is to introduce and initiate a study of the well known *Schatten r-class* operators and Trace Class operators in p-adic Hilbert spaces. We introduce the *Schatten r -class* operators on the p-adic Hilbert space E_ω. For the most part, the statements of the results are inspired by their classical counterparts, however, their proofs may depend heavily on the ultrametric nature of E_ω and the ground field $\mathbb{K}$. So as a result, though we are able to establish many of the known results in this new set up many a times we arrive at conclusions entirely different from the classical counterparts.We introduce the p-adic *Schatten r-class* operators. We prove that the adjoint of a *Schatten r-class* operator exists. We also prove that the set of all *Schatten r-class* operators is a left-sided ideal in the ring $\mathcal{B}_0(E_\omega)$ of all bounded operators which have adjoints. We show that a *Schatten r-class* operator is completely continuous in the sense that it is a limit, in $\mathcal{B}(E_\omega)$, of a sequence of operators of finite rank. The trace of an operator in p-adic Hilbert space was defined in [**2**]. We show that a *Schatten r-class* operator has a trace.We also show that the set of completely continuous operators in p-adic Hilbert is proper subset of the trace class operators which is in sharp contrast to the theory of operators in Classical Hilbert spaces. This gives a totally new perspective to the space of compact operators in p-adic Hilbert spaces. Apart from its intrinsic interest as discussed in this paper, p-adic functional analysis is a fast growing research area and has numerous applications in differential equations, statistics, quantum physics, dynamical systems, cognitive sciences, psychology and sociology, to name a few. For more on applications of p-adic functional analysis and other related issues see [**9**], [**6**], [**7**], [**8**] and [**14**].

2010 *Mathematics Subject Classification.* Primary 46S10.

Key words and phrases. p-adic Hilbert Space, Scahtten-class operators, Hilbert-Schimdt operators, Trace-class Operators.

2. Background

Let $\mathbb{K}$ be a complete ultrametric valued field. Classical examples of such a field include the field $\mathbb{Q}_p$ of p-adic numbers where p is a prime, and its various extensions, for example $\mathbb{C}_p$ the p-adic complex field and the field of formal Laurent series with coefficients in a given field ([**4**]). An ultrametric Banach space E over $\mathbb{K}$ is said to be a free Banach space if there exists a family $(e_i)_{i\in I}$ of elements of E such that each element $x \in E$ can be written uniquely as $x = \sum_{i\in I} x_i e_i$, that is, $\lim_{i\in I} x_i e_i = 0$ and $\|x\| = \sup_{i\in I} |x_i| \, \|e_i\|$. Such $(e_i)_{i\in I}$ is called an "orthogonal base" for E, and if $\|e_i\| = 1$, for all $i \in I$, then $(e_i)_{i\in I}$ is called an "orthonormal base".

Throughout this paper, we consider free Banach spaces over $\mathbb{K}$ and we shall assume that the index set to be countable and equal to the set of natural integers $\mathbb{N}$. For a detailed description and properties of these spaces, see [**3**],[**12**].

For a free Banach space E, let E'denote its topological dual and $\mathcal{B}(E)$ the set of all bounded linear operators on E. Both E' and $\mathcal{B}(E)$ are endowed with their respective usual norms.

For $(u, v) \in E \times E'$, we define $(v \otimes u)$ by:

$$\forall x \in E, \ (v \otimes u)(x) = v(x)\, u$$

then $(v \otimes u) \in \mathcal{B}(E)$ and $\|v \otimes u\| = \|v\| \,.\, \|u\|$.

Let $(e_i)_{i\in\mathbb{N}}$ be an orthogonal base for E, then one can define $e_i' \in E'$ by:

$$x = \sum_{i\in\mathbb{N}} x_i e_i, \ \ e_i'(x) = x_i.$$

It turns out that $\|e_i'\| = \dfrac{1}{\|e_i\|}$, furthermore, every $x' \in E'$ can be expressed as a pointwise convergent sum, $x' = \sum_{i\in\mathbb{N}} x'(e_i) e_i'$ and $\|x'\| = \sup_{i\in\mathbb{N}} \dfrac{|x'(e_i)|}{\|e_i\|}$.

Each operator A on E can be expressed as a pointwise convergent series, that is, there exists an infinite matrix $(a_{ij})_{(i,j)\in\mathbb{N}\times\mathbb{N}}$with coefficients in $\mathbb{K}$, such that, one has the pointwise convergent sum:

$$A = \sum_{i,j} a_{ij}(e_j' \otimes e_i), \text{ and for any } j \in \mathbb{N}, \ \lim_{i\to\infty} |a_{ij}| \, \|e_i\| = 0. \tag{2.1}$$

Moreover, for any $s \in \mathbb{N}$, $Ae_s = \sum_{i\in\mathbb{N}} a_{is} e_i$ and $\|A\| = \sup_{i,j} \dfrac{|a_{ij}| \, \|e_i\|}{\|e_j\|}$.

In this paper we shall work on a p-adic Hilbert space whose definition is given below. For more details, see [**1**] [**3**], [**6**].

Let $\omega = (\omega_i)_{i\in\mathbb{N}}$ be a sequence of non-zero elements in the valued field $\mathbb{K}$ and

$$E_\omega = \left\{ x = (x_i)_{i\in\mathbb{N}} \mid \forall i, \ \ x_i \in \mathbb{K} \ \text{ and } \lim_{i\to\infty} |x_i| \, |\omega_i|^{1/2} = 0 \right\}.$$

Then, it is easy to see that $x = (x_i)_{i\in\mathbb{N}} \in E_\omega$ if and only if $\lim_{i\to\infty} x_i^2 \omega_i = 0$. The space E_ω is a Banach space over $\mathbb{K}$, with the norm given by:

$$x = (x_i)_{i\in\mathbb{N}} \in E_\omega, \ \|x\| = \sup_{i\in\mathbb{N}} |x_i| \, |\omega_i|^{1/2}.$$

It is a free Banach space and it admits a *canonical orthogonal base*, namely, $(e_i)_{i\in\mathbb{N}}$, where e_i is the sequence all of whose terms are 0 except the i-th term which is 1, in other words, $e_i = (\delta_{ij})_{j\in\mathbb{N}}$, where δ_{ij} is the usual Kronecker symbol. We shall make extensive use of this canonical orthogonal base. We note that for each i, $\|e_i\| = |\omega_i|^{1/2}$. If $|\omega_i| = 1$, we shall refer to $(e_i)_{i\in\mathbb{N}}$ as the *canonical orthonormal base.*

Let $\langle , \rangle : E_\omega \times E_\omega \to \mathbb{K}$ be defined by:

$$\forall x, y \in E_\omega,\ x = (x_i)_{i\in\mathbb{N}},\ y = (y_i)_{i\in N},\ \langle x, y\rangle = \sum_{i\in\mathbb{N}} \omega_i x_i y_i.$$

Then, $\langle , \rangle$ is a symmetric, bilinear, non-degenerate form on E_ω, with value in $\mathbb{K}$, and it satisfies the Cauchy-Schwarz-Bunyakovsky inequality:

$$x, y \in E_\omega,\ |\langle x, y\rangle| \le \|x\| . \|y\| .$$

On the canonical orthogonal base, we have:

$$\langle e_i, e_j\rangle = \omega_i \delta_{ij} = \begin{cases} 0, & if\ i \neq j \\ \omega_i, & if\ i = j. \end{cases}$$

The space E_ω endowed with this form $\langle , \rangle$ is called a *p-adic Hilbert space.*
Let us notice that in a p-adic Hilbert space, it may happen that,

$$|\langle x, x\rangle| = 0 \text{ while } \|x\| \neq 0.$$

For example: let $p \equiv 1\ (mod\ 4)$, $\mathbb{K} = \mathbb{Q}_p$, $\omega = (\omega_i)_{i\in\mathbb{N}}$, where

$$\omega_0 = 1,\ \omega_1 = 1,\ \omega_i = p^i \text{ for } i \ge 2.$$

If we consider the non-zero vector $x = (x_i)_{i\in\mathbb{N}} \in E_\omega$, where

$$x_0 = 1,\ x_1 = \sqrt{-1} \in \mathbb{K},\ x_i = 0,\ i \ge 2.$$

Then $\langle x, x\rangle = 0$ but $\|x\| = 1$.

It should also be observed that for every operator A in $\mathcal{B}(E_\omega)$, the domain $D(A)$ $(D(A) := \{x = (x_i) \in E_\omega :\ \lim_i |x_i|\|Ae_i\| = 0\})$, of A, is actually the whole of E_ω. This is due to the fact that for every $x = \sum x_s e_s$ in E_ω,

$$|x_s|\,\|Ae_s\| \le \|A\|\,|x_s|\,\|e_s\| = \|A\|\,|x_s|\,|\omega_s|^{1/2} .$$

Throughout the paper we denote by $\mathcal{B}_0(E_\omega)$ the set of all operators A in $\mathcal{B}(E_\omega)$ which have adjoints A^* with respect to the non-degenerate form $\langle , \rangle$ defined above.

It is well known that the operator $A = \sum_{ij} a_{ij}(e'_j \otimes e_i) \in \mathcal{B}(E_\omega)$ is actually in $\mathcal{B}_0(E_\omega)$ if and only if $\forall i$, $\lim_{j\to\infty} \frac{|a_{ij}|}{|\omega_j|^{1/2}} = 0$ [**4**]. The space $\mathcal{B}_0(E_\omega)$ is stable under the operation of taking an adjoint and for any $A \in \mathcal{B}_0(E_\omega) : (A^*)^* = A$ and $\|A\| = \|A^*\|$.

3. Schatten-class Operators in Classical Hilbert spaces

Let us now recall few well known results from the Schatten class operators in classical Hilbert spaces.

Let H be a Hilbert space and let $T : H \longrightarrow H$ be a compact operator. Then the operator $|T|$, defined as

$$|T| = (TT^*)^{\frac{1}{2}}$$

is also a compact operator hence its spectrum consists of at most countably many distinct Eigenvalues. Let $\sigma_j(|T|)$ denote the eigen values of $|T|$. These numbers are called the singular numbers of T. Let $(\sigma_j(|T|)$ denote the non increasing sequence of the singular numbers of T, every number counted according to its multiplicity as an eigenvalue of $|T|$.

DEFINITION 3.1. For $0 < r < \infty$, let $\mathcal{B}_r(H)$ denote the following

$$\mathcal{B}_r(H) = \{T \in \mathcal{B}(H) : \sum_j \sigma_j(|T|^r) < \infty|\}$$

Then $\mathcal{B}_r(H)$ is called the *Schatten r-class* operators of H.

To introduce Schatten r-class operators in p-adic Hilbert spaces, we first look at the following straightforward observation for $\mathcal{B}_r(H)$. For more details, see [**15**].

PROPOSITION 3.2. *Let $\mathcal{B}_r(H)$ denote the Schatten r-class operators of H and let $\{e_j\}$ be the eigenvectors corresponding to the eigenvalues$\{\sigma_j(|T|)\}$. Then $T \in \mathcal{B}_r(H)$ if and only if $\sum_j \|T(e_j)\|^r < \infty$.*

4. Schatten Class operators in p-adic Hilbert spaces

Now we are in a position to define the analogue of *Schatten r-class* Operators in p-adic Hilbert space set up.

DEFINITION 4.1. An operator $A \in \mathcal{B}(E_\omega)$ is said to be in *Schatten r-class* ($1 \leq r < \infty$) denoted by $\mathcal{B}_r(E_\omega)$ if

$$\|A\|_r = (\sum_{s\geq 0} \frac{\|A(e_s)\|^r}{|\omega_s|^{\frac{r}{2}}})^{\frac{1}{r}} < \infty.$$

REMARK 4.2. If $r = 2$, we get the p-adic Hilbert Schimdt operators as in [**2**]

We look at some examples :

EXAMPLE 4.3. Suppose $\mathbb{K} = \mathbb{Q}_p$. Let, $\omega_0 = 1, \omega_i = m_i p$ where $i = pq_i + m_i$ if $1 \leq m_i \leq p - 1$, or in other words m_i is the remainder,when i is divided by p, and $\omega_i = p$, otherwise.

We define an operator A on E_ω as, $A = \sum a_{ij} e_j^{'} \otimes e_i$, where $a_{ij} = p^{i+j}, 0 \leq i \leq j$ and $a_{ij} = 0, i > j$. Then, clearly $\lim_i |a_{ij}| \|e_i\| = 0$, hence $A \in \mathcal{B}(E_\omega)$. Also,

$$\|A(e_s)\| = \|\sum_{i=0}^{\infty} a_{is} e_i\| = \sup_{0 \leq i \leq s} p^{-(i+s)} = p^{-s}$$

For $s \geq 1$,

$$\frac{\|A(e_s)\|}{|\omega_s|^{\frac{1}{2}}} \leq p^{-(s+\frac{1}{2})}.$$

Therefore,

$$\sum_{s=0}^{\infty}\frac{\|A(e_s)\|^r}{|\omega_s|^{\frac{r}{2}}}=1+p^{-r/2}\sum_{s=1}^{\infty}p^{-rs}=1+\frac{p^{-3r/2}}{1-p^{-r}}=1+\frac{1}{p^{3r/2}-p^{r/2}}=\|A\|_r<\infty$$

Hence $A\in\mathcal{B}_r(E_\omega)$.

EXAMPLE 4.4. Suppose $\mathbb{K}=\mathbb{Q}_p$. .Let, $\omega_i=p^{i+1}$, hence $|\omega_i|=p^{-i-1}\to 0$ as $i\to\infty$.

We define an operator A on E_ω as, $A=\sum a_{ij}e_j^{'}\otimes e_i$ where, $a_{ij}=\omega_i^j$
Then, clearly $\lim_i |a_{ij}|\|e_i\|=0$, hence $A\in\mathcal{B}(E_\omega)$
Also,

$$\begin{aligned}\|A(e_s)\| &= \|\sum_{i=0}^{\infty}a_{is}e_i\|\\ &= \sup_i |a_{is}|\|e_i\|\\ &= \sup_i |\omega_i|^s|\omega_i|^{1/2}\\ &= \sup_i |\omega_i|^{s+1/2}\\ &= \sup_i |p^{(2s+1)(i+1)/2}|\\ &= p^{-(2s+1)/2}\end{aligned}$$

Hence, for $s\geq 1$,

$$\frac{\|A(e_s)\|}{|\omega_s|^{\frac{1}{2}}}=\frac{p^{\frac{s+1}{2}}}{p^{\frac{2s+1}{2}}}=p^{\frac{-s}{2}}$$

Therefore,

$$\sum_{s=0}^{\infty}\frac{\|A(e_s)\|^r}{|\omega_s|^{\frac{r}{2}}}=\sum p^{-\frac{rs}{2}}=\frac{1}{1-p^{\frac{r}{2}}}<\infty.$$

Hence $A\in\mathcal{B}_r(E_\omega)$.

PROPOSITION 4.5. $\mathcal{B}_r(E_\omega)\subseteq\mathcal{B}_q(E_\omega)$for $r\leq q$ and $r,q\geq 1$.

PROOF. The proof follows from the definition of $\mathcal{B}_r(E_\omega)$ and the fact that $\ell_r\subseteq\ell_q$, for $r\leq q$ and $r,q\geq 1$. □

PROPOSITION 4.6. *If $A\in\mathcal{B}_r(E_\omega)$, then*

(1) *A has an adjoint A^*. i.e., $\mathcal{B}_r(E_\omega)\subseteq\mathcal{B}_0(E_\omega)$.*
(2) $\|A\|\leq\|A\|_r$

PROOF. We know from the general theory of operators on ultrametric Hilbert spaces (see [**4**],[**5**]and [**10**],) A^* exists if and only if

$$\lim_j\frac{|a_{ij}|}{|\omega_j|^{\frac{1}{2}}}=0\quad\text{for all}\quad i$$

Again because of the convergence of the real series,

$$\sum_{s\geq 0}\frac{\|A(e_s)\|^r}{|\omega_s|^{\frac{r}{2}}},$$

we have

$$\lim_s \frac{\|A(e_s)\|}{|\omega_s|} = 0.$$

Since

$$\|A(e_s)\| = \sup_i |a_{is}||\omega_i|^{\frac{1}{2}},$$

it follows that

$$\lim_s \frac{|a_{is}|}{|\omega_s|^{\frac{1}{2}}} = 0,$$

for all i. Hence $\mathcal{B}_r(E_\omega) \subseteq \mathcal{B}_0(E_\omega)$.

By definition,

$$\|A\|_r^r = \sum_s \frac{\|A(e_s)\|^r}{|\omega_s|^{\frac{r}{2}}} = \sum_s \frac{\sup_i |a_{is}|^r |\omega_i|^{\frac{r}{2}}}{|\omega_s|^{\frac{r}{2}}} \geq \sup_s \sup_i \frac{|a_{is}|^r |\omega_i|^{\frac{r}{2}}}{|\omega_s|^{\frac{r}{2}}} = \|A\|^r$$

Hence we have the inequality in part (2). □

REMARK 4.7. If $r = 2$, this is the Hilbert Schimdt norm of A and this norm satisfies the Cauchy-Schwarz-Bunyakovsky inequality where the inner product is given by $< A, B >= \sum_s < Ae_s, Be_s > /\omega_s$. ([**2**]).

PROPOSITION 4.8. *For any p-adic Hilbert space E_ω, the following is true*

(1) *$\mathcal{B}_r(E_\omega)$ is a left ideal of $\mathcal{B}(E_\omega)$.*
(2) *If $A, B \in \mathcal{B}_r(E\omega)$, then $AB \in \mathcal{B}_r(E_\omega)$ and $\|AB\|_r \leq \|A\|_r \ \|B\|_r$, i.e. $\mathcal{B}_r(E_\omega)$ is a normed algebra with respect to the Schatten r-norm.*
(3) *$\mathcal{B}_r(E_\omega)$ is complete with respect to $\| \ \|_r$.*

PROOF. Suppose $A, B \in \mathcal{B}_r(E_\omega)$, then

$$\begin{aligned}
\|A+B\|_r &= \left(\sum_{s\geq 0} \frac{\|(A+B)(e_s)\|^r}{|\omega_s|^{\frac{r}{2}}}\right)^{1/r} \\
&\leq \left(\left(\sum_{s\geq 0} \frac{\|A(e_s)\|}{|\omega_s|^{\frac{1}{2}}} + \frac{\|B(e_s)\|}{|\omega_s|^{\frac{1}{2}}}\right)^r\right)^{1/r} \\
&\leq \left(\sum_{s\geq 0} \frac{\|A(e_s)\|^r}{|\omega_s|^{\frac{r}{2}}}\right)^{1/r} + \left(\sum_{s\geq 0} \frac{\|B(e_s)\|^r}{|\omega_s|^{\frac{r}{2}}}\right)^{1/r} \\
&\leq \|A\|_r + \|B\|_r
\end{aligned}$$

Also, for $A \in \mathcal{B}(E_\omega)$, and $B \in \mathcal{B}_r(E_\omega)$,

$$\|AB\|_r^r = \sum_{s\geq 0} \frac{\|AB(e_s)\|^r}{|\omega_s|^{\frac{r}{2}}} \leq \|A\|^r \sum_{s\geq 0} \frac{\|B(e_s)\|^r}{|\omega_s|^{\frac{r}{2}}} \leq \|A\|^r \|B\|_r^r < \infty.$$

Hence $AB \in \mathcal{B}_r(E_\omega)$ and $\mathcal{B}_r(E_\omega)$ is a left ideal of $\mathcal{B}(E_\omega)$ and also of $\mathcal{B}_0(E_\omega)$.

Now $A, B \in \mathcal{B}_r(E_\omega)$.Since $\|AB\|_r \leq \|A\| \ \|B\|_r$, from Proposition 4.6, it follows that

$$\|AB\|_r \leq \|A\| \ \|B\|_r \leq \|A\|_r \ \|B\|_r,$$

and one obtains (2).

To prove $\mathcal{B}_r(E_\omega)$ is complete with respect to $\| \|_r$, suppose $\{T_n\}$ is a Cauchy sequence in $\mathcal{B}_r(E_\omega)$ with respect to $\|\|_r$. Hence $\|T_m - T_n\|_r \to 0$ as $m, n \to \infty$. Since $\|T_m - T_n\| \leq \|T_m - T_n\|_r$, it follows that $\{T_n\}$ is a Cauchy sequence in $\mathcal{B}_0(E_\omega)$ and hence there exists $T \in \mathcal{B}_0(E_\omega)$, such that $\|T_n - T\| \to 0$ as $n \to \infty$. We will show $T \in \mathcal{B}_r(E_\omega)$. Since $\{T_n\}$ is a Cauchy sequence in $\mathcal{B}_r(E_\omega)$, for any $\epsilon > 0$, there exists $m, n \geq N$ such that $(\sum_{s=0}^{\infty} \|T_m(e_s) - T_n(e_s)\|^r/|\omega_s|^{r/2}|)^1/r < \epsilon$. Since $\|T_n - T\| \to 0$ as $n \to \infty$, it follows that $\|T_n(e_s) - T(e_s)\| \to 0$ as $n \to \infty$, for all s. So for all $m, n \geq N$, $\sum_{s=0}^{k} \|T_m(e_s) - T_n(e_s)\|^r/|\omega_s|^{r/2}| < \epsilon^r$. Letting $n \to \infty$, we obtain $\sum_{s=0}^{k} \|T_m(e_s) - T(e_s)\|^r/|\omega_s|^{r/2}| < \epsilon^r$. Now letting $k \to \infty$, we have $\sum_{s=0}^{\infty} \|T_m(e_s) - T(e_s)\|^r/|\omega_s|^{r/2}| < \epsilon^r$, for all $m \geq N$. This implies $(T - T_m) \in \mathcal{B}_r(E_\omega)$, since $\mathcal{B}_r(E_\omega)$ is a linear subspace of $\mathcal{B}_0(E_\omega)$, it follows that $T = T_m + (T - T_m) \in \mathcal{B}_r(E_\omega)$ Hence $\mathcal{B}_r(E_\omega)$ is complete with respect to the $\| \|_r$ norm. □

5. Completely Continuous operators and Schatten-class operators

Let us first introduce a few notations.

Notations. (1) We let

$$c_0(E_\omega', (|\omega_s|)_{s \in N}) = \{(a_s)_{s \in N} \in (E_\omega')^N \mid \lim_{s \to \infty} \|a_s\| \, |\omega_s|^{1/2} = 0\}.$$

$c_0(E_\omega^*)$ is a Banach space over $\mathbb{K}$ with the norm $\|(a_s)_s\| = \sup_s \|a_s\| \, |\omega_s|^{1/2}$.

(2) We let

$$F_\omega = l^\infty(N, \frac{1}{|\omega_s|}_{s \in N}) = \{(x_s)_{s \in N} \in (\mathbb{K})^N \mid \sup_s \frac{|x_s|}{|\omega_s|^{1/2}} < \infty\}.$$

$l^\infty(N, \frac{1}{|\omega_s|})_{s \in N}$ is a Banach space over $\mathbb{K}$ with the norm $\|(x_s)_s\| = \sup_s \frac{|x_s|}{|\omega_s|^{1/2}}$.

PROPOSITION 5.1. As Banach spaces over $\mathbb{K}$, E_ω^* is isomorphic to F_ω.

PROOF. As in [**12**], the mapping $a \to (a_s(x))_s$ gives the desired isomorphism. □

DEFINITION 5.2. An operator $A \in \mathcal{B}(E_\omega)$ is *completely continuous* if it is the limit, in $\mathcal{B}(E_\omega)$, (i.e., a uniform limit) of a sequence of operators of *finite rank*. We denote by $\mathcal{C}(E_\omega)$ the subspace of all completely continuous operators on E_ω.

DEFINITION 5.3. For any operator $A \in \mathcal{B}(E_\omega)$, we define A_s, the s-component of A by the formula:

$$x \in E_\omega, \ Ax = \sum_{s \geq 0} a_s(x) e_s.$$

It is clear that $(a_s(x))_s \in E_\omega^*$.

PROPOSITION 5.4. *The mapping $A \to (A_s)_s$ is an isomorphism of the Banach space $\mathcal{C}(E_\omega)$ onto the Banach space $c_0(E_\omega, |\omega_s|, s \geq 0)$.*

PROOF. The proof is similar to that of [Proposition 4, [**12**]]. □

The following result was proved by Diarra in [**5**]. We include it for completeness.

PROPOSITION 5.5. [**5**] *The space* $\mathcal{C}(E_\omega)$ *of completely continuous operators which admits an adjoint is a closed two sided ideal of the unitary ultrametric Banach algebra* $\mathcal{B}_0(E_\omega)$. *Moreover* $\mathcal{C}(E_\omega)$ *can be identified isometrically to the topological tensor product space* $E'_{\omega,0}\hat{\otimes}E_\omega$, *where* $E'_{\omega,0}$ *is the closed linear subspace of* E'_ω *with orthogonal basis* $(e'_j)_{j\in I}$.

REMARK 5.6. It is proved in [**14**] that $\mathcal{C}(E_\omega)$ is isometrically isomorphic to the topological tensor product $E'_\omega\hat{\otimes}E_\omega$.

PROPOSITION 5.7. *Every Schatten r-class operator is completely continuous, i.e,* $\mathcal{B}_r(E_\omega)\subset\mathcal{C}(E_\omega)$.

PROOF. Let A be a *Schatten r-class* operator. Let $A=\sum_{(i,j)\in\mathbb{N}\times\mathbb{N}} a_{ij}\left(e'_j\otimes e_i\right)$

Since $A\in\mathcal{B}_r(E_\omega)$ it follows that $\lim_s\sup_i\frac{|a_{si}||\omega_i|^{\frac{1}{2}}}{|\omega_s|^{\frac{1}{2}}}=0$. Also since A is bounded, $\lim_i|a_{is}||\omega_i|^{\frac{1}{2}}=0$, for all s. Then, arguing as in [**5**], it follows that,

$$\lim_{i,s}\frac{|a_{is}||\omega_i|^{\frac{1}{2}}}{|\omega_s|^{\frac{1}{2}}}=0.$$

Hence, by Proposition 5.5, it follows that the family $(a_{ij}e'_j\otimes e_i)$ is norm summable and furthermore, $A\in E'_{\omega,0}\hat{\otimes}E_\omega=\mathcal{C}(E_\omega)$. □

6. Trace and Schatten-class operators

One important notion in the classical theory is that of *trace*. The *trace* of an operator in the p-adic case was defined in [**2**].

DEFINITION 6.1. For $A\in\mathcal{B}(E_\omega)$, we define the trace of A to be

$$TrA=\sum_s\frac{\langle Ae_s,e_s\rangle}{\omega_s}$$

if this series converges in $\mathbb{K}$. We denote by $\mathcal{TC}(E_\omega)$ the subspace of all Trace class operators, namely, those bounded operators for which the trace exists.

The following remark is easily verified.

REMARK 6.2. Let $A\in\mathcal{B}(E_\omega)$. Then $A\in\mathcal{TC}(E_\omega)$, if and only if $\lim_{s\to\infty}a_{ss}=0$. Moreover $Tr(A)=\sum_{s\geq0}a_{ss}$.

It is known that for any ultrametric Banach space E, any completely continuous operator on E, has a trace. We verify this fact here using matricial form of bounded operators.

THEOREM 6.3. $\mathcal{C}(E_\omega)\subseteq\mathcal{TC}(E_\omega)$, *i.e., every completely continuous operator has a trace.*

PROOF. Let $A\in C(E_\omega)$. By Proposition 5.4, it follows that $A\in c_0\left(E^*_\omega\right)$. Hence

$$\lim_s\|A_s\|\,|\omega_s|^{1/2}=\lim_s\sup_i\frac{|a_{si}|\,|\omega_s|^{1/2}}{|\omega_i|^{1/2}}=0.$$

Therefore,

$$\lim_s|a_{ss}|=\lim_s\frac{|a_{ss}||\omega_s|^{\frac{1}{2}}}{|\omega_s|^{\frac{1}{2}}}\leq\lim_s\sup_i\frac{|a_{si}|\,|\omega_s|^{1/2}}{|\omega_i|^{1/2}}=0.$$

This implies $A \in \mathcal{TC}\,(E_\omega)$. □

COROLLARY 6.4. *$\mathcal{B}_r(E_\omega) \subset \mathcal{TC}\,(E_\omega)$, i.e., every Schatten r-class operator has a trace.*

PROOF. Follows from the fact that

$$\mathcal{B}_r(E_\omega) \subseteq \mathcal{C}(E_\omega) \subseteq \mathcal{TC}\,(E_\omega).$$

Let us notice that setting

$$|Tr(A)|_r = \left(\sum_{s\geq 0} |a_{ss}|^r\right)^{\frac{1}{r}},$$

since

$$|a_{ss}| \leq \frac{\|A(e_s\|}{|\omega_s|^{\frac{1}{2}}},$$

one easily sees that, $|Tr(A)|_r \leq \|A\|_r$. □

REMARK 6.5. The above theorem is in sharp contrast to the classical case, since in this case, the trace class operators are strictly contained in the class of completely continuous operators. In fact, for a classical Hilbert space H, the trace class operator $B_1(H)$ is a subset of the algebra of Hilbert Schimdt operators in H. In the p-adic case however, we definitely have $B_1(E_\omega) \subseteq B_2(E_\omega)$ but the class of operators with a trace is a larger class.

To illustrate, $\mathcal{C}(E_\omega)$ is a proper subspace of $\mathcal{TC}(E_\omega)$, we have the following example.

EXAMPLE 6.6. Suppose $\mathbb{K} = \mathbb{Q}_p$ and suppose $|\omega_s| = 1$ for all s. Let

$$\begin{aligned} a_{ij} &= 1 \quad if \quad i < j \\ &= p^i \quad if \quad i = j \\ &= 0 \quad if \quad i > j \end{aligned}$$

For $s > 0$, one has $A(e_s) = \sum_{i=0}^{s-1} e_i + p^s e_s$, since $a_{is} = 0$ for all $i > s$, one has $\lim_i |a_{is}||\omega_i|^{\frac{1}{2}} = 0$. Moreover, $\|A(e_s)\| = max(1, |p^s|) = 1, \forall s \geq 0$ and this implies $\|A\| = 1$. Since $a_{ss} = p^s$, one has

$$Tr(A) = \sum_{s\geq 0} a_{ss} = \sum_{s\geq 0} p^s = \frac{1}{1-p}$$

and $A \in \mathcal{TC}\,(E_\omega)$. On the other hand, for any $s \geq 0$, one has

$$\sup_{i\geq 0} \frac{|a_{si}||\omega_i|}{|\omega_s|^{\frac{1}{2}}} = \sup_{i\geq s} |a_{si}| = 1.$$

Hence

$$\lim_{s\to\infty} \sup_{i\geq 0} \frac{|a_{si}||\omega_i|}{|\omega_s|^{\frac{1}{2}}} = 1$$

and $A \notin \mathcal{C}(E_\omega)$.

PROPOSITION 6.7. *$\mathcal{TC}\,(E_\omega)$ is a closed linear subspace of $\mathcal{B}\,(E_\omega)$ and the map $Tr : \mathcal{TC}\,(E_\omega) \to \mathbb{K}$ is a continuous linear functional with $|Tr(A)| \leq \|A\|$.*

PROOF. Let $A, B \in \mathcal{TC}(E_\omega)$. Then

$$\begin{aligned} Tr(A+B) &= \sum_s \frac{\langle (A+B)e_s, e_s \rangle}{\omega_s} \\ &= \sum_s \frac{\langle (Ae_s + Be_s), e_s \rangle}{\omega_s} \\ &= \sum_s \frac{\langle Ae_s, e_s \rangle}{\omega_s} + \sum_s \frac{\langle Be_s, e_s \rangle}{\omega_s} \\ &= Tr(A) + Tr(B). \end{aligned}$$

Also for $\lambda \in \mathbb{K}$ one has ,

$$Tr(\lambda A) = \sum_s \frac{\langle (\lambda A)e_s, e_s \rangle}{\omega_s} == \lambda \sum_s \frac{\langle A_s, e_s \rangle}{\omega_s} = \lambda Tr(A)$$

Now,

$$\|A\| = \sup_{i,j} \frac{|a_{ij}| \, \|e_i\|}{\|e_j\|} \geq \sup_i \frac{|a_{ii}| \, \|e_i\|}{\|e_i\|} = \sup_i |a_{ii}| \geq |Tr(A)|.$$

Also, let $A_n \in \mathcal{TC}(E_\omega)$, and $A_n \to A$, where $A \in \mathcal{B}(E_\omega)$. Suppose,

$$A_n = \sum_{i,j} a_{ij}(n) e_j^{'} \otimes e_i$$

and

$$A = \sum_{i,jd} a_{ij} e_j^{'} \otimes e_i.$$

Then, since

$$\begin{aligned} |a_{ii} - a_{ii}(n)| &\leq \sup_{i,j} \frac{|a_{ij} - a_{ij}(n)| \|e_i\|}{\|e_j\|} \\ &= \|A - A_n\| \end{aligned}$$

and for each integer $n \geq 1$, $\lim_{i \to +\infty} a_{ii}(n) = 0$, one sees first, that for any $i \in \mathbb{N}$, the sequence $(a_{ii}(n))_{n \geq 1}$ converges towards a_{ii} and, second, that the convergence is uniform with respect to i. Hence the limits $a_{ii} = \lim_{n \to \infty} a_{ii}(n)$ are such that $\lim_{i \to \infty} a_{ii} = 0$. It follows that the $Tr(A) = \sum_i a_{ii}$ exists with $Tr(A) = \lim_{n \to \infty} Tr(A_n)$.

Hence $\mathcal{TC}(E_\omega)$ is a linear closed subspace of $\mathcal{B}(E_\omega)$. From the above arguments, it is clear that $Tr : \mathcal{TC}(E_\omega) \to \mathbb{K}$ is a continuous linear functional. □

The following theorem is known for general ultrametric Banach spaces, free or not. Here we provide a proof based on the matrix expansion of bounded operators on the ultrametric Hilbert space E_ω.

THEOREM 6.8. *Suppose $A, B \in C(E_\omega)$, then $Tr(AB) = Tr(BA)$.*

PROOF. Let $A = \sum_{i,j} a_{ij} \left(e_j' \otimes e_i\right)$ and $B = \sum_{i,j} b_{ij} \left(e_j' \otimes e_i\right)$, then

$$\begin{aligned} Tr(AB) &= \sum_i \sum_j a_{ij} b_{ji}, \\ Tr(BA) &= \sum_j \sum_i a_{ij} b_{ji}. \end{aligned}$$

Since A and B are in $\mathcal{C}(E_\omega)$,using the identification from Proposition 5.4, it follows that, for every $\epsilon > 0$, there exists N_0 such that for $i, j \geq N_0$ and for any $k, l \in \mathbb{N}$,

$$\frac{|a_{kj}||\omega_k|}{|\omega_j|} < \sqrt{\epsilon} \quad \text{and} \quad \frac{|b_{li}||\omega_l|}{|\omega_i|} < \sqrt{\epsilon}.$$

We put $k = i$ and $l = j$, and we get, after a multiplication:

$$\text{For} \quad i, j \geq N_0, |a_{ij}||b_{ji}| < \epsilon.$$

Therefore, we can commute the double summation. □

Acknowledgments

The author is deeply grateful to the referee for deep insights for improvement, innumerable and valuable suggestions, comments and corrections.

References

[1] S. Albeverio, J. M. Bayod, C. Perez-Garcia, R. Cianci, and A. Khrennikov, *Non-Archimedean analogues of orthogonal and symmetric operators and p-adic quantization*, Acta Appl. Math. **57** (1999), no. 3, 205–237, DOI 10.1023/A:1006219101760. MR1722049

[2] S. Basu, T. Diagana, and F. Ramaroson, *A p-adic version of Hilbert-Schmidt operators and applications*, J. Anal. Appl. **2** (2004), no. 3, 173–188. MR2092641

[3] B. Diarra, *An operator on some ultrametric Hilbert spaces*, J. Anal. **6** (1998), 55–74. MR1671148

[4] B. Diarra, *Geometry of the p-adic Hilbert Spaces*, Preprint (1999).

[5] B. Diarra, *Bounded linear operators on ultrametric Hilbert spaces*, Afr. Diaspora J. Math. (N.S.) **8** (2009), no. 2, 173–181. MR2538713

[6] A. Yu. Khrennikov, *Mathematical methods in non-Archimedean physics* (Russian), Uspekhi Mat. Nauk **45** (1990), no. 4 (274), 79–110, 192, DOI 10.1070/RM1990v045n04ABEH002378; English transl., Russian Math. Surveys **45** (1990), no. 4, 87–125. MR1075387

[7] A. Yu. Khrennikov, *Generalized functions on a non-Archimedean superspace* (Russian), Izv. Akad. Nauk SSSR Ser. Mat. **55** (1991), no. 6, 1257–1286; English transl., Math. USSR-Izv. **39** (1992), no. 3, 1209–1238. MR1152212

[8] A. Yu. Khrennikov, *p-adic quantum mechanics with p-adic valued functions*, J. Math. Phys. **32** (1991), no. 4, 932–937, DOI 10.1063/1.529353. MR1097779

[9] H. A. Keller and H. Ochsenius A., *Algebras of bounded operators on nonclassical orthomodular spaces*, Proceedings of the International Quantum Structures Association, Part III (Castiglioncello, 1992), Internat. J. Theoret. Phys. **33** (1994), no. 1, 1–11, DOI 10.1007/BF00671609. MR1263295

[10] S. Ludkovsky and B. Diarra, *Spectral integration and spectral theory for non-Archimedean Banach spaces*, Int. J. Math. Math. Sci. **31** (2002), no. 7, 421–442, DOI 10.1155/S016117120201150X. MR1926812

[11] H. Ochsenius and W. H. Schikhof, *Banach spaces over fields with an infinite rank valuation*, p-adic functional analysis (Poznań, 1998), Lecture Notes in Pure and Appl. Math., vol. 207, Dekker, New York, 1999, pp. 233–293. MR1703500

[12] J.-P. Serre, *Endomorphismes complètement continus des espaces de Banach p-adiques* (French), Inst. Hautes Études Sci. Publ. Math. **12** (1962), 69–85. MR0144186

[13] J. R. Retherford, *Hilbert space: compact operators and the trace theorem*, London Mathematical Society Student Texts, vol. 27, Cambridge University Press, Cambridge, 1993. MR1237405

[14] A. C. M. van Rooij, *Non-Archimedean functional analysis*, Monographs and Textbooks in Pure and Applied Math., vol. 51, Marcel Dekker, Inc., New York, 1978. MR512894

[15] J. Weidmann, *Linear operators in Hilbert spaces*, Graduate Texts in Mathematics, vol. 68, Springer-Verlag, New York-Berlin, 1980. Translated from the German by Joseph Szücs. MR566954

Department of Mathematics, George Washington University, Washington DC 20052
Email address: sbasu@gwu.edu
Email address: sudeshna66@gmail.com

Contemporary Mathematics
Volume **704**, 2018
http://dx.doi.org/10.1090/conm/704/14160

Calculus on a non-Archimedean field extension of the real numbers: inverse function theorem, intermediate value theorem and mean value theorem

Gidon Bookatz and Khodr Shamseddine

ABSTRACT. In this paper, we introduce the concept of weakly locally uniformly differentiable functions (WLUD) on $\mathcal{N}$, a non-Archimedean field extension of the real numbers that is real closed and Cauchy complete in the topology induced by the order. We show that WLUD functions are C^1 and they form an $\mathcal{N}$-algebra that is closed under composition and contains all polynomial functions.

We formulate and prove a version of the inverse function theorem as well as a local intermediate value theorem for these functions. Then we generalize the WLUD concept to higher orders of differentiability and study WLUD^n functions at a point or on a subset of $\mathcal{N}$. In particular, we study the properties of WLUD^2 functions and we formulate and prove a local mean value theorem for such functions.

1. Introduction

We start this section by reviewing some basic terminology and facts about non-Archimedean fields. So let F be an ordered non-Archimedean field extension of $\mathbb{R}$. We introduce the following terminology.

DEFINITION 1 ($\sim$, $\approx$, $\ll$, S_F, λ). For $x, y \in F^* := F \setminus \{0\}$, we say that x is of the same order as y and write $x \sim y$ if there exist $n, m \in \mathbb{N}$ such that $n|x| > |y|$ and $m|y| > |x|$, where $|\cdot|$ denotes the ordinary absolute value on F: $|x| = \max\{x, -x\}$. For nonnegative $x, y \in F$, we say that x is infinitely smaller than y and write $x \ll y$ if $nx < y$ for all $n \in \mathbb{N}$, and we say that x is infinitely small if $x \ll 1$ and x is finite if $x \sim 1$; finally, we say that x is approximately equal to y and write $x \approx y$ if $x \sim y$ and $|x - y| \ll |x|$. We also set $\lambda(x) = [x]$, the class of x under the equivalence relation $\sim$.

The set of equivalence classes S_F (under the relation $\sim$) is naturally endowed with an addition via $[x] + [y] = [x \cdot y]$ and an order via $[x] < [y]$ if $|y| \ll |x|$ (or $|x| \gg |y|$), both of which are readily checked to be well-defined. Note that we use $+$ instead of $\cdot$ for the operation in S_F because, for the fields discussed in this paper, S_F is isomorphic to an additive subgroup of $\mathbb{R}$. It follows that $(S_F, +, <)$ is an ordered group, often referred to as the Hahn group or skeleton group, whose neutral element is $[1]$, the class of 1. It follows from the above that the projection λ from F^* to S_F is a valuation.

The theorem of Hahn [**2**] provides a complete classification of non-Archimedean extensions of $\mathbb{R}$ in terms of their skeleton groups. In fact, invoking the axiom of choice it is shown that the elements of any such ordered field F can be written as (generalized) formal power series (also called Hahn series) over its skeleton group S_F with real coefficients, and the set of appearing exponents forms a well-ordered subset of S_F. That is, for all $x \in F$, we have

$$x = \sum_{q \in S_F} a_q d^q; \tag{1.1}$$

with $a_q \in \mathbb{R}$ for all q, d a positive infinitely small element of F, and the support of x, given by

$$\operatorname{supp}(x) := \{q \in S_F : a_q \neq 0\},$$

forming a well-ordered subset of S_F. With the representation of elements of F as in Equation (1.1) it follows that, for $x \neq 0$ in F,

$$\lambda(x) = \min\left(\operatorname{supp}(x)\right),$$

which exists since $\operatorname{supp}(x)$ is well-ordered. Moreover, we set $\lambda(0) = \infty$.

Addition, multiplication and order on the Hahn series are defined as follows. Given $x = \sum_{q \in \operatorname{supp}(x)} a_q d^q$ and $y = \sum_{t \in \operatorname{supp}(y)} b_t d^t$, then

$$x + y = \sum_{r \in \operatorname{supp}(x) \cup \operatorname{supp}(y)} (a_r + b_r) d^r; \text{ and}$$

$$x \cdot y = \sum_{r \in \operatorname{supp}(x) \oplus \operatorname{supp}(y)} \left(\sum_{\substack{q \in \operatorname{supp}(x),\, t \in \operatorname{supp}(y) \\ q + t = r}} a_q \cdot b_t \right) d^r. \tag{1.2}$$

Note that, since $\operatorname{supp}(x)$ and $\operatorname{supp}(y)$ are well-ordered, only finitely many terms contribute to the sum

$$\sum_{\substack{q \in \operatorname{supp}(x),\, t \in \operatorname{supp}(y) \\ q + t = r}} a_q \cdot b_t,$$

in Equation (1.2), for each $r \in \operatorname{supp}(x) \oplus \operatorname{supp}(y)$.

Given a nonzero $x = \sum_{q \in \operatorname{supp}(x)} a_q d^q$, then $x > 0$ if and only if $a_{\lambda(x)} > 0$.

From general properties of formal power series fields [**6**, **8**], it follows that if S_F is divisible then F is real closed; that is, every positive element of F is a square in F and every polynomial of odd degree over F has at least one root in F. For a general overview of the algebraic properties of formal power series fields, we refer to the comprehensive overview by Ribenboim [**9**], and for an overview of the related valuation theory the book by Krull [**3**]. A thorough and complete treatment of ordered structures can also be found in [**7**].

Throughout this paper, we will denote by $\mathcal{N}$ any totally ordered non-Archimedean field extension of $\mathbb{R}$ that is real closed and complete in the order topology and whose skeleton group $S_{\mathcal{N}}$ is Archimedean, i.e. a subgroup of $\mathbb{R}$. The coefficient a_q of the qth power in the Hahn representation of a given x will be denoted by $x[q]$, and hence the number d is given by $d[1] = 1$ and $d[q] = 0$ for $q \neq 1$. It is easy to check that, for $q \in S_{\mathcal{N}}$, $0 < d^q \ll 1$ if and only if $q > 0$, and $d^q \gg 1$ if and only if $q < 0$; moreover, $x \approx x[\lambda(x)] d^{\lambda(x)}$ for all $x \neq 0$.

The smallest such field $\mathcal{N}$ is the Levi-Civita field $\mathcal{R}$, first introduced in [**4**,**5**]. In this case $S_{\mathcal{R}} = \mathbb{Q}$, and for any element $x \in \mathcal{R}$, $\text{supp}(x)$ is a left-finite subset of $\mathbb{Q}$, i.e. below any rational bound r there are only finitely many exponents in the Hahn representation of x. The Levi-Civita field $\mathcal{R}$ is of particular interest because of its practical usefulness. Since the supports of the elements of $\mathcal{R}$ are left-finite, it is possible to represent these numbers on a computer [**1**]. Having infinitely small numbers allows for many computational applications; one such application is the computation of derivatives of real functions representable on a computer [**13**], where both the accuracy of formula manipulators and the speed of classical numerical methods are achieved. For a review of the Levi-Civita field $\mathcal{R}$, see [**12**,**14**] and references therein.

In the wider context of valuation theory, it is interesting to note that the topology induced by the order on $\mathcal{N}$ is the same as that introduced via the valuation λ, as shown in Remark 1 below. It follows therefore that the field $\mathcal{N}$ is just a special case of the class of fields discussed in [**11**].

REMARK 1. *The mapping $\Lambda : \mathcal{N} \times \mathcal{N} \to \mathbb{R}$, given by $\Lambda(x, y) = \exp(-\lambda(x - y))$, is an ultrametric distance (and hence a metric); the valuation topology it induces is equivalent to the order topology (we will use τ_v to denote either one of the two topologies in this paper). For if A is an open set in the order topology and $a \in A$, then there exists $r > 0$ in $\mathcal{N}$ such that, for all $x \in \mathcal{N}$, $|x - a| < r \Rightarrow x \in A$. Let $l = \exp(-\lambda(r))$, then we also have that, for all $x \in \mathcal{N}$, $\Lambda(x, a) < l \Rightarrow x \in A$; and hence A is open with respect to the valuation topology. The other direction of the equivalence of the topologies follows analogously.*

It follows from Remark 1 that $\mathcal{N}$ which is complete in the order topology is also complete in the valuation topology induced by the ultrametric Λ.

REMARK 2. *Contrary to the field $\mathbb{R}^*$ of Nonstandard Analysis* [**10**,**18**], *the field $\mathcal{N}$ is an ordered field extension of the field of real numbers $\mathbb{R}$; and the enmbedding of $\mathbb{R}$ in $\mathcal{N}$ is compatible with the orders in $\mathbb{R}$ and $\mathcal{N}$. While in Nonstandard Analysis there is a generally valid transfer principle that allows the transformation of known results of conventional analysis, here all relevant calculus theorems are developed separately. Moreover, besides being non-Archimedeanly valued, the fact that the Levi-Civita field $\mathcal{R}$ has a total order (which is also non-Archimedean) gives the field a richer structure, thus opening up new possibilities of study, like monotonicity, which are not available in other non-Archimedean valued fields like the p-Adic fields for example* [**11**]. *This makes $\mathcal{N}$ an outstanding example, worthy to be studied in detail in its own right.*

The following results were proved in [**17**]; they show that the topological structure of $\mathcal{N}$ is different from that of $\mathbb{R}$ or $\mathbb{C}$, and that makes doing Calculus on the field more difficult.

- $(\mathcal{N}, \tau_v)$ is a totally disconnected topological space. It is Hausdorff and nowhere locally compact. There are no countable bases. The topology induced to $\mathbb{R}$ is the discrete topology. As an immediate consequence of the fact that $(\mathcal{N}, \tau_v)$ is totally disconnected, it follows that, for any $x_0 \in \mathcal{N}$, the connected component of x_0 is $\{x_0\}$; moreover, the topology is zero-dimensional, that is, there is a base of clopen sets for the topology.
- If we view $\mathcal{N}$ as an infinite dimensional vector space over $\mathbb{R}$ then τ_v is not a vector topology; that is, $(\mathcal{N}, \tau_v)$ is not a linear topological space.

- If A is compact in $(\mathcal{N}, \tau_v)$ then A is closed and bounded and it has an empty interior in $(\mathcal{N}, \tau_v)$, that is,
$$\text{int}(A) := \{a \in A : \exists r > 0 \text{ in } \mathcal{N} \ni (a-r, a+r) \subset A\} = \emptyset.$$
The converse is not true: the set $A = [0,1] \cap \mathbb{Q}$ is a (countably infinite) closed and bounded subset of $\mathcal{N}$ with an empty interior; but A is not compact in $(\mathcal{N}, \tau_v)$ [**17**].
- Given a sequence (x_n) of elements of $\mathcal{N}$, the series $\sum_{n=1}^{\infty} x_n$ converges if and only if the sequence (x_n) converges to zero.

In [**17**] we studied the properties of locally uniformly differentiable (LUD) functions on $\mathcal{N}$. In particular, we showed that this class of functions is closed under addition, multiplication and composition of functions. Then we stated and proved local versions of the inverse function theorem and the intermediate value theorem for $\mathcal{N}$-valued LUD functions on $\mathcal{N}$. The stronger condition (local uniform differentiability) on the function than that of the real case was needed for the proofs of both theorems because of the total disconnectedness of the field $\mathcal{N}$ in the order topology. Then in [**15**], we generalized the definition of local uniform differentiability to any order. Then we studied the properties of n-times locally uniformly differentiable (LUD^n) functions and we formulated and proved a local mean value theorem for $\mathcal{N}$-valued functions that are LUD^2 at a point of $\mathcal{N}$.

In this paper, we introduce a new smoothness criterion which we call weak local uniform differentiability (WLUD) which is strictly weaker than local uniform differentiability and strictly stronger than continuous differentiability (C^1), we study the properties of $\mathcal{N}$-valued WLUD and WLUD^n functions and we show that this weaker criterion is sufficient to get all the nice calculus results obtained in [**15, 17**].

2. WLUD Functions

DEFINITION 2. Let $A \subseteq \mathcal{N}$ be open, let $f : A \to \mathcal{N}$, and let $x_0 \in A$ be given. We say that f is weakly locally uniformly differentiable (abbreviated as WLUD) at x_0 if f is differentiable in a neighbourhood of x_0 in A and if for every $\epsilon > 0$ in $\mathcal{N}$ there exists $\delta > 0$ in $\mathcal{N}$ such that for every $x, y \in (x_0 - \delta, x_0 + \delta) \cap A$ we have that $|f(y) - f(x) - f'(x)(y-x)| < \epsilon |y - x|$. Moreover, we say that f is WLUD on A if f is WLUD at every point in A.

One may notice, that the definition of WLUD is very similar to that of LUD in [**15, 17**], and indeed it is easy to see that LUD implies WLUD. However, as we will prove later, the two concepts are not equivalent. Nonetheless, similar to the case with LUD functions, we get that the class of WLUD functions is contained in the class of C^1 functions, and is an $\mathcal{N}$-algebra that is closed under composition of functions.

NOTATION 1. Throughout the rest of the paper, we will use $B(x_0, r)$ to denote the open interval $(x_0 - r, x_0 + r)$, for $x_0 \in \mathcal{N}$ and $r > 0$ in $\mathcal{N}$.

PROPOSITION 1. *Let $A \subseteq \mathcal{N}$ be open and let $f : A \to \mathcal{N}$ be WLUD at $x_0 \in A$. Then f is C^1 at x_0.*

PROOF. Let $\epsilon > 0$ in $\mathcal{N}$ be given. Then there exists $\delta > 0$ in $\mathcal{N}$ such that, for every $x, y \in B(x_0, \delta) \cap A$, we have that
$$|f(y) - f(x) - f'(x)(y-x)| < \frac{\epsilon}{2} |y - x| .$$

It follows that, for all $x \in B(x_0, \delta) \cap A$, we have that

$$\begin{aligned} |f'(x) - f'(x_0)| &\leq \left| f'(x) - \frac{f(x_0) - f(x)}{x_0 - x} \right| + \left| \frac{f(x) - f(x_0)}{x - x_0} - f'(x_0) \right| \\ &< \frac{\epsilon}{2} + \frac{\epsilon}{2} = \epsilon. \end{aligned}$$

□

PROPOSITION 2. *Let $f, g : A \to \mathcal{N}$ be WLUD at $x_0 \in A$ and let $\alpha \in \mathcal{N}$ be given. Then $(f + \alpha g)$ is WLUD at x_0.*

PROOF. Without loss of generality, we may assume $\alpha \neq 0$. Let $\epsilon > 0$ in $\mathcal{N}$ be given. Then there exists $\delta > 0$ in $\mathcal{N}$ such that, for every $x, y \in B(x_0, \delta) \cap A$, we have that

$$|f(y) - f(x) - f'(x)(y - x)| < \frac{\epsilon}{2}|y - x|$$

and

$$|g(y) - g(x) - g'(x)(y - x)| < \frac{\epsilon}{2|\alpha|}|y - x|.$$

Hence, for every $x, y \in B(x_0, \delta) \cap A$, we have that

$$\begin{aligned} &|(f{+}\alpha g)(y) - (f + \alpha g)(x) - (f' + \alpha g')(x)(y - x)| \\ &\quad \leq |f(y) - f(x) - f'(x)(y - x)| + |\alpha|\,|g(y) - g(x) - g'(x)(y - x)| \\ &\quad < \frac{\epsilon}{2}|y - x| + |\alpha|\frac{\epsilon}{2|\alpha|}|y - x| \\ &\quad = \epsilon|y - x|. \end{aligned}$$

□

PROPOSITION 3. *Let $f, g : A \to \mathcal{N}$ be WLUD at $x_0 \in A$. Then fg is WLUD at x_0.*

PROOF. Let $\epsilon > 0$ in $\mathcal{N}$ be given. By proposition 1, f and g are C^1 at x_0, and so there exists a $\delta_c > 0$ in $\mathcal{N}$ such that $|f(x) - f(x_0)| < 1$, $|g(x) - g(x_0)| < 1$, and $|f'(x) - f'(x_0)| < 1$ on $B(x_0, \delta_c) \cap A$. Moreover, there exist $\delta_f, \delta_g, \delta_0 > 0$ in $\mathcal{N}$ such that

$$|f(y) - f(x) - f'(x)(y - x)| < \frac{\epsilon}{3(|g(x_0)| + 1)}|y - x|$$

if $x, y \in B(x_0, \delta_f) \cap A$;

$$|g(y) - g(x) - g'(x)(y - x)| < \frac{\epsilon}{3(|f(x_0)| + 1)}|y - x|$$

if $x, y \in B(x_0, \delta_g) \cap A$; and

$$|g(y) - g(x)| < \frac{\epsilon}{3(|f'(x_0)| + 1)}$$

if $x, y \in B(x_0, \delta_0) \cap A$.

Let $\delta = \min\{\delta_c, \delta_f, \delta_g, \delta_0\}$. Then it follows that, for every $x, y \in B(x_0, \delta) \cap A$, we have that

$$\begin{aligned}
&\left|f(y)g(y) - f(x)g(x) - \Big(f'(x)g(x) + f(x)g'(x)\Big)(y - x)\right| \\
&\le |f(y)g(y) - f(x)g(y) - f'(x)g(y)(y - x)| \\
&\quad + |f(x)g(y) - f(x)g(x) - f(x)g'(x)(y - x)| \\
&\quad + |f'(x)g(y)(y - x) - f'(x)g(x)(y - x)| \\
&= |g(y)|\,|f(y) - f(x) - f'(x)(y - x)| \\
&\quad + |f(x)|\,|g(y) - g(x) - g'(x)(y - x)| \\
&\quad + |f'(x)|\,|g(y) - g(x)|\,|y - x| \\
&< \frac{|g(y)|}{3(|g(x_0)| + 1)}\epsilon|y - x| + \frac{|f(x)|}{3(|f(x_0)| + 1)}\epsilon|y - x| + \frac{|f'(x)|}{3(|f'(x_0)| + 1)}\epsilon|y - x| \\
&< \epsilon|y - x|.
\end{aligned}$$

□

PROPOSITION 4. *Let $g : A \to B$ be WLUD at $x_0 \in A$ and $f : B \to \mathcal{N}$ be WLUD at $g(x_0) \in B$. Then $f \circ g : A \to \mathcal{N}$ is WLUD at x_0.*

PROOF. Let $\epsilon > 0$ in $\mathcal{N}$ be given. As g is WLUD at x_0, and f is WLUD at $g(x_0)$, there exist $\delta_1, \delta_2 > 0$ in $\mathcal{N}$ such that

$$|f(y) - f(x) - f'(x)(y - x)| < \frac{\epsilon}{2\,(1 + |g'(x_0)|)}|y - x|$$

for every $x, y \in B \cap B(g(x_0), \delta_1)$; and

$$|g(y) - g(x) - g'(x)(y - x)| < \frac{\epsilon}{2\Big(|f'(g(x_0))| + 1\Big)}|y - x|$$

for every $x, y \in A \cap B(x_0, \delta_2)$.

Moreover, as f is C^1 at $g(x_0)$ and g is C^1 at x_0, there exists a $\delta_3 > 0$ in $\mathcal{N}$ such that for every $x, y \in A \cap B(x_0, \delta_3)$ we have that $|g(y) - g(x)| < \delta_1$ and $|f'(g(x)) - f'(g(x_0))| < 1$. Finally, since g is WLUD at x_0, there exists $\delta_4 > 0$ in $\mathcal{N}$ such that for every $x, y \in A \cap B(x_0, \delta_4)$ we have that

$$|g(y) - g(x)| < (1 + |g'(x_0)|)\,|y - x|.$$

Let $\delta = \min\{\delta_1, \delta_2, \delta_3, \delta_4\}$. Then for every $x, y \in A \cap B(x_0, \delta)$ we have that

$$\begin{aligned}
&|f(g(y)) - f(g(x)) - g'(x)f'(g(x))(y - x)| \\
&\le \left|f(g(y)) - f(g(x)) - f'(g(x))\Big(g(y) - g(x)\Big)\right| \\
&\quad + |f'(g(x))|\,|g(y) - g(x) - g'(x)(y - x)| \\
&< \frac{\epsilon}{2(1 + |g'(x_0)|)}\,|g(y) - g(x)| + \frac{|f'(g(x))|}{2\Big(|f'(g(x_0))| + 1\Big)}\epsilon|y - x| \\
&< \epsilon|y - x|.
\end{aligned}$$

□

In the following, we will extend the WLUD concept to higher orders of differentiability and we will define WLUD^n analogously to how LUD^n was defined in [**15**].

DEFINITION 3. Let $A \subseteq \mathcal{N}$ be open, let $f : A \to \mathcal{N}$, let $x_0 \in A$, and let $n \in \mathbb{N}$ be given. We say that f is WLUD^n at x_0 if f is n times differentiable in a neighbourhood of x_0 in A and if for every $\epsilon > 0$ in $\mathcal{N}$ there exists $\delta > 0$ in $\mathcal{N}$ such that for every $x, y \in B(x_0, \delta) \cap A$ we have that

$$\left| f(y) - \sum_{k=0}^{n} \frac{f^{(k)}(x)}{k!}(y-x)^k \right| < \epsilon \left| y - x \right|^n .$$

Moreover, we say that f is WLUD^n on A if f is WLUD^n at every point in A.

REMARK 3. *If we use this definition to make an analogous concept of WLUD^0, we get the condition that for every $\epsilon > 0$ in $\mathcal{N}$ there exists $\delta > 0$ in $\mathcal{N}$ such that for every $x, y \in B(x_0, \delta) \cap A$ we have that $|f(y) - f(x)| < \epsilon$, which can be fairly easily seen to be an equivalent statement to that of f being continuous at x_0.*

Just as it is the case for LUD^n, in the real case WLUD^n is equivalent to C^n. Moreover, as we will see (in Proposition 5 below), this definition of WLUD^n implies WLUD^{n-1} (for $n \in \mathbb{N}$).

LEMMA 1. *Let $A \subseteq \mathcal{N}$ be open and let $f : A \to \mathcal{N}$ be WLUD^n at $x_0 \in A$. Then $f^{(n)}$ is locally bounded at x_0; that is, there exist a neighborhood U of x_0 in A and an $M > 0$ in $\mathcal{N}$ such that, for every $x \in U$, we have that $\left|f^{(n)}(x)\right| \leq M$.*

PROOF. As f is WLUD^n at x_0, there exists $\delta_1 > 0$ in $\mathcal{N}$ such that $B(x_0, \delta_1) \subset A$ and, for every $x, y \in B(x_0, \delta_1)$, we have that

$$\left| f(y) - \sum_{k=0}^{n} \frac{f^{(k)}(x)}{k!}(y-x)^k \right| < \left| y - x \right|^n .$$

For each $k \in \{0, 1, \ldots, n-1\}$, $f^{(k)}$ is continuous at x_0, and so there exist a neighborhood V_k of x_0 in A and a number $M_k > 0$ in $\mathcal{N}$ such that for every $x \in V_k$ we have that $\left|f^{(k)}(x)\right| \leq M_k$. Let $\delta > 0$ in $\mathcal{N}$ be such that $B(x_0, \delta) \subseteq B(x_0, \delta_1) \cap \left(\cap_{k=0}^{n-1} V_k\right)$ and let $U = B(x_0, \delta)$. Now let

$$M = n! \left(\frac{2}{\delta}\right)^n \left(\left(\frac{\delta}{2}\right)^n + M_0 + \sum_{k=0}^{n-1} \frac{M_k}{k!} \left(\frac{\delta}{2}\right)^k \right),$$

and let $x \in U$ be given. Choose

$$y = \begin{cases} x + \frac{\delta}{2} & \text{if } x + \frac{\delta}{2} \in B(x_0, \delta) \\ x - \frac{\delta}{2} & \text{otherwise.} \end{cases}$$

Then we have that $x, y \in U \subseteq B(x_0, \delta_1)$ and thus,

$$\begin{aligned} \left|f^{(n)}(x)\right| &\leq \frac{n!}{|y-x|^n} \left(\left| f(y) - \sum_{k=0}^{n} \frac{f^{(k)}(x)}{k!}(y-x)^k \right| + |f(y)| + \sum_{k=0}^{n-1} \left| \frac{f^{(k)}(x)}{k!}(y-x)^k \right| \right) \\ &< \frac{n!}{|y-x|^n} \left(|y-x|^n + M_0 + \sum_{k=0}^{n-1} \frac{M_k}{k!} |y-x|^k \right) \\ &= n! \left(\frac{2}{\delta}\right)^n \left(\left(\frac{\delta}{2}\right)^n + M_0 + \sum_{k=0}^{n-1} \frac{M_k}{k!} \left(\frac{\delta}{2}\right)^k \right) = M. \end{aligned}$$

□

PROPOSITION 5. *Let $A \subseteq \mathcal{N}$ be open and let $f : A \to \mathcal{N}$ be WLUD^n at $x_0 \in A$ for some $n \in \mathbb{N}$. Then f is WLUD^{n-1} at x_0.*

PROOF. By Lemma 1, there exists a neighborhood U of x_0 in A such that $f^{(n)}$ is locally bounded by some $M > 0$ on U. Let $\epsilon > 0$ in $\mathcal{N}$ be given. As f is WLUD^n at x_0, there exists $\delta_1 > 0$ in $\mathcal{N}$ such that, for every $x, y \in B(x_0, \delta_1)$, we have that

$$\left| f(y) - \sum_{k=0}^{n} \frac{f^{(k)}(x)}{k!}(y-x)^k \right| < \frac{\epsilon}{2} |y-x|^n .$$

Moreover, we may assume without loss of generality, that $B(x_0, \delta_1) \subseteq U$. Let $\delta = \min\{\delta_1, 1, n!\epsilon/(2M)\}$. Then, for every $x, y \in B(x_0, \delta)$, we have that

$$\begin{aligned} \left| f(y) - \sum_{k=0}^{n-1} \frac{f^{(k)}(x)}{k!}(y-x)^k \right| &\leq \left| f(y) - \sum_{k=0}^{n} \frac{f^{(k)}(x)}{k!}(y-x)^k \right| \\ &\quad + \left| \frac{f^{(n)}(x)}{n!}(y-x)^n \right| \\ &< \frac{\epsilon}{2} |y-x|^n + \left| \frac{f^{(n)}(x)}{n!} \right| \delta \, |y-x|^{n-1} \\ &\leq \frac{\epsilon}{2} |y-x| \, |y-x|^{n-1} + \frac{M}{n!} \frac{n!\epsilon}{2M} |y-x|^{n-1} \\ &< \epsilon \, |y-x|^{n-1} . \end{aligned}$$

□

PROPOSITION 6. *Let $A \subseteq \mathcal{N}$ be open and let $f : A \to \mathcal{N}$ be WLUD^2 at $x_0 \in A$. Then f is C^2 at x_0.*

PROOF. Let $\epsilon > 0$ in $\mathcal{N}$ be given. Then there exists $\delta_1 > 0$ in $\mathcal{N}$ such that, for every $x, y \in B(x_0, \delta_1)$, we have that

$$\left| f(y) - f(x) - f'(x)(y-x) - \frac{1}{2} f''(x)(y-x)^2 \right| < \frac{\epsilon}{6}(y-x)^2.$$

As f' is differentiable at x_0, there exists $\delta_2 > 0$ in $\mathcal{N}$ such that, for every $x \in A$ satisfying $0 < |x - x_0| < \delta_2$, we have that

$$\left| \frac{f'(x) - f'(x_0)}{x - x_0} - f''(x_0) \right| < \frac{\epsilon}{6}.$$

Let $\delta = \min\{\delta_1, \delta_2\}$. Then it follows that if $0 < |x - x_0| < \delta$ then we have that

$$
\begin{aligned}
|f''(x) - f''(x_0)| &\le 2\left|\frac{1}{2}f''(x) + \frac{f'(x)}{x_0 - x} + \frac{f(x) - f(x_0)}{(x_0 - x)^2}\right| \\
&\quad + 2\left|\frac{f'(x) - f'(x_0)}{x - x_0} - f''(x_0)\right| \\
&\quad + 2\left|\frac{1}{2}f''(x_0) + \frac{f'(x_0)}{x - x_0} + \frac{f(x_0) - f(x)}{(x - x_0)^2}\right| \\
&= 2\left|\frac{f(x_0) - f(x)}{(x_0 - x)^2} - \frac{f'(x)}{x_0 - x} - \frac{1}{2}f''(x)\right| \\
&\quad + 2\left|\frac{f'(x) - f'(x_0)}{x - x_0} - f''(x_0)\right| \\
&\quad + 2\left|\frac{f(x) - f(x_0)}{(x - x_0)^2} - \frac{f'(x_0)}{x - x_0} - \frac{1}{2}f''(x_0)\right| \\
&< 2\frac{\epsilon}{6} + 2\frac{\epsilon}{6} + 2\frac{\epsilon}{6} = \epsilon.
\end{aligned}
$$

□

PROPOSITION 7. *Let $A \subseteq \mathcal{N}$ be open and let $f : A \to \mathcal{N}$ be WLUD^2 at $x_0 \in A$. Then f' is WLUD at x_0.*

PROOF. Let $\epsilon > 0$ in $\mathcal{N}$ be given. As f is WLUD^2 at x_0, there exists $\delta_1 > 0$ in $\mathcal{N}$ such that $B(x_0, \delta_1) \subset A$ and, for any distinct $x, y \in B(x_0, \delta_1)$, we have that

$$
\left|\frac{f(y) - f(x)}{y - x} - f'(x) - \frac{1}{2}f''(x)(y - x)\right| < d\epsilon|y - x|.
$$

Moreover, by Proposition 6, f is C^2 at x_0, and hence there exists $\delta_2 > 0$ in $\mathcal{N}$ such that $B(x_0, \delta_2) \subset A$ and, for every $x, y \in B(x_0, \delta_2)$ we have that

$$
|f''(y) - f''(x)| \le |f''(y) - f''(x_0)| + |f''(x_0) - f''(x)| < d\epsilon.
$$

Let $\delta = \min\{\delta_1, \delta_2\}$. Then it follows that for distinct $x, y \in B(x_0, \delta)$ we have that

$$
\begin{aligned}
|f'(y) - f'(x) - f''(x)(y - x)| &\le \left|\frac{f(y) - f(x)}{y - x} - f'(x) - \frac{1}{2}f''(x)(y - x)\right| \\
&\quad + \left|\frac{1}{2}f''(y)(x - y) + f'(y) - \frac{f(y) - f(x)}{y - x}\right| \\
&\quad + \frac{1}{2}\left|f''(y) - f''(x)\right|\left|y - x\right| \\
&< d\epsilon|y - x| + d\epsilon|y - x| + \frac{1}{2}d\epsilon|y - x| < \epsilon|y - x|.
\end{aligned}
$$

□

Just as with the $n = 1$ case, WLUD^n functions form an $\mathcal{N}$-algebra that is closed under composition, for any $n \in \mathbb{N}$. As this paper will focus primarily on functions that are WLUD^2, we mainly present the proofs for the $n = 2$ case here. The proofs for the general case are similar. We start with the following proposition whose proof (for any $n \ge 2$) is very similar to that of the case $n = 1$ (Proposition 2) and will thus be omitted here.

PROPOSITION 8. *Let $A \subseteq \mathcal{N}$ be open, let $f, g : A \to \mathcal{N}$ be WLUD^n at $x_0 \in A$ and let $\alpha \in \mathcal{N}$ be given. Then $(f + \alpha g)$ is WLUD^n at x_0.*

PROPOSITION 9. *Let $A \subseteq \mathcal{N}$ be open and let $f, g : A \to \mathcal{N}$ be WLUD^2 at $x_0 \in A$. Then fg is WLUD^2 at x_0.*

PROOF. Let $\epsilon > 0$ in $\mathcal{N}$ be given. By proposition 6, f and g are C^2 at x_0, and so there exists a $\delta_c > 0$ in $\mathcal{N}$ such that $B(x_0, \delta_c) \subset A$ and $|f(x) - f(x_0)| < 1$, $|g(x) - g(x_0)| < 1$, $|f'(x) - f'(x_0)| < 1$, and $|f''(x) - f''(x_0)| < 1$ on $B(x_0, \delta_c)$. Moreover, there exist $\delta_f, \delta_g, \delta_0, \delta_1 > 0$ in $\mathcal{N}$ such that

$$\left|f(y) - f(x) - f'(x)(y - x) - \frac{1}{2}f''(x)(y - x)^2\right| < \frac{\epsilon}{4(|g(x_0)| + 1)}(y - x)^2$$

if $x, y \in B(x_0, \delta_f)$;

$$\left|g(y) - g(x) - g'(x)(y - x) - \frac{1}{2}g''(x)(y - x)^2\right| < \frac{\epsilon}{4(|f(x_0)| + 1)}(y - x)^2$$

if $x, y \in B(x_0, \delta_g)$;

$$|g(y) - g(x) - g'(x)(y - x)| < \frac{\epsilon}{4(|f'(x_0)| + 1)}|y - x|$$

if $x, y \in B(x_0, \delta_0)$; and

$$|g(y) - g(x)| < \frac{\epsilon}{2(|f''(x_0)| + 1)}$$

if $x, y \in B(x_0, \delta_1)$.

Let $\delta = \min\{\delta_c, \delta_f, \delta_g, \delta_0, \delta_1\}$. Then it follows that, for every $x, y \in B(x_0, \delta)$, we have that

$$\begin{aligned}
&\left|(fg)(y) - (fg)(x) - (fg)'(x)(y-x) - \frac{1}{2}(fg)''(x)(y-x)^2\right| \\
&= \Big| f(y)g(y) - f(x)g(x) - \Big(f'(x)g(x) + f(x)g'(x)\Big)(y-x) \\
&\qquad - \frac{1}{2}\Big(f''(x)g(x) + 2f'(x)g'(x) + f(x)g''(x)\Big)(y-x)^2\Big| \\
&\le \left| f(y)g(y) - f(x)g(y) - f'(x)g(y)(y-x) - \frac{1}{2}f''(x)g(y)(y-x)^2\right| \\
&\qquad + \left| f(x)g(y) - f(x)g(x) - f(x)g'(x)(y-x) - \frac{1}{2}f(x)g''(x)(y-x)^2\right| \\
&\qquad + \left| f'(x)g(y)(y-x) - f'(x)g(x)(y-x) - f'(x)g'(x)(y-x)^2\right| \\
&\qquad + \frac{1}{2}\left| f''(x)g(y) - f''(x)g(x)\right|(y-x)^2 \\
&= |g(y)|\left| f(y) - f(x) - f'(x)(y-x) - \frac{1}{2}f''(x)(y-x)^2\right| \\
&\qquad + |f(x)|\left| g(y) - g(x) - g'(x)(y-x) - \frac{1}{2}g''(x)(y-x)^2\right| \\
&\qquad + |f'(x)|\,|y-x|\,|g(y) - g(x) - g'(x)(y-x)| \\
&\qquad + \frac{1}{2}|f''(x)|\,|g(y) - g(x)|\,(y-x)^2 \\
&< \frac{|g(y)|}{4(|g(x_0)|+1)}\epsilon(y-x)^2 + \frac{|f(x)|}{4(|f(x_0)|+1)}\epsilon(y-x)^2 \\
&\qquad + \frac{|f'(x)|}{4(|f'(x_0)|+1)}\epsilon(y-x)^2 + \frac{|f''(x)|}{4(|f''(x_0)|+1)}\epsilon(y-x)^2 \\
&< \epsilon(y-x)^2.
\end{aligned}$$

□

COROLLARY 1. *All polynomials are WLUD^2 on $\mathcal{N}$.*

PROOF. Using Proposition 8 and Proposition 9, it suffices to show that the function $f(x) = x$ is WLUD^2 on $\mathcal{N}$. But that follows readily from the fact that, for all $x, y \in \mathcal{N}$, we have that

$$\left| f(y) - f(x) - f'(x)(y-x) - \frac{1}{2}f''(x)(y-x)^2\right| = |y - x - (y-x)| = 0.$$

□

PROPOSITION 10. *Let $A, B \subseteq \mathcal{N}$ be open and let $g : A \to B$ be WLUD^2 at $x_0 \in A$ and $f : B \to \mathcal{N}$ be WLUD^2 at $g(x_0)$. Then $f \circ g : A \to \mathcal{N}$ is WLUD^2 at x_0.*

PROOF. Let $\epsilon > 0$ in $\mathcal{N}$ be given. As g and g^2 are WLUD^2 at x_0, and f is WLUD^2 at $g(x_0)$, there exist $\delta_1, \delta_2, \delta_3 > 0$ in $\mathcal{N}$ such that

$$\left| f(y) - f(x) - f'(x)(y-x) - \frac{1}{2}f''(x)(y-x)^2\right| < \frac{\epsilon}{3\left(1 + |g'(x_0)|\right)^2}(y-x)^2$$

for every $x, y \in B \cap B(g(x_0), \delta_1)$;

$$\left|g(y) - g(x) - g'(x)(y-x) - \frac{1}{2}g''(x)(y-x)^2\right| < \frac{\epsilon}{3\Big(\left|f'(g(x_0))\right| + \left|f''(g(x_0))g(x_0)\right| + 2\Big)}(y-x)^2$$

for every $x, y \in A \cap B(x_0, \delta_2)$; and

$$\left|g^2(y) - g^2(x) - 2g(x)g'(x)(y-x) - \Big(g'(x)^2 + g(x)g''(x)\Big)(y-x)^2\right| < \frac{\epsilon}{3\Big(\left|f''(g(x_0))\right| + 1\Big)}(y-x)^2.$$

for every $x, y \in A \cap B(x_0, \delta_3)$.

Moreover, as f is C^2 at $g(x_0)$ and g is C^2 at x_0, there exists a $\delta_4 > 0$ in $\mathcal{N}$ such that for every $x, y \in A \cap B(x_0, \delta_4)$ we have that $|g(y) - g(x)| < \delta_1$, $|f'(g(x)) - f'(g(x_0))| < 1$, and $|f''(g(x))g(x) - f''(g(x_0))g(x_0)| < 1$. Finally, since g is WLUD at x_0, there exists $\delta_5 > 0$ in $\mathcal{N}$ such that for every $x, y \in A \cap B(x_0, \delta_5)$ we have that

$$|g(y) - g(x)| < (1 + |g'(x_0)|)\,|y - x|.$$

Let $\delta = \min\{\delta_1, \delta_2, \delta_3, \delta_4, \delta_5\}$. Then for every $x, y \in A \cap B(x_0, \delta)$ we have that

$$\begin{aligned}
&\left|f(g(y)) - f(g(x)) - g'(x)f'(g(x))(y-x) - \frac{1}{2}\Big(g''(x)f'(g(x)) + g'(x)^2 f''(g(x))\Big)(y-x)^2\right| \\
&\le \left|f(g(y)) - f(g(x)) - f'(g(x))\Big(g(y) - g(x)\Big) - \frac{1}{2}f''(g(x))\Big(g(y) - g(x)\Big)^2\right| \\
&\quad + \left|f'(g(x)) - f''(g(x))g(x)\right| \left|g(y) - g(x) - g'(x)(y-x) - \frac{1}{2}g''(x)(y-x)^2\right| \\
&\quad + \frac{1}{2}\left|f''(g(x))\right| \left|g(y)^2 - g(x)^2 - 2g(x)g'(x)(y-x) - \Big(g'(x)^2 + g(x)g''(x)\Big)(y-x)^2\right| \\
&< \frac{\epsilon}{3\left(1 + |g'(x_0)|\right)^2}\left(g(y) - g(x)\right)^2 + \frac{\left|f'(g(x)) - f''(g(x))g(x)\right|}{3\Big(\left|f'(g(x_0))\right| + \left|f''(g(x_0))g(x_0)\right| + 2\Big)}\epsilon(y-x)^2 \\
&\quad + \frac{\left|f''(g(x))\right|}{3\Big(\left|f''(g(x_0))\right| + 1\Big)}\epsilon(y-x)^2 \\
&< \epsilon(y-x)^2.
\end{aligned}$$

□

So far we have shown that the LUD class of functions is contained in the WLUD class of functions, which in turn is contained in the C^1 class of functions. To complete our discussion of the basic properties of WLUD functions, we show that these inclusions are strict, meaning that LUD, WLUD, and C^1 are indeed distinct concepts in the non-Archimedean case while they are equivalent in the classical case.

NOTATION 2. For the next two examples we will use the following notation for convenience: Given $x, y \in \mathcal{R}\setminus\{0\}$, define $z = x - x[\lambda(x)]d^{\lambda(x)}$, $w = y - y[\lambda(y)]d^{\lambda(y)}$, and let $\lambda(x) = l/m$ and $\lambda(y) = j/k$ in lowest terms.

We begin with a function that is C^1 and WLUD but is not LUD.

EXAMPLE 1. Let $f : (-1,1) \to \mathcal{R}$ be given by

$$f(x) = \begin{cases} xd^m & m > 2\lambda(x),\ z \geq d^{2\lambda(x)},\text{ and } \lambda\left(z - d^{2\lambda(x)}\right) \leq m \\ 0 & \text{else.} \end{cases}$$

We note that for any $r \in \mathbb{R}$ and $q \in \mathbb{Q}$, the set $\{x \in (-1,1) \mid \lambda(x) = q, x[q] = r, z \geq d^{2q}, \lambda\left(z - d^{2q}\right) \leq m\}$ is clopen in $\mathcal{R}$, and thus f is locally linear everywhere but zero. Hence we have that f is differentiable for $x \neq 0$ with

$$f'(x) = \begin{cases} d^m & m > 2\lambda(x),\ z \geq d^{2\lambda(x)},\text{ and } \lambda\left(z - d^{2\lambda(x)}\right) \leq m \\ 0 & \text{else,} \end{cases}$$

which is similarly continuous. Moreover, for $x \neq 0$, we have

$$\begin{aligned}\frac{f(x) - f(0)}{x} = \frac{f(x)}{x} &= \begin{cases} d^m & m > 2\lambda(x),\ z \geq d^{2\lambda(x)},\text{ and } \lambda\left(z - d^{2\lambda(x)}\right) \leq m \\ 0 & \text{else} \end{cases} \\ &= f'(x) \ll d^{2\lambda(x)}.\end{aligned}$$

Therefore $\lim_{x\to 0} f(x)/x = \lim_{x\to 0} f'(x) = 0 = f'(0)$ and hence f is C^1.

Moreover, as we will show, f is WLUD as well. For $x_0 \neq 0$, this is trivial, as f is locally linear. For $x_0 = 0$, let $\epsilon > 0$ be given, choose $\delta = d^2\epsilon$, and let $x, y \in (-\delta, \delta)$ be given. We have 3 cases.

Case 1. *If $f(x) = 0$, $f(y) \neq 0$, then we must have $d^k \ll |y - x|$, and so*

$$|f(y) - f(x) - f'(x)(y-x)| = |yd^k| \ll \delta|y - x| \ll \epsilon|y - x|.$$

Case 2. *If $f(y) = 0$, $f(x) \neq 0$, then we must have $d^m \ll |y - x|$, and so*

$$|f(y) - f(x) - f'(x)(y-x)| = |xd^m + d^m(y - x)| = |y|d^m \ll \delta|y - x| \ll \epsilon|y - x|.$$

Case 3. *If $f(x), f(y) \neq 0$, then*

$$\begin{aligned}|f(y) - f(x) - f'(x)(y-x)| = |y||d^k - d^m| &\ll \delta d^{-1}\left|d^{2\lambda(y)} - d^{2\lambda(x)}\right| \\ &\ll \delta d^{-1}|y - x| \ll \epsilon|y - x|.\end{aligned}$$

We will now show f is not LUD at 0. Let a basic open neighbourhood $(-d^n, d^n)$ of 0 be given. Choose $\epsilon = d^{2n}$ and let $\delta > 0$ be given. Let $N = \max\{\lambda(\delta), 4n\}$, $q = (2nN-1)/N$, $x = d^q + d^{2q}$, and $y = x + d^m$. Then $x, y \in U$, $m = N \geq 4n > 2n$, and

$$|f(y) - f(x) - f'(x)(y-x)| = yd^m = y|y - x| > \epsilon|y - x|.$$

Thus, f is C^1 but not LUD. In fact, f', being locally constant away from zero with $\lim_{x\to 0} f'(x) = 0$, is differentiable with derivative zero everywhere, and so f is C^∞ but not LUD.

This shows that LUD is a strictly stronger condition than WLUD. In the next example, we will show that WLUD is a strictly stronger condition than C^1.

EXAMPLE 2. Let $f : (-1,1) \to \mathcal{R}$ be given by

$$f(x) = \begin{cases} z^2 & z < d^{2\lambda(x)} \text{ or } \left|z - d^{2\lambda(x)}\right| \ll d^m \\ z^2 + 2xz & z > d^{2\lambda(x)} \text{ and } \lambda\left(z - d^{2\lambda(x)}\right) \leq m \\ 0 & x = 0. \end{cases}$$

First we will show that f is C^1. Just as in the previous example, the pieces on which f is defined (other than $\{0\}$ of course) are clopen sets. Thus, for $x_0 \neq 0$, f is locally a polynomial, and we have that f is C^1 on $(-1,1)\setminus\{0\}$ with

$$f'(x) = \begin{cases} 2z & z < d^{2\lambda(x)} \text{ or } \left|z - d^{2\lambda(x)}\right| \ll d^m \\ 4z + 2x & z > d^{2\lambda(x)} \text{ and } \lambda\left(z - d^{2\lambda(x)}\right) \leq d^m. \end{cases}$$

For $x_0 = 0$, we have

$$\begin{aligned} \left|\frac{f(x)}{x}\right| &= \begin{cases} \left|\frac{z}{x}z\right| & z < d^{2\lambda(x)} \text{ or } \left|z - d^{2\lambda(x)}\right| \ll d^m \\ \left|\frac{z}{x}(z+2x)\right| & z > d^{2\lambda(x)} \text{ and } \lambda\left(z - d^{2\lambda(x)}\right) \leq d^m \end{cases} \\ &\leq 3|x|, \end{aligned}$$

and hence $f'(0) = \lim_{x\to 0} f(x)/x = 0$. Moreover, by our expression for f', we have $\lim_{x\to 0} f'(x) = 0$, and so f is C^1 on $(-1,1)$.

Now we will show that f is not WLUD at 0. Let $\epsilon = d$, let $\delta > 0$ be given, let $n \in \mathbb{N}$ be such that $n \geq 2$ and $d^n < \delta$, let $m = 4n$, and let $l = 4n^2 + 1$. Then l and m are co-prime and

$$\frac{m}{3} = \frac{4}{3}n > n + \frac{1}{4n} = \frac{4n^2+1}{4n} = \frac{l}{m} > n.$$

Let $x = d^{l/m} + d^{2l/m}$, $y = d^{l/m} + d^{2l/m} + d^m$. Then, $|x|$, $|y| < 2d^{l/m} \ll d^n < \delta$, and so $x, y \in B(0,\delta)$. Moreover,

$$f(y) - f'(x)y = w^2 + 2yw - 2zy \geq 2y(w-z) = 2yd^m \geq 0,$$

$$f'(x)x - f(x) = 2zx - z^2 = 2z(x-z) = 2d^{3l/m} \geq 0,$$

and so,

$$\begin{aligned} |f(y) - f(x) - f'(x)(y-x)| &= f(y) - f'(x)y + f'(x)x - f(x) \\ &\geq f'(x)x - f(x) = 2d^{3l/m} \gg d^m \gg \epsilon|y-x|. \end{aligned}$$

3. Calculus Theorems

The following lemma is a trivial consequence of the fact that a WLUD function is C^1.

LEMMA 2. *Let $A \subseteq \mathcal{N}$ be open and let $f : A \to \mathcal{N}$ be WLUD at $x_0 \in A$. Then for every $\epsilon > 0$ in $\mathcal{N}$, there exists $\delta > 0$ in $\mathcal{N}$ such that $B(x_0,\delta) \subset A$ and, for every $x, y \in B(x_0,\delta)$, we have that*

$$|f(y) - f(x) - f'(x_0)(y-x)| < \epsilon\,|y-x|\,.$$

REMARK 4. *As the preceding lemma will be the primary tool we will use to prove our major calculus theorems with WLUD, it is noteworthy to point out that the converse of the preceding lemma is also true for differentiable functions.*

PROPOSITION 11. *Let $A \subseteq \mathcal{N}$ be open, let $x_0 \in A \subset \mathcal{N}$ and let $f : A \to \mathcal{N}$ be a differentiable function such that for every $\epsilon > 0$ in $\mathcal{N}$, there exists $\delta > 0$ in $\mathcal{N}$ such that, for every $x, y \in B(x_0,\delta) \cap A$, we have that $|f(y) - f(x) - f'(x_0)(y-x)| < \epsilon\,|y-x|$. Then f is WLUD at x_0.*

PROOF. It suffices to prove that f' is C^1 at x_0. Let $\epsilon > 0$ in $\mathcal{N}$ be given. Then there exists $\delta > 0$ in $\mathcal{N}$ such that $B(x_0, \delta) \subset A$ and, for every $x, y \in B(x_0, \delta)$, we have

$$|f(y) - f(x) - f'(x_0)(y - x)| < \frac{1}{2}\epsilon|y - x|.$$

Let $x \in B(x_0, \delta)$ be given. Then as f is differentiable at x, there exists $y \in B(x_0, \delta)$ such that

$$\left|\frac{f(y) - f(x)}{y - x} - f'(x)\right| < \frac{1}{2}\epsilon,$$

and so,

$$|f'(x) - f'(x_0)| \leq \left|f'(x) - \frac{f(y) - f(x)}{y - x}\right| + \left|\frac{f(y) - f(x)}{y - x} - f'(x_0)\right| < \epsilon.$$

□

The following lemma, which is unrelated to the properties of WLUD functions, was proved in [**16**] (Lemma 4.1) .

LEMMA 3. *Let $\delta_1 > 0$ in $\mathcal{N}$ be given and let $\phi : B(0, \delta_1) \to \mathcal{N}$ be such that $|\phi(t)| \leq c|t|$ for every $t \in B(0, \delta_1)$, where $0 < c \ll 1$. For $m \in \mathbb{N}$ let $\phi^{[m]} = \underbrace{\phi \circ \cdots \circ \phi}_{m \text{ times}}$ and set $\phi^{[0]}$ to be the identity map. Let $\delta \in \mathcal{N}$ be such that $0 < \delta \leq (1 - c)\delta_1$ and let $\psi(t) = \sum_{m=0}^{\infty} \phi^{[m]}(t)$, for every $t \in B(0, \delta)$. Then*

(i) $|\psi(t)| \leq \frac{|t|}{1-c}$; *and*
(ii) $\psi(t) - \phi(\psi(t)) = t$.

LEMMA 4. *Let $A \subseteq \mathcal{N}$ be open and let $f : A \to \mathcal{N}$ be WLUD on A with $f'(x_0) \neq 0$ for some $x_0 \in A$ and with $f(x_0) = y_0$. Then there exist $\delta, \eta > 0$ in $\mathcal{N}$ and a function F defined on $B(y_0, \eta)$ such that*

(i) $B(x_0, \delta) \subseteq A$;
(ii) $f|_{B(x_0,\delta)}$ *is injective;*
(iii) $B(y_0, \eta) \subseteq f(B(x_0, \delta))$ *and* $F(B(y_0, \eta) \subseteq B(x_0, \delta)$;
(iv) $f(F(x)) = x$ *for every* $x \in B(y_0, \eta)$; *and*
(v) F *is WLUD on* $B(y_0, \eta)$ *with* $F' = 1/f' \circ F$.

PROOF. Without loss of generality, we may assume that $x_0 = 0$ and $y_0 = 0$, for if this is not the case, then we can replace $f(x)$ with $\tilde{f}(x) = f(x + x_0) - y_0$. Moreover, without loss of generality, we may assume $f'(x_0) > 0$, for if $f'(x_0) < 0$ we could apply this proof to $(-f)$ and get the desired result.

By Proposition 1, f is C^1, and so there exists $\omega > 0$ in $\mathcal{N}$ such that $f'(x) \geq \frac{1}{2}f'(0) > 0$ for every $x \in B(x_0, \omega) \cap A$. Let $L = f'(0)$. Let $\phi(x) = x - \frac{1}{L}f(x)$. It follows that $\phi'(x) = 1 - \frac{1}{L}f'(x)$ and so $\phi'(0) = 0$. Let $c \in \mathcal{N}$ be such that $0 < c \ll 1$. As ϕ is WLUD at 0, then by Lemma 2, there exists $\delta_0 > 0$ in $\mathcal{N}$ such that $B(0, \delta_0) \subseteq A$ and, for every $s, t \in B(0, \delta_0)$, we have that

$$|\phi(s) - \phi(t) - \phi'(0)(s - t)| < c\,|s - t|\,.$$

Thus,

$$|\phi(s) - \phi(t)| < c\,|s - t|\,. \tag{3.1}$$

Let $s,t\in B(0,\delta_0)$ be such that $f(s)=f(t)$. Then

$$|\phi(s)-\phi(t)|=|s-t|\le c\,|s-t|\,.$$

As $c\ll 1$, it follows that $s=t$, and thus $f|_{B(0,\delta_0)}$ is injective. By Lemma 2, there exists $\delta_f>0$ in $\mathcal{N}$ such that for every $s,t\in B(0,\delta_f)$ we have that

$$|f(s)-f(t)-L(s-t)|<\frac{L}{2}\,|s-t|\,.$$

Let $\delta=\min\{(1-c)\delta_0,\omega,\delta_f\}$. Then $B(0,\delta)\subset B(0,\delta_0)\subset A$ and thus $f|_{B(0,\delta)}$ is injective. This shows (i) and (ii).

By Equation (3.1) with $t=0$, we have that $|\phi(s)|<c\,|s|$ for every $s\in B(0,\delta)$, and so we have a function ψ with properties of that in Lemma 3. Let $\eta=L(1-c)\delta$ and define $F(x)=\psi(\frac{x}{L})$ for every $x\in B(0,\eta)$. Thus for every $x\in B(0,\eta)$ we have that

$$|F(x)|=\left|\psi\left(\frac{x}{L}\right)\right|\le\frac{|x|}{L(1-c)}<\frac{\eta}{L(1-c)}=\delta.$$

Thus $F(B(0,\eta))\subseteq B(0,\delta)$. Furthermore, for every $x\in B(0,\delta)$, we have that

$$x-\phi(x)=\frac{f(x)}{L}.$$

Let $x\in B(0,\eta)$. Then

$$\frac{|x|}{L}<(1-c)\delta<\delta.$$

Thus $\frac{x}{L}\in B(0,\delta)$ and hence

$$\frac{x}{L}-\phi\left(\frac{x}{L}\right)=\frac{1}{L}f\left(\frac{x}{L}\right).$$

Moreover, we have by Lemma 3 that

$$\psi\left(\frac{x}{L}\right)-\phi\left(\psi\left(\frac{x}{L}\right)\right)=\frac{x}{L}$$

and thus

$$\frac{1}{L}f\left(\psi\left(\frac{x}{L}\right)\right)=\frac{x}{L}.$$

It follows that for every $x\in B(0,\eta)$,

$$f(F(x))=f\left(\psi\left(\frac{x}{L}\right)\right)=x$$

and hence $B(0,\eta)\subseteq f(B(0,\delta))$, as $F(x)\in B(0,\delta)$ for every $x\in B(0,\eta)$. This shows (iii) and (iv).

Now, for any $x\in B(0,\delta)$ we have

$$|f(s)-f(t)-L(s-t)|<\frac{L}{2}\,|s-t|$$

and thus

$$|f(s)-f(t)|\ge L\,|s-t|-\frac{L}{2}\,|s-t|=\frac{L}{2}\,|s-t|\,.$$

Let $\epsilon>0$ be given. As f is WLUD at 0, there exists $\delta_2>0$ such that for every $x,y\in B(0,\delta_2)$ we have

$$|f(y)-f(x)-f'(x)(y-x)|<\frac{L^2}{4}\epsilon|y-x|.$$

Let $\delta_3 = \min\{L\delta_2/2, \eta, \delta\}$, let $x, y \in B(0, \delta_3)$ be given, and let $t_x = F(x)$ and $t_y = F(y)$. Then

$$|t_y - t_x| \le \frac{2}{L}\,|f(t_y) - f(t_x)| = \frac{2}{L}\,|y - x| < \delta_2$$

Thus,

$$\begin{aligned}\left|F(y) - F(x) - \frac{1}{f'(F(x))}(y - x)\right| &= \left|\frac{1}{f'(F(x))}\right| |y - x - f'(F(x))(F(y) - F(x))| \\ &\le \frac{2}{L}\,|f(t_y) - f(t_x) - f'(t_x)(t_y - t_x)| \\ &< \epsilon\frac{L}{2}|t_y - t_x| \le \epsilon|y - x|,\end{aligned}$$

which shows (v). □

THEOREM 3.1. *Let $A \subseteq \mathcal{N}$ be open and let $f : A \to \mathcal{N}$ be WLUD on A with $f'(x_0) \neq 0$ for some $x_0 \in A$. Then there exists a neighborhood U of x_0 in A such that*

(i) *$f|_U$ is injective;*
(ii) *$f(U)$ is open; and*
(iii) *f^{-1} exists and is WLUD on $f(U)$ with $(f^{-1})' = 1/f' \circ f^{-1}$.*

PROOF. By Lemma 4, there exists a neighborhood U_0 of x_0 in A such that f is injective on U_0. As f is C^1 and $f'(x_0) \neq 0$, there exists a neighborhood U_1 of x_0 in A such that $f'(x) \neq 0$ for every $x \in U_1$. Let $U = U_0 \cap U_1$. Then U is a neighborhood of x_0 and $f|_U$ is injective.

Let $x \in U$ and let $y = f(x)$. Lemma 4 applied to $f|_U$ at x gives a δ, η, and F as stated in the lemma, for which $B(y, \eta) \subseteq f(B(x, \delta)) \subseteq f(U)$. As this holds for every $x \in U$, we have that $f(U)$ is open.

As f is injective f^{-1} exists on $f(U)$ and for any $y \in f(U)$,

$$f(f^{-1}(y)) = y = f(F(y))$$

and so we have that $F = f^{-1}$, which is WLUD at y with $(f^{-1})' = 1/f' \circ f^{-1}$ by lemma 4.

□

THEOREM 3.2 (local intermediate value theeorem). *Let $A \subseteq \mathcal{N}$ be open, let $f : A \to \mathcal{N}$ be WLUD on A and let $x_0 \in A$ be such that $f'(x_0) \neq 0$. Then there exists a neighborhood U of x_0 in A such that f has the intermediate value property on U. That is, for every $a, b \in U$ with $a < b$, if c is between $f(a)$ and $f(b)$, then there exists $x \in (a, b)$ such that $f(x) = c$.*

PROOF. Without loss of generality, we may assume $f'(x_0) > 0$. By Lemma 2, there exists $\delta > 0$ in $\mathcal{N}$ such that $B(x_0, \delta) \subset A$ and, for all $x \neq y$ in $B(x_0, \delta)$, we have that

$$|f(y) - f(x) - f'(x_0)(y - x)| < \frac{f'(x_0)}{2}\,|y - x|$$

and thus

$$\frac{f(y) - f(x)}{y - x} > f'(x_0) - \frac{f'(x_0)}{2} = \frac{f'(x_0)}{2} > 0.$$

Hence f is strictly increasing on $B(x_0, \delta)$. Applying Theorem 3.1 to f gives a neighborhood $U_0 \subseteq B(x_0, \delta)$ of x_0 such that $f(U_0)$ is open. Let $\epsilon > 0$ in $\mathcal{N}$ be such that $B(f(x_0),\epsilon) \subseteq f(U_0)$ and let $U = f^{-1}(B(f(x_0),\epsilon))$, which is an open neighborhood of x_0. Let $a, b \in U$ be such that $a < b$ and let $c \in (f(a), f(b))$ be given. As $f(a), f(b) \in B(f(x_0),\epsilon)$ and $B(f(x_0),\epsilon)$ is a convex set, we have that $c \in B(f(x_0),\epsilon)$. Thus there exists $x \in U = f^{-1}(B(f(x_0),\epsilon))$ such that $f(x) = c$. As f is strictly increasing on U, it follows that $x \in (a, b)$. □

THEOREM 3.3 (local mean value theorem). *Let $A \subseteq \mathcal{N}$ be open, let $f : A \to \mathcal{N}$ be WLUD^2 at $x_0 \in A$ and assume that $f''(x_0) \neq 0$. Then there exists a neighborhood U of x_0 in A such that f has the mean value property on U. That is, for every $a, b \in U$ with $a < b$, there exists $c \in (a, b)$ such that*

$$f'(c) = \frac{f(b) - f(a)}{b - a}.$$

PROOF. We may assume without loss of generality that $f''(x_0) > 0$. By Proposition 6, f'' is continuous at x_0. Thus there exists $\delta_1 > 0$ such that $U_1 = B(x_0, \delta_1) \subset A$ and, for every $x \in U_1$ we have that

$$|f''(x) - f''(x_0)| < \frac{1}{4} f''(x_0).$$

As f is WLUD^2 at x_0, there exists $\delta_2 > 0$ in $\mathcal{N}$ such that, for every $x, y \in U_1$ with $0 < |y - x| < \delta_2$, we have that

$$\left| f(y) - f(x) - f'(x)(y - x) - \frac{1}{2} f''(x)(y - x)^2 \right| < \frac{1}{4} f''(x_0)(y - x)^2.$$

Let $\delta = \min\{\delta_1, \delta_2\}$. Then it follows that, for every $x, y \in B(x_0, \delta)$, we have that

$$\begin{aligned} f(y) - f(x) - f'(x)(y - x) &> \left(\frac{1}{2} f''(x) - \frac{1}{4} f''(x_0) \right) (y - x)^2 \\ &> \frac{1}{8} f''(x_0)(y - x)^2 > 0. \end{aligned} \tag{3.2}$$

Applying Theorem 3.2 to f' at x_0 gives a neighborhood U_0 of x_0 in A such that f' has the intermediate value property in U_0. Let $U = U_0 \cap B(x_0, \delta)$ and let $a, b \in U$ with $a < b$ be given. By Equation (3.2) we have that $f(b) > f(a) + f'(a)(b - a)$, and thus

$$f'(a) < \frac{f(b) - f(a)}{b - a}.$$

Similarly, we have that $f(a) > f(b) + f'(b)(a - b)$, and thus

$$f'(b) > \frac{f(b) - f(a)}{b - a}$$

Thus, by Theorem 3.2, there exists $c \in (a, b)$ such that

$$f'(c) = \frac{f(b) - f(a)}{b - a}$$

□

References

[1] M. Berz. Calculus and numerics on Levi-Civita fields. In M. Berz, C. Bischof, G. Corliss, and A. Griewank, editors, *Computational Differentiation: Techniques, Applications, and Tools*, pages 19–35, Philadelphia, 1996. SIAM.

[2] H. Hahn. Über die nichtarchimedischen Größensysteme. *Sitzungsbericht der Wiener Akademie der Wissenschaften Abt. 2a*, 117:601–655, 1907.

[3] W. Krull, *Allgemeine Bewertungstheorie* (German), J. Reine Angew. Math. **167** (1932), 160–196, DOI 10.1515/crll.1932.167.160. MR1581334

[4] T. Levi-Civita. Sugli infiniti ed infinitesimi attuali quali elementi analitici. *Atti Ist. Veneto di Sc., Lett. ed Art.*, 7a, 4:1765, 1892.

[5] T. Levi-Civita. Sui numeri transfiniti. *Rend. Acc. Lincei*, 5a, 7:91,113, 1898.

[6] S. MacLane, *The universality of formal power series fields*, Bull. Amer. Math. Soc. **45** (1939), 888–890, DOI 10.1090/S0002-9904-1939-07110-3. MR0000610

[7] S. Prieß-Crampe, *Angeordnete Strukturen* (German), Ergebnisse der Mathematik und ihrer Grenzgebiete [Results in Mathematics and Related Areas], vol. 98, Springer-Verlag, Berlin, 1983. Gruppen, Körper, projektive Ebenen. [Groups, fields, projective planes]. MR704186

[8] F. J. Rayner, *Algebraically closed fields analogous to fields of Puiseux series*, J. London Math. Soc. (2) **8** (1974), 504–506, DOI 10.1112/jlms/s2-8.3.504. MR0349642

[9] P. Ribenboim, *Fields: algebraically closed and others*, Manuscripta Math. **75** (1992), no. 2, 115–150, DOI 10.1007/BF02567077. MR1160093

[10] A. Robinson, *Non-standard analysis*, North-Holland Publishing Co., Amsterdam, 1966. MR0205854

[11] W. H. Schikhof, *Ultrametric calculus. An introduction to p-adic analysis*, Cambridge Studies in Advanced Mathematics, vol. 4, Cambridge University Press, Cambridge, 1984. MR791759

[12] K. Shamseddine, *Analysis on the Levi-Civita field and computational applications*, Appl. Math. Comput. **255** (2015), 44–57, DOI 10.1016/j.amc.2014.04.108. MR3316582

[13] K. Shamseddine and M. Berz, *Exception handling in derivative computation with non-Archimedean calculus*, Computational differentiation (Santa Fe, NM, 1996), SIAM, Philadelphia, PA, 1996, pp. 37–51. MR1431040

[14] K. Shamseddine and M. Berz, *Analysis on the Levi-Civita field, a brief overview*, Advances in p-adic and non-Archimedean analysis, Contemp. Math., vol. 508, Amer. Math. Soc., Providence, RI, 2010, pp. 215–237, DOI 10.1090/conm/508/10002. MR2597696

[15] K. Shamseddine and G. Bookatz, *A local mean value theorem for functions on non-Archimedean field extensions of the real numbers*, p-Adic Numbers Ultrametric Anal. Appl. **8** (2016), no. 2, 160–175, DOI 10.1134/S2070046616020059. MR3503302

[16] K. Shamseddine, T. Rempel, and T. Sierens, *The implicit function theorem in a non-Archimedean setting*, Indag. Math. (N.S.) **20** (2009), no. 4, 603–617, DOI 10.1016/S0019-3577(09)80028-1. MR2776902

[17] K. Shamseddine and T. Sierens, *On locally uniformly differentiable functions on a complete non-Archimedean ordered field extension of the real numbers*, ISRN Math. Anal. (2012), Art. ID 387053, 20. MR2910742

[18] N. Vakil, *Real analysis through modern infinitesimals*, Encyclopedia of Mathematics and its Applications, vol. 140, Cambridge University Press, Cambridge, 2011. MR2752815

Department of Physics and Astronomy, University of Manitoba, Winnipeg, Manitoba R3T 2N2, Canada

Email address: bookatzg@myumanitoba.ca

Department of Physics and Astronomy, University of Manitoba, Winnipeg, Manitoba R3T 2N2, Canada

Email address: khodr.shamseddine@umanitoba.ca

Contemporary Mathematics
Volume **704**, 2018
http://dx.doi.org/10.1090/conm/704/14161

p-adic meromorphic functions $f'P'(f)$, $g'P'(g)$ sharing a small function, ignoring multiplicity

Kamal Boussaf and Alain Escassut

ABSTRACT. Let $\mathbb{K}$ be a complete algebraically closed p-adic field of characteristic zero. Let f, g be two transcendental meromorphic functions in the whole field $\mathbb{K}$ or meromorphic functions in an open disk that are not quotients of bounded analytic functions. Let P be a polynomial of uniqueness for meromorphic functions in $\mathbb{K}$ or in an open disk and let α be a small meromorphic function with regard to f and g. If $f'P'(f)$ and $g'P'(g)$ share α ignoring multiplicity orders, then we show that $f = g$ provided that the multiplicity order of zeros of P' satisfies certain inequalities. If α is a Moebius function or a non-zero constant, we can obtain more general results on P and if f is an analytic function in $\mathbb{K}$ or in the disk, we also obtain more precise results. All results follow previous ones obtained for similar meromorphic functions sharing a small function, counting multiplicity. That comes after similar results obtained by Buy Thi Kieu Oanh and Ngo Thi Thu Thuy for complex functions and results obtained by the present authors with Jacqueline Ojeda, counting multiplicity.

1. Introduction and Main Results

Let f, g be two meromorphic functions in a p-adic field. Here we study polynomials P such that, when $f'P'(f)$ and $g'P'(g)$ share a small function α, then $f = g$. Problems of uniqueness on meromorphic functions were examined first in $\mathbb{C}$ [**3**], [**11**], [**14**], [**16**], [**17**], [**18**], [**23**], [**24**], [**25**] and next in a p-adic field [**1**], [**2**], [**5**], [**7**], [**8**], [**12**], [**13**], [**19**], [**20**], [**21**], [**22**]. After examining problems of the form $P(f) = P(g)$, several studies were made on the equality $f'P'(f) = g'P'(g)$, or value sharing questions: if $f'P'(f)$ and $g'P'(g)$ share a value, or a small function *counting multiplicity*, do we have $f = g$? Here we will try to generalize results previously obtained in [**2**] by supposing that $f'P'(f)$ and $g'P'(g)$ share a value, or a small function *ignoring multiplicity*. That is similar to a work by Buy Thi Kieu Oanh and Ngo Thi Thu concerning complex meromorphic functions. But in the present paper as in [**2**] we can also examine the situation for meromorphic functions inside an open disk.

Let $\mathbb{K}$ be an algebraically closed field of characteristic zero, complete for an ultrametric absolute value denoted by $|\ .\ |$. We denote by $\mathcal{A}(\mathbb{K})$ the $\mathbb{K}$-algebra of

2010 *Mathematics Subject Classification.* Primary 12J25, 30D35, 30G06.
Key words and phrases. Meromorphic, Nevanlinna, ultrametric, sharing value, unicity, distribution of values.

entire functions in $\mathbb{K}$, by $\mathcal{M}(\mathbb{K})$ the field of meromorphic functions in $\mathbb{K}$, i.e. the field of fractions of $\mathcal{A}(\mathbb{K})$ and by $\mathbb{K}(x)$ the field of rational functions.

Let $a \in \mathbb{K}$ and $R \in]0, +\infty[$. We denote by $d(a, R)$ the closed disk $\{x \in \mathbb{K} : |x - a| \leq R\}$ and by $d(a, R^-)$ the "open" disk $\{x \in \mathbb{K} : |x - a| < R\}$. We denote by $\mathcal{A}(d(a, R^-))$ the set of analytic functions in $d(a, R^-)$, i.e. the $\mathbb{K}$-algebra of power series $\sum_{n=0}^{\infty} a_n(x-a)^n$ converging in $d(a, R^-)$ and by $\mathcal{M}(d(a, R^-))$ the field of meromorphic functions inside $d(a, R^-)$, i.e. the field of fractions of $\mathcal{A}(d(a, R^-))$. Moreover, we denote by $\mathcal{A}_b(d(a, R^-))$ the $\mathbb{K}$ - subalgebra of $\mathcal{A}(d(a, R^-))$ consisting of the bounded analytic functions in $d(a, R^-)$, i.e. which satisfy $\sup_{n \in \mathbb{N}} |a_n| R^n < +\infty$. And we denote by $\mathcal{M}_b(d(a, R^-))$ the field of fractions of $\mathcal{A}_b(d(a, R^-))$. Finally, we denote by $\mathcal{A}_u(d(a, R^-))$ the set of unbounded analytic functions in $d(a, R^-)$, i.e. $\mathcal{A}(d(a, R^-)) \setminus \mathcal{A}_b(d(a, R^-))$. Similarly, we set $\mathcal{M}_u(d(a, R^-)) = \mathcal{M}(d(a, R^-)) \setminus \mathcal{M}_b(d(a, R^-))$.

The problem of value sharing a small function by functions of the form $f'P'(f)$, counting multiplicity, was examined first when P was just of the form x^n [**17**], [**19**]. Next, it was examined when P was a polynomial such that P' had exactly two distinct zeros [**16**], [**18**], [**21**], both in complex analysis and in p-adic analysis. In [**16**], [**18**] the functions were meromorphic on $\mathbb{C}$, with a small function that was a constant or the identity. In [**21**], the problem was considered for analytic functions in the field $\mathbb{K}$: on the one hand for entire functions and on the other hand for unbounded analytic functions in an open disk.

Actually solving a value sharing problem involving $f'P'(f)$, $g'P'(g)$ requires to know polynomials of uniqueness P for meromorphic functions.

In [**21**] its author studied several problems of uniqueness and particularly the following:

Let $f, g \in \mathcal{A}(\mathbb{K})$ be transcendental (resp. Let $f, g \in \mathcal{A}_u(d(0, R^-))$) and $\alpha \in \mathcal{A}(\mathbb{K})$ (resp. $\alpha \in \mathcal{A}_u(d(0, R^-))$) be a small function, such that $f^n(f-a)^k f'$ and $g^n(g-a)^k g'$ share α, counting multiplicity, with $n, k \in \mathbb{N}$ and $a \in \mathbb{K} \setminus \{0\}$. Do we have $f = g$?

Here we consider functions $f,\ g \in \mathcal{M}(\mathbb{K})$ or $f,\ g \in \mathcal{M}(d(a, R^-))$ and polynomials of uniqueness P: we must only assume certain hypotheses on the multiplicity order of the zeros of P'. The method for the various theorems we will show is the following: assuming that $f'P'(f)$ and $g'P'(g)$ share a small function, we first prove that $f'P'(f) = g'P'(g)$. Next, we derive $P(f) = P(g)$. And then, when P is a polynomial of uniqueness for the functions we consider, we can conclude $f = g$. Similar reasonings were made in [**2**], where the major hypothesis was that $f'P'(f)$ and $g'P'(g)$ share a small function *counting multiplicity*. Here we erase that hypothesis and only suppose that $f'P'(f)$ and $g'P'(g)$ shared a small function *ignoring multiplicity*.

Now, in order to define small functions, we have to briefly recall definitions of the classical Nevanlinna theory in the field $\mathbb{K}$ and a few specific properties of ultrametric analytic or meromorphic functions.

Let log be a real logarithm function of base $b > 1$ and let $f \in \mathcal{M}(\mathbb{K})$ (resp. $f \in \mathcal{M}(d(0, R^-))$) having no zero and no pole at 0. Let $r \in]0, +\infty[$ (resp. $r \in]0, R[$) and let $\gamma \in d(0, r)$. If f has a zero of order n at γ, we put $\omega_\gamma(f) = n$. If f has a pole of order n at γ, we put $\omega_\gamma(f) = -n$ and finally, if $f(\gamma) \neq 0, \infty$, we set $\omega_\gamma(f) = 0$.

We denote by $Z(r, f)$ the *counting function of zeros of f* in $d(0, r)$, counting multiplicity, i.e. we set

$$Z(r, f) = \sum_{\omega_\gamma(f)>0,\ |\gamma|\le r} \omega_\gamma(f)(\log r - \log|\gamma|).$$

Similarly, we denote by $\overline{Z}(r, f)$ the counting function of zeros of f in $d(0, r)$, ignoring multiplicity, and set

$$\overline{Z}(r, f) = \sum_{\omega_\gamma(f)>0,\ |\gamma|\le r} (\log r - \log|\gamma|).$$

In the same way, we set $N(r, f) = Z\Big(r, \dfrac{1}{f}\Big)$ $\Big($resp. $\overline{N}(r, f) = \overline{Z}\Big(r, \dfrac{1}{f}\Big)\Big)$ to denote the *counting function of poles of f* in $d(0, r)$, counting multiplicity (resp. ignoring multiplicity).

For $f \in \mathcal{M}(d(0, R^-))$ having no zero and no pole at 0, the *Nevanlinna function* is defined by $T(r, f) = \max\big\{Z(r, f) + \log|f(0)|, N(r, f)\big\}$.

Now, we must recall the definition of a *small function* with respect to a meromorphic function and some pertinent properties.

DEFINITION. Let $f \in \mathcal{M}(\mathbb{K})$ $\big($resp. let $f \in \mathcal{M}(d(0, R^-))\big)$ such that $f(0) \neq 0, \infty$. A function $\alpha \in \mathcal{M}(\mathbb{K})$ $\big($resp. $\alpha \in \mathcal{M}(d(0, R^-))\big)$ having no zero and no pole at 0 is called *a small function with respect to f*, if it satisfies $\lim\limits_{r\to+\infty} \dfrac{T(r, \alpha)}{T(r, f)} = 0$ $\Big($resp. $\lim\limits_{r\to R^-} \dfrac{T(r, \alpha)}{T(r, f)} = 0\Big)$.

If 0 is a zero or a pole of f or α, we can make a change of variable such that the new origin is not a zero or a pole for both f and α. Thus it is easily seen that the last relation does not really depend on the origin.

We denote by $\mathcal{M}_f(\mathbb{K})$ $\big($resp. $\mathcal{M}_f(d(0, R^-))\big)$ the set of small meromorphic functions with respect to f in $\mathbb{K}$ $\big($resp. in $d(0, R^-)\big)$.

REMARK 1. Thanks to classical properties of the Nevanlinna function $T(r, f)$ with respect to the operations in a field of meromorphic functions, such as $T(r, f + g) \le T(r, f) + T(r, g)$ and $T(r, fg) \le T(r, f) + T(r, g)$, for $f, g \in \mathcal{M}(\mathbb{K})$ and $r > 0$, it is easily proved that $\mathcal{M}_f(\mathbb{K})$ $\big($resp. $\mathcal{M}_f(d(0, R^-))\big)$ is a subfield of $\mathcal{M}(\mathbb{K})$ $\big($resp. $\mathcal{M}(d(0, R^-))\big)$ and that $\mathcal{M}(\mathbb{K})$ $\big($resp. $\mathcal{M}(d(0, R))\big)$ is a transcendental extension of $\mathcal{M}_f(\mathbb{K})$ $\big($resp. of $\mathcal{M}_f(d(0, R^-))\big)$ **[9]**.

Let us remember the following definition.

DEFINITION. Let $f, g, \alpha \in \mathcal{M}(\mathbb{K})$ $\big($resp. let $f, g, \alpha \in \mathcal{M}(d(0, R^-))\big)$. We say that f and g *share the function α I.M.*, if $f - \alpha$ and $g - \alpha$ have the same zeros in $\mathbb{K}$ $\big($resp. in $d(0, R^-)\big)$. Similarly, we say f and g *share the function α C.M.*, if $f - \alpha$ and $g - \alpha$ have the same zeros in $\mathbb{K}$ $\big($ resp. in $d(0, R^-)\big)$ with the same multiplicity, in $\mathbb{K}$ (resp. in $d(0, R^-)$).

Recall that a polynomial $P \in \mathbb{K}[x]$ is called *a polynomial of uniqueness* for a class of functions $\mathcal{F}$ if for any two functions $f,\ g \in \mathcal{F}$ the property $P(f) = P(g)$ implies $f = g$.

The definition of polynomials of uniqueness was introduced in [**17**] by P. Li and C. C. Yang and was studied in many papers [**10**], [**11**] for complex functions and [**1**], [**5**], [**7**], [**12**], [**13**], [**15**], [**22**], for p-adic functions.

Actually, in a p-adic field, we can obtain various results, not only for functions defined in the whole field $\mathbb{K}$ but also for functions defined inside an open disk because the p-adic Nevanlinna Theory works inside a disk, for functions of $\mathcal{M}_u(d(0,R^-))$.

Let us recall Theorem A [**7**], [**9**], [**22**]:

THEOREM A. *Let $P \in \mathbb{K}[x]$ be such that P' has exactly two distinct zeros a_1 of order k_1 and a_2 of order k_2. Then P is a polynomial of uniqueness for $\mathcal{A}(\mathbb{K})$. Moreover, if $\min\{k_1,k_2\} \geq 2$, then P is a polynomial of uniqueness for $\mathcal{M}(\mathbb{K})$.*

Theorem A was first proved in [**22**] with the addiditional hypothesis $P(a_1) \neq P(a_2)$. Actually this hypothesis is useless because, as showed in Lemma 10 [**7**], (see also [**9**]) the equality $P(a_1) = P(a_2)$ is impossible since P' only has two distinct zeros.

NOTATION. Let $P \in \mathbb{K}[x] \setminus \mathbb{K}$ and let $\Xi(P)$ be the set of zeros c of P' such that $P(c) \neq P(b)$ for every zero b of P' other than c. We denote by $\Phi(P)$ the cardinal of $\Xi(P)$.

REMARK 2. If $\deg(P) = q$ then $\Phi(P) \leq q-1$.

From [**7**] we have the following results:

THEOREM B. *Let $d(a,R^-)$ be an open disk in $\mathbb{K}$ and $P \in \mathbb{K}[x]$.*
If $\Phi(P) \geq 2$ then P is a polynomial of uniqueness for $\mathcal{A}(\mathbb{K})$.
If $\Phi(P) \geq 3$ then P is a polynomial of uniqueness for both $\mathcal{A}_u(d(a,R^-))$ and $\mathcal{M}(\mathbb{K})$. If $\Phi(P) \geq 4$ then P is a polynomial of uniqueness for $\mathcal{M}_u(d(a,R^-))$.

And from [**21**] we have:

THEOREM C. *Let $P \in \mathbb{K}[x]$ be of degree $n \geq 6$ be such that P' only has two distinct zeros, one of them being of order 2. Then P is a polynomial of uniqueness for $\mathcal{M}_u(d(0,R^-))$.*

We can now state our main theorems.

THEOREM 1. *Let P be a polynomial of uniqueness for $\mathcal{M}(\mathbb{K})$, let $P' = b(x-a_1)^n \prod_{i=2}^{l}(x-a_i)^{k_i}$ with $b \in \mathbb{K}^*$, $l \geq 2$, $k_i \geq k_{i+1}$, $2 \leq i \leq l-1$ and let $k = \sum_{i=2}^{l} k_i$. Suppose P satisfies the following conditions:*

$$n \geq 18 + \max(0, 10-k_2) + \sum_{i=3}^{l} \max(0, 7-k_i),$$

$$n \geq k+2,$$

if $l=2$, then $n \neq 2k,\ 2k+1,\ 3k+1$,
if $l=3$, then $n \neq 2k+1,\ 3k_i - k\ \forall i = 2,3$.

Let $f, g \in \mathcal{M}(\mathbb{K})$ be transcendental and let $\alpha \in \mathcal{M}_f(\mathbb{K}) \cap \mathcal{M}_g(\mathbb{K})$ be non-identically zero. If $f'P'(f)$ and $g'P'(g)$ share α I.M., then $f = g$.

By Theorem B, we have Corollary 1.1:

COROLLARY 1.1. *Let $P \in \mathbb{K}[x]$ satisfy $\Phi(P) \geq 3$, let $P' = b(x-a_1)^n \prod_{i=2}^{l}(x-a_i)^{k_i}$ with $b \in \mathbb{K}^*$, $l \geq 3$, $k_i \geq k_{i+1}$, $2 \leq i \leq l-1$ and let $k = \sum_{i=2}^{l} k_i$. Suppose P satisfies the following conditions:*

$n \geq 18 + \max(0, 10-k_2) + \sum_{i=3}^{l} \max(0, 7-k_i)$,

$n \geq k+2$,

if $l = 3$, *then* $n \neq 2k+1,\ 3k_i - k\ \forall i = 2, 3$.

Let $f, g \in \mathcal{M}(\mathbb{K})$ be transcendental and let $\alpha \in \mathcal{M}_f(\mathbb{K}) \cap \mathcal{M}_g(\mathbb{K})$ be non-identically zero. If $f'P'(f)$ and $g'P'(g)$ share α I.M., then $f = g$.

EXAMPLE. Let $P'(x) = x^{19}(x-1)^{10}(x-2)^7$. Consider the polynomial function $P(t)$ defined in $[0, +\infty[$. Clearly, $P'(t) < 0\ \forall t \in]0,1[$ hence $P(1) < 0$. At the point 1, P' has a zero of even order, hence P remains decreasing in $]1,2[$ hence $P(2) < P(1) < P(0)$. Therefore, $\Phi(P) = 3$ and hence, P is a polynomial of uniqueness for $\mathcal{M}(\mathbb{K})$. Now, consider two functions $f,\ g \in \mathcal{M}(\mathbb{K})$ and a small function α. If $f'P'(f)$ and $g'P'(g)$ share α I.M., then $f = g$.

And by Theorem A we also have Corollary 1.2.

COROLLARY 1.2. *Let $P \in \mathbb{K}[x]$ be such that P' is of the form $b(x-a_1)^n(x-a_2)^k$ with $k \geq 2$ and*

$n \geq 18 + \max(0, 10-k)$,

$n \geq k+2$,

$n \neq 2k,\ 2k+1,\ 3k+1$.

Let $f, g \in \mathcal{M}(\mathbb{K})$ be transcendental and let $\alpha \in \mathcal{M}_f(\mathbb{K}) \cap \mathcal{M}_g(\mathbb{K})$ be non-identically zero. If $f'P'(f)$ and $g'P'(g)$ share α I.M., then $f = g$.

EXAMPLE. Let $P'(x) = x^{18}(x-1)^{10}$. By Theorem A, P is a polynomial of uniqueness for $\mathcal{M}(\mathbb{K})$. Now, consider two functions $f,\ g \in \mathcal{M}(\mathbb{K})$ and a small function $\alpha \in \mathcal{M}_f(\mathbb{K}) \cap \mathcal{M}_g(\mathbb{K})$. If $f'P'(f)$ and $g'P'(g)$ share α I.M., then $f = g$.

THEOREM 2. *Let P be a polynomial of uniqueness for $\mathcal{M}(\mathbb{K})$, let*

$$P' = b(x-a_1)^n \prod_{i=2}^{l}(x-a_i)^{k_i}$$

with $b \in \mathbb{K}^*$, $l \geq 2$, $k_i \geq k_{i+1}$, $2 \leq i \leq l-1$ and let $k = \sum_{i=2}^{l} k_i$. Suppose P satisfies the following conditions:

$n \geq 17 + \max(0, 10-k_2) + \sum_{i=3}^{l} \max(0, 7-k_i)$,

$n \geq k+2$,

if $l = 2$, *then* $n \neq 2k,\ 2k+1,\ 3k+1$,

if $l = 3$, *then* $n \neq 2k+1,\ 3k_i - k\ \forall i = 2, 3$.

Let $f, g \in \mathcal{M}(\mathbb{K})$ be transcendental and let α be a Moebius function. If $f'P'(f)$ and $g'P'(g)$ share α I.M., then $f = g$.

By Theorem B, we have Corollary 2.1.

COROLLARY 2.1. *Let $P \in \mathbb{K}[x]$ satisfy $\Phi(P) \geq 3$, let* $P' = b(x-a_1)^n \prod_{i=2}^{l}(x-a_i)^{k_i}$ *with $b \in \mathbb{K}^*$, $l \geq 3$, $k_i \geq k_{i+1}$, $2 \leq i \leq l-1$ and let $k = \sum_{i=2}^{l} k_i$. Suppose P satisfies the following conditions:*

$$n \geq 17 + \max(0, 10 - k_2) + \sum_{i=3}^{l} \max(0, 7 - k_i),$$

$n \geq k+2$,

if $l = 3$, then $n \neq 2k+1,\ 3k_i - k\ \forall i = 2, 3$.

Let $f, g \in \mathcal{M}(\mathbb{K})$ be transcendental and let α be a Moebius function. If $f'P'(f)$ and $g'P'(g)$ share α I.M., then $f = g$.

EXAMPLE. Let $P'(x) = x^{18}(x-1)^{10}(x-2)^6$. Consider the polynomial function $P(t)$ defined in $[0, +\infty[$. Clearly, $P'(t) > 0\ \forall t \in]0, 1[\cup]1, 2[$ hence $P(2) > P(1) > P(0)$. Therefore, $\Phi(P) = 3$ and hence, P is a polynomial of uniqueness for $\mathcal{M}(\mathbb{K})$. On the other hand, P satisfies the hypotheses of Corollary 2.1. Now, consider two functions $f,\ g \in \mathcal{M}(\mathbb{K})$ and a Moebius function α. If $f'P'(f)$ and $g'P'(g)$ share α I.M., then $f = g$.

And by Theorem A, we have Corollary 2.2.

COROLLARY 2.2. *Let $P \in \mathbb{K}[x]$ be such that P' is of the form $b(x-a_1)^n(x-a_2)^k$ with $\min(k, n) \geq 2$ and with $b \in \mathbb{K}^*$. Suppose P satisfies the following conditions:*

$n \geq 17 + \max(0, 10 - k)$,

$n \geq k+2$,

$n \neq 2k,\ 2k+1,\ 3k+1$,

Let $f, g \in \mathcal{M}(\mathbb{K})$ be transcendental and let α be a Moebius function. If $f'P'(f)$ and $g'P'(g)$ share α I.M., then $f = g$.

EXAMPLE. Let $P'(x) = x^{17}(x-1)^{10}$. By Theorem A, P is a polynomial of uniqueness for $\mathcal{M}(\mathbb{K})$ and it satisfies all hypotheses of Corollary 2.2. Now, consider two functions $f,\ g \in \mathcal{M}(\mathbb{K})$ and a Moebius function α. If $f'P'(f)$ and $g'P'(g)$ share α I.M., then $f = g$.

THEOREM 3. *Let P be a polynomial of uniqueness for $\mathcal{M}(\mathbb{K})$, let*

$$P' = b(x-a_1)^n \prod_{i=2}^{l}(x-a_i)^{k_i}$$

with $b \in \mathbb{K}^*$, $l \geq 2$, $k_i \geq k_{i+1}$, $2 \leq i \leq l-1$ and let $k = \sum_{i=2}^{l} k_i$. Suppose P satisfies the following conditions:

$n \geq k+2$,

$$n \geq 17 + \max(0, 10 - k_2) + \sum_{i=3}^{l} \max(0, 7 - k_i).$$

Let $f, g \in \mathcal{M}(\mathbb{K})$ be transcendental and let α be a non-zero constant. If $f'P'(f)$ and $g'P'(g)$ share α I.M., then $f = g$.

By Theorem B, we have Corollary 3.1.

COROLLARY 3.1. *Let $P \in \mathbb{K}[x]$ satisfy $\Phi(P) \geq 3$, let* $P' = b(x-a_1)^n \prod_{i=2}^{l}(x-a_i)^{k_i}$ *with $b \in \mathbb{K}^*$, $l \geq 3$, $k_i \geq k_{i+1}$, $2 \leq i \leq l-1$ and let $k = \sum_{i=2}^{l} k_i$. Suppose P satisfies the following conditions:*

$n \geq k+2,$

$n \geq 17 + \max(0, 10-k_2) + \sum_{i=3}^{l} \max(0, 7-k_i).$

Let $f, g \in \mathcal{M}(\mathbb{K})$ be transcendental and let α be a non-zero constant. If $f'P'(f)$ and $g'P'(g)$ share α I.M., then $f = g$.

EXAMPLE. Let $P'(x) = x^{23}(x-1)^6(x-2)^5$. Consider the polynomial function $P(t)$ defined in $[0, +\infty[$. Clearly, $P'(t) < 0$ $\forall t \in]0,1[\cup]1,2[$ hence $P(2) < P(1) < P(0)$. Therefore, $\Phi(P) = 3$ and hence, P is a polynomial of uniqueness for $\mathcal{M}(\mathbb{K})$. On the other hand, P satisfies the hypotheses of Corollary 3.1. Now, consider two functions f, $g \in \mathcal{M}(\mathbb{K})$ and a constant α. If $f'P'(f)$ and $g'P'(g)$ share α I.M., then $f = g$. Here we notice that $n = 2k+1 = 23$ but this hypothesis is not excluded in Theorem 3.

And by Theorem A, we have Corollary 3.2

COROLLARY 3.2. *Let $P \in \mathbb{K}[x]$ be such that P' is of the form $b(x-a_1)^n(x-a_2)^k$ with $k \geq 2$ and with $b \in \mathbb{K}^*$. Suppose P satisfies the following conditions:*

$n \geq 17 + \max(0, 10-k),$

$n \geq k+2,$

Let $f, g \in \mathcal{M}(\mathbb{K})$ be transcendental and let α be a non-zero constant. If $f'P'(f)$ and $g'P'(g)$ share α I.M., then $f = g$.

EXAMPLE. Let $P'(x) = x^{18}(x-1)^9$. By Theorem A, P is a polynomial of uniqueness for $\mathcal{M}(\mathbb{K})$ and it satisfies all hypotheses of Corollary 3.2. Now, consider two functions f, $g \in \mathcal{M}(\mathbb{K})$ and a constant $\alpha \neq 0$. If $f'P'(f)$ and $g'P'(g)$ share α I.M., then $f = g$. Here we notice that $n = 2k$ by this hypothesis is not excluded in Corollary 3.2.

THEOREM 4. *Let $a \in \mathbb{K}$ and $R > 0$. Let P be a polynomial of uniqueness for $\mathcal{M}_u(d(a, R^-))$ and let*

$$P' = b(x-a_1)^n \prod_{i=2}^{l}(x-a_i)^{k_i}$$

with $b \in \mathbb{K}^*$, $l \geq 2$, $k_i \geq k_{i+1}$, $2 \leq i \leq l-1$ and let $k = \sum_{i=2}^{l} k_i$. Suppose P satisfies the following conditions:

$n \geq 18 + \max(0, 10-k_2) + \sum_{i=3}^{l} \max(0, 7-k_i),$

$n \geq k+3,$

if $l = 2$, then $n \neq 2k,\ 2k+1,\ 3k+1,$

if $l = 3$, then $n \neq 2k+1,\ 3k_i - k$ $\forall i = 2, 3.$

Let $f, g \in \mathcal{M}_u(d(a, R^-))$ and let $\alpha \in \mathcal{M}_f(d(a, R^-)) \cap \mathcal{M}_g(d(a, R^-))$ be non-identically zero. If $f'P'(f)$ and $g'P'(g)$ share α I.M., then $f = g$.

By Theorem B we can state Corollary 4.1.

COROLLARY 4.1. *Let $a \in \mathbb{K}$ and $R > 0$. Let $P \in \mathbb{K}[x]$ satisfy $\Phi(P) \geq 4$, let $P' = b(x - a_1)^n \prod_{i=2}^{l}(x - a_i)^{k_i}$ with $b \in \mathbb{K}^*$, $l \geq 4$, $k_i \geq k_{i+1}$, $2 \leq i \leq l - 1$ and let $k = \sum_{i=2}^{l} k_i$. Suppose P satisfies the following conditions:*

$$n \geq 18 + \max(0, 10 - k_2) + \sum_{i=3}^{l} \max(0, 7 - k_i),$$

$$n \geq k + 3,$$

Let $f, g \in \mathcal{M}_u(d(a, R^-))$ and let $\alpha \in \mathcal{M}_f(d(a, R^-)) \cap \mathcal{M}_g(d(a, R^-))$ be non-identically zero. If $f'P'(f)$ and $g'P'(g)$ share α I.M., then $f = g$.

EXAMPLE. Let $P'(x) = x^{23}(x-1)^{10}(x-2)^4(x-3)^5$. Consider the polynomial function $P(t)$ defined in $[0, +\infty[$. Clearly, $P'(t) < 0\ \forall t \in]0, 1[\cup]1, 2[\cup]2, 3[$ hence $P(3) < P(2) < P(1) < P(0)$. Therefore, $\Phi(P) = 4$ and hence, P is a polynomial of uniqueness for $\mathcal{M}(d(a, R^-))$. On the other hand, we check that P satisfies the hypotheses of Corollary 4.1. Now, consider two functions $f,\ g \in \mathcal{M}_u(d(a, R^-))$ and a constant α. If $f'P'(f)$ and $g'P'(g)$ share α I.M., then $f = g$.

And by Theorem C we have Corollary 4.2:

COROLLARY 4.2. *Let $a \in \mathbb{K}$ and $R > 0$. Let $P \in \mathbb{K}[x]$ be such that P' is of the form $b(x - a_1)^n(x - a_2)^2$ with $b \in \mathbb{K}^*$. Suppose P satisfies*

$$n \geq 18 + \max(0, 10 - k).$$

Let $f, g \in \mathcal{M}_u(d(a, R^-))$ and let $\alpha \in \mathcal{M}_f(d(a, R^-)) \cap \mathcal{M}_f(d(a, R^-))$ be non-identically zero. If $f'P'(f)$ and $g'P'(g)$ share α I.M., then $f = g$.

EXAMPLE. Let $P'(X) = x^{26}(x-1)^2$. By Theorem A, P is a polynomial of uniqueness for $\mathcal{M}(d(a, R^-))$ and it satisfies all hypotheses of Corollary 4.2. Now, consider two functions $f,\ g \in \mathcal{M}(d(a, R^-))$ and a function $\alpha \in \mathcal{M}_f(d(a, R^-)) \cap \mathcal{M}_g(d(a, R^-))$. If $f'P'(f)$ and $g'P'(g)$ share α I.M., then $f = g$.

THEOREM 5. *Let P be a polynomial of uniqueness for $\mathcal{M}(\mathbb{K})$ such that P' is of the form $b(x - a_1)^n \prod_{i=2}^{l}(x - a_i)$ with $l \geq 3$, $b \in \mathbb{K}^*$, satisfying $n \geq 10 + 4l$. Let $f, g \in \mathcal{M}(\mathbb{K})$ be transcendental and let $\alpha \in \mathcal{M}_f(\mathbb{K}) \cap \mathcal{M}_g(\mathbb{K})$ be non-identically zero. If $f'P'(f)$ and $g'P'(g)$ share α I.M., then $f = g$.*

By Theorem B, we have Corollary 5.1:

COROLLARY 5.1. *Let $P \in \mathbb{K}[x]$ satisfy $\Phi(P) \geq 3$ and be such that P' is of the form*

$b(x - a_1)^n \prod_{i=2}^{l}(x - a_i)$ with $l \geq 3$, $b \in \mathbb{K}^*$ satisfying:

$$n \geq 10 + 4l.$$

Let $f, g \in \mathcal{M}(\mathbb{K})$ be transcendental and let $\alpha \in \mathcal{M}_f(\mathbb{K}) \cap \mathcal{M}_g(\mathbb{K})$ be non-identically zero. If $f'P'(f)$ and $g'P'(g)$ share α I.M., then $f = g$.

EXAMPLE. Let $P'(x) = x^{22}(x^2 - 1)$. Consider the polynomial function $P(t)$ defined in $[-1, +1]$. Clearly, $P'(t) < 0\ \forall t \in] - 1, 0[\cup]0, 1[$ hence $P(-1) < P(0) < P(1)$. Therefore, $\Phi(P) = 3$ and hence, P is a polynomial of uniqueness for $\mathcal{M}(\mathbb{K})$. On the other hand, we check that P satisfies the hypotheses of Corollary 5.1. Now, consider two functions $f,\ g \in \mathcal{M}(\mathbb{K})$ and a function $\alpha \in \mathcal{M}_f(\mathbb{K}) \cap \mathcal{M}_g(\mathbb{K})$. If $f'P'(f)$ and $g'P'(g)$ share α I.M., then $f = g$.

THEOREM 6. *Let $a \in \mathbb{K}$ and $R > 0$. Let P be a polynomial of uniqueness for $\mathcal{M}_u(d(a, R^-))$ such that P' is of the form $P' = b(x - a_1)^n \prod_{i=2}^{l}(x - a_i)$ with $l \geq 3$, $b \in \mathbb{K}^*$ satisfying:*
$n \geq 10 + 4l$.
Let $f, g \in \mathcal{M}_u(d(a, R^-))$ and let $\alpha \in \mathcal{M}_f(d(a, R^-)) \cap \mathcal{M}_g(d(a, R^-))$ be non-identically zero. If $f'P'(f)$ and $g'P'(g)$ share α I.M., then $f = g$.

By Theorem B, we have Corollary 6.1:

COROLLARY 6.1. *Let $a \in \mathbb{K}$ and $R > 0$. Let $P \in \mathbb{K}[x]$ satisfy $\Phi(P) \geq 4$ and be such that P' is of the form $P' = b(x - a_1)^n \prod_{i=2}^{l}(x - a_i)$ with $l \geq 4$, $b \in \mathbb{K}^*$ and $n \geq 10 + 4l$.*

Let $f, g \in \mathcal{M}_u(d(a, R^-))$ and let $\alpha \in \mathcal{M}_f(d(a, R^-)) \cap \mathcal{M}_g(d(a, R^-))$ be non-identically zero. If $f'P'(f)$ and $g'P'(g)$ share α I.M., then $f = g$.

EXAMPLE. Let $P(x) = \frac{x^{30}}{30} - \frac{2x^{29}}{29} - \frac{x^{28}}{28} + \frac{2x^{27}}{27}$. Then $P'(x) = x^{29} - 2x^{28} - x^{27} + 2x^{26} = x^{26}(x - 1)(x + 1)(x - 2)$. We check that:
$P(0) = 0$,
$P(1) = \frac{1}{30} - \frac{2}{29} - \frac{1}{28} + \frac{2}{27} \neq 0$,
$P(-1) = \frac{1}{30} + \frac{2}{29} - \frac{1}{28} - \frac{2}{27} \neq 0, P(1)$, and

$$P(2) = \frac{2^{30}}{30} - \frac{2^{30}}{29} - \frac{2^{28}}{28} + \frac{2^{28}}{27} \neq 0, P(1), P(-1).$$

Then $\Phi(P) = 4$. So, P is a polynomial of uniqueness for both $\mathcal{M}(\mathbb{K})$ and $\mathcal{M}(d(0, R^-))$. Moreover, we have $n = 26, l = 4$, hence we can apply Corollaries 5.1 and 6.1.

Given $f,\ g \in \mathcal{M}(\mathbb{K})$ transcendental or $f,\ g \in \mathcal{M}_u(d(0, R^-))$ such that $f'P'(f)$ and $g'P'(g)$ share a small function α I.M., we have $f = g$.

THEOREM 7. *Let P be a polynomial of uniqueness for $\mathcal{M}(\mathbb{K})$ such that P' is of the form $P' = b(x - a_1)^n \prod_{i=2}^{l}(x - a_i)$ with $l \geq 3$, $b \in \mathbb{K}^*$ satisfying $n \geq 9 + 4l$.*

Let $f, g \in \mathcal{M}(\mathbb{K})$ be transcendental and let α be a Moebius function or a non-zero constant. If $f'P'(f)$ and $g'P'(g)$ share α I.M., then $f = g$.

EXAMPLE. Let $P'(x) = x^{21}(x^2 - 1)$. Then as in the previous example, P is a uniqueness polynomial for $\mathcal{M}(\mathbb{K})$. Let $f,\ g \in \mathcal{M}(\mathbb{K})$ be transcendental and let $\alpha \in \mathcal{M}(\mathbb{K})$ be a Moebius function or a non-zero constant such that $f'P'(f)$, $g'P'(g)$ share α I.M. Then, $f = g$.

THEOREM 8. *Let $f,\ g \in \mathcal{M}(\mathbb{K})$ be transcendental and let $\alpha \in \mathcal{M}_f(\mathbb{K}) \cap \mathcal{M}_g(\mathbb{K})$ be non-identically zero. Let $a \in \mathbb{K} \setminus \{0\}$. If $f'f^n(f - a)$ and $g'g^n(g - a)$ share the function α I.M. and if $n \geq 18$, then either $f = g$ or there exists $h \in \mathcal{M}(\mathbb{K})$ such* *that* $f = \dfrac{a(n+2)}{n+1}\Big(\dfrac{h^{n+1}-1}{h^{n+2}-1}\Big)h$ *and* $g = \dfrac{a(n+2)}{n+1}\Big(\dfrac{h^{n+1}-1}{h^{n+2}-1}\Big)$. *Moreover, if* α *is a constant or a Moebius function, then the conclusion holds whenever* $n \geq 17$.

Inside an open disk, we have a version similar to the general case in the whole field.

THEOREM 9. *Let* $f,\ g \in \mathcal{M}_u(d(0,R^-))$, *and let* $\alpha \in \mathcal{M}_f(d(0,R^-)) \cap \mathcal{M}_g(d(0,R^-))$ *be non-identically zero. Let* $a \in \mathbb{K} \setminus \{0\}$. *If* $f'f^n(f-a)$ *and* $g'g^n(g-a)$ *share the function* α *C.M. and* $n \geq 18$, *then either* $f = g$ *or there exists* $h \in \mathcal{M}(d(0,R^-))$ *such that* $f = \dfrac{a(n+2)}{n+1}\Big(\dfrac{h^{n+1}-1}{h^{n+2}-1}\Big)h$ *and* $g = \dfrac{a(n+2)}{n+1}\Big(\dfrac{h^{n+1}-1}{h^{n+2}-1}\Big)$.

REMARK 3. As already noticed in [**2**], in Theorems 8 and 9, the second conclusion does occur. Indeed, let $h \in \mathcal{M}(\mathbb{K})$ (resp. let $h \in \mathcal{M}_u(d(0,R^-))$). Now, let us precisely define f and g as: $g = (\dfrac{n+2}{n+1})\Big(\dfrac{h^{n+1)}-1}{h^{n+2}-1}\Big)$ and $f = hg$. Then we can see that the polynomial $P(y) = \dfrac{1}{n+2}y^{n+2} - \dfrac{1}{n+1}y^{n+1}$ satisfies $P(f) = P(g)$, hence $f'P'(f) = g'P'(g)$, therefore $f'P'(f)$ and $g'P'(g)$ trivially share any function.

Consider now the situation with analytic functions.

THEOREM 10. *Let* $P(x) \in \mathbb{K}[x]$ *be a polynomial of uniqueness for* $\mathcal{A}(\mathbb{K})$ *(resp. for* $\mathcal{A}(d(a,R^-))$*) and let* $P'(x) = b(x-a_1)^n \prod_{i=2}^{l}(x-a_i)^{k_i}$ *with* $b \in \mathbb{K}^*$, $f,\ g \in \mathcal{A}(\mathbb{K})$ *(resp.* $f,\ g \in \mathcal{A}(d(0,R^-))$*) and be such that* $f'P'(f)$ *and* $g'P'(g)$ *share a function* $\alpha \in \mathcal{A}_f(\mathbb{K}) \cap \mathcal{A}_g(\mathbb{K})$ *I.M. (resp.* $\alpha \in \mathcal{A}_f(d(a,R^-)) \cap \mathcal{A}_g(d(a,R^-))$ *I.M.). If* $n \geq \sum_{i=2}^{l} k_i + 2$ *(resp.* $n \geq \sum_{i=2}^{l} k_i + 3$*) and* $n \geq 8 + \max(0, 10-k_2) + \sum_{i=3}^{l} \max(0, 7-k_i)$, *then* $f = g$.

EXAMPLE. Let $P'(x) = x^{20}(x-1)^{10}(x-2)^7$. Consider the polynomial function $P(t)$ defined in $[0,+\infty[$, vanishing at 0. Clearly, $P'(t) > 0\ \forall t \in]0,1[\cup]1,2[$ hence $P(2) > P(1) > P(0) = 0$. Therefore, $\Phi(P) = 3$ and hence, P is a polynomial of uniqueness for $\mathcal{A}(d(a,R^-))$. On the other hand, P satisfies the hypotheses of Theorem 8. Now, consider two functions $f,\ g \in \mathcal{A}(d(a,R^-))$ such that $f'P'(f)$ and $g'P'(g)$ share a function $\alpha \in \mathcal{A}_f(d(a,R^-)) \cap \mathcal{A}_g(d(a,R^-))$ I.M. Then $f = g$.

By Theorem A, we can derive this corollary:

COROLLARY 10.1. *Let* $P(x) \in \mathbb{K}[x]$ *and let* $P'(x) = x^n(x-a)^k$. *Let* $f,\ g \in \mathcal{A}(\mathbb{K})$ *be such that* $f'P'(f)$ *and* $g'P'(g)$ *share a function* $\alpha \in \mathcal{A}_f(\mathbb{K}) \cap \mathcal{A}_g(\mathbb{K})$ *I.M. If* $n \geq k+2$ *and* $n \geq 8 + \max(0, 10-k)$, *then* $f = g$.

THEOREM 11. *Let* $P(x) \in \mathbb{K}[x]$ *and let* $P'(x) = b(x-a_1)^n \prod_{i=2}^{l}(x-a_i)^{k_i}$ *with* $b \in \mathbb{K}^*$, $f,\ g \in \mathcal{A}(\mathbb{K})$ *be a polynomial of uniqueness for* $\mathcal{A}(\mathbb{K})$ *and be such that* $f'P'(f)$ *and* $g'P'(g)$ *share an affine function* α *or a non-zero constant* α *I.M. If* $n \geq \sum_{i=2}^{l} k_i + 1$ *and* $n \geq 7 + \max(0, 10-k_2) + \sum_{i=3}^{l} \max(0, 7-k_i)$, *then* $f = g$.

2. Basic Results

All lemmas used in [**2**] apply again here, even those whose hypotheses involve sharing values because they only concern sharing values IM, except Lemma 8 that we will thoroughly check.

NOTATION. Given two meromorphic functions $f,\ g \in \mathcal{M}(\mathbb{K})$ (resp. $f,\ g \in \mathcal{M}(d(0,R^-))$), we will denote by $\Theta_{f,g}$ the function

$$\frac{f''}{f'} - \frac{2f'}{f-1} - \frac{g''}{g'} + \frac{2g'}{g-1}.$$

LEMMA 1. *Let $f \in \mathcal{M}(\mathbb{K})$ (resp. Let $f \in \mathcal{M}(d(0,R^-))$), let $a \in \mathbb{K}$, let $Q(x) \in \mathbb{K}[x]$ of degree s. Then $T(r,Q(f)) = sT(r,f) + O(1)$ and $T(r,f'Q(f)) \geq sT(r,f) + O(1)$.*

LEMMA 2. *Let $f \in \mathcal{M}(\mathbb{K})$ (resp. Let $f \in \mathcal{M}(d(0,R^-))$). Then $N(r,f') = N(r,f) + \overline{N}(r,f)$, $Z(r,f') \leq Z(r,f) + \overline{N}(r,f) - \log r + O(1)$. Moreover, $T(r,f) - Z(r,f) \leq T(r,f') - Z(r,f') + O(1)$.*

LEMMA 3. *Let $f \in \mathcal{M}(\mathbb{K})$ (resp. Let $f \in \mathcal{M}(d(0,R^-))$) and let $\phi = \dfrac{f'}{f}$. Then $Z(r,\phi) \leq N(r,\phi) - \log r + O(1)$.*

LEMMA 4. *The function $\Theta_{f,g}$ satisfies $Z(r,\Theta_{f,g}) \leq N(r,\Theta_{f,g}) - \log r + O(1)$.*

LEMMA 5. *Let $f,\ g \in \mathcal{M}(\mathbb{K})$ (resp. Let $f,\ g \in \mathcal{M}(d(0,R^-))$). If a is a simple zero of $f-1$ and $g-1$, it is a zero of $\Theta_{f,g}$.*

In order to state the next lemma, we must recall the definition of quasi-exceptional values.

(i) Let $f \in \mathcal{M}(\mathbb{K}) \setminus \mathbb{K}$. (resp. let $f \in \mathcal{M}_u(d(0,R^-))$). Then b will be said to be *a Picard exceptional value* of f (or just *an exceptional value*) if $f(x) \neq b\ \forall x \in \mathbb{K}$ (resp $f(x) \neq b\ \forall x \in d(0,R^-)$).
(ii) Let $f \in \mathcal{M}(\mathbb{K}) \setminus \mathbb{K}(x)$. (resp. let $f \in \mathcal{M}_u(d(0,R^-))$) and let $b \in \mathbb{K}$. Then b will be said to be *a quasi-exceptional value* of f if the function $f-b$ has a finite number of zeros in $\mathbb{K}$ (resp. in $d(0,R^-)$).

LEMMA 6. *Let $f \in \mathcal{A}(\mathbb{K}) \setminus \mathbb{K}$ $\big($resp. let $f \in \mathcal{A}_u(d(0,R^-))\big)$. Then f has no exceptional value. If f is transcendental, it has no quasi-exceptional value. Let $f \in \mathcal{M}(\mathbb{K}) \setminus \mathbb{K}$ $\big($resp. let $f \in \mathcal{M}_u(d(0,R^-))\big)$. Then f has at most one exceptional value in $\mathbb{K}$. Let $f \in \mathcal{M}(\mathbb{K})$ be transcendental $\big($resp. let $f \in \mathcal{M}_u(d(0,R^-))\big)$. Then f has at most one quasi-exceptional value in $\mathbb{K}$.*

We now have to recall the *ultrametric Nevanlinna Second Main Theorem* in a basic form which we will frequently use.

Let $f \in \mathcal{M}(\mathbb{K})$ $\big($resp. $f \in \mathcal{M}(d(0,R^-))\big)$ satisfy $f'(0) \neq 0, \infty$. Let S be a finite subset of $\mathbb{K}$ and $r \in]0,+\infty[$ $\big($resp. $r \in]0,R[\big)$. We denote by $Z_0^S(r,f')$ *the counting function of zeros of f' in $d(0,r)$ which are not zeros of any $f-s$ for $s \in S$*. This is, if $(\gamma_n)_{n\in\mathbb{N}}$ is the finite or infinite sequence of zeros of f' in $d(0,r)$ that are not zeros of $f-s$ for $s \in S$, with multiplicy order q_n respectively, we set

$$Z_0^S(r,f') = \sum_{|\gamma_n|\leq r} q_n(\log r - \log|\gamma_n|).$$

THEOREM N ([**4**], [**5**]). *Let* $a_1, ..., a_q \in \mathbb{K}$ *with* $q \geq 2, q \in \mathbb{N}$, *and let* $f \in \mathcal{M}(\mathbb{K})$ *(resp. let* $f \in \mathcal{M}(d(0, R^-))$*). Let* $S = \{a_1, ..., a_q\}$*. Assume that none of* f, f' *and* $f - a_j$ *with* $1 \leq j \leq q$, *equals* 0 *or* ∞ *at the origin. Then, for* $r > 0$ $\big($*resp. for* $r \in]0, R[\big)$, *we have*

$$(q-1)T(r,f) \leq \sum_{j=1}^{q} \overline{Z}(r, f - a_j) + \overline{N}(r,f) - Z_0^S(r, f') - \log r + O(1).$$

3. Specific Lemmas

The following Lemma is proven in [**2**].

LEMMA 7. *Let* $f,\ g \in \mathcal{M}(\mathbb{K})$ *be transcendental (resp.* $f,\ g \in \mathcal{M}_u(d(0, R^-))$*). Let* $P(x) = x^{n+1}Q(x)$ *be a polynomial such that* $n \geq deg(Q) + 2$ *(resp.* $n \geq deg(Q) + 3$*). If* $P'(f)f' = P'(g)g'$ *then* $P(f) = P(g)$.

Lemma 8 is proven in [**2**] under the assumption that F, G share 1 value C.M. But actually, the hypothesis IM is sufficient to conclude.

LEMMA 8. *Let* $F, G \in \mathcal{M}(\mathbb{K})$ $\big($*resp. let* $F, G \in \mathcal{M}(d(0, R^-))\big)$ *be non-constant, having no zero and no pole at* 0 *and sharing the value* 1 *I.M.*

If $\Theta_{F,G} = 0$ *and if*

$$\lim_{r \to +\infty} \Big(T(r,F) - [\overline{Z}(r,F) + \overline{N}(r,F) + \overline{Z}(r,G) + \overline{N}(r,G)] \Big) = +\infty$$

(resp.

$$\lim_{r \to R^-} \Big(T(r,F) - [\overline{Z}(r,F) + \overline{N}(r,F) + \overline{Z}(r,G) + \overline{N}(r,G)] \Big) = +\infty)$$

then either $F = G$ *or* $FG = 1$.

The following Lemma is proven in [**2**] and in [**9**], (Theorem 56.3). Here, we just have to add the last sentence which actually is immediate: f, g *can not belong to* $\mathcal{A}(\mathbb{K})$ *(resp.* $f,\ g \in \mathcal{A}_u(d(0, R^-))$*).*

LEMMA 9. *Let* $Q(x) = (x - a_1)^n \prod_{i=2}^{l} (x - a_i)^{k_i} \in \mathbb{K}[x]$ $(a_i \neq a_j,\ \forall i \neq j)$ *with* $l \geq 2$ *and* $n \geq \max\{k_2, .., k_l\}$ *and let* $k = \sum_{i=2}^{l} k_i$*. Let* $f,\ g \in \mathcal{M}(\mathbb{K})$ *be transcendental (resp.* $f,\ g \in \mathcal{M}_u(d(0, R^-))$*) such that* $\theta = Q(f)f'Q(g)g'$ *is a small function with respect to* f *and* g*. We have the following :*

If $l = 2$ *then* n *belongs to* $\{k,\ k+1,\ 2k,\ 2k+1,\ 3k+1\}$.

If $l = 3$ *then* n *belongs to* $\{\frac{k}{2}, k+1, 2k+1, 3k_2 - k, .., 3k_l - k\}$.

If $l \geq 4$ *then* $n = k + 1$.

If θ *is a constant and* $f,\ g \in \mathcal{M}(\mathbb{K})$ *then* $n = k + 1$.

Moreover, f, g *can not belong to* $\mathcal{A}(\mathbb{K})$ *(resp.* $f,\ g \in \mathcal{A}_u(d(0, R^-))$*).*

Lemma 10 is known and easily checked [**6**], [**24**]:

LEMMA 10. *Let* $f,\ g \in \mathcal{M}(\mathbb{K})$ *be transcendental (resp. let* $f,\ g \in \mathcal{M}_u(d(0, R^-))$*) satisfy* $(f - a)f^n = (g - a)g^n$ *with* $a \in \mathbb{K}$ *and let* $h = \dfrac{f}{g}$*. If* h *is not identically* 1, *then*

$$g = \frac{h^n - 1}{h^{n+1} - 1},\ f = \frac{h^{n+1} - h}{h^{n+1} - 1}.$$

NOTATION. Let $f,g \in \mathcal{M}(\mathbb{K})$ (resp. Let $f \in \mathcal{M}(d(0,R^-))$) be such that $f(0) \neq 0, \infty$ We denote by $Z_{1)}(r,f)$ the counting function of simple zeros of f, we denote by $Z_{(2}(r,f)$ the counting function of multiple zeros of f, each counted with multiplcity and we denote by $\overline{Z}_{(2}(r,f)$ the counting function of multiple zeros of f, each counted without multiplcity.

In the same way, we set
$N_{1)}(r,f) = Z_{1)}(r,\frac{1}{f})$, $N_{(2}(r,f) = Z_{(2}(r,\frac{1}{f}), \overline{N}_{(2}(r,f) = \overline{Z}_{(2}(r,\frac{1}{f})$.
Consequently, by definition, one has $\overline{Z}(r,f) = Z_{1)}(r,f) + \overline{Z}_{(2}(r,f)$, $\overline{N}(r,f) = N_{1)}(r,f) + \overline{N}_{(2}(r,f)$.

Finally, we denote by $Z_{[2]}(r,f)$ the counting fiunction of the zeros of f each counted with multiplicity when it is at most 2 and with multiplicity 2 when it is bigger and we set $N_{[2]}(r,f) = Z_{[2]}(r,\frac{1}{f})$.

Moreover, here we denote by $Z_0(r,f')$ the counting function of the zeros of f' that are not zeros of $f(f-1)$.

Given $f,g \in \mathcal{M}(\mathbb{K})$ (resp. Let $f,g \in \mathcal{M}(d(0,R^-)))$, we denote by $\Theta_{f,g}$ the function

$$\frac{f''}{f'} - \frac{2f'}{f-1} - \frac{g''}{g'} + \frac{2g'}{g-1}.$$

The following Lemma 11 is proven in [**24**] for complex functions. Here we check that it holds for p-adic meromorphic functions. Moreover, we enjoy an additional negative function in $-\log r$. This is a generalization published in [**16**].

LEMMA 11. *Let $F,G \in \mathcal{M}(\mathbb{K})$ (resp. Let F, $G \in \mathcal{M}(d(0,R^-))$) be distinct and share the value* 1 *I.M. If $\Theta_{F,G}$ is not identically zero, then,*

$$T(r,F) \leq N_{[2]}(r,F)) + Z_{[2]}(r,F) + N_{[2]}(r,G) + Z_{[2]}(r,G)$$
$$+2\big(\overline{Z}(r,F) + 2\overline{N}(r,F)\big) + \overline{Z}(r,G) + \overline{N}(r,G) - 4\log(r) + O(1).$$

PROOF. By hypotheses, we have $\overline{Z}(r,F-1) = Z_{1)}(r,F-1) + \overline{Z}_{(2}(r,F-1)$.

Next, we can easilly check that each simple zero of $F-1$ and $G-1$ is a zero of $\Theta_{F,G}$. Therefore
(2) $Z_{1)}(r,F-1) \leq Z(r,\Theta_{F,G}) + \overline{Z}_{(2}(r,G-1)$ and similarly
(3) $Z_{1)}(r,G-1) \leq Z(r,\Theta_{F,G}) + \overline{Z}_{(2}(r,F-1)$.

On the other hand, by Lemma 4, we have $Z(r,\Theta_{F,G}) \leq N(r,\Theta_{F,G}) - \log r + O(1)$. Consequently, by (2) we obtain
(4) $Z_{1)}(r,F-1) \leq N(r,\Theta_{F,G}) + \overline{Z}_{(2}(r,G-1) - \log r + O(1)$
and similarly by (3),
(5) $Z_{1)}(r,G-1) \leq N(r,\Theta_{F,G}) + \overline{Z}_{(2}(r,F-1) - \log r + O(1)$.

Let $Z_0(r,F')$ denote the counting function of zeros of F' when $F(x) \neq 0$ and 1. Of course, each pole of $\Theta_{F,G}$ is of order 1. Next, if F has a pole of order 1, it is not a pole of $\frac{F''}{F'} - \frac{2F'}{F-1}$ and if F has a zero of order ≥ 2 it is not a pole of $\frac{F''}{F'} - \frac{2F'}{F-1}$, and similarly for $\frac{G''}{G'} - \frac{2G'}{G-1}$.

Consequently, a pole of $\Theta_{F,G}$ can only occur either when $F-1$ or $G-1$ has a zero of order ≥ 2 or when F' or G' has a zero that is not a zero of $F-1$ or when F or G has a pole of order ≥ 2 or when $F-1$ or $G-1$ has a zero of order ≥ 2.

On the other hand, we can dissociate zeros of F' when $F(x) \neq 0$ and 1 and other zeros of F' and similarly for G'. Therefore

$$(6)\ N(r,\Theta_{F,G}) \leq \overline{Z}_0(r,F') + \overline{Z}_0(r,G') + Z_{(2}(r,F) + Z_{(2}(r,G) + \overline{N}_{(2}(r,F) + \overline{N}_{(2}(r,G) + \big(\overline{Z}_{(2}(r,F-1) + \overline{Z}_{(2}(r,G-1)\big)$$

And hence by (4) we have

$$(7)\ Z_{1)}(r,F-1) \leq \overline{Z}_0(r,F') + \overline{Z}_0(r,G') + \overline{N}_{(2}(r,F) + \overline{N}_{(2}(r,G) + \overline{Z}_{(2}(r,F-1) + 2\overline{Z}_{(2}(r,G-1) - \log(r) + O(1)$$

and similarly by (5),

$$(8)\ Z_{1)}(r,G-1) \leq \overline{Z}_0(r,F') + \overline{Z}_0(r,G') + \overline{N}_{(2}(r,F) + \overline{N}_{(2}(r,G) + \overline{Z}_{(2}(r,G-1) + 2\overline{Z}_{(2}(r,F-1) - \log(r) + O(1).$$

Moreover, we have

$$(9)\ Z_{1)}(r,F-1) \leq \overline{Z}_0(r,F') + \overline{Z}_0(r,G') + \overline{N}_{(2}(r,F) + \overline{N}_{(2}(r,G) + \overline{Z}_{(2}(r,F-1) + \overline{Z}_{(2}(r,G-1) - \log(r) + O(1)$$

and similarly

$$(10)\ Z_{1)}(r,G-1) \leq \overline{Z}_0(r,F') + \overline{Z}_0(r,G') + \overline{N}_{(2}(r,F) + \overline{N}_{(2}(r,G) + \overline{Z}_{(2}(r,G-1) + \overline{Z}_{(2}(r,F-1) - \log(r) + O(1).$$

Now, since each zero of $F-1$ of order ≥ 2 is a zero of F' that is not a zero of F, we have

$$Z(r,F') = Z_0(r,F') + Z_{(2}(r,F-1) - \overline{Z}_{(2}(r,F-1) + Z_{(2}(r,F) - \overline{Z}_{(2}(r,F)$$

hence

$$(11)\ \overline{Z}_0(r,F') + \overline{Z}_{(2}(r,F-1) + Z_{(2}(r,F) - \overline{Z}_{(2}(r,F) \leq Z(r,F')$$

and similarly

$$(12)\ \overline{Z}_0(r,G') + \overline{Z}_{(2}(r,G-1) + Z_{(2}(r,G) - \overline{Z}_{(2}(r,G) \leq Z(r,G').$$

Hence by Lemma 3 and by (11) and (12) we have

$$(13)\ \overline{Z}_{(2}(r,F-1) \leq -\overline{Z}_0(r,F') + \overline{Z}_{(2}(r,F) - Z_{(2}(r,F) + Z(r,F) + \overline{N}(r,F) - \log r + O(1)$$

and similarly

$$(14)\ \overline{Z}_{(2}(r,G-1) \leq -\overline{Z}_0(r,G') + \overline{Z}_{(2}(r,G) - Z_{(2}(r,G) + Z(r,G) + \overline{N}(r,G) - \log r + O(1).$$

Now $Z(r,F) + \overline{Z}_{(2}(r,F) - Z_{(2}(r,F)$ is just $\overline{Z}(r,F)$. Consequently, by (13) we obtain

$$(15)\ \overline{Z}_{(2}(r,F-1) \leq \overline{N}(r,F) + \overline{Z}(r,F) - \overline{Z}_0(r,F') - \log r + O(1)$$

and similarly by (14),

$$(16)\ \overline{Z}_{(2}(r,G-1) \leq \overline{N}(r,G) + \overline{Z}(r,G) - \overline{Z}_0(r,G') - \log r + O(1).$$

Consequently, by (7), (8), (15) and (16) we obtain

$$Z_{1)}(r,F-1) \leq \overline{Z}_0(r,F') + \overline{Z}_0(r,G') + \overline{N}_{(2}(r,F) + \overline{N}_{(2}(r,G) + \overline{Z}_{(2}(r,F) + \overline{Z}_{(2}(r,G) + (\overline{N}(r,F) + \overline{Z}(r,F) - \overline{Z}_0(r,F')) + 2((\overline{N}(r,G) + \overline{Z}(r,G) - \overline{Z}_0(r,G') - \overline{Z}_0(r,F') - 2\log(r) + O(1)$$

i.e.

$$(17)\ Z_{1)}(r, F-1) \leq N_{[2]}(r, F) + N_{[2]}(r, G) + Z_{[2]}(r, F) + Z_{[2]}(r, G)$$
$$+\overline{N}(r, G) + \overline{Z}(r, G) - 2\log(r) + O(1)$$

and similarly

$$(18)\ Z_{1)}(r, G-1) \leq N_{[2]}(r, F) + N_{[2]}(r, G) + Z_{[2]}(r, F) + Z_{[2]}(r, G)$$
$$+\overline{N}(r, F) + \overline{Z}(r, F) - 2\log(r) + O(1)$$

Morever, by (9) and (10) we have

$$(19)\ Z_{1)}(r, F-1) \leq N_{[2]}(r, F) + N_{[2]}(r, G) + Z_{[2]}(r, F) + Z_{[2]}(r, G)$$
$$-\log(r) + O(1)$$

and similarly

$$(20)\ Z_{1)}(r, G-1) \leq N_{[2]}(r, F) + N_{[2]}(r, G) + Z_{[2]}(r, F) + Z_{[2]}(r, G) - \log(r) + O(1)$$

Now, Theorem N lets us write

$$T(r, F) \leq \overline{N}(r, F) + \overline{Z}(r, F) + \overline{Z}(r, F-1) - Z_0(r, F') - \log r + O(1)$$

hence

$$T(r, F) \leq \overline{N}(r, F) + \overline{Z}(r, F) + Z_{1)}(r, F-1) + \overline{Z}_{(2}(r, F-1) - \log r + O(1)$$

and hence, by (14) and (16), we can derive

$$T(r, F) \leq N_{[2]}(r, F) + Z_{[2]}(r, F) + N_{[2]}(r, G) + Z_{[2]}(r, G)$$
$$+2\big(\overline{Z}(r, F) + \overline{N}(r, F)\big) + \overline{N}(r, G) + \overline{Z}(r, G) - 4\log r + O(1)$$

and similarly

$$T(r, G) \leq N_{[2]}(r, F) + Z_{[2]}(r, F) + N_{[2]}(r, G) + Z_{[2]}(r, G)$$
$$+2\big(\overline{Z}(r, G) + \overline{N}(r, G)\big) + \overline{N}(r, F) + \overline{Z}(r, F) - 4\log r + O(1).$$

□

The following Lemma is immediate:

LEMMA 12. *Let $f \in \mathcal{M}(\mathbb{K})$ (resp. Let $f \in \mathcal{M}(d(0, R^-))$) be non-constant. Let $P(x) = \prod_{i=1}^{l}(x-a_i)^{k_i} \in \mathbb{K}[x]$ be of degree q. Then $\overline{Z}(r, P(f)) \leq \sum_{i=1}^{l} \overline{Z}(r, f-a_i) \leq lT(r, f)$.*

4. Proof of Theorems

PROOF OF THE THEOREMS. The polynomial P is considered in theorems 1, 2, 3, 4, 5, 6, 7, 10, 11 and we can assume $a_1 = 0$. In Theorems 8, and 9, we call P the polynomial such that $P'(x) = x^n(x-a)^k$ and $P(0) = 0$. Set $F = \dfrac{f'P'(f)}{\alpha}$ and $G = \dfrac{g'P'(g)}{\alpha}$. Clearly F and G share the value 1 C.M.

Since f, g are transcendental (resp. unbounded), we notice that so are F and G. Recall that

$$\Theta_{F,G} = \frac{F''}{F'} - \frac{2F'}{F-1} - \frac{G''}{G'} + \frac{2G'}{G-1}$$

We will prove that under the hypotheses of each theorem, $\Theta_{F,G}$ is identically zero.

Set $\widehat{F} = P(f)$, $\widehat{G} = P(g)$. We notice that $P(x)$ is of the form $x^{n+1}Q(x)$ with $Q \in K[x]$ of degree k. Now, with help of Lemma 2, we can check that we have

$$T(r, \widehat{F}) - Z(r, \widehat{F}) \le T(r, \widehat{F}') - Z(r, \widehat{F}') + O(1)$$

Consequently, since $(\widehat{F})' = \alpha F$, we have

$$T(r, \widehat{F}) \le T(r, F) + Z(r, \widehat{F}) - Z(r, F) + T(r, \alpha) + O(1), \tag{1}$$

hence, by (1), we obtain

(1)

$$T(r, \widehat{F}) \le T(r, F) + (n+1)Z(r, f) + Z\big(r, Q(f)\big) - nZ(r, f) - \sum_{i=2}^{l} k_i Z(r, f - a_i)$$
$$- Z(r, f') + T(r, \alpha) + O(1).$$

i.e.

(2)

$$T(r, \widehat{F}) \le T(r, F) + Z(r, f) + Z\big(r, Q(f)\big) - \sum_{i=2}^{l} k_i Z(r, f - a_i) - Z(r, f') + T(r, \alpha) + O(1).$$

and similarly,

(3)

$$T(r, \widehat{G}) \le T(r, G) + Z(r, g) + Z\big(r, Q(g)\big) - \sum_{i=2}^{l} k_i Z(r, g - a_i) - Z(r, g') + T(r, \alpha) + O(1).$$

Now, using Lemma 13, it follows from the definition of F and G that

(4)

$$Z_{[2]}(r, F) + N_{[2]}(r, F) \le 2Z(r, f) + 2\sum_{i=2}^{l} Z(r, f - a_i) + Z(r, f') + 2\overline{N}(r, f) + T(r, \alpha) + O(1)$$

and similarly

(5)

$$Z_{[2]}(r, G) + N_{[2]}(r, G) \le 2Z(r, g) + 2\sum_{i=2}^{l} Z(r, g - a_i) + Z(r, g') + 2\overline{N}(r, g) + T(r, \alpha) + O(1)$$

And particularly, if $k_i = 1$, $\forall i \in \{2, .., l\}$, then

(6)

$$Z_{[2]}(r, F) + N_{[2]}(r, F) \le 2Z(r, f) + \sum_{i=2}^{l} Z(r, f - a_i) + Z(r, f') + 2\overline{N}(r, f) + T(r, \alpha) + O(1)$$

and similarly

(7)

$$Z_{[2]}(r, G) + N_{[2]}(r, G) \le 2Z(r, g) + \sum_{i=2}^{l} Z(r, g - a_i) + Z(r, g') + 2\overline{N}(r, g) + T(r, \alpha) + O(1).$$

Suppose now that $\Theta_{F,G}$ is not identically zero. Now, by Lemma 11, we have

$$T(r, F) \le Z_{[2]}(r, F) + N_{[2]}(r, F) + Z_{[2]}(r, G) + N_{[2]}(r, G)$$
$$+2\big(\overline{N}(r, F) + \overline{Z}(r, F)\big) + \overline{N}(r, G) + \overline{Z}(r, G) - 4\log r + O(1)$$

hence by (2), we obtain

$$T(r,\widehat{F}) \leq Z_{[2]}(r,F) + N_{[2]}(r,F) + Z_{[2]}(r,G) + N_{[2]}(r,G) + Z(r,f) + Z(r,Q(f)) \\ +2\big(\overline{N}(r,F) + \overline{Z}(r,F)\big) + \overline{N}(r,G) + \overline{Z}(r,G) \\ -\sum_{i=2}^{l} k_i Z(r,f-a_i) - Z(r,f') + T(r,\alpha) - 4\log r + O(1)$$

and hence by (4) and (5):

$$T(r,\widehat{F}) \leq 2Z(r,f) + 2\sum_{i=2}^{l} Z(r,f-a_i) + Z(r,f') + 2\overline{N}(r,f) + 2Z(r,g) \\ + 2\sum_{i=2}^{l} Z(r,g-a_i) + Z(r,g') \\ +2\big(\overline{N}(r,F) + \overline{Z}(r,F)\big) + \overline{N}(r,G) + \overline{Z}(r,G) \\ +2\overline{N}(r,g)+Z(r,f)+Z(r,Q(f))-\sum_{i=2}^{l} k_i Z(r,f-a_i)-Z(r,f')+2T(r,\alpha)-4\log r+O(1) \tag{8}$$

and similarly,

$$T(r,\widehat{G}) \leq 2Z(r,g) + 2\sum_{i=2}^{l} Z(r,g-a_i) + Z(r,g') + 2\overline{N}(r,g) + 2Z(r,f) \\ + 2\sum_{i=2}^{l} Z(r,f-a_i) + Z(r,f') \\ +2\big(\overline{N}(r,G) + \overline{Z}(r,G)\big) + \overline{N}(r,F) + \overline{Z}(r,F) \\ +2\overline{N}(r,f)+Z(r,g)+Z(r,Q(g))-\sum_{i=2}^{l} k_i Z(r,g-a_i)-Z(r,g')+2T(r,\alpha)-4\log r+O(1). \tag{9}$$

Consequently,

$$T(r,\widehat{F})+T(r,\widehat{G}) \leq 7\big(Z(r,f)+Z(r,g)\big)+7\sum_{i=2}^{l}\big(Z(r,f-a_i)+Z(r,f-a_i)\big)+Z(r,g')+Z(r,f') \\ +4\big(\overline{N}(r,f) + 2\overline{N}(r,g)\big) + Z(r,Q(f)) + Z(r,Q(g)) \\ -\sum_{i=2}^{l} k_i Z(r,f-a_i) - Z(r,f') + 2T(r,\alpha) - 4\log r + O(1) \tag{10}$$

Particularly, if $k_i = 1\ \forall i = 2, ..., l$, then we have

$$T(r,\widehat{F}) \leq 2Z(r,f)+\sum_{i=2}^{l} Z(r,f-a_i)+Z(r,f')+2\overline{N}(r,f)+2Z(r,g)+\sum_{i=2}^{l} Z(r,g-a_i)+Z(r,g') \\ +2\big(\overline{N}(r,F) + \overline{Z}(r,F)\big) + \overline{N}(r,G) + \overline{Z}(r,G) \\ +2\overline{N}(r,g)+Z(r,f)+Z(r,Q(f))-\sum_{i=2}^{l} Z(r,f-a_i)-Z(r,f')+2T(r,\alpha)-4\log r+O(1) \tag{11}$$

and similarly,

$$T(r,\widehat{G}) \le 2Z(r,g)+\sum_{i=2}^{l} Z(r,g-a_i)+Z(r,g')+2\overline{N}(r,g)+2Z(r,f)+\sum_{i=2}^{l} Z(r,f-a_i)+Z(r,f')$$
$$+2\big(\overline{N}(r,G)+\overline{Z}(r,G)\big)+\overline{N}(r,F)+\overline{Z}(r,F)$$

(12)

$$+2\overline{N}(r,f)+Z(r,g)+Z(r,Q(g))-\sum_{i=2}^{l} Z(r,g-a_i)-Z(r,g')+2T(r,\alpha)-4\log r+O(1).$$

therefore by (11) and (12) we can adapt (10) and obtain

$$T(r,\widehat{F})+T(r,\widehat{G}) \le 7\big(Z(r,f)+Z(r,g)\big)+4\sum_{i=2}^{l}\big(Z(r,f-a_i)+Z(r,f-a_i)\big)+Z(r,g')+Z(r,f')$$

$$(13)\quad +4\big(\overline{N}(r,f)+\overline{N}(r,g)\big)+Z(r,Q(f))+Z(r,Q(g))+4T(r,\alpha)-8\log r+O(1)$$

Now, let us go back to the general case. By (10) we have

$$T(r,\widehat{F})+T(r,\widehat{G}) \le 5(Z(r,f)+Z(r,g))+\sum_{i=2}^{l}(4-k_i)(Z(r,f-a_i)+Z(r,g-a_i))+(Z(r,f')+Z(r,g'))$$

$$+3\big(\overline{Z}(r,F)+\overline{Z}(r,G)\big)+7(\overline{N}(r,f)+\overline{N}(r,g))+(Z(r,Q(f))+Z(r,Q(g)))+4T(r,\alpha)$$

(14)

$$+3\Big(Z(r,f')+\sum_{i=2}^{l}(Z(r,f-a_i)+Z(r,g-a_i))+Z(r,f)+Z(r,g)+Z(r,g')\Big)+4T(r,\alpha)-8\log r+O(1).$$

Here, by Lemma 2, we may notice that we can give $Z(r,f')$ the following upper bound $Z(r,f') \le Z(r,f-a_2)+\overline{N}(r,f)-\log(r)$ and similarly $Z(r,g') \le Z(r,g-a_2)+\overline{N}(r,g-\log(r))$. Consequently, by (14) we obtain

$$T(r,\widehat{F})+T(r,\widehat{G}) \le 8\big(Z(r,f)+Z(r,g)\big)+10\big(\overline{N}(r,f)+\overline{N}(r,g)\big)+(10-k_2)\big(Z(r,f-a_2)+Z(r,g-a_2)\big)$$

$$+\sum_{i=3}^{l}(7-k_i)\big(Z(r,f-a_i)+Z(r,g-a_i)\big)+Z(r,Q(f))+Z(r;Q(g))+4T(r\alpha)-10\log(r)+O(1)$$

therefore

$$T(r,\widehat{F})+T(r,\widehat{G}) \le 8\big(Z(r,f)+Z(r,g)\big)+10\big(\overline{N}(r,f)+\overline{N}(r,g)\big)+(10-k_2)\big(Z(r,f-a_2)+Z(r,g-a_2)\big)$$

$$(15)\quad +\sum_{i=3}^{l}(7-k_i)\big(Z(r,f-a_i)+Z(r,g-a_i)\big)+Z(r,Q(f))+Z(r;Q(g))+4T(r,\alpha)-10\log(r)+O(1).$$

Now, we notice that given $m \in \mathbb{Z}$ and $h \in \mathcal{M}(\mathbb{K})$ (resp. $h \in \mathcal{M}(d(0,R^-))$ we have $mZ(r,h) \le \max(0, mT(r,h)$. Consequently, by (15) we obtain

$$T(r,\widehat{F})+T(r,\widehat{G}) \le 8\big(Z(r,f)+Z(r,g)\big)+10\big(\overline{N}(r,f)+\overline{N}(r,g)\big)+\max\big(0,(10-k_2)\big)\big)\big(T(r,f)+T(r,g)\big)\big)$$

$$(16)\quad +\sum_{i=3}^{l}\max\big(0,(7-k_i)\big)\big(T(r,f)+T(r,g)\big)+Z(r,Q(f))+Z(r,Q(g))+4T(r,\alpha)-10\log(r)+O(1).$$

Now, since $\deg(Q)=k$, we have $T(r,Q(f))=kT(r,f)+O(1)$ and $T(r,Q(g))=kT(r,g)+O(1)$. On the other hand, since $\widehat{F}$ is a polynomial in f of degree $n+k+1$, we have $T(r,\widehat{F}) = (n+1+k)T(r,f)+O(1)$ and $T(r,\widehat{G}) = (n+1+k)T(r,f)+O(1)$. Consequently, by (16) we obtain

$$n\big((T(r,f)+T(r,g)\big) \leq 17\big(T(r,f)+T(r,g)\big) + \max(0,10-k_2)\big((T(r,f)+T(r,g)\big)$$

$$+\sum_{i=3}^{l}(\max(7-k_i)\big((T(r,f)+T(r,g)\big) + 4T(r,\alpha) - 10\log(r) + O(1). \tag{17}$$

Thus, in Theorems 1 and 4, if $\Theta_{F,G}$ is not identically zero, Relation (17) is impossible. In Theorem 2, $T(r,\alpha)$ is of the form $\log(r)+O(1)$ and in Theorem 3,$T(r,\alpha)=0$, hence Relation (17) is impossible again if $\Theta_{F,G}$ is not identically zero. So, we have proven that $\Theta_{F,G}$ is identically zero in Theorems 1, 2, 3, 4.

Suppose now that $k_i = 1$, $\forall i \in \{2,..,l\}$. Then $k = l$ and then, by (13) we can derive

$$(n+l+1)\big(T(r,f)+T(r,g)\big) \leq (7+4(l-1))\big(T(r,f)+T(r,g)\big) + T(r,f)+T(r,g)$$

$$+(4+1)\big(\overline{N}(r,f)+\overline{N}(r,g)\big)+l\big(T(r,f)+T(r,g)\big)+4T(r,\alpha)-10\log(r)+O(1) \tag{18}$$

therefore $n \leq 9+4l$.

Further, assuming again $k_i = 1$, $\forall i \in \{2,..,l\}$, suppose now that α is a Moebius function. Then by (18) we now have $n \leq 8+4l$.

Thus, we can check that in Theorems 5 and 6, if $n \geq 10+4l$, $\Theta_{F,G}$ must be identically zero. And in Theorems 7, if $n \geq 9+4l$, $\Theta_{F,G}$ must be identically zero.

Similarly, in Theorems 8 and 9, if $n \geq 18$, $\Theta_{F,G}$ must be identically zero and in Theorem 8, if α is a constant or a Moebius function, the conclusion holds for $n \geq 17$

Consider now the situation in Theorems 10 and 11. Since $N(r,f) = N(r,g) = 0$, by (14) we have

$$T(r,\widehat{F})+T(r,\widehat{G}) \leq 8\big(Z(r,f)+Z(r,g)\big) + (10-k_2)\big(Z(r,f-a_2)+Z(r,g-a_2)\big)$$

$$+\sum_{i=3}^{l}(7-k_i)\big(Z(r,f-a_i)+Z(r,g-a_i)\big)+Z(r,Q(f))+Z(r,Q(g))+4T(r\alpha)-10\log(r)+O(1)$$

therefore

$$T(r,\widehat{F})+T(r,\widehat{G}) \leq 8\big(Z(r,f)+Z(r,g)\big) + (10-k_2)\big(Z(r,f-a_2)+Z(r,g-a_2)\big)$$

$$+\sum_{i=3}^{l}(7-k_i)\big(Z(r,f-a_i)+Z(r,g-a_i)\big)+Z(r,Q(f))+Z(r;Q(g))+4T(r,\alpha)-10\log(r)+O(1) \tag{19}$$

As in Relation (15), given $m \in \mathbb{Z}$ and $h \in \mathcal{M}(\mathbb{K})$ (resp. $h \in \mathcal{M}(d(0,R^-))$ we have $mZ(r,h) \leq \max(0,mT(r,h)$. Consequently, by (19) we obtain

$$T(r,\widehat{F})+T(r,\widehat{G}) \leq 8\big(Z(r,f)+Z(r,g)\big) + \max\big(0,(10-k_2)\big)\big)\big(T(r,f)+T(r,g)\big)\big)$$

$$+\sum_{i=3}^{l}\max\big(0,(7-k_i)\big)\big(T(r,f)+T(r,g)\big)+Z(r,Q(f))+Z(r,Q(g))+4T(r,\alpha)-10\log(r)+O(1). \tag{20}$$

Now, since $\deg(Q) = k$, we have $T(r,Q(f)) = kT(r,f)+O(1)$ and $T(r,Q(g)) = kT(r,g)+O(1)$. On the other hand, since $\widehat{F}$ is a polynomial in f of degree $n+k+1$, we have $T(r,\widehat{F}) = (n+1+k)T(r,f)+O(1)$ and $T(r,\widehat{G}) = (n+1+k)T(r,f)+O(1)$.

Consequently, by (20) we obtain

$$n\big(T(r,f)+T(r,g)\big) \le 7\big(T(r,f)+T(r,g)\big)+\max(0,10-k_2)\big((T(r,f)+T(r,g)\big)$$

$$+\sum_{i=3}^{l}(\max(7-k_i)\big((T(r,f)+T(r,g)\big)+4T(r,\alpha)-10\log(r)+O(1). \tag{21}$$

Thus, in Theorems 10, if $\Theta_{F,G}$ is not identically zero, Relation (21) is impossible. In Theorem 11, $T(r,\alpha)$ is of the form $-4\log(r)+O(1)$ hence Relation (21) is impossible again if $\Theta_{F,G}$ is not identically zero. So, we have proven that $\Theta_{F,G}$ is identically zero in Theorems 10 and 11 too.

Consequently, we can assume that $\Theta_{F,G}=0$ in each hypothesis of all theorems. Therefore, we can perform the end of the proof exactly as in [**2**] or [**9**] (chapter 59) for Theorems 1, 2, 3, 4, 5, 6, 7, 8, 9, 10, 11 because we check that the reasonning never uses the hypothesis: *f and g share α C.M.* That ends the proof of all theorems. □

Acknowledgement

The authors thank the referee for pointing out to us several orthographic and typographic corrections.

References

[1] T. T. H. An, J. T.-Y. Wang, and P.-M. Wong, *Unique range sets and uniqueness polynomials in positive characteristic. II*, Acta Arith. **116** (2005), no. 2, 115–143, DOI 10.4064/aa116-2-2. MR2110391

[2] K. Boussaf, A. Escassut, and J. Ojeda, *p-adic meromorphic functions $f'P'(f)$, $g'P'(g)$ sharing a small function*, Bull. Sci. Math. **136** (2012), no. 2, 172–200, DOI 10.1016/j.bulsci.2011.06.006. MR2889135

[3] K. Boussaf, A. Escassut, and J. Ojeda, *Complex meromorphic functions $f'P'(f)$, $g'P'(g)$ sharing a small function*, Indag. Math. (N.S.) **24** (2013), no. 1, 15–41, DOI 10.1016/j.indag.2012.06.003. MR2997749

[4] A. Boutabaa, *Théorie de Nevanlinna p-adique* (French, with English summary), Manuscripta Math. **67** (1990), no. 3, 251–269, DOI 10.1007/BF02568432. MR1046988

[5] A. Boutabaa and A. Escassut, *Urs and ursims for p-adic meromorphic functions inside a disc*, Proc. Edinb. Math. Soc. (2) **44** (2001), no. 3, 485–504, DOI 10.1017/S0013091599000759. MR1875765

[6] B. T. K. Oanh and N. T. T. Thuy, *Uniqueness of differential polynomials of meromorphic functions sharing a small function without counting multiplicity*, Fasc. Math. **57** (2016), 121–135, DOI 10.1515/fascmath-2016-0020. MR3590222

[7] A. Escassut, *Meromorphic functions of uniqueness*, Bull. Sci. Math. **131** (2007), no. 3, 219–241, DOI 10.1016/j.bulsci.2006.05.004. MR2309863

[8] A. Escassut, J. Ojeda, and C. C. Yang, *Functional equations in a p-adic context*, J. Math. Anal. Appl. **351** (2009), no. 1, 350–359, DOI 10.1016/j.jmaa.2008.10.006. MR2472948

[9] A. Escassut, *p-adic value distribution*, Some topics on value distribution and differentiability in complex and p-adic analysis, Math. Monogr. Ser., vol. 11, Sci. Press Beijing, Beijing, 2008, pp. 42–138. MR2732024

[10] G. Frank and M. Reinders, *A unique range set for meromorphic functions with* 11 *elements*, Complex Variables Theory Appl. **37** (1998), no. 1-4, 185–193. MR1687880

[11] H. Fujimoto, *On uniqueness of meromorphic functions sharing finite sets*, Amer. J. Math. **122** (2000), no. 6, 1175–1203. MR1797660

[12] Nguyen Trong Hoa, *On the functional equation $P(f)=Q(g)$ in non-Archimedean field*, Acta Math. Vietnam. **31** (2006), no. 2, 167–180. MR2255701

[13] P.-C. Hu and C.-C. Yang, *Meromorphic functions over non-Archimedean fields*, Mathematics and its Applications, vol. 522, Kluwer Academic Publishers, Dordrecht, 2000. MR1794326

[14] C.-C. Yang and X. Hua, *Uniqueness and value-sharing of meromorphic functions*, Ann. Acad. Sci. Fenn. Math. **22** (1997), no. 2, 395–406. MR1469799

[15] Ha Huy Khoai and Ta Thi Hoai An, *On uniqueness polynomials and bi-URs for p-adic meromorphic functions*, J. Number Theory **87** (2001), no. 2, 211–221, DOI 10.1006/jnth.2000.2591. MR1824143

[16] I. Lahiri and N. Mandal, *Uniqueness of nonlinear differential polynomials sharing simple and double 1-points*, Int. J. Math. Math. Sci. **12** (2005), 1933–1942, DOI 10.1155/IJMMS.2005.1933. MR2176445

[17] P. Li and C.-C. Yang, *Some further results on the unique range sets of meromorphic functions*, Kodai Math. J. **18** (1995), no. 3, 437–450, DOI 10.2996/kmj/1138043482. MR1362919

[18] W. Lin and H. Yi, *Uniqueness theorems for meromorphic functions concerning fixed-points*, Complex Var. Theory Appl. **49** (2004), no. 11, 793–806, DOI 10.1080/02781070412331298624. MR2097218

[19] J. Ojeda, *Applications of the p-adic Nevanlinna theory to problems of uniqueness*, Advances in p-adic and non-Archimedean analysis, Contemp. Math., vol. 508, Amer. Math. Soc., Providence, RI, 2010, pp. 161–179, DOI 10.1090/conm/508/09998. MR2597692

[20] J. Ojeda, *Zeros of ultrametric meromorphic functions* $f'f^n(f-a)^k-\alpha$, Asian-Eur. J. Math. **1** (2008), no. 3, 415–429, DOI 10.1142/S1793557108000357. MR2452254

[21] J. Ojeda, *Uniqueness for ultrametric analytic functions*, Bull. Math. Soc. Sci. Math. Roumanie (N.S.) **54(102)** (2011), no. 2, 153–165. MR2817219

[22] J. T.-Y. Wang, *Uniqueness polynomials and bi-unique range sets for rational functions and non-Archimedean meromorphic functions*, Acta Arith. **104** (2002), no. 2, 183–200, DOI 10.4064/aa104-2-7. MR1914252

[23] Y. Xu and H. Qu, *Entire functions sharing one value IM*, Indian J. Pure Appl. Math. **31** (2000), no. 7, 849–855. MR1779444

[24] C.-C. Yang and X.-h. Hua, *Unique polynomials of entire and meromorphic functions* (English, with English, Russian and Ukrainian summaries), Mat. Fiz. Anal. Geom. **4** (1997), no. 3, 391–398. MR1615796

[25] H.X. Yi and C.C. Yang Uniqueness theorems of meromorphic functions. Science Press, China (1995).

Université Clermont Auvergne, Laboratoire de Mathématiques Blaise Pascal, BP 10448 63000 Clermont-Ferrand, France

Email address: `kamal.boussaf@math.univ-bpclermont.fr`

Université Clermont Auvergne, Laboratoire de Mathématiques Blaise Pascal, BP 10448 63000 Clermont-Ferrand, France

Email address: `alain.escassut@math.univ-bpclermont.fr`

Contemporary Mathematics
Volume **704**, 2018
http://dx.doi.org/10.1090/conm/704/14162

Spectra of commutative non-unital Banach rings

Chi-Wai Leung and Chi-Keung Ng

ABSTRACT. In this note, we give a brief study of the spectrum of commutative non-unital Banach rings. In particular, we transplant several results in the case of complex Banach algebras to this general situation. We also give some examples that show the different and difficulty in this general situation. Furthermore, we associate with a commutative Banach ring S, a continuous bundle of Banach algebras with compact base space $\mathbf{Mult}(\mathbb{Z}_{\|\cdot\|_1})$ whose fibres are of the form $S\widehat{\otimes}_{\mathbb{Z}}K$ for some complete valued fields K such that elements in S can be represented as continuous sections on this bundle and that continuous local sections of this bundle are "close to those generated by the images of elements in S".

1. Introduction, notation and preliminary

In the study of Banach rings, it is very often to come across non-unital ones. For example, a closed ideal of a unital Banach ring is in general non-unital. However, the "non-unitalness" creates some problem in the study (e.g. their spectra need not be non-emtpy). Therefore, it seems useful to have a look at the non-unital case. The aims of the note is to give a first study on this.

Throughout this note, by a *ring*, we mean an additive group together with a compatible multiplication, but we do not assume that this multiplication has an identity. In order words, "rings" in this article are "rngs" in some literatures.

Let T be a commutative ring. A function $\alpha : T \to \mathbb{R}_+$ is called a *semi-norm* if

$$\alpha(0) = 0, \quad \alpha(x+y) \leq \alpha(x) + \alpha(y) \quad \text{and} \quad \alpha(-x) = \alpha(x) \qquad (x, y \in T).$$

A semi-norm α is said to be *ultrametric* if $\alpha(x+y) \leq \max\{\alpha(x), \alpha(y)\}$ $(x, y \in T)$. Notice that some of the semi-norms that we consider in this article are "non-ultrametric". Moreover, α is said to be *submultiplicative* (respectively, *multiplicative*) if $\alpha(xy) \leq \alpha(x)\alpha(y)$ (respectively, $\alpha(xy) = \alpha(x)\alpha(y)$) for every $x, y \in T$. On the other hand, α is called a *norm* if the additive subgroup

$$\ker \alpha := \{x \in T : \alpha(x) = 0\}$$

is zero. A norm α is said to be *complete* if T is complete under the metric defined by $d(x, y) := \|x - y\|$ $(x, y \in T)$. We denote

$$\mathbf{Mult}(T) := \big\{\alpha : \alpha \text{ is a multiplicative semi-norm on } T \text{ with } \alpha(T) \neq \{0\}\big\}.$$

2010 *Mathematics Subject Classification.* Primary 46S10, 47S10.

The second named author is supported by the National Natural Science Foundation of China (11471168).

If T is a unital ring and $1 \in T$ is the identity, then $\alpha(T) \neq \{0\}$ is equivalent to $\alpha(1) = 1$.

If S is a ring equipped with a complete submultiplicative norm $\|\cdot\|$, then $S_{\|\cdot\|}$ (or simply S if no confusion occurs) is called a *Banach ring*. A Banach ring S is called a *unital Banach ring* if S has an identity 1 and $\|1\| = 1$. If R is another Banach ring, we denote by $\mathcal{C}(R, S)$ the set of all non-zero contractive ring homomorphisms from R to S.

If K is a field as well as a unital Banach ring such that the norm is multiplicative, then K is called a *complete valued field*, and will be abbreviated as CVF. The norm on K will be denoted by $|\cdot|$. A CVF K is said to be *non-Archimedean* if its norm $|\cdot|$ is ultrametric. Note that some CVFs in this article are "Archimedean" (i.e., they are isomorphic to either $\mathbb{R}$ or $\mathbb{C}$).

An algebra A over a CVF K is called a *Banach K-algebra* if A is also a Banach ring such that $\|ta\| = |t|\|a\|$ $(t \in K; a \in A)$. If A is ultrametric, then K has to be non-Archimedean. However, unlike some literatures, we may consider "non-ultrametric" Banach algebra over non-Archimedean CVF. In case when A is unital and $1 \in A$ is the identity, we will identify $t \in K$ with $t1 \in A$.

The *spectrum* of a commutative Banach ring S is the set

$$\mathbf{Mult}(S_{\|\cdot\|}) := \{\alpha \in \mathbf{Mult}(S) : \alpha(a) \leq \|a\| \text{ for any } a \in S\}$$

endowed with the topology of pointwise convergence. Clearly, $\mathbf{Mult}(S_{\|\cdot\|})$ is Hausdorff, and we will see in Corollary 2.5(a) that $\mathbf{Mult}(S_{\|\cdot\|})$ is locally compact. For each $\alpha \in \mathbf{Mult}(S_{\|\cdot\|})$, it is easy to see that $\ker \alpha$ is a closed prime ideal of S.

REMARK 1.1. Let S be a commutative unital Banach ring, and B be a commutative unital Banach algebra over a CVF K.
(a) It follows from the proof of [**1**, Theorem 1.2.1] that every maximal ideal of S equals $\ker \beta$, for some element β in $\mathbf{Mult}(S_{\|\cdot\|})$. The same is true when we weaken the unital assumption of S to $S^2 = S$, because of Proposition 2.6 below and the proof of [**2**, Proposition 3.1] (see the paragraph following Proposition 2.6).
(b) For each $\alpha \in \mathbf{Mult}(B_{\|\cdot\|})$ and $t \in K \setminus \{0\}$, since $\alpha(t) \leq \|t\| = |t|$ and $\alpha(t)^{-1} = \alpha(t^{-1}) \leq |t^{-1}| = |t|^{-1}$, we actually have $\alpha(t) = |t|$. Thus, we automatically have

$$\alpha(tb) = |t|\alpha(b) \qquad (b \in B; t \in K).$$

When B is non-unital, we will establish the same fact in Remark 2.4(b) below.
(c) If K is non-Archimedean, then for any $\alpha \in \mathbf{Mult}(B_{\|\cdot\|})$, one has $\alpha(b^n) = \alpha(b)^n$ and $\alpha(n) \leq 1$ for every $b \in B$ and $n \in \mathbb{N}$ (where n is identified with its image in $K \subseteq B$), from which we know that α is ultrametric (see e.g. the argument of Lemma 8.2(ii) in Chapter 1 of [**11**]), whether or not the norm on B is ultrametric. When B is non-unital, we will establish the same fact in Remark 2.4(c) below.

Spectra of commutative unital ultrametric Banach algebras were first studied by A. Escassut ([**3**]), G. Garandel ([**5**]) and B. Guennebaud ([**6**]). It was then generalized to commutative unital Banach ring by V.G. Berkovich ([**1**]). For a more recent account on this subject, one may consult [**4**].

Our first lemma is easy to verify.

LEMMA 1.2. *Let R be a unital Banach ring and K be a CVF.*
(a) Any bounded unital ring homomorphism ϕ from R to K is contractive.
(b) Any element in $\mathcal{C}(K, R)$ is isometric.

Our second lemma is well-known in the case of unital Banach ring. The proof for the non-unital case is the same as the one for the unital case, but we will repeat it here for completeness.

LEMMA 1.3. *Let T be a commutative ring and $\alpha \in \mathbf{Mult}(T)$. There exist a CVF $\mathcal{H}_\alpha$ and a ring homomorphism $\varphi_\alpha : T \to \mathcal{H}_\alpha$ with*

- *the subfield generated by $\varphi_\alpha(T)$ being dense in $\mathcal{H}_\alpha$;*
- *$\alpha(x) = |\varphi_\alpha(x)|$ $(x \in T)$.*

Moreover, such a pair $(\mathcal{H}_\alpha, \varphi_\alpha)$ is unique up to isometric isomorphism.

PROOF. Since α is non-zero and multiplicative, we know that $T_\alpha := T/\ker\alpha$ has no non-zero zero divisor, and we denote by $\tilde{\alpha}$ the multiplicative norm on T_α induced by α. Let F_α be the field of fraction of T_α (note that this can be constructed even though T_α may not be unital). We set

$$|xy^{-1}|_\alpha := \tilde{\alpha}(x)\tilde{\alpha}(y)^{-1} \qquad (x \in T; y \in T \setminus \{0\}).$$

It is easy to check that $|\cdot|_\alpha$ is a well-defined multiplicative norm on F_α, and we denote by $\mathcal{H}_\alpha$ its completion. If $\varphi_\alpha : T \to F_\alpha \subseteq \mathcal{H}_\alpha$ is the canonical map, then it is easy to see that the two required properties are satisfied.

Suppose now that K is another CVF and $\phi : T \to K$ is a map satisfying the corresponding properties in the statement. Then

$$|\phi(x)| = \alpha(x) = |\varphi_\alpha(x)| \qquad (x \in T).$$

This shows that $\Psi_0 : \phi(x) \mapsto \varphi_\alpha(x)$ is a well-defined isometric ring homomorphism from $\phi(T)$ onto $\varphi_\alpha(T)$. As $\{\phi(x)\phi(y)^{-1} : x, y \in T; \phi(y) \neq 0\}$ is dense in K, we know that Ψ_0 extends to an isometric ring homomorphism Ψ from K to $\mathcal{H}_\alpha$ such that $\varphi_\alpha = \Psi \circ \phi$. Moreover, Ψ is clearly surjective. □

2. Double centalizer algebras and unitalizations

From now on, S is a commutative Banach ring (not necessarily unital). We denote by $\mathcal{B}(S)$ the set of all bounded additive maps from S to itself, equipped with the usual addition, composition as well as the norm:

$$\|m\| := \inf\{\kappa > 0 : \|m(x)\| \leq \kappa\|x\|, \text{ for every } x \in S\} \qquad (m \in \mathcal{B}(S)).$$

It is well-known that $\mathcal{B}(S)$ is a unital Banach ring. If we set

$$\mathcal{D}(S) := \{m \in \mathcal{B}(S) : xm(y) = m(xy) = m(x)y, \forall x, y \in S\},$$

then $\mathcal{D}(S)$ is a unital Banach subring of $\mathcal{B}(S)$. Note, however, that $\mathcal{D}(S)$ needs not be commutative. For example if the product on S is defined as $xy := 0$ for every $x, y \in S$, then $\mathcal{D}(S) = \mathcal{B}(S)$.

REMARK 2.1. (a) The map $\iota : S \to \mathcal{D}(S)$ given by $\iota(a)(x) := ax$ $(a, x \in S)$ is a contractive ring homomorphism with its image being an ideal of $\mathcal{D}(S)$. Moreover, if S has identity, then ι is unital and injective.
(b) Suppose that

$$S^2 := \Big\{\sum\nolimits_{k=1}^{n} x_k y_k : n \in \mathbb{N}; x_1, \dots, x_n, y_1, \dots, y_n \in S\Big\}.$$

is dense in S. Then $\mathcal{D}(S)$ is commutative and the intersection of $\iota(S)$ with any non-zero ideal of $\mathcal{D}(S)$ is non-zero.

LEMMA 2.2. *For any CVF K and $\phi \in \mathcal{C}(S,K)$, there is a unique $\bar{\phi} \in \mathcal{C}(\mathcal{D}(S),K)$ satisfying*

$$\phi(m(a)) = \bar{\phi}(m)\phi(a) \qquad (m \in \mathcal{D}(S), a \in S). \tag{2.1}$$

Moreover, $\ker\bar{\phi} = \{m \in \mathcal{D}(S) : m(S) \subseteq \ker\phi\}$.

PROOF. Observe, first of all, that the uniqueness of $\bar{\phi}$ and the equality concerning $\ker\bar{\phi}$ follow directly from Relation (2.1). In the following, we will establish the existence of $\bar{\phi}$.

Fix any $m \in \mathcal{D}(S)$ and $x_0 \in S$ with $\phi(x_0) \neq 0$. We set

$$\bar{\phi}(m) := \phi(m(x_0))\phi(x_0)^{-1}.$$

As ϕ is multiplicative, it is clear that $\bar{\phi}$ is well-defined and that Relation (2.1) holds when $\phi(a) \neq 0$. If $\phi(a) = 0$, then $\phi(m(a))\phi(x_0) = \phi(m(x_0))\phi(a) = 0$, which implies that $\phi(m(a)) = 0$. Hence, Relation (2.1) is established.

Obviously, $\bar{\phi}$ is an additive group homomorphism. For any $m, n \in \mathcal{D}(S)$, Relation (2.1) implies

$$\bar{\phi}(mn) = \phi(m(n(x_0)))\phi(x_0)^{-1} = \bar{\phi}(m)\phi(n(x_0))\phi(x_0)^{-1} = \bar{\phi}(m)\bar{\phi}(n).$$

Furthermore, since

$$|\bar{\phi}(m)| = |\phi(m(x_0))\phi(x_0)^{-1}| \leq \|m(x_0)\|\,|\phi(x_0)|^{-1} \leq \|x_0\|\,|\phi(x_0)|^{-1}\,\|m\|,$$

we know that $\bar{\phi}$ is bounded. It now follows from Lemma 1.2(a) that $\bar{\phi}$ is contractive. $\square$

PROPOSITION 2.3. *Let S be a commutative Banach ring and $I \subseteq S$ be a closed ideal.*
(a) If K is a CVF and $\phi \in \mathcal{C}(I;K)$, one can find an element $\check{\phi} \in \mathcal{C}(S;K)$ satisfying

$$\check{\phi}(a)\phi(x) = \phi(ax) \qquad (a \in S; x \in I). \tag{2.2}$$

(b) If the spectrum of $I_{\|\cdot\|}$ is non-empty, then there is a homeomorphism τ from $\mathbf{Mult}(I_{\|\cdot\|})$ onto an open subset of $\mathbf{Mult}(S_{\|\cdot\|})$ such that $\tau(\alpha)|_I = \alpha$ $(\alpha \in \mathbf{Mult}(I))$. Moreover, an element $\beta \in \mathbf{Mult}(S_{\|\cdot\|})$ belongs to the image of τ if and only if $\beta(I) \neq \{0\}$.
(c) The spectrum of $(S/I)_{\|\cdot\|}$ is homeomorphic to $\mathbf{Mult}(S_{\|\cdot\|}) \setminus \tau(\mathbf{Mult}(I_{\|\cdot\|}))$ in the canonical way.

PROOF. (a) Define $\iota_0 : S \to \mathcal{B}(I)$ by $\iota_0(a)(x) := ax$ $(a \in S; x \in I)$. It is easy to check that ι_0 is a contractive ring homomorphism with $\iota_0(S) \subseteq \mathcal{D}(I)$. If $\bar{\phi} \in \mathcal{C}(\mathcal{D}(I),K)$ is the homomorphism as in Lemma 2.2, then $\check{\phi} := \bar{\phi} \circ \iota_0$ will satisfy the required relation.
(b) Let $\alpha \in \mathbf{Mult}(I_{\|\cdot\|})$. If $(\mathcal{H}_\alpha, \varphi_\alpha)$ is as in Lemma 1.3 and $\check{\varphi}_\alpha$ is as in part (a), we define $\tau(\alpha)(a) := |\check{\varphi}_\alpha(a)|$ $(a \in S)$. It is clear that $\tau(\alpha) \in \mathbf{Mult}(S_{\|\cdot\|})$. As $\tau(\alpha)|_I = \alpha$, we know that τ is injective. Moreover, it follows from Relation (2.2) that

$$\tau(\alpha)(a)\alpha(x) = \alpha(ax) \qquad (a \in S; x \in I). \tag{2.3}$$

Consider now any $\beta \in \mathbf{Mult}(S_{\|\cdot\|})$ satisfying $\beta(I) \neq \{0\}$. If we set $\alpha := \beta|_I$, then $\alpha \in \mathbf{Mult}(I_{\|\cdot\|})$. Choose any $x_0 \in I \setminus \ker\alpha$. It follows from $\beta(a) = \alpha(ax_0)\alpha(x_0)^{-1}$ $(a \in S)$ and Relation (2.3) that $\beta = \tau(\alpha)$. These establish the second statement. The second statement clearly implies the image of τ being an open subset of $\mathbf{Mult}(S_{\|\cdot\|})$. Thus, it remains to show that τ is a homeomorphism from $\mathbf{Mult}(I_{\|\cdot\|})$

onto this image. In fact, let $\{\alpha_i\}_{i\in\mathfrak{I}}$ be a net in $\mathbf{Mult}(I_{\|\cdot\|})$ and $\alpha_0 \in \mathbf{Mult}(I_{\|\cdot\|})$. Clearly, if $\tau(\alpha_i) \to \tau(\alpha_0)$, then $\alpha_i \to \alpha_0$. Conversely, suppose $\alpha_i \to \alpha_0$. Pick an element $y_0 \in I \setminus \ker \alpha_0$. Then $\alpha_i(y_0) \neq 0$ when i is large enough. For any $a \in S$, if i is large enough, then

$$\tau(\alpha_i)(a) = \alpha_i(ay_0)\alpha_i(y_0)^{-1} \to \alpha_0(ay_0)\alpha_0(y_0)^{-1} = \tau(\alpha_0)(a).$$

(c) It is clear that the quotient map from S to S/I induces a homeomorphism from $\mathbf{Mult}((S/I)_{\|\cdot\|})$ to its image in $\mathbf{Mult}(S_{\|\cdot\|})$. Since any element β in the spectrum of $S_{\|\cdot\|}$ with $\beta(I) = \{0\}$ induces an element in the spectrum of $(S/I)_{\|\cdot\|}$, we know from the second statement of part (b) that the image of $\mathbf{Mult}((S/I)_{\|\cdot\|})$ in the spectrum of $S_{\|\cdot\|}$ is precisely $\mathbf{Mult}(S_{\|\cdot\|}) \setminus \tau\big(\mathbf{Mult}(I_{\|\cdot\|})\big)$. □

Let us set $\hat{S} := S \oplus \mathbb{Z}$ and equip it with the ring structure such that $(0,1)$ is the identity of $\hat{S}$ and that S becomes a subring of $\hat{S}$ under the canonical embedding. We endow $\hat{S}$ with the norm $\|(a,k)\|_{\hat{S}} := \|a\|_S + |k|_1$ $(a \in S, k \in \mathbb{Z})$, where $|\cdot|_1$ is the Euclidean norm. Then $\hat{S}$ is a unital Banach ring so that any bounded (respectively, contractive) ring homomorphism from S to a unital Banach ring extends to a bounded (respectively, contractive) unital ring homomorphism on $\hat{S}$ to the same Banach ring.

Since $\hat{S}/S \cong \mathbb{Z}_{|\cdot|_1}$, we know from parts (b) and (c) of Proposition 2.3 that the spectrum of $\hat{S}_{\|\cdot\|}$ is a union of the open subset $\mathbf{Mult}(S_{\|\cdot\|})$ with the closed subset $\mathbf{Mult}(\mathbb{Z}_{|\cdot|_1})$. In particular, if S is already unital, then the spectrum of $S_{\|\cdot\|}$ is a compact Hausdorff space and $\mathbf{Mult}(\hat{S}_{\|\cdot\|})$ equals the disjoint union of the above two compact subsets.

REMARK 2.4. Suppose that $A_{\|\cdot\|}$ is a commutative Banach algebra over a CVF K. Let $A^{\mathbf{1}}_{\|\cdot\|}$ be the Banach K-algebra unitalization of A; i.e. $A^{\mathbf{1}} := A \oplus K$, equipped with the norm $\|(a,t)\| := \|a\| + |t|$ as well as the unital K-algebra structure such that $(0,1)$ is the identity and the canonical embedding from A to $A^{\mathbf{1}}$ is an algebra homomorphism. Notice that the Banach K-algebra unitalization of (0) is K itself. Clearly, A is a closed ideal of $A^{\mathbf{1}}$. Proposition 2.3(b) tells us that every element $\beta \in \mathbf{Mult}(A_{\|\cdot\|})$ extends to an element in $\tau(\beta) \in \mathbf{Mult}(A^{\mathbf{1}}_{\|\cdot\|})$.
(a) By Proposition 2.3(b)&(c), the spectrum of $A^{\mathbf{1}}_{\|\cdot\|}$ is the one-point compactification of $\mathbf{Mult}(A_{\|\cdot\|})$, where the extra point in $\mathbf{Mult}(A^{\mathbf{1}}_{\|\cdot\|})$ is given by the canonical map in $\mathcal{C}(A^{\mathbf{1}}, K)$ (we use the convention that the one-point compactification of an empty set is the one-point space).
(b) Remark 1.1(b) tells us that $\beta(ta) = \tau(\beta)(ta) = |t|\beta(a)$ $(a \in A; t \in K)$. In particular, $\ker \beta$ is a K-subspace of A (on top of being a prime ideal), even though A is not assumed to be unital.
(c) If K is non-Archimedean (but the norm on A is not assumed to be ultrametric), it follows from Remark 1.1(c) that $\tau(\beta)$ is ultrametric and hence so is β.

COROLLARY 2.5. *Let S be a commutative Banach ring.*
(a) If $\mathbf{Mult}(S_{\|\cdot\|})$ is non-empty, then $\mathbf{Mult}(S_{\|\cdot\|})$ is a locally compact space.
(b) If S contains a non-zero idempotent e, then $\mathbf{Mult}(S_{\|\cdot\|}) \neq \emptyset$.

PROOF. (a) By [**1**, Theorem 1.2.1], the Hausdorff space $\mathbf{Mult}(\hat{S}_{\|\cdot\|})$ is compact. It then follows from Proposition 2.3(b) that $\mathbf{Mult}(S_{\|\cdot\|})$ is locally compact.

(b) Let $I := eS$. Then I is a closed ideal of S and is a unital Banach ring. Thus, it follows from Remark 2.1(a), [**1**, Theorem 1.2.1] and Proposition 2.3(b) that $\mathbf{Mult}(S_{\|\cdot\|}) \neq \emptyset$. □

Part (b) above gives a situation when $\mathbf{Mult}(S_{\|\cdot\|})$ is non-empty. In general, it is possible for S to be empty (see Example 2.7(a) below). The following is another situation when $\mathbf{Mult}(S_{\|\cdot\|}) \neq \emptyset$. This proposition can be regarded as a generalization of Remark 1.1(a) (note that in the unital case, all maximal ideals are closed).

PROPOSITION 2.6. *Let S be a commutative Banach ring with the subring S^2 (see Remark 2.1) being norm-dense in S. For any closed maximal ideal $I \subsetneq S$, there is $\alpha \in \mathbf{Mult}(S_{\|\cdot\|})$ with $\ker\alpha = I$.*

PROOF. Let $T := S/I$. The hypothesis implies that T^2 is dense in T. Hence, one can find $x_0 \in T$ with $x_0T \neq (0)$. Since T contains no non-trivial ideal, we know that

$$x_0T = T$$

and there exists $e \in T$ with $x_0e = x_0$. From this, we conclude that $(x_0y)e = x_0y$ ($y \in T$), and e is the identity of T. Now, Remark 2.1(a) and [**1**, Theorem 1.2.1] ensures the existence of $\beta \in \mathbf{Mult}(T_{\|\cdot\|})$. As $\ker\beta$ is a proper ideal of T, one has $\ker\beta = (0)$. If $q : S \to T$ is the quotient map, then $\alpha := \beta \circ q$ will satisfy the requirement. □

Notice that if $S^2 = S$, then it follows from the argument of [**2**, Proposition 3.1] that all maximal ideals of S are closed. In general, it is possible to have non-closed maximal ideal even in the case of non-unital complex Banach algebras.

As in the paragraph preceding Remark 2.4, the spectrum of $\hat{S}_{\|\cdot\|}$ is a union of the open subset $\mathbf{Mult}(S_{\|\cdot\|})$ and the closed subset $\mathbf{Mult}(\mathbb{Z}_{|\cdot|_1})$. However, unlike the situation of Banach algebra unitalizations, there is no universal way to describe how these two sets are combined together to form the topology of the spectrum $\mathbf{Mult}(\hat{S}_{\|\cdot\|})$.

Let us recall that the spectrum of $\mathbb{Z}_{|\cdot|_1}$ is the following topological subspace of $\mathbb{R}^2$:

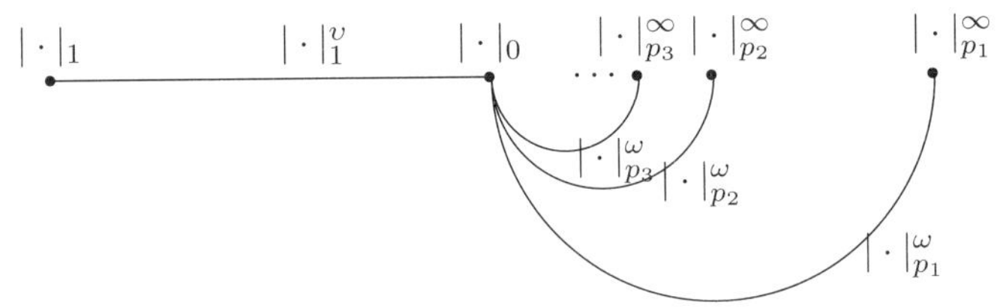

,

where v is in $(0,1)$, ω is in $(0,\infty)$, $\{p_1, p_2, \dots\}$ is the set of all prime numbers, $|\cdot|_1^v$ is the Euclidean norm on $\mathbb{Z}$ raised to the power v, $|\cdot|_0$ is the trivial norm on $\mathbb{Z}$, $|\cdot|_p^\omega$ is the usual p-adic norm on $\mathbb{Z}$ raised to the power ω and $|\cdot|_p^\infty$ is the semi-norm on $\mathbb{Z}$ induced by the trivial norm on

$$\mathbb{F}_p := \mathbb{Z}/p\mathbb{Z}. \tag{2.4}$$

EXAMPLE 2.7. (a) Suppose that every element in a commutative Banach ring S is *nilpotent*; i.e. for any $a \in S$, there exists $n_a \in \mathbb{N}$ with $a^{n_a} = 0$. Then the spectrum of $S_{\|\cdot\|}$ is empty. In fact, if, on the contrary, there exists $\alpha \in \mathbf{Mult}(S_{\|\cdot\|})$, then one can find $b \in S$ with $\alpha(b) \neq 0$, but we will arrive at the contradiction that $\alpha(b)^{n_b} = \alpha(b^{n_b}) = 0$.

In particular, let K be a CVF, r be a positive real number and $n \in \mathbb{N}$. Consider X to be the K-vector subspace of $K\langle r^{-1}\mathbf{t}\rangle$ (see [**1**, Example 1.1.1(v)]) generated by $\{\mathbf{t}^k : 0 \leq k \leq n\}$ (which is necessarily closed). Suppose that $D \in \mathcal{B}(X)$ is the operator defined by

$$D\left(\sum\nolimits_{k=0}^{n} s_k \mathbf{t}^k\right) := \sum\nolimits_{k=1}^{n} k s_k \mathbf{t}^{k-1}.$$

Then $D^n = 0$. If we set S to be the (automatically closed) non-unital K-subalgebra of $\mathcal{B}(X)$ generated by D, then $\mathbf{Mult}(S_{\|\cdot\|}) = \emptyset$.

(b) Let $\mathbb{P}$ be the set of all prime numbers, $R := \bigoplus_{p\in\mathbb{P}} \mathbb{F}_p$ and $\|\cdot\|_0$ be the trivial norm on R. Then the spectrum of $R_{\|\cdot\|_0}$ is a countable infinite set equipped with the discrete topology. Moreover, the spectrum of $\hat{R}_{\|\cdot\|}$ is a subspace of $\mathbb{R}^2$ consisting of the set $\mathbf{Mult}(\mathbb{Z}_{|\cdot|_1})$ and a sequence outside $\mathbf{Mult}(\mathbb{Z}_{|\cdot|_1})$ that converges to $|\cdot|_0 \in \mathbf{Mult}(\mathbb{Z}_{|\cdot|_1})$.

(c) Fix a number $q \in \mathbb{P}$. Let $\mathbb{Q}_q$ be the field of q-adic numbers, endowed with the norm $|\cdot|_q$. Set S to be the Banach $\mathbb{Q}_q$-algebra

$$\left\{(r_n)_{n\in\mathbb{N}} \in \prod\nolimits_{n\in\mathbb{N}} \mathbb{Q}_q : \lim |r_n|_q = 0\right\},$$

equipped with the norm $\|(r_n)_{n\in\mathbb{N}}\|_S := \sup_{n\in\mathbb{N}} |r_n|_q$. Again, $\mathbf{Mult}(S_{\|\cdot\|_S})$ is a countable infinite set with the discrete topology. Unlike part (b), the spectrum of $\hat{S}_{\|\cdot\|}$ is the union of $\mathbf{Mult}(\mathbb{Z}_{|\cdot|_1})$ with a sequence outside $\mathbf{Mult}(\mathbb{Z}_{|\cdot|_1})$ that converges to $|\cdot|_q$

(d) Consider A to be the $\mathbb{R}$-algebra of all continuous real functions on $\mathbb{R}$ that vanish at infinity, equipped with the usual sup-norm. Let R and S be as in parts (b) and (c), respectively. If $T := A \oplus R \oplus S$, equipped with direct sum norm, then $\mathbf{Mult}(T_{\|\cdot\|})$ equals the disjoint union of $\mathbb{R}$ with two distinct copies of $\mathbb{N}$. The spectrum of $\hat{T}_{\|\cdot\|}$ is a subspace of $\mathbb{R}^2$ consisting of $\mathbf{Mult}(\mathbb{Z}_{|\cdot|_1})$, a circle that intersects $\mathbf{Mult}(\mathbb{Z}_{|\cdot|_1})$ at $|\cdot|_1$ as well as two convergent sequences with limits being $|\cdot|_0$ and $|\cdot|_q$ respectively.

3. Semi-simplicity

In the following, we consider the case when $\mathbf{Mult}(S_{\|\cdot\|})$ is big enough in the sense that it "separates points of S". More precisely, we denote by $\Theta : S \to \prod_{\alpha\in\mathbf{Mult}(S_{\|\cdot\|})} \mathcal{H}_\alpha$ the Gelfand transform as in [**1**]; namely,

$$\Theta(a) = (\varphi_\alpha(a))_{\alpha\in\mathbf{Mult}(S_{\|\cdot\|})} \qquad (a \in S).$$

We say that S is *Gelfand semi-simple* if Θ is injective. It is easy to see that the spectrum of $S/\ker\Theta_{\|\cdot\|}$ equals that of $S_{\|\cdot\|}$, and hence $\mathbf{Mult}(\ker\Theta_{\|\cdot\|}) = \emptyset$. Consider the "Jacobson radical" (observe that if S is unital, maximal ideals are automatically closed):

$$\mathbf{J}(S_{\|\cdot\|}) := \bigcap\{M \subseteq S : M \text{ is a closed maximal ideal of } S\}$$

(we use the convention that $\mathbf{J}(S_{\|\cdot\|})$ equals S when S has no closed maximal ideal). It is clear that if S is unital and $\mathbf{J}(S_{\|\cdot\|}) = \{0\}$, then S is Gelfand semi-simple. However, there exists a Gelfand semi-simple unital Banach algebra S with the spectrum of $\mathbf{J}(S_{\|\cdot\|})_{\|\cdot\|}$ being non-empty (see Example 4.4(a) below).

It is easy to see that if R is a Gelfand semi-simple commutative Banach ring and there exists a contractive ring monomorphism from S to R, then S is also Gelfand semi-simple. In particular, closed subrings of Gelfand semi-simple Banach rings are Gelfand semi-simple. The following proposition tells us that it is also preserved under "essential extension".

PROPOSITION 3.1. *Let S be a commutative Banach ring and $I \subseteq S$ be an* essential closed ideal, *in the sense that every non-zero closed ideal of S intersects I non-trivially. If I is Gelfand semi-simple, then so is S.*

PROOF. Let $\iota_0 \in \mathcal{C}(S, \mathcal{D}(I))$ be as in the proof of Proposition 2.3(a). Assume that $\ker \iota_0 \neq (0)$. Since $\ker \iota_0$ is a closed ideal of S (because ι_0 is contractive), the hypothesis produces $a \in I \cap \ker \iota_0 \setminus \{0\}$. However, the relation $aI = \{0\}$ will imply that $a^2 = 0$. Thus, $\alpha(a)^2 = \alpha(a^2) = 0$ for every $\alpha \in \mathbf{Mult}(I_{\|\cdot\|})$ which contradicts with the assumption that I is Gelfand semi-simple. Consequently, ι_0 is injective and it suffices to show that $\mathcal{D}(I)$ is a Gelfand semi-simple commutative Banach algebra.

We denote $\iota := \iota_0|_I$. For every $\alpha \in \mathbf{Mult}(I_{\|\cdot\|})$, let φ_α be as in Lemma 1.3, and define a ring homomorphism $\bar{\Theta} : \mathcal{D}(I) \longrightarrow \big\{(t_\alpha)_{\alpha \in \mathbf{Mult}(I_{\|\cdot\|})} \in \prod_{\alpha \in \mathbf{Mult}(I_{\|\cdot\|})} \mathcal{H}_\alpha : \sup_{\alpha \in \mathbf{Mult}(I_{\|\cdot\|})} |t_\alpha| < \infty\big\}$ by

$$\bar{\Theta}(m) := (\bar{\varphi}_\alpha(m))_{\alpha \in \mathbf{Mult}(I_{\|\cdot\|})} \qquad (m \in \mathcal{D}(I)),$$

where $\bar{\varphi}_\alpha$ is as in Lemma 2.2. If $m \in \ker \bar{\Theta}$, then one knows that $m(I) \subseteq \ker \varphi_\alpha$ for all $\alpha \in \mathbf{Mult}(I_{\|\cdot\|})$ (because of the last statement of Lemma 2.2), and the semi-simplicity assumption implies $m(I) = (0)$; in other words, $m = 0$. Consequently, $\mathcal{D}(I)$ is commutative and ι is injective (because $\bar{\Theta} \circ \iota$ coincides with the Gelfand transform of I). □

However, quotient rings of Gelfand semi-simple Banach rings need not be Gelfand semi-simple (in fact, there are examples of commutative Gelfand semi-simple complex Banach algebras with quotients not being Gelfand semi-simple). Let I be a proper closed ideal of S (not assumed to be Gelfand semi-simple) and

$$\check{I} := \bigcap \big\{ \ker \alpha : \alpha \in \mathbf{Mult}(S_{\|\cdot\|}) \text{ with } \alpha(I) = \{0\}\big\}.$$

Then S/I is Gelfand semi-simple if and only if $\check{I} = I$.

4. Comparison with the prime spectra

Next, we would like to investigate the relation between the sepctrum of $S_{\|\cdot\|}$ and the prime spectrum, $\mathbf{Spec}(S)$, of S, equipped with the Zariski topology (although the Zariski topology usually defined for the prime spectra of unital rings, it can also be defined for non-unital rings). It is clear that the map

$$\Lambda : \mathbf{Mult}(S_{\|\cdot\|}) \to \mathbf{Spec}(S_{\|\cdot\|})$$

given by $\Lambda(\alpha) := \ker \alpha$ ($\alpha \in \mathbf{Mult}(S_{\|\cdot\|})$) is continuous. Let us set $\mathbf{Spec}_0(S_{\|\cdot\|}) := \Lambda\big(\mathbf{Mult}(S_{\|\cdot\|})\big)$. Clearly, the topology on $\mathbf{Spec}_0(S_{\|\cdot\|})$ as induced by the Zariski topology on $\mathbf{Spec}(S)$ is the one with closed subsets being precisely subsets of the form $\big\{P \in \mathbf{Spec}_0(S_{\|\cdot\|}) : I \subseteq P\big\}$ for some closed ideal $I \subseteq S$. By abuse of notation, we call this topology *the Zariski topology on* $\mathbf{Spec}_0(S_{\|\cdot\|})$.

The following easy proposition is a generalization of a well-known result in the case of complex Banach algebras (see e.g. [**10**, Theorem 3.2.10]).

PROPOSITION 4.1. *Let S be a commutative Banach ring.*
(a) The Zariski topology on $\mathbf{Spec}_0(S_{\|\cdot\|})$ is the finest topology under which Λ is continuous if and only if for any closed subset $\Omega \subseteq \mathbf{Mult}(S_{\|\cdot\|})$ satisfying $\Omega = \Lambda^{-1}(\Lambda(\Omega))$ and any $\alpha \in \mathbf{Mult}(S_{\|\cdot\|}) \setminus \Omega$, one can find $a \in S$ such that $\alpha(a) \neq 0$ and $\omega(a) = 0$ for all $\omega \in \Omega$.
(b) If S is unital, then each one of following conditions is stronger than the next one:

(1) *$\mathbf{Spec}_0(S_{\|\cdot\|})$ is T_1 and Λ is an open map;*
(2) *$\mathbf{Spec}_0(S_{\|\cdot\|})$ is Hausdorff;*
(3) *The Zariski topology is the finest topology under which Λ is continuous.*

PROOF. (a) $\Rightarrow$). Let Ω be a non-empty closed subset satisfying $\Omega = \Lambda^{-1}(\Lambda(\Omega))$ and $\alpha \notin \Omega$. We set $\Xi := \Lambda(\Omega)$. The assumption implies that Ξ is closed in the Zariski topology. In other words, if $I := \bigcap \Xi$, then $\Xi = \{P \in \mathbf{Spec}_0(S_{\|\cdot\|}) : I \subseteq P\}$. It follows that $I \nsubseteq \ker \alpha$ and any element $a \in I \setminus \ker \alpha$ will satisfy the requirement.
$\Leftarrow$). Let $\Xi \subsetneq \mathbf{Spec}_0(S_{\|\cdot\|})$ be a proper subset with $\Omega := \Lambda^{-1}(\Xi)$ being closed in $\mathbf{Mult}(S_{\|\cdot\|})$. We need to show that Ξ is closed in the Zariski topology. Set $I := \bigcap \Xi$. Assume on the contrary that one can find $P \in \mathbf{Spec}_0(S_{\|\cdot\|}) \setminus \Xi$ satisfying $I \subseteq P$. Consider $\alpha \in \mathbf{Mult}(S_{\|\cdot\|})$ with $P = \ker \alpha$. Since $\alpha \notin \Omega$, the assumption produces $a \in I$ with $\alpha(a) \neq 0$. However, this will contradict $I \subseteq \ker \alpha$.
(b) (1) $\Rightarrow$ (2). Let P_1 and P_2 be two distinct points in $\mathbf{Spec}_0(S_{\|\cdot\|})$. Then $\Lambda^{-1}(P_1)$ and $\Lambda^{-1}(P_2)$ are two disjoint closed subsets of the compact Hausdorff space $\mathbf{Mult}(S_{\|\cdot\|})$. Hence, we can find two disjoint open subsets V_1 and V_2 of $\mathbf{Mult}(S_{\|\cdot\|})$ with $\Lambda^{-1}(P_i) \subseteq V_i$ $(i = 1, 2)$. Now, $\Lambda(V_1)$ and $\Lambda(V_2)$ are two disjoint open subsets containing P_1 and P_2 respectively.
(2) $\Rightarrow$ (3). Suppose that $\Xi \subseteq \mathbf{Spec}_0(S_{\|\cdot\|})$ with $\Lambda^{-1}(\Xi)$ being closed. Then $\Lambda^{-1}(\Xi)$ is compact and hence $\Xi = \Lambda(\Lambda^{-1}(\Xi))$ is compact. As $\mathbf{Spec}_0(S_{\|\cdot\|})$ is Hausdorff, we know that Ξ is closed as required. □

If Λ is injective, then $\Omega = \Lambda^{-1}(\Lambda(\Omega))$ for any subset $\Omega \subseteq \mathbf{Mult}(S_{\|\cdot\|})$, and the three conditions in part (b) are equivalent. Hence, for a complex Banach algebra A, the corresponding condition in part (a) is the same as A being "completely regular" in the sense of [**10**, Definition 3.2.9].

On the other hand, if Λ is an open map, then the Zariski topology is the finest topology under which Λ is continuous (since $\Lambda^{-1}(\Xi)$ being open will imply that Ξ is open). In general, it is possible to have Λ open, but $\mathbf{Spec}_0(S_{\|\cdot\|})$ not being T_1. We will give such an example in the following.

Fix a positive integer N and a prime number p. We denote by $\mathbb{Z}_p[\mathbf{t}]$ the ring of polynomials (with one variable $\mathbf{t}$) over the ring $\mathbb{Z}_p$ of p-adic integers, and consider

$$Q : \mathbb{Z}_p[\mathbf{t}] \to \mathbb{F}_p[\mathbf{t}]$$

to be canonical map (see (2.4)). The following lemma is more or less well-known, but we present its argument here for completeness.

LEMMA 4.2. *Let $P \subseteq \mathbb{Z}_p[\mathbf{t}]$ be a non-zero prime ideal.*
(a) $P \cap \mathbb{Z}_p \neq \{0\}$ if and only if $P = Q^{-1}(J)$ for a (not necessarily non-zero) prime ideal J of $\mathbb{F}_p[\mathbf{t}]$.
(b) If $P \cap \mathbb{Z}_p = \{0\}$, there is an irreducible element $\mathbf{p}_0 \in \mathbb{Z}_p[\mathbf{t}]$ such that $P = \mathbf{p}_0 \cdot \mathbb{Z}_p[\mathbf{t}]$.

PROOF. (a) Suppose that $I := P \cap \mathbb{Z}_p$ is non-zero. Since I is a prime ideal of $\mathbb{Z}_p$, it is the unique maximal ideal of $\mathbb{Z}_p$. Thus, $\ker Q = I[\mathbf{t}] \subseteq P$, and we know that $P = Q^{-1}(Q(P))$. It is clear that $J := Q(P)$ is a prime ideal of $\mathbb{F}_p[\mathbf{t}]$. The converse is obvious.
(b) Set $m_0 := \min_{\mathbf{p}\in P\setminus\{0\}} \deg \mathbf{p}$ and

$$\mathcal{S} := \{\mathbf{q} \in P : \deg \mathbf{q} = m_0\}.$$

For any $\mathbf{p} \in \mathbb{Z}_p[\mathbf{t}]$, we denote by $l(\mathbf{p}) \in \mathbb{Z}_p$ the leading coefficient of $\mathbf{p}$. We pick an element $\mathbf{p}_0 \in \mathcal{S}$ satisfying $|l(\mathbf{p}_0)|_p \geq |l(\mathbf{q})|_p$ for all $\mathbf{q} \in \mathcal{S}$ (this can be done because the value group of $\mathbb{Z}_p$ is $\{p^{-m} : m \in \mathbb{Z}_+\}$).

Suppose $\mathbf{p}_1, \mathbf{p}_2 \in \mathbb{Z}_p[\mathbf{t}]$ satisfying $\mathbf{p}_1\mathbf{p}_2 = \mathbf{p}_0$. As P is prime, we may assume $\mathbf{p}_1 \in P$. Thus,

$$\deg \mathbf{p}_1 \geq m_0 = \deg \mathbf{p}_0 \qquad \text{and} \qquad \deg \mathbf{p}_1 + \deg \mathbf{p}_2 = \deg \mathbf{p}_0.$$

This forces $\mathbf{p}_1 \in \mathcal{S}$ and $\mathbf{p}_2 \in \mathbb{Z}_p$. Furthermore, since

$$|l(\mathbf{p}_0)|_p = |l(\mathbf{p}_1)|_p|\mathbf{p}_2|_p \leq |l(\mathbf{p}_1)|_p,$$

we know from the choice of $\mathbf{p}_0$ that $|\mathbf{p}_2|_p = 1$ and hence $\mathbf{p}_2$ is an invertible element in $\mathbb{Z}_p$. In other words, $\mathbf{p}_0$ is irreducible in $\mathbb{Z}_p[\mathbf{t}]$.

Clearly, we have $\mathbf{p}_0 \cdot \mathbb{Z}_p[\mathbf{t}] \subseteq P$. Conversely, pick an arbitrary element $\mathbf{p} \in P$. There exist $a \in \mathbb{Z}_p\setminus\{0\}$ and $\mathbf{r}_0, \dots, \mathbf{r}_n \in \mathbb{Z}_p[\mathbf{t}]$ satisfying $\deg \mathbf{r}_0, \dots, \deg \mathbf{r}_n < \deg \mathbf{p}_0$ as well as

$$a\mathbf{p} = \mathbf{r}_0 + \mathbf{r}_1\mathbf{p}_0 + \cdots + \mathbf{r}_n\mathbf{p}_0^n \in \mathbf{p}_0 \cdot \mathbb{Z}_p[\mathbf{t}].$$

Since $\mathbf{p}_0 \cdot \mathbb{Z}_p[\mathbf{t}]$ is a prime ideal of $\mathbb{Z}_p[\mathbf{t}]$ (notice that $\mathbb{Z}_p[\mathbf{t}]$ is a UFD and $\mathbf{p}_0$ is irreducible), we know that either $a \in \mathbf{p}_0 \cdot \mathbb{Z}_p[\mathbf{t}]$ or $\mathbf{p} \in \mathbf{p}_0 \cdot \mathbb{Z}_p[\mathbf{t}]$. However, the fact that $(\mathbf{p}_0 \cdot \mathbb{Z}_p[\mathbf{t}]) \cap \mathbb{Z}_p \subseteq P \cap \mathbb{Z}_p = \{0\}$ rules out the possibility of $a \in \mathbf{p}_0 \cdot \mathbb{Z}_p[\mathbf{t}]$. In other words, $P \subseteq \mathbf{p}_0 \cdot \mathbb{Z}_p[\mathbf{t}]$. □

Let G be a finite cyclic group with a generator $u \in G$. We denote by $\mathbb{Z}_p[G]$ the group ring of G over $\mathbb{Z}_p$, and endow it with the norm $\|\sum_{g\in G} a_g g\| := \max_{g\in G} |a_g|_p$. Clearly, $\mathbb{Z}_p[G]$ is a commutative unital Banach ring.

For $l = 1, ..., N$, we consider the following element in $\mathbb{Z}_p[\mathbf{t}]$:

$$\mathbf{q}_l := \mathbf{t}^{p^{l-1}(p-1)} + \mathbf{t}^{p^{l-1}(p-2)} + \cdots + \mathbf{t}^{p^{l-1}} + 1$$

and we denote $\mathbf{q}_0 := \mathbf{t} - 1$. It can be shown that the polynomials $\mathbf{q}_0, \dots, \mathbf{q}_N$ are all irreducible (see e.g. [**8**, Lemma 2.14]). We consider

$$q_G : \mathbb{Z}_p[\mathbf{t}] \to \mathbb{Z}_p[G]$$

to be the ring homomorphism that sends $\mathbf{t}$ to u.

The following result could also be known, but we will again present the complete treatment here.

PROPOSITION 4.3. *Let G be a cyclic group of order p^N (where $N \in \mathbb{N}$) and $u \in G$ be a generator. Then* $\mathbf{Spec}(\mathbb{Z}_p[G]) = \{J_\infty, I_0, I_1, \dots, I_N\}$, *where*

$$J_\infty := p \cdot \mathbb{Z}_p[G] + (u-1) \cdot \mathbb{Z}_p[G] \quad \text{and} \quad I_k := q_G(\mathbf{q}_k) \cdot \mathbb{Z}_p[G].$$

Furthermore, a non-empty subset of $\mathbf{Spec}(\mathbb{Z}_p[G])$ *is closed if and only if it contains* J_∞.

PROOF. Obviously, q_G is surjective and it induces a bijection from $\mathbf{Spec}(\mathbb{Z}_p[G])$ onto the set of prime ideals of $\mathbb{Z}_p[\mathbf{t}]$ that contain $(\mathbf{t}^{p^N}-1)\cdot\mathbb{Z}_p[\mathbf{t}]$. Notice that

$$\mathbf{t}^{p^N}-1=\mathbf{q}_0\cdot\mathbf{q}_1\cdot\mathbf{q}_2\cdots\mathbf{q}_N \tag{4.1}$$

is the prime factorization of $\mathbf{t}^{p^N}-1$ in the UFD $\mathbb{Z}_p[\mathbf{t}]$.

Since $(\mathbf{t}-1)\cdot\mathbb{F}_p[\mathbf{t}]$ is a prime ideal of $\mathbb{F}_p[\mathbf{t}]$, we know that

$$p\cdot\mathbb{Z}_p[\mathbf{t}]+\mathbf{q}_0\cdot\mathbb{Z}_p[\mathbf{t}]=Q^{-1}\big((\mathbf{t}-1)\cdot\mathbb{F}_p[\mathbf{t}]\big)$$

is a prime ideal of $\mathbb{Z}_p[\mathbf{t}]$, and hence $J_\infty\in\mathbf{Spec}(\mathbb{Z}_p[G])$. Moreover, Equality (4.1) and the irreducibility of $\mathbf{q}_k$ implies that $I_k\in\mathbf{Spec}(\mathbb{Z}_p[G])$ $(k=0,\dots,N)$.

Conversely, consider $I\subseteq\mathbb{Z}_p[\mathbf{t}]$ to be an arbitrary prime ideal containing $(\mathbf{t}^{p^N}-1)\cdot\mathbb{Z}_p[\mathbf{t}]$. Suppose that $I\cap\mathbb{Z}_p\neq\{0\}$. By Lemma 4.2(a), there is an irreducible polynomial $\mathbf{p}_0\in\mathbb{F}_p[\mathbf{t}]$ such that $I=Q^{-1}\big(\mathbf{p}_0\cdot\mathbb{F}_p[\mathbf{t}]\big)$. Since $(\mathbf{t}^{p^N}-1)\cdot\mathbb{Z}_p[\mathbf{t}]\subseteq I$, one knows that $\mathbf{p}_0$ divides the polynomial $\mathbf{t}^{p^N}-1\in\mathbb{F}_p[\mathbf{t}]$. However, as $\mathbf{t}^{p^N}-1=(\mathbf{t}-1)^{p^N}$ in $\mathbb{F}_p[\mathbf{t}]$, we conclude that $\mathbf{p}_0=\mathbf{t}-1$ and $q_G(I)=J_\infty$. On the other hand, suppose that $I\cap\mathbb{Z}_p=\{0\}$. Lemma 4.2(b) implies that I is an ideal generated by an irreducible element in $\mathbb{Z}_p[\mathbf{t}]$ that divides $\mathbf{t}^{p^N}-1$. Thus, the prime factorization of $\mathbf{t}^{p^N}-1$ (see Equality (4.1)) tells us $I=\mathbf{q}_k\cdot\mathbb{Z}_p[\mathbf{t}]$, for a unique integer $k\in\{0,1,2,\dots,N\}$.

Next, we consider the second assertion. For $0\leq k\leq N-1$, since $\mathbf{q}_0\cdots\mathbf{q}_k=\mathbf{t}^{p^k}-1$, we have the equality

$$\mathbf{q}_{k+1}=\mathbf{q}_0\cdots\mathbf{q}_k\big(\mathbf{t}^{p^k(p-2)}+2\mathbf{t}^{p^k(p-3)}+\cdots+(p-2)\mathbf{t}^{p^k}+p-1\big)+p,$$

which implies $I_{k+1}\subseteq J_\infty$. This ensures that J_∞ belongs to every non-empty closed subset. Conversely, let $k_1,\dots,k_n$ be distinct integers in $\{0,1,2,\dots,N\}$, since all $\mathbf{q}_0,\mathbf{q}_1,\dots,\mathbf{q}_N$ are irreducible elements in the UFD $\mathbb{Z}_p[\mathbf{t}]$, we know that if $k_0\in\{0,1,2,\dots,N\}$ with I_{k_0} containing $I_{k_1}\cap\cdots\cap I_{k_n}$, then k_0 equals k_i for a unique number i. This shows that $\{J_\infty,I_{k_1},\dots,I_{k_n}\}$ is a closed subset in the Zariski topology. □

EXAMPLE 4.4. Let $p\in\mathbb{P}$, $N\in\mathbb{N}$ and G be the cyclic group of order p^N with a generator u.

(a) Let $S:=\mathbb{Z}_p[G]$. Consider $\mathbb{C}_p$ to be the completion of the algebraic closure of $\mathbb{Q}_p$, with the induced norm on $\mathbb{C}_p$ being denoted by $|\cdot|_p$. We put $r_0:=1$ and pick a root $r_k\in\mathbb{C}_p$ for $\mathbf{q}_k$ for each $k=1,\dots,N$. If $\omega\in[1,\infty)$ and $a_0,\dots,a_{p^N-1}\in\mathbb{Z}_p$, we set

$$\gamma_k^\omega\Big(\sum_{l=0}^{p^N-1}a_lu^l\Big):=\Big|\sum_{l=0}^{p^N-1}a_lr_k^l\Big|_p^\omega\quad\textit{and}\quad\beta\Big(\sum_{l=0}^{p^N-1}a_lu^l\Big):=\Big|\sum_{l=0}^{p^N-1}Q(a_l)\Big|_{\mathbb{F}_p}.$$

It was shown in [8, Example 2.15] that γ_k^ω is independent of the choice of r_k, and that

$$\mathbf{Mult}(S_{\|\cdot\|})=\{\gamma_k^\omega:\omega\in[1,\infty);k=0,\dots,N\}\cup\{\beta\}.$$

Moreover, $\mathbf{Mult}(S_{\|\cdot\|})$ is a subspace of $\mathbb{R}^2$ of the form:

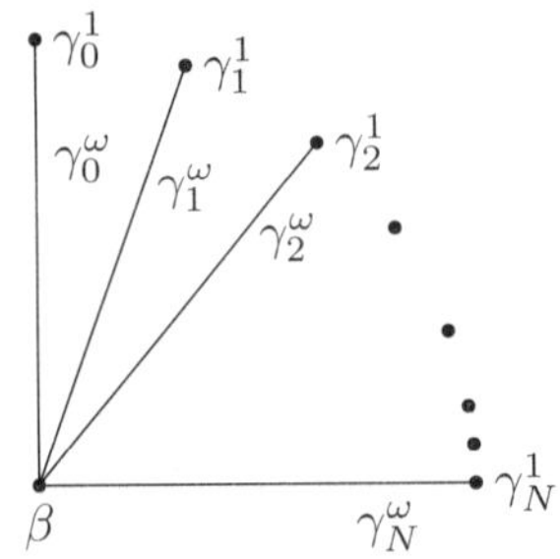

It is easy to see that $\ker\beta = J_\infty$ and that $\ker\gamma_k^\omega = I_k$ for $k \in \{0, \ldots, N\}$ and $\omega \in [1, \infty)$. Hence, $\mathbf{Spec}_0(S_{\|\cdot\|}) = \mathbf{Spec}(S)$.

Since any subset of $\mathbf{Spec}(S)$ that does not contains J_∞ is open and that Λ will send any open subset of $\mathbf{Mult}(S_{\|\cdot\|})$ containing β to the whole space $\mathbf{Spec}(S)$, we know that Λ is an open map, but $\mathbf{Spec}(S)$ is non-Hausdorff.

Furthermore, the relation $J_\infty \cap \bigcap_{k=0}^N I_k = (0)$ implies that S is Gelfand semi-simple. On the other hand, as J_∞ is the only maximal ideal of S, the Jacobson radical $\mathbf{J}(S_{\|\cdot\|})$ equals J_∞. However, since $S/J_\infty \cong \mathbb{F}_p[G]$, we know from Proposition 2.3(b)&(c) that the spectrum of $\mathbf{J}(S_{\|\cdot\|})_{\|\cdot\|}$ is homeomorphic to the disjoint union of the $N+1$ intervals $\{\gamma_k^\omega : \omega \in [1, \infty)\}$ for $k = 0, ..., N$ in the above picture.
(b) Let $\varepsilon : S \to \mathbb{Z}_p$ be the map given by $\varepsilon\big(\sum_{k=0}^{p^N-1} a_k u^k\big) := \sum_{k=0}^{p^N-1} a_k$. Then we know from Proposition 2.3(b) that

$$\mathbf{Mult}\big((\ker\varepsilon)_{\|\cdot\|}\big) := \big\{\gamma_k^\omega|_{\ker\varepsilon} : k = 1, ..., N; \omega \in [1, \infty)\big\}.$$

Hence, the spectrum of $(\ker\varepsilon)_{\|\cdot\|}$ is homeomorphic to the disjoint union of N intervals of the form $[1, \infty)$. On the other hand,

$$\mathbf{Spec}(\ker\varepsilon) := \{I_k \cap \ker\varepsilon : k = 1, ..., N\},$$

and the Zariski topology on the prime spectrum of $\ker\varepsilon$ is the discrete one. Thus, the canonical map from $\mathbf{Mult}((\ker\varepsilon)_{\|\cdot\|})$ to $\mathbf{Spec}(\ker\varepsilon)$ is open and the Zariski topology is the finest topology on $\mathbf{Spec}(\ker\varepsilon)$ such that this map is continuous.

5. Representations on continuous sections of bundles of Banach algebras

Direct sums of Banach algebras is a special class of Banach rings on which there exist many useful tools (thanks to the tools available to Banach algebras). A boarder class of Banach rings that are interesting and easier to handle will be that of continuous sections of "continuous bundles of Banach algebras". In the following, we associate a general Banach ring with a ring of continuous sections on a canonical "Banach algebra bundle" over $\mathbf{Mult}(\mathbb{Z}_{|\cdot|_1})$, and constructs a contractive ring homomorphism from the former ring to the later such that continuous local sections of this Banach algebra bundle are closely related to those in the image of this ring homomorphism.

Let us begin with some more notation. Suppose that $R_{\|\cdot\|}$ and $S_{\|\cdot\|}$ are two commutative Banach rings. We consider the following seminorm on the algebraic tensor product $R\otimes_{\mathbb{Z}}S$:

$$\|z\|:=\inf\Big\{\sum\nolimits_{k=1}^{n}\|a_k\|\|b_k\|:n\in\mathbb{N};a_1,...,a_n\in R;b_1,...,b_n\in S;z=\sum\nolimits_{k=1}^{n}a_k\otimes b_k\Big\}.$$

We denote by $R\widehat{\otimes}_{\mathbb{Z}}S$ the completion of $R\otimes_{\mathbb{Z}}S/\ker\|\cdot\|$ under the quotient norm. When $R\widehat{\otimes}_{\mathbb{Z}}S\neq(0)$, it is a commutative Banach ring under the above norm and the canonical ring structure. Notice that $\mathcal{H}_\alpha\widehat{\otimes}_{\mathbb{Z}}\mathcal{H}_\beta=(0)$ when α and β are two distinct elements in $\mathbf{Mult}(\mathbb{Z}_{|\cdot|_1})$.

The following lemma is easy to verify.

LEMMA 5.1. *Let $\beta\in\mathbf{Mult}(\mathbb{Z}_{|\cdot|_1})$. The algebra $\hat{S}\widehat{\otimes}_{\mathbb{Z}}\mathcal{H}_\beta$ is the Banach $\mathcal{H}_\beta$-algebra unitalization $(S\widehat{\otimes}_{\mathbb{Z}}\mathcal{H}_\beta)^{\mathbf{1}}$ of the Banach $\mathcal{H}_\beta$-algebra $S\widehat{\otimes}_{\mathbb{Z}}\mathcal{H}_\beta$ (see Remark 2.4).*
Hence, $\mathbf{Mult}\big((\hat{S}\widehat{\otimes}_{\mathbb{Z}}\mathcal{H}_\beta)_{\|\cdot\|}\big)$ is the one-point compactification of $\mathbf{Mult}\big((S\widehat{\otimes}_{\mathbb{Z}}\mathcal{H}_\beta)_{\|\cdot\|}\big)$.

In the following, $\eta:\mathbf{Mult}(\hat{S}_{\|\cdot\|})\to\mathbf{Mult}(\mathbb{Z}_{|\cdot|_1})$ is the continuous map given by the canonical embedding from $\mathbb{Z}$ to $\hat{S}$. Note that η is surjective, thanks to the existence of the quotient map from $\hat{S}$ to $\mathbb{Z}$. On the other hand, for every $\beta\in\mathbf{Mult}(\mathbb{Z}_{|\cdot|_1})$, we consider

$$\tilde{\Psi}_\beta:\mathbf{Mult}\big((\hat{S}\widehat{\otimes}_{\mathbb{Z}}\mathcal{H}_\beta)_{\|\cdot\|}\big)\to\mathbf{Mult}(\hat{S}_{\|\cdot\|})$$

to be the map induced by the canonical homomorphism $\Psi_\beta\in\mathcal{C}(\hat{S},\hat{S}\widehat{\otimes}_{\mathbb{Z}}\mathcal{H}_\beta)$.

PROPOSITION 5.2. *Let $S_{\|\cdot\|}$ be a commutative Banach ring and $\beta\in\mathbf{Mult}(\mathbb{Z}_{|\cdot|_1})$. The subspace $\eta^{-1}(\beta)$ is homeomorphic to the one-point compactification of the spectrum of $(S\widehat{\otimes}_{\mathbb{Z}}\mathcal{H}_\beta)_{\|\cdot\|}$.*

PROOF. By Lemma 5.1, it suffices to show that

$$\eta^{-1}(\beta)=\tilde{\Psi}_\beta\big(\mathbf{Mult}\big((\hat{S}\widehat{\otimes}_{\mathbb{Z}}\mathcal{H}_\beta)_{\|\cdot\|}\big)\big)$$

and that $\tilde{\Psi}_\beta$ is injective.

In fact, for any $\gamma\in\mathbf{Mult}\big((\hat{S}\widehat{\otimes}_{\mathbb{Z}}\mathcal{H}_\beta)_{\|\cdot\|}\big)$, since φ_γ restricts to an isometry on the image of $\mathcal{H}_\beta$ (see Lemma 1.2(b)), we know that

$$\gamma\big((0,k)\otimes 1\big)=\gamma\big((0,1)\otimes k\big)=\beta(k)\qquad(k\in\mathbb{Z}).$$

Therefore, $\tilde{\Psi}_\beta(\gamma)\in\eta^{-1}(\beta)$. Conversely, suppose that $\alpha\in\eta^{-1}(\beta)$. It is clear that $\mathcal{H}_\beta$ is the minimal CVF contained in $\mathcal{H}_\alpha$ and hence $\mathcal{H}_\alpha\widehat{\otimes}_{\mathbb{Z}}\mathcal{H}_\beta\cong\mathcal{H}_\alpha$. Thus, the map $\varphi_\alpha:\hat{S}\to\mathcal{H}_\alpha$ induces an element in $\Phi_\alpha\in\mathcal{C}(\hat{S}\widehat{\otimes}_{\mathbb{Z}}\mathcal{H}_\beta,\mathcal{H}_\alpha)$ with $\Phi_\alpha\circ\Psi_\beta=\varphi_\alpha$. From this, one knows that $\hat{S}\widehat{\otimes}_{\mathbb{Z}}\mathcal{H}_\beta\neq(0)$. If we define $\bar{\alpha}\in\mathbf{Mult}\big((\hat{S}\widehat{\otimes}_{\mathbb{Z}}\mathcal{H}_\beta)_{\|\cdot\|}\big)$ by $\bar{\alpha}(x):=|\Phi_\alpha(x)|$ $(x\in\hat{S}\widehat{\otimes}_{\mathbb{Z}}\mathcal{H}_\beta)$, then $\alpha=\bar{\alpha}\circ\Psi_\beta$. These show that the image of $\tilde{\Psi}_\beta$ is precisely $\eta^{-1}(\beta)$.

In order to verify the injectivity of $\tilde{\Psi}_\beta$, let us assume that

$$\gamma,\gamma'\in\mathbf{Mult}\big((\hat{S}\widehat{\otimes}_{\mathbb{Z}}\mathcal{H}_\beta)_{\|\cdot\|}\big)$$

satisfying $\tilde{\Psi}_\beta(\gamma)=\tilde{\Psi}_\beta(\gamma')$. Then there exists an isometric ring isomomorphism Υ from the normed subring $\varphi_\gamma(\hat{S}\otimes 1)$ (of $\mathcal{H}_\gamma$) to $\varphi_{\gamma'}(\hat{S}\otimes 1)$ such that

$$\Upsilon(\varphi_\gamma(x\otimes 1))=\varphi_{\gamma'}(x\otimes 1)\qquad(x\in\hat{S}).$$

When $\mathcal{H}_\beta$ is of characteristic zero (respectively, characteristic p), the closed subfield of $\mathcal{H}_\gamma$ generated by $\varphi_\gamma(\hat{S}\otimes 1)$ will contain $\varphi_\gamma(\hat{S}\otimes\mathbb{Q})$ (respectively, $\varphi_\gamma(\hat{S}\otimes\mathbb{F}_p)$).

As $\mathbb{Q}$ (respectively, $\mathbb{F}_p$) is dense in $\mathcal{H}_\beta$, the continuity of φ_γ tells us that $\varphi_\gamma(\hat{S}\otimes 1)$ generates $\mathcal{H}_\gamma$, and the same is true for γ'. Consequently, Υ extends to an isometric isomorphism from $\mathcal{H}_\gamma$ onto $\mathcal{H}_{\gamma'}$, whose composition with φ_γ equals $\varphi_{\gamma'}$. This shows that $\gamma' = \gamma$ as required. □

The above tells us that the set $\mathbf{Mult}(\hat{S}_{\|\cdot\|})$ is a union of the spectrum of $(\hat{S}\widehat{\otimes}_{\mathbb{Z}}\mathcal{H}_\beta)_{\|\cdot\|}$ when β runs through all elements in $\mathbf{Mult}(\mathbb{Z}_{|\cdot|_1})$. Moreover, if we regard $\mathbf{Mult}(\mathbb{Z}_{|\cdot|_1})$ as a subspace of $\mathbf{Mult}(\hat{S}_{\|\cdot\|})$ through the quotient map from $\hat{S}$ to $\mathbb{Z}$, then $\tilde{\Psi}_\beta\big(\mathbf{Mult}\big((\hat{S}\widehat{\otimes}_{\mathbb{Z}}\mathcal{H}_\beta)_{\|\cdot\|}\big)\big) = \tilde{\Psi}_\beta\big(\mathbf{Mult}\big((S\widehat{\otimes}_{\mathbb{Z}}\mathcal{H}_\beta)_{\|\cdot\|}\big)\big)\cup\{\beta\}$. Consequently, the spectrum of $S_{\|\cdot\|}$ can be identified with

$$\bigcup\nolimits_{\beta\in\mathbf{Mult}(\mathbb{Z}_{|\cdot|_1})} \mathbf{Mult}\big((S\widehat{\otimes}_{\mathbb{Z}}\mathcal{H}_\beta)_{\|\cdot\|}\big)$$

(where some of the sets in the union could be empty), when this union is equipped with certain topology compatible with the one on each $\mathbf{Mult}\big((S\widehat{\otimes}_{\mathbb{Z}}\mathcal{H}_\beta)_{\|\cdot\|}\big)$.

As a "lifting" of the above picture to the "rings level", we will define, in the following, a T_1 topology on the disjoint union:

$$\mathcal{A}^{S_{\|\cdot\|}} := \bigcup\nolimits_{\beta\in\mathbf{Mult}(\mathbb{Z}_{|\cdot|_1})} S\widehat{\otimes}_{\mathbb{Z}}\mathcal{H}_\beta$$

so that it becomes a "continuous bundle of Banach algebras" over the compact Hausdorff space $\mathbf{Mult}(\mathbb{Z}_{|\cdot|_1})$ and that

- $\alpha\mapsto a\otimes\varphi_\alpha(1)$ is a continuous section from $\mathbf{Mult}(\mathbb{Z}_{|\cdot|_1})$ to $\mathcal{A}^{S_{\|\cdot\|}}$, for every $a\in S$;
- a section $\rho:\mathbf{Mult}(\mathbb{Z}_{|\cdot|_1})\to\mathcal{A}^{S_{\|\cdot\|}}$ is continuous at α_0 if and only if "ρ is close to a map of the form $\alpha\mapsto a\otimes\varphi_\alpha(k)^{-1}$ near α_0" (see Lemma 5.4(b) below).

We will do this by first introducing a topology on $\bigcup_{\beta\in\mathbf{Mult}(\mathbb{Z}_{|\cdot|_1})}\hat{S}\widehat{\otimes}_{\mathbb{Z}}\mathcal{H}_\beta$ and then consider $\mathcal{A}^{S_{\|\cdot\|}}$ as subspace of it. Because of this, it makes no harm to assume that S is unital directly.

From now on, S is a commutative unital Banach ring. Moreover, we denote by $\mathbb{R}_+^*$ the set of all non-zero positive real numbers, and by $\pi:\mathcal{A}^{S_{\|\cdot\|}}\to\mathbf{Mult}(\mathbb{Z}_{|\cdot|_1})$ the canonical map.

The idea of the topology on $\mathcal{A}^{S_{\|\cdot\|}}$ comes from the alternative description of the topology on the "Gelfand-Berkovich spectrum" of commutative unital Banach rings as in [**9**, Proposition 3.4]. More precisely, let us fix an element $x_0\in\mathcal{A}^{S_{\|\cdot\|}}$. For any $\epsilon>0$, we set

$$\mathcal{S}_\epsilon^{x_0} := \big\{(a,k)\in S\times\mathbb{N} : \pi(x_0)(k)>0 \text{ and } \big\|x_0 - a\otimes\varphi_{\pi(x_0)}(k)^{-1}\big\|<\epsilon\big\}.$$

Given any $(\delta,n,\epsilon)\in\mathbb{R}_+^*\times\mathbb{Z}\times\mathbb{R}_+^*$ as well as $(a,k)\in\mathcal{S}_\epsilon^{x_0}$, we put

$$U^{x_0}_{\delta,n,\epsilon,a,k} := \big\{y\in\mathcal{A}^{S_{\|\cdot\|}} : |\pi(y)(n)-\pi(x_0)(n)|<\delta,\ \pi(y)(k)>0 \text{ and } \big\|y-a\otimes\varphi_{\pi(y)}(k)^{-1}\big\|<\epsilon\big\}.$$

Proposition 5.3. *Let S be a commutative unital Banach ring. There exists a T_1 topology $\mathcal{P}$ on $\mathcal{A}^{S_{\|\cdot\|}}$ such that intersections of finite numbers of elements in collection*

$$\mathcal{U}^{x_0} := \big\{U^{x_0}_{\delta,n,\epsilon,a,k} : (\delta,n,\epsilon)\in\mathbb{R}_+^*\times\mathbb{Z}\times\mathbb{R}_+^*; (a,k)\in\mathcal{S}_\epsilon^{x_0}\big\}$$

form a local open neighborhood basis for $x_0\in\mathcal{A}^{S_{\|\cdot\|}}$, and that π becomes an open map.

PROOF. Let $\mathcal{P}$ be the collection of all subsets $V \subseteq \mathcal{A}^{S_{\|\cdot\|}}$ satisfying the following: for any $x_0 \in V$ there exist a finite number of sets $U_1, \ldots, U_m \in \mathcal{U}^{x_0}$ such that $\bigcap_{k=1}^m U_m \subseteq V$. It is easy to see that $\mathcal{P}$ is a topology on $\mathcal{A}^{S_{\|\cdot\|}}$, and that π is an open map under this topology.

For $x_0 \in \mathcal{A}^{S_{\|\cdot\|}}$, $(\delta, n, \epsilon) \in \mathbb{R}_+^* \times \mathbb{Z} \times \mathbb{R}_+^*$ and $(a, k) \in \mathcal{S}_\epsilon^{x_0}$, we claim that the set $U_{\delta,n,\epsilon,a,k}^{x_0}$ belongs to $\mathcal{P}$ (and hence intersections of finite numbers of elements in $\mathcal{U}^{x_0}$ form a local open neighborhood base for x_0). In fact, consider an arbitrary element $x_1 \in U_{\delta,n,\epsilon,a,k}^{x_0}$. Clearly, $(a, k) \in \mathcal{S}_\epsilon^{x_1}$. If we set

$$\gamma := \delta - |\pi(x_1)(n) - \pi(x_0)(n)| \in (0, \delta],$$

then we have $U_{\gamma,n,\epsilon,a,k}^{x_1} \subseteq U_{\delta,n,\epsilon,a,k}^{x_0}$. This implies that $U_{\delta,n,\epsilon,a,k}^{x_0} \in \mathcal{P}$.

Next, we will show that $\mathcal{P}$ is T_1. Let x_1 and x_2 be two distinct points in $\mathcal{A}^{S_{\|\cdot\|}}$. If $\pi(x_1) \neq \pi(x_2)$, one can find $n_0 \in \mathbb{Z}$ and $\delta_0 > 0$ satisfying

$$\{\alpha \in \mathbf{Mult}(\mathbb{Z}_{|\cdot|_1}) : |\alpha(n_0) - \pi(x_1)(n_0)| < \delta_0\} \cap \{\alpha \in \mathbf{Mult}(\mathbb{Z}_{|\cdot|_1}) : |\alpha(n_0) - \pi(x_2)(n_0)| < \delta_0\} = \emptyset.$$

Hence, if we fix $\epsilon > 0$ and $(a_i, k_i) \in \mathcal{S}_\epsilon^{x_i}$ $(i = 1, 2)$ arbitrarily, then $U_{\delta_0,n_0,\epsilon,a_1,k_1}^{x_1} \cap U_{\delta_0,n_0,\epsilon,a_2,k_2}^{x_2} = \emptyset$. Secondly, we consider the case when there exists $\alpha_0 \in \mathbf{Mult}(\mathbb{Z}_{|\cdot|_1})$ with $x_1, x_2 \in S\widehat{\otimes}_{\mathbb{Z}}\mathcal{H}_{\alpha_0}$ (i.e., $\pi(x_1) = \pi(x_2)$). Since $\|x_1 - x_2\| > 0$ and

$$\{a \otimes \varphi_{\alpha_0}(k)^{-1} : a \in S \text{ and } k \in \mathbb{N} \text{ with } \alpha_0(k) > 0\}$$

is dense in $S\widehat{\otimes}_{\mathbb{Z}}\mathcal{H}_{\alpha_0}$ (recall that $\mathcal{H}_{\alpha_0}$ is a minimal CVF), one can find $\epsilon_0 > 0$ and $(a_1, k_1), (a_2, k_2) \in S \times \mathbb{N}$ with $\alpha_0(k_j) > 0$ and $\|x_j - a_j \otimes \varphi_{\alpha_0}(k_j)^{-1}\| < \epsilon_0$ $(j = 1, 2)$ such that

$$\{y \in S\widehat{\otimes}_{\mathbb{Z}}\mathcal{H}_{\alpha_0} : \|y - a_1 \otimes \varphi_{\alpha_0}(k_1)^{-1}\| < \epsilon_0\} \cap \{y \in S\widehat{\otimes}_{\mathbb{Z}}\mathcal{H}_{\alpha_0} : \|y - a_2 \otimes \varphi_{\alpha_0}(k_2)^{-1}\| < \epsilon_0\} = \emptyset.$$

Thus, for arbitrary chosen $\delta > 0$ and $n \in \mathbb{N}$, one has $x_1 \notin U_{\delta,n,\epsilon_0,a_2,k_2}^{x_2}$ and $x_2 \notin U_{\delta,n,\epsilon_0,a_1,k_1}^{x_1}$. □

We will see in Example 5.7(a) below that $\mathcal{P}$ needs not be Hausdorff in general.

LEMMA 5.4. *(a) Let $\alpha_0 \in \mathbf{Mult}(\mathbb{Z}_{|\cdot|_1})$. For $a \in S$ and $k \in \mathbb{N}$, if $\alpha_0(k) > 0$, then the function $\alpha \mapsto \|a \otimes \varphi_\alpha(k)^{-1}\|$ is well-defined and upper semi-continuous in an open neighborhood of α_0.*
(b) Suppose that $\{x_i\}_{i\in\mathfrak{I}}$ is a net in $\mathcal{A}^{S_{\|\cdot\|}}$ and $x_0 \in \mathcal{A}^{S_{\|\cdot\|}}$. Then $x_i \xrightarrow{\mathcal{P}} x_0 \in \mathcal{A}^{S_{\|\cdot\|}}$ if and only if $\pi(x_i) \to \pi(x_0)$ and for any $\gamma > 0$, there exist $(a_\gamma, k_\gamma) \in \mathcal{S}_\gamma^{x_0}$ and $i_1 \in \mathfrak{I}$ such that $\pi(x_i)(k_\gamma) > 0$ and $\|x_i - a_\gamma \otimes \varphi_{\pi(x_i)}(k_\gamma)^{-1}\| < \gamma$ whenever $i \geq i_1$.

PROOF. (a) As $\|a \otimes \varphi_\alpha(k)^{-1}\| = \alpha(k)^{-1}\|a \otimes \varphi_\alpha(1)\|$, we know from [**7**, Lemma 4.4] and the continuity of $\alpha \mapsto \alpha(k)$ that $\alpha \mapsto \|a \otimes \varphi_\alpha(k)^{-1}\|$ is well-defined and upper semi-continuous in a neighborhood of α_0.
(b) $\Rightarrow$). Consider any $\gamma \in \mathbb{R}_+^*$. As

$$\{a \otimes \varphi_{\pi(x_0)}(k)^{-1} : (a, k) \in S \times \mathbb{N} \text{ with } \pi(x_0)(k) > 0\}$$

is dense in $S\widehat{\otimes}_{\mathbb{Z}}\mathcal{H}_{\pi(x_0)}$, we know that $\mathcal{S}_\gamma^{x_0} \neq \emptyset$ and we pick a pair $(a_\gamma, k_\gamma) \in \mathcal{S}_\gamma^{x_0}$. Since $U_{1,1,\gamma,a_\gamma,k_\gamma}^{x_0}$ is open in $\mathcal{P}$, one can find $i_1 \in \mathfrak{I}$ so that $x_i \in U_{1,1,\gamma,a_\gamma,k_\gamma}^{x_0}$ for every $i \geq i_1$. In this case,

$$\|x_i - a_\gamma \otimes \varphi_{\pi(x_i)}(k_\gamma)^{-1}\| < \gamma.$$

On the other hand, for every $(\delta, n) \in \mathbb{R}_+^* \times \mathbb{Z}$, the openness of $U_{\delta,n,1,a_1,k_1}^{x_0}$ implies that $x_i \in U_{\delta,n,1,a_1,k_1}^{x_0}$ for large enough i. This shows that $\pi(x_i) \to \pi(x_0)$.

$\Leftarrow$). Suppose that $(\delta, n, \epsilon) \in \mathbb{R}_+^* \times \mathbb{Z} \times \mathbb{R}_+^*$ and $(a,k) \in \mathcal{S}_\epsilon^{x_0}$. As $\pi(x_i) \to \pi(x_0)$, there exists $i_2 \in \mathfrak{I}$ with $|\pi(x_i)(n) - \pi(x_0)(n)| < \delta$ whenever $i \geq i_2$. On the other hand, we set

$$\kappa := \|x_0 - a \otimes \varphi_{\pi(x_0)}(k)^{-1}\|.$$

Since $\kappa < \epsilon$, one can find $\gamma > 0$ with $\kappa + 2\gamma < \epsilon$. Let $(a_\gamma, k_\gamma) \in \mathcal{S}_\gamma^{x_0}$ be as in the statement. Then

$$\pi(x_0)(k_\gamma k)^{-1}\|(ka_\gamma - k_\gamma a) \otimes \varphi_{\pi(x_0)}(1)\| = \|a_\gamma \otimes \varphi_{\pi(x_0)}(k_\gamma)^{-1} - a \otimes \varphi_{\pi(x_0)}(k)^{-1}\| < \kappa + \gamma.$$

Since $\pi(x_i)(k_\gamma k) \to \pi(x_0)(k_\gamma k)$, we know from part (a) that there exists $i_3 \in \mathfrak{I}$ such that

$$\|a_\gamma \otimes \varphi_{\pi(x_i)}(k_\gamma)^{-1} - a \otimes \varphi_{\pi(x_i)}(k)^{-1}\| < \kappa + \gamma \qquad (i \geq i_3).$$

If i_1 is as in the statement and $i_0 \in \mathfrak{I}$ satisfying $i_0 \geq i_1, i_2, i_3$, then for any $i \geq i_0$, one has $\|x_i - a \otimes \varphi_{\pi(x_i)}(k)^{-1}\| < \kappa + 2\gamma < \epsilon$ and $|\pi(x_i)(n) - \pi(x_0)(n)| < \delta$; in other words, $x_i \in U_{\delta,n,\epsilon,a,k}^{x_0}$. □

THEOREM 5.5. *Let S be a commutative unital Banach ring.*
(a) The function $x \mapsto \|x\|$ on $\mathcal{A}^{S_{\|\cdot\|}}$ is upper semicontinuous.
(b) If $\{x_i\}_{i\in\mathfrak{I}}$ and $\{y_i\}_{i\in\mathfrak{I}}$ are two nets in $\mathcal{A}^{S_{\|\cdot\|}}$ with $\pi(x_i) = \pi(y_i)$ (for all $i \in \mathfrak{I}$) such that $x_i \xrightarrow{\mathcal{P}} x_0$ and $y_i \xrightarrow{\mathcal{P}} y_0$, then $x_i y_i \xrightarrow{\mathcal{P}} x_0 y_0$ as well as $x_i + y_i \xrightarrow{\mathcal{P}} x_0 + y_0$.
(c) For any $b \in S$, the map $\Delta(b) : \mathbf{Mult}(\mathbb{Z}_{|\cdot|_1}) \to \mathcal{A}^{S_{\|\cdot\|}}$ given by $\Delta(b)(\alpha) := b \otimes \varphi_\alpha(1)$ is continuous.

PROOF. (a) Let $\kappa \in \mathbb{R}_+^*$ and set $\mathcal{A}_\kappa^{S_{\|\cdot\|}} := \{x \in \mathcal{A}^{S_{\|\cdot\|}} : \|x\| < \kappa\}$. Consider any $x_0 \in \mathcal{A}_\kappa^{S_{\|\cdot\|}}$. If we pick $\epsilon \in \mathbb{R}_+^*$ with $\|x_0\| + 2\epsilon < \kappa$, and choose an arbitrary element $(a,k) \in \mathcal{S}_\epsilon^{x_0}$, then

$$\|a \otimes \varphi_{\pi(x_0)}(k)^{-1}\| < \|x_0\| + \epsilon.$$

By Lemma 5.4(a), one can find $\delta > 0$ and $n \in \mathbb{Z}$ such that whenever $\alpha \in \mathbf{Mult}(\mathbb{Z}_{|\cdot|_1})$ satisfying $|\alpha(n) - \pi(x_0)(n)| < \delta$, one has $\|a \otimes \varphi_\alpha(k)^{-1}\| < \|x_0\| + \epsilon$. Thus, for any $y \in U_{\delta,n,\epsilon,a,k}^{x_0}$, we know that

$$\|a \otimes \varphi_{\pi(y)}(k)^{-1}\| < \|x_0\| + \epsilon \quad \text{and} \quad \|y - a \otimes \varphi_{\pi(y)}(k)^{-1}\| < \epsilon.$$

This tells us that $\|y\| < \|x_0\| + 2\epsilon < \kappa$. Consequently, $\mathcal{A}_\kappa^{S_{\|\cdot\|}} \in \mathcal{P}$.
(b) By part (a), there exist $\kappa > 0$ and $i_2 \in \mathfrak{I}$ such that $\|x_0\|, \|y_0\|, \|x_i\|, \|y_i\| < \kappa$ when $i \geq i_2$. Therefore, we may assume that $\|x_0\|, \|y_0\|, \|x_i\|, \|y_i\| < \kappa$ for all $i \in \mathfrak{I}$.

Put $\alpha_i := \pi(x_i)$ $(i \in \mathfrak{I} \cup \{0\})$, and fix a positive number $\epsilon \in (0,1)$. By Lemma 5.4(b), we know that $\alpha_i \to \alpha_0$. Moreover, one can find $(a,k) \in \mathcal{S}_{\frac{\epsilon}{2\kappa+1}}^{x_0}$, $(b,l) \in \mathcal{S}_{\frac{\epsilon}{2\kappa+1}}^{y_0}$ as well as $i_1 \in \mathfrak{I}$ such that for $i \geq i_1$, we have $\alpha_i(k) > 0$, $\alpha_i(l) > 0$,

$$\|x_i - a \otimes \varphi_{\alpha_i}(k)^{-1}\| < \epsilon/(2\kappa+1) \quad \text{and} \quad \|y_i - b \otimes \varphi_{\alpha_i}(l)^{-1}\| < \epsilon/(2\kappa+1).$$

Hence, $\|x_0 y_0 - ab \otimes \varphi_{\alpha_0}(kl)^{-1}\| < \frac{\kappa\epsilon}{2\kappa+1} + \frac{\epsilon}{2\kappa+1}\left(\kappa + \frac{1}{2\kappa+1}\right) < \epsilon$ and $\|x_i y_i - ab \otimes \varphi_{\alpha_i}(kl)^{-1}\| < \epsilon$ for any $i \geq i_1$. Consequently, Lemma 5.4(b) tells us that $x_i y_i \xrightarrow{\mathcal{P}} x_0 y_0$. By replacing $\epsilon/(2\kappa+1)$ in the above by $\epsilon/2$, we can also show that $x_i + y_i \xrightarrow{\mathcal{P}} x_0 + y_0$.
(c) Let $\{\alpha_i\}_{i\in\mathfrak{I}}$ be a net in $\mathbf{Mult}(\mathbb{Z}_{|\cdot|_1})$ that converges to α_0. Denote $x_0 := b \otimes \varphi_{\alpha_0}(1)$. Suppose that $(\delta, n, \epsilon) \in \mathbb{R}_+^* \times \mathbb{Z} \times \mathbb{R}_+^*$. Clearly, $(b,1) \in \mathcal{S}_\epsilon^{x_0}$. For every $i \in \mathfrak{I}$, one has $\alpha_i(1) = 1$ and $\|b \otimes \varphi_{\alpha_i}(1) - b \otimes \varphi_{\alpha_i}(1)\| = 0$. Thus, Lemma 5.4(b) tells us that $b \otimes \varphi_{\alpha_i}(1) \to x_0$ as required. □

We will see in Example 5.7(a) below that, in general, the norm function on $\mathcal{A}^{S_{\|\cdot\|}}$ may not be continuous.

A map $\rho : \mathbf{Mult}(\mathbb{Z}_{|\cdot|_1}) \to \mathcal{A}^{S_{\|\cdot\|}}$ is called a *continuous section on* $\mathcal{A}^{S_{\|\cdot\|}}$ if it is a continuous map satisfying $\rho(\alpha) \in S\widehat{\otimes}_{\mathbb{Z}}\mathcal{H}_\alpha$. We denote by $\Gamma(\mathcal{A}^{S_{\|\cdot\|}})$ the set of all continuous sections on $\mathcal{A}^{S_{\|\cdot\|}}$.

PROPOSITION 5.6. *(a) $\Gamma(\mathcal{A}^{S_{\|\cdot\|}})$ is a commutative unital Banach ring, under the canonical ring structure as well as the norm defined by $\|\rho\| := \sup_{\alpha\in\mathbf{Mult}(\mathbb{Z}_{|\cdot|_1})} \|\rho(\alpha)\|$ $(\rho \in \Gamma(\mathcal{A}^{S_{\|\cdot\|}}))$.*
(b) The map $\Delta : S \to \Gamma(\mathcal{A}^{S_{\|\cdot\|}})$ as given by Theorem 5.5(c) is a contractive unital ring homomorphism.

PROOF. (a) It follows from Theorem 5.5(b) that $\Gamma(\mathcal{A}^{S_{\|\cdot\|}})$ is a commutative normed ring. Moreover, this ring is unital because of Theorem 5.5(c). Suppose now that $\{\rho_n\}_{n\in\mathbb{N}}$ is a Cauchy sequence in $\Gamma(\mathcal{A}^{S_{\|\cdot\|}})$. For each $\beta \in \mathbf{Mult}(\mathbb{Z}_{|\cdot|_1})$, since $S\widehat{\otimes}_{\mathbb{Z}}\mathcal{H}_\beta$ is complete, we know that $\{\rho_n(\beta)\}_{n\in\mathbb{N}}$ converges to an element $\rho(\beta) \in S\widehat{\otimes}_{\mathbb{Z}}\mathcal{H}_\beta$. Consider $\alpha_0 \in \mathbf{Mult}(\mathbb{Z}_{|\cdot|_1})$. Let $\{\alpha_i\}_{i\in\mathfrak{I}}$ be a net in $\mathbf{Mult}(\mathbb{Z}_{|\cdot|_1})$ converging to α_0 and $\gamma \in \mathbb{R}_+^*$. There exists $n_1 \in \mathbb{N}$ such that for any $n \geq n_1$, one has

$$\sup_{\beta\in\mathbf{Mult}(\mathbb{Z}_{|\cdot|_1})}\|\rho_n(\beta) - \rho(\beta)\| < \gamma/2.$$

As ρ_{n_1} is continuous, one may find, through Lemma 5.4(b), elements $(a_1, k_1) \in \mathcal{S}_{\gamma/2}^{\rho_{n_1}(\alpha_0)}$ and $i_1 \in \mathfrak{I}$ such that $\alpha_i(k_1) > 0$ and

$$\|\rho_{n_1}(\alpha_i) - a_1 \otimes \varphi_{\alpha_i}(k_1)^{-1}\| < \gamma/2$$

whenever $i \geq i_1$. Clearly, $(a_1, k_1) \in \mathcal{S}_\gamma^{\rho(\alpha_0)}$ and $\|\rho(\alpha_i) - a_1 \otimes \varphi_{\alpha_i}(k_1)^{-1}\| < \gamma$ for every $i \geq i_1$. Thus, Lemma 5.4(b) tells us that $\rho(\alpha_i) \to \rho(\alpha_0)$. Consequently, $\rho \in \Gamma(\mathcal{A}^{S_{\|\cdot\|}})$ as required.
(b) This part is obvious. □

We say that S is *quasi-regular* when Δ is an isometric bijection. Obviously, the underlying Banach ring of a Banach algebra is quasi-regular, and quasi-regular rings are very closed to being Banach algebras. Since there are a lots of tools available for Banach algebras, quasi-regular Banach rings should be easier to handle. Unfortunately, not all Banach rings are quasi-regular (see part (b) of the following example).

EXAMPLE 5.7. (a) Let S be the ring $\mathbb{R}$ equipped with the usual Euclidean norm $|\cdot|_1$ and $R := \hat{S}$. In this case, the fiber of $\mathcal{A}^{R_{\|\cdot\|}}$ over $|\cdot|_1 \in \mathbf{Mult}(\mathbb{Z}_{|\cdot|_1})$ is the Banach $\mathbb{R}$-algebra unitalization $\mathbb{R} \oplus \mathbb{R}$ of $\mathbb{R}$, while the fiber over $\alpha \in \mathbf{Mult}(\mathbb{Z}_{|\cdot|_1}) \setminus \{|\cdot|_1\}$ is $\mathbb{R}$. If e is the identity of S and $t \in \mathbb{R}$, then

$$\Delta(te)(\alpha) = \begin{cases} 0_\alpha & \text{when } \alpha \neq |\cdot|_1 \\ (t,0) & \text{when } \alpha = |\cdot|_1 \end{cases},$$

where 0_α is the zero element in the fiber over α. Suppose that $\{\alpha_i\}_{i\in\mathfrak{I}}$ is a net in $\mathbf{Mult}(\mathbb{Z}_{|\cdot|_1}) \setminus \{|\cdot|_1\}$ that converges to $|\cdot|_1$. We learn from the definition of $\mathcal{P}$ that $\{0_{\alpha_i}\}_{i\in\mathfrak{I}}$ will $\mathcal{P}$-converge to every element in $\{(t,0) : t \in \mathbb{R}\}$. Hence, $\mathcal{P}$ is non-Hausdorff. Furthermore, as $|0_{\alpha_i}| = 0$ $(i \in \mathfrak{I})$ and $\|(1,0)\| = 1$, we know that the norm function on $\mathcal{A}^{R_{\|\cdot\|}}$ is not conitnuous. On the other hand, it is clear that R is quasi-regular.

(b) It is easy to see that $\mathbb{Z}_{|\cdot|_1}\widehat{\otimes}_{\mathbb{Z}}\mathcal{H}_\alpha = \mathcal{H}_\alpha$ for every $\alpha \in \mathbf{Mult}(\mathbb{Z}_{|\cdot|_1})$. In this case, $\mathcal{A}^{\mathbb{Z}_{|\cdot|_1}}$ is the topological subspace of $\mathbb{A}^1_{\mathbb{Z}_{|\cdot|_1}}$ consisting of images of elements from all minimal CVFs. It was shown in [**9**] that $\Gamma(\mathcal{A}^{\mathbb{Z}_{|\cdot|_1}})$ is the ring

$$\{f \in C([0,1];\mathbb{R}^1) : \lim_{v\to 0}|f(v) - n|_1^v = 0, \text{ for some } n \in \mathbb{Z}\}$$

equipped with the norm: $\|f\| := \sup_{v\in(0,1]}|f(v)|_1^v$. In this case, Δ is not surjective.
(c) Let $\|\cdot\|_0$ be the trivial norm on $\mathbb{Z}$. Then

$$\mathbf{Mult}(\mathbb{Z}_{\|\cdot\|_0}) = \mathbf{Mult}(\mathbb{Z}_{|\cdot|_1}) \setminus \{|\cdot|_1^v : v \in (0,1]\}.$$

Moreover, $\mathbb{Z}_{\|\cdot\|_0}\widehat{\otimes}_{\mathbb{Z}}\mathcal{H}_\alpha = \mathcal{H}_\alpha$ for $\alpha \in \mathbf{Mult}(\mathbb{Z}_{\|\cdot\|_0})$, while $\mathbb{Z}_{\|\cdot\|_0}\widehat{\otimes}_{\mathbb{Z}}\mathcal{H}_\alpha = (0)$ for $\alpha \in \{|\cdot|_1^v : v \in (0,1]\}$. Furthermore, $\mathcal{A}^{\mathbb{Z}_{\|\cdot\|_0}}$ is a topological subspace $\mathbb{A}^1_{\mathbb{Z}_{|\cdot|_1}}$ consisting of images of elements from all minimal non-Archimedean CVFs, and Δ is an isometric isomorphism from S onto $\Gamma(\mathcal{A}^{\mathbb{Z}_{\|\cdot\|_0}})$; in other words, $\mathbb{Z}_{\|\cdot\|_0}$ is quasi-regular. The details can be found in [**9**].

Acknowledgement

The authors thank the referee for some helpful comments in making a better presentation of this article. This work is supported by the National Natural Science Foundation of China (11471168).

References

[1] V. G. Berkovich, *Spectral theory and analytic geometry over non-Archimedean fields*, Mathematical Surveys and Monographs, vol. 33, American Mathematical Society, Providence, RI, 1990. MR1070709
[2] G. Dales, Maximal ideals in commutative Banach algebras, preprint (arXiv:1408.3815).
[3] A. Escassut, *Propriétés spectrales en analyse non archimédienne* (French), Journées Arithmétiques de Bordeaux (Conf., Univ. Bordeaux, Bordeaux, 1974), Soc. Math. France, Paris, 1975, pp. 157–167. Astérisque, Nos. 24-25. MR0374915
[4] A. Escassut, *Ultrametric Banach algebras*, World Scientific Publishing Co., Inc., River Edge, NJ, 2003. MR1978961
[5] G. Garandel, *Les semi-normes multiplicatives sur les algèbres d'éléments analytiques au sens de Krasner* (French), Nederl. Akad. Wetensch. Proc. Ser. A **78**=Indag. Math. **37** (1975), no. 4, 327–341. MR0390286
[6] B. Guennebaud, *Semi-normes multiplicatives sur une algèbre de Banach ultramétrique* (French), Séminaire de Théorie des Nombres, 1972–1973 (Univ. Bordeaux I, Talence), Exp. No. 13, Lab. Théorie des Nombres, Centre Nat. Recherche Sci., Talence, 1973, pp. 3. MR0405110
[7] C.W. Leung and C.K. Ng, Berkovich spectra of elements in Banach Rings, preprint. arXiv:1410.5893
[8] C.W. Leung and C.K. Ng, The topology on Berkovich affine lines over complete valuation rings, Quart. J. Math., to appear.
[9] C.K. Ng, The Gelfand functor for unital Banach algebras over general fields, in preparation.
[10] T. W. Palmer, *Banach algebras and the general theory of *-algebras. Vol. I*, Encyclopedia of Mathematics and its Applications, vol. 49, Cambridge University Press, Cambridge, 1994. Algebras and Banach algebras. MR1270014
[11] W. H. Schikhof, *Ultrametric calculus: An introduction to p-adic analysis*, Cambridge Studies in Advanced Mathematics, vol. 4, Cambridge University Press, Cambridge, 1984. MR791759

DEPARTMENT OF MATHEMATICS, THE CHINESE UNIVERSITY OF HONG KONG, HONG KONG
Email address: cwleung@math.cuhk.edu.hk

CHERN INSTITUTE OF MATHEMATICS AND LPMC, NANKAI UNIVERSITY, TIANJIN 300071, CHINA
Email address: ckng@nankai.edu.cn

Contemporary Mathematics
Volume **704**, 2018
http://dx.doi.org/10.1090/conm/704/14163

Ultrametric Calkin algebras

Bertin Diarra

ABSTRACT. Let K be a rank one complete ultrametric valued field and let E be an ultrametric K-Banach space. By a completely continuous linear operator on E one means a linear endomorphism of E which is the limit, in the norm-topology of bounded linear operators, of a sequence of continuous linear endomorphisms of E of finite rank. The set $\mathcal{C}(E)$ of the completely continuous operators on E is a closed two-sided ideal of the Banach algebra $\mathcal{L}(E)$ of bounded linear operators. The closed two-sided ideal structure of the algebra $\mathcal{L}(E)$ for general complex or ultrametric Banach spaces E, is far to be well understood. However for classical Hilbert spaces and for ultrametric Banach spaces over complete discrete valuation field, this structure is well known since the sixties of the last century.

If the ultrametric Banach space E is of countable type, then $\mathcal{C}(E)$ is the unique closed non-trivial two-sided ideal of $\mathcal{L}(E)$, property that is the same as in separable classical real or complex Hilbert spaces. The quotient Banach algebra $\mathfrak{B}(E) = \mathcal{L}(E)/\mathcal{C}(E)$ is called the Calkin algebra of E. As done by Calkin in the classical case of separable complex Hilbert spaces, we shall construct, but using complete ultrametric ultraproduct theory, a faithful linear representation of the Banach algebra $\mathfrak{B}(E)$, when the ultrametric Banach space E is of countable type.

Also, when the Banach space E is pseudoreflexive, we shall prove in a classical way, an ultrametric version of Atkinson theorem.

1. The completely continuous linear operators

Let K be a rank one complete ultrametric valued field and let E and F be two ultrametric K-Banach spaces. One considers on the vector space $\mathcal{L}(E, F)$ of the continuous linear operators of E in F the norm $\|u\| = \sup\limits_{0 \neq x \in E} \dfrac{\|u(x)\|}{\|x\|}$. With this norm the K-linear vector space $\mathcal{L}(E, F)$ becomes an ultrametric Banach space. When $E = F$, one writes $\mathcal{L}(E, E) = \mathcal{L}(E)$, which together with the law of composition of applications, is a unitary Banach algebra.

DEFINITION 1.1. A continuous linear map $u : E \longrightarrow F$ is said to be *completely continuous* if there exists a sequence of continuous linear maps of finite rank u_n : $E \longrightarrow F$ such that $\lim\limits_{n \to +\infty} \|u - u_n\| = 0$.

2010 *Mathematics Subject Classification.* Primary 47S10, 46B08; 47B10, 46S10, 47A53.
Key words and phrases. Banach spaces, completely continuous linear operators, Calkin algebras, ultraproduits, Fredholm operators.

In classical analysis a continuous linear map $u : E \longrightarrow F$ between Banach spaces is said to be compact if $u(B_E)$ is relatively compact in F, where B_E is the unit ball of E. The definition of completely continuous operators is the same as the previous one. It is easily seen that any completely continuous operator is compact, the converse is not always true and linked to the famous approximation property. However, on real and complex Hilbert spaces the two definitions coincide. In ultrametric Banach space theory, replacing relatively compact by compactoid, one says that an operator u is compact if $u(B_E)$ is a compactoid. Then in ultrametric Banach space theory completely continuous linear operators coincide with compact operators (cf. [**13**] or [**16**], Theorem 4.39-).

We shall denote by $\mathcal{C}(E, F)$ the space of completely continuous linear operators of E in F. It is a closed subspace of $\mathcal{L}(E, F)$. For $E = F$, one sets $\mathcal{C}(E, E) = \mathcal{C}(E)$. Let E_1 and F_1 be two other ultrametric K-Banach spaces. Then for the following linear operators $v \in \mathcal{L}(E_1, E)$, $u \in \mathcal{C}(E, F)$, and $w \in \mathcal{L}(F, F_1)$, one sees that $u \circ v$ belongs to $\mathcal{C}(E_1, F)$, (resp. $w \circ u$ belongs to $\mathcal{C}(E, F_1)$) and $w \circ u \circ v$ belongs to $\mathcal{C}(E_1, F_1)$. In particular $\mathcal{C}(E)$ is a closed two-sided ideal of the Banach algebra $\mathcal{L}(E)$ and is the closure of the two-sided ideal $\mathcal{L}_f(E)$ of the continuous linear endomorphisms of E of finite rank.

Let α be a real number such that $0 < \alpha \leq 1$. A sequence $(e_n)_{n\geq 1}$ of elements of the Banach space E is said to be *α-orthogonal* if for any finite sum $x = \sum_{n\in J} \lambda_n e_n$, $\lambda_n \in K, J$ a finite subset of $\mathbb{N} \setminus \{0\}$, one has $\alpha \max_{n\in J} |\lambda_n| \|e_n\| \leq \|x\| \leq \max_{n\in J} |\lambda_n| \|e_n\|$. Furthermore for any sequence $(\lambda_n)_{n\geq 1} \subset K$ such that $\lim_{n\to+\infty} |\lambda_n| \|e_n\| = 0$, the series $x = \sum_{n\geq 1} \lambda_n e_n$ converges in E and $\alpha \max_{n\geq 1} |\lambda_n| \|e_n\| \leq \|x\| \leq \max_{n\geq 1} |\lambda_n| \|e_n\|$. The closure $\overline{Span}[e_n, n \geq 1]$ in E of the subspace spanned by the sequence $(e_n)_{n\geq 1}$ is equal to the space of elements x of E which can be written in the (unique) form of convergent sum $x = \sum_{n\geq 1} \lambda_n e_n$. One then says that $(e_n)_{n\geq 1}$ is an α-orthogonal basis of $\overline{Span}[e_n, n \geq 1]$.

One says that an ultrametric Banach space E is of *countable type*, if it is the closure of a subspace V which has a countable algebraic basis.

PROPOSITION 1.2. *Let α be a real number such that $0 < \alpha < 1$.*
Any ultrametric Banach space E of countable type has an α-orthogonal basis.

PROOF. See [**13**] or [**16**].

When it is not specified the Banach spaces need not to be of countable type.

Now, let us consider on the tensor product $E \otimes F$ of the ultrametric Banach spaces E and F the ultrametric (semi)-norm $\|z\| = \|z\|_\otimes = \inf_{\sum x_i \otimes y_i = z} \max_i \|x_i\| \|y_i\|$. The completion $E \widehat{\otimes} F$ of the normed $(E \otimes F, \| \|_\otimes)$ is called the *topological tensor product* of the Banach spaces E and F. □

PROPOSITION 1.3. *Let α be a real number such that $0 < \alpha < 1$ and let $z \in E\widehat{\otimes}F$, there exists a countable α-orthogonal family $(e_n)_{n\geq 1} \subset E$ such that $z = \sum_{n\geq 1} e_n \otimes y_n$, with $y_n \in F$ and $\lim_{n\to+\infty} \|e_n\| \|y_n\| = 0$.*

Furthermore $\alpha \sup_{n\geq 1} \|e_n\| \|y_n\| \leq \|z\| \leq \sup_{n\geq 1} \|e_n\| \|y_n\|$.

PROOF. Let $z = \sum_{\nu=1}^{m} x_\nu \otimes \zeta_\nu \in E \otimes F$ and let V be the subspace of E spanned by the finite sequence $(x_\nu)_{1\le\nu\le m}$. Then, since V is of finite dimension $q \le m$, it is of countable type and for $0<\alpha<1$, it admits an α-orthogonal basis $(e_n)_{1\le n\le q}$. It follows that $x_\nu = \sum_{n=1}^{q} \lambda_{n\nu} e_n$, with $\alpha \max_{1\le n\le q} |\lambda_{n\nu}|\|e_n\| \le \|x_\nu\| \le \max_{1\le n\le q} |\lambda_{n\nu}|\|e_n\|$. Furthermore $z = \sum_{\nu=1}^{m}\sum_{n=1}^{q} \lambda_{n\nu} e_n \otimes \zeta_\nu = \sum_{n=1}^{q} e_n \otimes y_n$ where $y_n = \sum_{\nu=1}^{m} \lambda_{n\nu}\zeta_\nu$ and $\|z\| \le \max_{1\le n\le q} \|e_n\|\|y_n\|$. On the other hand let $e'_n : V \to K$ be the linear form defined for $x = \sum_{n=1}^{q} \lambda_n e_n \in V$, by setting $< e'_n, x >= \lambda_n$. Then $\frac{1}{\|e_n\|} = \frac{|< e'_n, e_n >|}{\|e_n\|} \le \|e'_n\|$. Since $\alpha \max_{\le 1\le n\le q} |\lambda_n|\|e_n\| \le \|x\|$, one has $| < e'_n, x > | = |\lambda_n| = |\lambda_n|\|e_n\|\frac{1}{\|e_n\|} \le \frac{\|x\|}{\alpha}\cdot\frac{1}{\|e_n\|}$. Hence $\|e'_n\| \le \frac{1}{\alpha}\frac{1}{\|e_n\|}$, and $\alpha\|e'_n\| \le \frac{1}{\|e_n\|} \le \|e'_n\|$.
Let us consider the linear operator $e'_n \otimes 1_F : V \otimes F \to F$ such that $(e'_n \otimes 1_F)(x\otimes y) = \,=< e'_n, x > y$. One has $(e'_n \otimes 1_F)(z) = y_n$. Therefore $\|y_n\| = \|(e'_n \otimes 1_F)(z)\| \le \|e'_n \otimes 1_F\|\|z\| = \|e'_n\|\|z\|$. It follows that $\|e_n\|\|y_n\| \le \|e_n\|\|e'_n\|\|z\| \le \frac{1}{\alpha}\|z\|$ and $\alpha \max_{1\le n\le q} \|e_n\|\|y_n\| \le \|z\| \le \max_{1\le n\le q} \|e_n\|\|y_n\|$.

Any $z \in E\widehat{\otimes}F$ is a limit of a sequence of elements $z_m \in E \otimes F$. Selecting sums of the form $z_m = \sum_{\nu_m=1}^{\ell_m} x_{\nu_m} \otimes \zeta_{\nu_m} \in E \otimes F$, one obtains a countable subset $A = \bigcup_{m\ge1} \{x_{\nu_m},\ 1 \le \nu_m \le \ell_m\}$ of E. The closed subspace W of E spanned by A admits an α-orthogonal basis $(e_n)_{\ge1}$ such that for any $m \ge 1$ there exists q_m such that z_m belongs to the finite dimensional subpace $V_{q_m} = Span[e_1, e_2, \cdots, e_{q_m}]$ of W. Therefore there exists $y_n \in F, 1 \le n \le q_m$ such that $z_m = \sum_{n=1}^{q_m} e_n \otimes y_n$, with $\alpha \max_{1\le n\le q_m} \|e_n\|\|y_n\| \le \|z\| \le \max_{1\le n\le q_m} \|e_n\|\|y_n\|$. Since $z = \lim_{m\to+\infty} z_m$, one concludes the proof by completion. □

REMARK 1.4. In the same way for $z \in E\widehat{\otimes}F$, there exists a countable α-orthogonal family $(f_n)_{n\ge1} \subset F$ such that $z = \sum_{n\ge1} x_n \otimes f_n$, where $x_n \in E$ and $\lim_{n\to+\infty} \|x_n\|\|f_n\| = 0$.

Moreover $\alpha \sup_{n\ge1} \|x_n\|\|f_n\| \le \|z\| \le \sup_{n\ge1} \|x_n\|\|f_n\|$

Let E' be the continuous dual of E, that is the space of continuous linear forms $x' : E \longrightarrow K$. Considering on E' the norm $\|x'\| = \sup_{0\ne x\in E} \frac{|< x', x >|}{\|x\|}$, one sees that E' is a Banach space, Let $\mathcal{L}_f(E, F)$ be the space of continuous linear operators of E in F of finite rank. One has an identification of $\mathcal{L}_f(E, F)$ with the tensor product

$E' \otimes F$. Indeed, one associates to $x' \in E'$ and $y \in F$ the continuous linear map of rank 1 by setting $\psi(x' \otimes y)(x) = < x', x > y$, for $x \in E$.

One immediately sees that $\|\psi(x' \otimes y)\| = \|x'\|\|y\|$. And one obtains by linearity, a continuous linear map, again noticed ψ of $E' \otimes F$ in $\mathcal{L}(E, F)$ with range included in $\mathcal{L}_f(E, F)$. On the other hand, if $u \in \mathcal{L}_f(E, F)$, then $u(E)$ is a subspace of F of finite dimension m. Let $(f_j)_{1 \le j \le m}$ be a given basis of $u(E)$ and its dual basis $(f'_j)_{1 \le j \le m}$. For any $x \in E$, one has $u(x) = \sum_{j=1}^{m} \lambda_j(x) f_j$ and for $1 \le j \le m$ each map $\lambda_j : x \longrightarrow \lambda_j(x) = < f'_j, u(x) >$ is a continuous linear form. Hence, considering $v = \sum_{j=1}^{m} \lambda_j \otimes f_j \in E' \otimes F$, one has $\psi(v) = u$.

Assume now $(f_j)_{1 \le j \le m}$ to be an α-orthogonal basis of the finite dimensional space $u(E)$. For any $x \in E$, one has $u(x) = \sum_{j=1}^{m} \lambda_j(x) f_j$, with $\alpha \max_{1 \le j \le m} |\lambda_j(x)|\|f_j\| \le \|u(x)\| \le \max_{1 \le j \le m} |\lambda_j(x)|\|f_j\|$.

Hence $\alpha \max_{1 \le j \le m} \|\lambda_j\|\|f_j\| \le \|u\| \le \max_{1 \le j \le m} \|\lambda_j\|\|f_j\|$.

On the other hand, the element v of $E' \otimes F$ is such that $\alpha \max_{1 \le j \le m} \|\lambda_j\|\|f_j\| \le \|v\|_{\otimes} \le \max_{1 \le j \le m} \|\lambda_j\|\|f_j\|$.

It follows that $\alpha\|u\| \le \|v\|_{\otimes} \le \frac{1}{\alpha}\|u\|$, $\forall \alpha,\ 0 < \alpha < 1$; then $\|v\|_{\otimes} = \|u\|$.

Therefore the map ψ is an isometric linear map of $E' \otimes F$ onto $\mathcal{L}_f(E, F)$.

From this observation one deduces the following proposition.

PROPOSITION 1.5. *Let E and F be two ultrametric Banach spaces. The topological tensor product $E' \widehat{\otimes} F$ is isometrically isomorphic to the Banach space $\mathcal{C}(E, F)$ of the completely continuous linear operators of E in F. For $u \in \mathcal{C}(E, F)$ and $0 < \alpha < 1$, there exists an α-orthogonal sequence $(f_n)_{n \ge 1} \subset F$ such that $u = \sum_{n \ge 1} x'_n \otimes f_n$, where $x'_n \in E'$ and $\lim_{n \to +\infty} \|x'_n\|\|f_n\| = 0$.*

Moreover $\alpha \sup_{n \ge 1} \|x'_n\|\|f_n\| \le \|u\| \le \sup_{n \ge 1} \|x'_n\|\|f_n\|$.

2. Ultrametric Calkin algebras

When the Banach space E is of finite dimension m, then the algebra $\mathcal{L}(E)$ of continuous linear endomorphisms of E is isomorphic to the algebra of $m \times m$-square matrices $\mathrm{Mat}_m(K)$. It is well known that the only two-sided ideal of $\mathrm{Mat}_m(K)$ distinct from $\mathrm{Mat}_m(K)$ is $\{0\}$. In other words, $\mathrm{Mat}_m(K)$ is a simple algebra.

The situation is more contrasted when the space E is of infinite dimension and the two-sided ideal structure of $\mathcal{L}(E)$ may be difficult to handle.

Henceforth if not specified, the Banach spaces are supposed to be infinite dimensional.

We have noticed above that if the Banach space E is infinite dimensional then the space $\mathcal{C}(E)$ of completely continuous linear endomorphisms of E is a closed two-sided ideal of $\mathcal{L}(E)$ and is isometrically isomorphic to the topological tensor product $E' \widehat{\otimes} E$. Observing that if the ground field K is **not** spherically complete, there exits K-Banach space E whose continuous dual $E' = \{0\}$, an example is

$\ell^\infty(\mathbb{N}, K)/c_0(\mathbb{N}, K)$ (see for instance [**16**]), then for such a Banach space E, one has $\mathcal{C}(E) = \{0\}$ and for any ultrametric K-Banach space F, one also has $\mathcal{C}(E, F) = \{0\}$.

Let us suppose that the complete valued field K is of discrete valuation. In this case, it is well known, see for instance [**18**], that any ultrametric K-Banach space E, such that $\|E\| \subset |K|$, has an orthonormal basis $(e_i)_{i\in I}$, that means that any $x \in E$ can be written in a unique form of summable family $x = \sum_{i\in I} \lambda_i e_i$, where $\lambda_i \in K$ and $\|x\| = \sup_{i\in I} |\lambda_i|$. Notice that when the complete field K is of discrete valuation the norm $\| \ \|$ of any Banach space E is equivalent to a norm $\| \ \|_0$ such that $\|E\|_0 \subset |K|$. Any closed subspace V of of E admits an orthogonal complement. In other words, there exists a continuous linear projection P of E onto V with norm $\|P\| = 1$. If V is a closed subspace of E, one puts $\dim_t V$ to be the *cardinal* of any orthonormal basis of V. If κ is a cardinal $\leq \dim_t(E)$, one sets $\mathcal{C}_\kappa(E)$ to be the smallest closed two-sided ideal of $\mathcal{L}(E)$ that contains the continuous projections P such that $\dim_t P(E) < \kappa$. If $\kappa = \aleph_0$ the cardinal of a countable set, then $\mathcal{C}_{\aleph_0}(E) = \mathcal{C}(E)$. With this, M. van der Put has proved in [**15**] the following ultrametric counterpart of the closed two-sided ideal structure of the Banach algebra of bounded operators on complex or real Hilbert spaces obtained in [**6**] and [**10**].

THEOREM 2.1. *Assume that K is a complete discrete valuation field. Let E be an ultrametric K-Banach space. The map, which associates to any cardinal $\kappa \leq \dim_t P(E)$ the closed two-sided ideal $\mathcal{C}_\kappa(E)$, is an increasing bijection of $\aleph_0 \leq \kappa \leq \dim_t E$ onto the set of non trivial closed two-sided ideals of $\mathcal{L}(E)$, ordered by inclusion.* □

DEFINITION 2.2. Assume that the complete valued field K is of discrete valuation. For $\aleph_0 \leq \kappa \leq \dim_t E$, let $\mathcal{C}_\kappa$ be the closed two-sided ideal of $\mathcal{L}(E)$ corresponding to κ. The quotient algebra $\mathfrak{B}_\kappa(E) = \mathcal{L}(E)/\mathcal{C}_\kappa(E)$, with the quotient norm is a Banach algebra, called the κ-Calkin algebra. of E □

If $\aleph_0 \leq \kappa \leq \kappa' \leq \dim_t E$, one has the canonical surjective algebra homomorphisms $\phi_\kappa : \mathcal{L}(E) \longrightarrow \mathfrak{B}_\kappa(E)$ and $\phi_{\kappa',\kappa} : \mathfrak{B}_\kappa(E) \longrightarrow \mathfrak{B}_{\kappa'}(E)$ such that $\phi_{\kappa',\kappa} \circ \phi_\kappa = \phi_{\kappa'}$. Considering the *opposite* order of the set of cardinals $[\aleph_0, \dim_t E] = \mathcal{J}(E)$, one obtains the projective system $(\mathfrak{B}_\kappa(E), \phi_{\kappa',\kappa})$ in the category of unitary Banach algebras, the morphisms having norms ≤ 1. Then a natural problem is to describe the relationship between $\mathcal{L}(E)$ and the projective limit $\varprojlim_n(\mathfrak{B}_\kappa(E), \phi_{\kappa',\kappa})$ in the category of unitary Banach algebras. In fact, since $\aleph_0$ is the greatest element of $\mathcal{J}(E)$, one sees that the projection of $\varprojlim_n(\mathfrak{B}_\kappa(E), \phi_{\kappa',\kappa})$ onto $\mathfrak{B}_{\kappa_0}(E)$ is an isomorphism.

From the theorem above, one deduces the next corollary.

COROLLARY 2.3. *Assume that K is a complete valued field of discrete valuation. If the K-Banach space is of* countable type, *then the closed two-sided ideal $\mathcal{C}(E)$ is the only proper closed two-sided ideal of $\mathcal{L}(E)$*

One can drop in this corollary the condition that K is of discrete valuation.

LEMMA 2.4. *Let K be any ultrametric complete rank one valued field. Let E be an ultrametric K-Banach space of countable type. Any closed subspace D of E*

has a topological complement in E. More precisely, for any $\varepsilon > 0$, there exists a continuous linear projection P of E onto D with $\|P\| < 1+\varepsilon$.

PROOF. See [**13**]-Theorem 2.3.13. □

Let us notice that a consequence of this lemma is that any closed subspace of a Banach space of countable type is also of countable type.

PROPOSITION 2.5. *Let K be any complete rank one valued field. Let E be an ultrametric K-Banach space of countable type. If $T \in \mathcal{L}(E) \setminus \mathcal{C}(E)$. Then the two-sided ideal of $\mathcal{L}(E)$ generated by T is equal to $\mathcal{L}(E)$.*

PROOF. It runs as in Calkin paper [**2**].

First, let us remind that u is a completely continuous linear map if and only if each closed subspace contained in $u(E)$ is finite dimensional (cf. [**16**]- Theorem 4.40, γ). Then, it follows that if T does not belong to $\mathcal{C}(E)$, then $T(E)$ contains an infinite dimensional subspace D closed in E.

Let $V = T^{-1}(\{0\})$ be the kernel of T which then is closed in E. According to the above lemma, giving $\varepsilon > 0$, there exists a continuous linear projection P of E onto V such that $\|P\| < 1+\varepsilon$. Let $Q = id_E - P$ and $W = Q(E)$, one has $E = V \oplus W$ and any x in E can be written in the unique form $x = x_1 + x_2$ with $(1+\varepsilon)^{-1}\max(\|x_1\|, \|x_2\|) \le \|x\| \le \max(\|x_1\|, \|x_2\|)$.

Furthermore, one has $T(x) = T(x_1) + T(x_2) = 0 + T(x_2) = T(x_2)$ and setting $T_1 = T_{|W}$ the restriction of T to W, one has an injective continuous linear map of W into E with $T_1(W) = T(E) \supseteq D$.

Let us set $F = T_1^{-1}(D) \subset W$. Since D is a closed subset of E and T_1 is continuous, one sees that F is a closed subspace of W and of E. Moreover $T(F) = T_1(F) = T_1(T_1^{-1}(D)) = D$.

Since E and F are infinite dimensional Banach space of countable type, they are isomorphic Banach spaces. Indeed, for $0 < \alpha < 1$, one can find an α-orthogonal basis $(e_n)_{n\ge1}$ in E (resp. $(f_n)_{n\ge1}$ in F) such that $\alpha < \|e_n\| \le 1$ (resp. $\alpha < \|f_n\| \le 1$). Then one defines a bijective linear map θ of E onto F by setting $\theta(\sum\limits_{n\ge1} a_n e_n) = \sum\limits_{n\ge1} a_n f_n$ with norm $\|\theta\| \le \alpha^{-2}$.

In the same way, one obtains a bijective continuous linear map ϖ of E onto D.

Set $S = \varpi^{-1} \circ T_1 \circ \theta$, one has $S(E) = \varpi^{-1} \circ T_1 \circ \theta(E) = \varpi^{-1} \circ T_1(F) = \varpi^{-1}(D) = E$, that is S is surjective. Since the maps $\varpi^{-1} \circ T_1$ and θ are injective, their composition $S = \varpi^{-1} \circ T_1 \circ \theta$ is also injective. It follows that S is an invertible element of the algebra $\mathcal{L}(E)$. Let i_F be the canonical injection of F into E and π_D be a continuous linear projection of E onto its closed subspace D, then $S = (\varpi^{-1} \circ i_F) \circ T \circ (\pi_D \circ \theta)$ is contained in the two-sided ideal of $\mathcal{L}(E)$ generated by T. But S is invertible, it follows that this two-sided ideal is equal to $\mathcal{L}(E)$. □

An obvious consequence of Proposition 2.4 is that if E is of countable type, then any proper two-sided ideal $\mathfrak{I}$ of $\mathcal{L}(E)$ is contained in $\mathcal{C}(E)$. On the other hand one can prove that every non zero two-sided ideal $\mathfrak{I}$ of $\mathcal{L}E)$ contains the two-sided ideal $\mathcal{L}_f(E)$ of the continuous linear operators of finite rank. With this all together one obtains :

COROLLARY 2.6. *If E is an ultrametric Banach space of countable type, then $\mathcal{C}(E)$ is the unique proper closed two-sided ideal of $\mathcal{L}(E)$.*

By applying Corollary 2.6, one has

PROPOSITION 2.7. *Let $\mathcal{A}$ be an ultrametric unitary Banach algebra. Assume that the ultrametric Banach space E is of countable type. If $\varphi : \mathcal{L}(E) \to \mathcal{A}$ is a continuous homomorphism of unitary Banach algebras, then or φ is injective, or ker $\varphi = \mathcal{C}(E)$.*

PROOF. Indeed if $\varphi : \mathcal{L}(E) \to \mathcal{A}$ is a continuous homomorphism of unitary Banach algebras, one has $\varphi(id_E) = 1_{\mathcal{A}}$, the unity of $\mathcal{A}$. However kerφ is a closed two-sided ideal of $\mathcal{L}(E)$ which does not contain id_E and therefore is different from $\mathcal{L}(E)$. Hence or ker$\varphi = \{0\}$ or ker$\varphi = \mathcal{C}(E)$. □

In his paper loc.cit., on classification of two-sided ideals of the Banach algebra of bounded linear operators on a separable complex Hilbert space, J.W. Calkin has given a continuous linear representation of the quotient Banach algebra $\mathcal{L}(H)/\mathcal{C}(H)$ in an algebra of bounded linear operators of a Hilbert space that he constructed by using *Banach limits* on the complex Banach space of bounded sequences. Banach limits are linear forms that extend the linear form lim of the limit of convergent sequences. Unlike to use general Banach limits, we have studied in [**3**] the ultrametric ultraproducts of complete fields and of some ultrametric Banach spaces. Ultrametric ultraproducts are constructed on base of ultrafilters.

3. Ultrametric ultraproducts of Banach spaces

Let $(E_i)_{i\in I}$ be a family of ultrametric K-Banach spaces Let us consider the K-normed space $\overline{\prod}_{i\in I} E_i = \{x = (x_i)_{i\in I} \ : \ \|x\|_\infty = \sup_{i\in I} \|x_i\| < +\infty\}$. It is readily seen and well known that $\overline{\prod}_{i\in I} E_i$ endowed with the norm $\| \ \|_\infty$ is complete.

Let $\mathcal{U}$ be an ultrafilter on I. For each $x = (x_i)_{i\in I} \in \overline{\prod}_{i\in I} E_i$, the family of real numbers $(\|x_i\|)_{i\in I}$ being bounded, one sees that the limit $\|x\| = \lim_{\mathcal{U}} \|x_i\|$ exists and one defines an ultrametric semi-norm on $\overline{\prod}_{i\in I} E_i$ such that $\|x\| \le \|x\|_\infty$. The subset $\mathcal{V}$ of the elements x of $\overline{\prod}_{i\in I} E_i$ such that $\|x\| = \lim_{\mathcal{U}} \|x_i\| = 0$ is a $\| \ \|_\infty$-closed subspace. Let us write again x the class of $x = (x_i)_{i\in I}$ modulo $\mathcal{V}$. On the quotient space $\overline{\prod}_{i\in I} E_i/\mathcal{V}$ one has the norms $\|x\|_\sim = \inf_{y\in\mathcal{V}} \|x - y\|_\infty$ and $\|x\| = \lim_{\mathcal{U}} \|x_i\| = \inf_{J\in\mathcal{U}} \sup_{i\in J} \|x_i\|$. One verifies immediately that these two norms are equal, because $\|x\| = \inf_{J\in\mathcal{U}} \sup_{i\in J} \|x_i\| = \inf_{y\in\mathcal{V}} \|x - y\|_\infty$. As a consequence, with respect to the norm $\| \ \|$, the quotient space $\overline{\prod}_{i\in I} E_i/\mathcal{V}$ is complete.

One says that the space $\overline{\prod}_{i\in I} E_i/\mathcal{V}$ is the *ultrametric ultraproduct* with respect to the ultrafilter $\mathcal{U}$ of the family of Banach spaces $(E_i)_{i\in I}$.

We put $\overline{\prod}_{i\in I} E_i/\mathcal{V} = \overline{\prod}_{i\in I} E_i/\mathcal{U}$.

An ultrafilter $\mathcal{U}$, on the set I is said to be *countably-incomplete* if there exists a sequence $(X_n)_{n\ge1} \subset \mathcal{U}$ such that $X_{n+1} \subset X_n$ and $\bigcap_{n\ge1} X_n = \emptyset$. If the set I is countable, every non-principal ultrafilter is countably-incomplete.

PROPOSITION 3.1. *The ultrametric ultraproduct* $\overline{\prod}_{i\in I} E_i/\mathcal{U}$ *is a* K*-Banach space.*

Moreover, if the ultrafilter is countably-incomplete, then the K*-vector space* $\overline{\prod}_{i\in I} E_i/\mathcal{U}$ *is spherically complete*

PROOF. A K-ultrametric normed E is said to be spherically complete, if any decreasing sequence of closed discs of E has a non empty intersection. For the spherical completeness of $\overline{\prod}_{i\in I} E_i/\mathcal{U}$ see the proof of Proposition 1 in [**3**]. $\square$

When $E_i = E, \forall i \in I$, one has $\overline{\prod}_{i\in I} E = \ell^\infty(I,E)$ and $\ell^\infty(I,E)/\mathcal{U}$ is called the *ultrapower* of E.

REMARK 3.2. As we know (cf. [**3**]), if the $E_i = K_i$ are complete valued fields, then $\overline{\prod}_{i\in I} K_i/\mathcal{U}$ is a compete valued field.. In particular the ultrapower $\ell^\infty(I,K)/\mathcal{U}$ is a complete valued field.

Let us set for any $\lambda = (\lambda_i)_{i\in I} \in \ell^\infty(I,K)/\mathcal{U}$ and $x = (x_i)_{i\in I} \in \overline{\prod}_{i\in I} E_i/\mathcal{U}$: $\lambda x = (\lambda_i x_i)_{i\in I}$, then $\overline{\prod}_{i\in I} E_i/\mathcal{U}$ becomes a $\ell^\infty(I,K)/\mathcal{U}$-Banach space.

Notice that if the fields K_i are distinct and E_i is an ultrametric K_i-Banach space, then one has in the same way a structure of $\overline{\prod}_{i\in I} K_i/\mathcal{U}$-Banach space on $\overline{\prod}_{i\in I} E_i/\mathcal{U}$. $\square$

Let $(E_i)_{i\in I}$ and $(F_i)_{i\in I}$ be two families of ultrametric K-Banach spaces and a bounded family $(u_i)_{i\in I}$ of continuous linear maps $u_i : E_i \longrightarrow F_i$, that is $\sup_{i\in I} \|u_i\| < +\infty$.

For $x = (x_i)_{i\in I} \in \overline{\prod}_{i\in I} E_i$, put $u(x) = (u_i(x_i))_{i\in I}$, one immediately sees that $u(x)$ belongs to $\overline{\prod}_{i\in I} E_i$, with $\|u_i(x_i)\| \le (\sup_{i\in I} \|u_i\|)\cdot \|x_i\|$ and $\lim_{\mathcal{U}} \|x_i\| = 0 \Longrightarrow \lim_{\mathcal{U}} \|u_i(x_i)\| = 0$. From what, one deduces a continuous linear map, again noticed u of $\overline{\prod}_{i\in I} E_i/\mathcal{U}$ into $\overline{\prod}_{i\in I} F_i/\mathcal{U}$.

PROPOSITION 3.3. *The continuous* K*-linear map* u *of* $\overline{\prod}_{i\in I} E_i/\mathcal{U}$ *into* $\overline{\prod}_{i\in I} F_i/\mathcal{U}$ *associated to the bounded family* $(u_i)_{i\in I}$ *of continuous linear maps* $u_i : E_i \longrightarrow F_i$, *is in fact a continuous linear map of* $\ell^\infty(I,K)/\mathcal{U}$*-Banach spaces.*

Moreover, one has $\|u\| = \lim_{\mathcal{U}} \|u_i\|$.

PROOF. The fact that $\|u\| = \lim_{\mathcal{U}} \|u_i\|$ is well known for ultraproducts of real or complex Banach spaces. (cf. [**9**]). Indeed, since $\sup_{i\in I} \|u_i\| < +\infty$, the limit $\lim_{\mathcal{U}} \|u_i\|$ exists. But for $x = (x_i)_{i\in I}$, one has $\|u_i(x_i)\| \le \|u_i\|\|x_i\|, \forall i \in I \Longrightarrow \lim_{\mathcal{U}} \|u_i(x_i)\| \le \lim_{\mathcal{U}} \|u_i\| \cdot \lim_{\mathcal{U}} \|x_i\|$, that is $\|u(x)\| \le \lim_{\mathcal{U}} \|u_i\|\|x\|$. Then $\|u\| \le \lim_{\mathcal{U}} \|u_i\|$. On the other hand, for $\varepsilon > 0$, there exists $J_\varepsilon \in \mathcal{U}$, such that $\dfrac{\|u_i(x_i)\|}{\|x_i\|} - \varepsilon \le \|u_i\| - \varepsilon \le \lim_{\mathcal{U}} \|u_i\|$, $\forall i \in J_\varepsilon$ and $\|x_i\| \ne 0$. Hence $\|u_i(x_i)\| - \varepsilon\|x_i\| \le (\lim_{\mathcal{U}} \|u_i\|)\cdot \|x_i\|$. Taking limits, one has $\|u(x)\| - \varepsilon\|x| \leq \lim_{\mathcal{U}} \|u_i\| \cdot \|x\|$. It follows that $\|u\| - \varepsilon \leq \lim_{\mathcal{U}} \|u_i\|$ and $\|u\| \leq \lim_{\mathcal{U}} \|u_i\|$.

Let $\lambda = (\lambda_i)_{i\in I} \in \ell^\infty(I, K)/\mathcal{U}$ and $(x_i)_{i\in I} \in \overline{\prod}_{i\in I} E_i/\mathcal{U}$, then one has $u(\lambda x) = (u_i(\lambda_i x_i))_{i\in I} = (\lambda_i \cdot u_i(x_i))_{i\in I} = ((\lambda_i)_{i\in I}) \cdot (u(x_i))_{i\in I}) = \lambda u(x)$. □

Notice that the proposition means that one has a natural linear isometric map of the ultrametric ultraproduct $\overline{\prod}_{i\in I} \mathcal{L}(E_i, F_i)/\mathcal{U}$ into $\mathcal{L}\left(\overline{\prod}_{i\in I} E_i/\mathcal{U}, \overline{\prod}_{i\in I} F_i/\mathcal{U}\right)$.

COROLLARY 3.4. *Let E be an ultrametric Banach space. There exists a canonical isometric unitary Banach algebra homomorphism of $\mathcal{L}(E)$ into $\mathcal{L}(\ell^\infty(I, E)/\mathcal{U})$.*

PROOF. Let $E_i = F_i = E$ and $u_i = u \in \mathcal{L}(E)$, $\forall i \in I$. As above one defines a continuous linear endomorphism $\varphi(u)$ of $\ell^\infty(I, E)/\mathcal{U}$ by setting $\varphi(u)(x) = (u(x_i))_{\in I}$, for $x = (x_i)_{i\in I}$. Then one has a linear map of $\mathcal{L}(E)$ into $\mathcal{L}(\ell^\infty(I, E)/\mathcal{U})$.
It is readily seen that $\varphi(id_E) = id_\infty$, $\varphi(u \circ v) = \varphi(u) \circ \varphi(v)$ with $\|\varphi(u)\| = \|u\|$. □

One has the canonical isometric injections: $\mathcal{L}(E) \longrightarrow \ell^\infty(I, \mathcal{L}(E))/\mathcal{U}$ and $\ell^\infty(I, \mathcal{L}(E))/\mathcal{U} \longrightarrow \mathcal{L}(\ell^\infty(I, E)/\mathcal{U})$.

Let $\ell_0^\infty(I, E)$ be the subspace of $\ell^\infty(I, E)$ of the bounded families $(x_i)_{i\in I}$ such that for any continuous linear form x' on E, one has $\lim_{\mathcal{F}} | < x', x_i > | = 0$ (the limit is with respect to the Fréchet filter $\mathcal{F}$ of I; that is the filter of the complements of the finite subsets of I). Any non-principal ultrafilter $\mathcal{U}$ of I contains $\mathcal{F}$.

The space $\ell_0^\infty(I, E)$ is a closed subspace of $\ell^\infty(I, E)$. Indeed, let $x = (x_i)_{i\in I}$ be in the closure of $\ell_0^\infty(I, E)$ and $(x_n)_{n\geq 1} = (x_{i,n})_{i,n}$ a sequence that converges towards x. For $\varepsilon > 0$, there exists n_ε such that $\|x_i - x_{i,n_\varepsilon}\| \leq \|x - x_{n_\varepsilon}\|_\infty < \varepsilon$. Hence for $x' \in E'$, one has $| < x', x_i > | \leq \max(| < x', x_i - x_{i,n_\varepsilon} > |, | < x', x_{i,n_\varepsilon} > |) \leq \max(\|x'\|\varepsilon, | < x', x_{i,n_\varepsilon} > |)$. It follows that $\lim_{\mathcal{F}} | < x', x_i > | = 0$ and x belongs to $\ell_0^\infty(I, E)$. The subspace $\mathcal{V}_0 = \mathcal{V} \bigcap \ell_0^\infty(I, E)$ is closed in $\ell_0^\infty(I, E)$ and the quotient space $\ell_0^\infty(I, E)/\mathcal{V}_0$ can be identified with a closed subspace $\ell_0^\infty(I, E)/\mathcal{U}$ of $\ell^\infty(I, E)/\mathcal{U}$. □.

4. A linear representation of the Calkin algebra when the Banach space E is infinite dimensional of countable type

For any complete utrametric valued field K, let E be an infinite dimensional ultrametric K-Banach space of countable type. We have seen that the closed two-sided ideal $\mathcal{C}(E)$ is the unique proper closed two-sided ideal of $\mathcal{L}(E)$ and it is the greatest proper two-sided ideal. We are now interested to the linear representation of $\mathfrak{B}_{\aleph_0}(E) = \mathfrak{B}(E) = \mathcal{L}(E)/\mathcal{C}(E)$ on appropriate Banach spaces.

For this purpose let $\mathcal{U}$ be a non-principal ultrafilter on the set $\mathbb{N}^\star$ of positive integers. With such ultrafilter is associated the Banach space $\ell^\infty(\mathbb{N}^\star, E)/\mathcal{U}$ an ultrapower of E and its closed subspace $\ell_0^\infty(\mathbb{N}^\star, E)/\mathcal{U}$.

THEOREM 4.1. *Let E be an infinite dimensional ultrametric Banach space of countable type.*
There exists a continuous injective unitary algebra homomorphism of the Calkin algebra $\mathfrak{B}(E)$ into the algebra of continuous linear operators $\mathcal{L}(\ell_0^\infty(\mathbb{N}^\star, E)/\mathcal{U})$

PROOF. -(i)- We have encountered above the map φ which associates to $u \in \mathcal{L}(E)$ the linear endomorphism $\varphi(u)$ of $\ell^\infty(\mathbb{N}^\star, E)$ such $\varphi(u)((x_i)_{i\geq 1}) = (u(x_i))_{i\geq 1}$. This map giving rise to an isometrical homomorphism of unitary algebras φ of $\mathcal{L}(E)$ into $\mathcal{L}(\ell^\infty(\mathbb{N}^\star, E)/\mathcal{U})$.

-(ii)- Now, let $x = (x_i)_{i\geq 1}$ be an element of $\ell_0^\infty(\mathbb{N}^\star, E)$. For any $u \in \mathcal{L}(E)$ and any continuous linear form $x' \in E'$, one has $\lim\limits_{i\to+\infty} < x', u(x_i) >= \lim\limits_{i\to+\infty} < {}^t u(x'), x_i >= 0$. It follows that $\varphi(u)(x) = (u(x_i))_{i\in I}$ belongs to $\ell_0^\infty(\mathbb{N}^\star, E)$.

Therefore, the restriction $\varphi_0(u)$ of $\varphi(u)$ to $\ell_0^\infty(\mathbb{N}^\star, E)$ is also a continuous linear endomorphism of $\ell_0^\infty(\mathbb{N}^\star, E)$.

-(iii)- Let $x = (x_i)_{i\geq 1}$ be an element of $\mathcal{V}$ and $u \in \mathcal{L}(E)$, since $\|u(x_i)\| \leq \|u\|\|x_i\|$, one has $\lim\limits_{\mathcal{U}} \|u(x_i)\| = 0$. Hence the closed subspace $\mathcal{V}$ of $\ell^\infty(\mathbb{N}^\star, E)$ is stable by $\varphi(u)$ for any element u of $\mathcal{L}(E)$. It follows that $\mathcal{V}_0 = \mathcal{V} \bigcap \ell_0^\infty(\mathbb{N}^\star, E)$ is stable by $\varphi_0(u)$. Taking quotient, one defines a continuous linear endomorphism again noticed $\varphi_0(u)$ of $\ell_0^\infty(\mathbb{N}^\star, E)/\mathcal{V}_0 = \ell_0^\infty(\mathbb{N}^\star, E)/\mathcal{U}$. Then one obtains the continuous unitary algebra homomorphism $\varphi_0 : \mathcal{L}(E) \longrightarrow \mathcal{L}(\ell_0^\infty(\mathbb{N}^\star, E)/\mathcal{U})$ such that $\|\varphi_0(u)\| \leq \|u\|$.

-(iv)- We will show that if u belongs to $\mathcal{C}(E)$, then for any element $x = (x_i)_{i\geq 1}$ of $\ell_0^\infty(\mathbb{N}^\star, E)$, one has $\varphi_0(u)(x) = (u(x_i))_{i\geq 1} \in \mathcal{V}_0$.

For this, let us remind that if u is an element of $\mathcal{C}(E)$, then for a real number $0 < \alpha < 1$, there exists an α-orthogonal sequence $(e_n)_{n\geq 1}$ in E and a sequence $(x'_n)_{n\geq 1} \subset E'$ such that $u = \sum\limits_{n\geq 1} x'_n \otimes e_n$, $\lim\limits_{n\to+\infty} \|x'_n\|\|e_n\| = 0$ and $\alpha \sup\limits_{n\geq 1} \|x'_n\|\|e_n\| \leq \|u\| \leq \sup\limits_{n\geq 1} \|x'_n\|\|e_n\|$.

For any real number $\varepsilon > 0$, there exists n_ε such that $\|u - \sum\limits_{n=1}^{n_\varepsilon} x'_n \otimes e_n\| = \|\sum\limits_{n>n_\varepsilon} x'_n \otimes e_n\| < \varepsilon$.

Now, let $x = (x_i)_{i\geq 1} \in \ell_0^\infty(\mathbb{N}^\star, E)$, since $\|u(x_i) - \sum\limits_{n=1}^{n_\varepsilon} < x'_n, x_i > e_n\| \leq \|\sum\limits_{n>n_\varepsilon} < x'_n, x_i > e_n\| < \|x_i\|\varepsilon \leq \|x\|_\infty \varepsilon$, one has

$$\|u(x_i)\| \leq \max(\|\sum_{n=1}^{n_\varepsilon} < x'_n, x_i > e_n\|, \|x\|_\infty \varepsilon) \leq$$

$$\max\Big(\max_{1\leq n\leq n_\varepsilon} | < x'_i, e_n > |\|e_n\|, \|x\|_\infty \varepsilon\Big) \leq \max\Big(\max_{1\leq n\leq n_\varepsilon} | < x'_i, e_n > |, \|x\|_\infty \varepsilon\Big).$$

However, for $1 \leq n \leq n_\varepsilon$, one has $\lim\limits_{i\to+\infty} | < x'_n, x_i > | = 0$. Then for $u \in \mathcal{C}(E)$ and $x = (x_i)_{i\geq 1} \in \ell_0^\infty(\mathbb{N}^\star, E)$, one has $\lim\limits_{i\to+\infty} \|u(x_i)\| = 0$ and $\lim\limits_{\mathcal{U}} \|u(x_i)\| = 0$.

Hence, we have proved that $\varphi_0(u)(x) = (u(x_i))_{i\geq 1}$ belongs $\mathcal{V}_0$.

-(v)- By passing to the quotient modulo $\mathcal{V}_0$, one sees that for any $u \in \mathcal{C}(E)$ one has $\varphi_0(u) = 0$ as endomorphism of $\ell_0^\infty(\mathbb{N}^\star, E)/\mathcal{U}$. This means that $\mathcal{C}(E) \subset ker\varphi_0$.

On the other hand $\varphi_0(id_E) = id_\infty$. Therefore the closed two-sided ideal $\ker\varphi_0$ is different from $\mathcal{L}(E)$ and containing the unique closed two-sided $\mathcal{C}(E)$ it coincides with it. Again, passing to the quotient but modulo $\mathcal{C}(E) = \ker\varphi_0$, one obtains a unitary algebra homomorphism $\overline{\varphi}_0 : \mathfrak{B}(E) \longrightarrow \mathcal{L}(\ell_0^\infty(E, \mathbb{N}^\star)/\mathcal{U})$.

If $\overline{u} \in \mathfrak{B}(E)$ is the class of u, one has by definition $\bar{\varphi}_0(\overline{u}) = \varphi_0(u)$.

Hence $\bar{\varphi}_0(\overline{u}) = 0 \Longleftrightarrow \overline{u} = 0$; it follows that $\overline{\varphi}_0$ is injective.

Notice that

$$\|\varphi_0(u)\| = \sup \left\{ \frac{\lim_{\mathcal{U}} \|u(x_i)\|}{\lim\limits_{\mathcal{U}} \|x_i\|},\ 0 \neq x = (x_i)_{\geq 1} \in \ell_0^\infty(\mathbb{N}^\star, E)/\mathcal{U} \right\} \leq \|u\|.$$

For any $u \in \mathcal{L}(E)$ and $v \in \mathcal{C}(E)$, one has $\overline{\varphi}_0(\overline{u}) = \varphi_0(u - v)$ and $\|\overline{\varphi}_0(\overline{u})\| = \|\varphi_0(u - v)\| \leq \|u - v\|$. Hence $\|\overline{\varphi}_0(\overline{u})\| \leq \inf\limits_{u \in \mathcal{C}(E)} \|u - v\| = \|\overline{u}\|$, that is $\overline{\varphi}_0$ is continuous. □

REMARK 4.2. Let $\sigma(E, E')$ be the weak topology induced on E by the duality between E and E'. What is proved in -(iv)- is that if $(x_i)_{i \geq 1}$ is a sequence in E which converges towards 0 in the weak topology and if u is a completely continuous operator then the sequence $(u(x_i))_{i \geq 1}$ converges towards 0 in the norm topology. A well known result for compact operators in complex and real analysis. □

REMARK 4.3. Let K be a complete discrete valuation field, If E is an ultrametric K-Banach space, considering if necessary an equivalent norm, one can assume that $\|E\| \subset |K|$ and E admits an orthonormal basis $(e_i)_{i \in I}$. By definition $\mathrm{card}(I) = \dim_t(E) = d_t(E)$. Then according to M. van der Put theorem loc.cit, the two-sided ideal $\mathcal{C}_{d_t(E)}(E)$ is the greatest closed two-sided ideal of $\mathcal{L}(E)$. Considering I in place of $\mathbb{N}^\star$, with the arguments in the proof of Theorm 4.1, one can only conclude that $\ker\overline{\varphi}_0$ contains $\mathcal{C}(E)$. □

REMARK 4.4. -(a)- Let us remind that the so-called ulrametric Hilbert spaces have very few properties of complex or real Hilbert spaces. Let $\omega = (\omega_j)_{j \in I}$ be a family of elements of $K^\star = K \setminus \{0\}$. One sets $E_\omega = \{x = (a_j)_{j \in I} \subset K$ such that $\lim\limits_{I} |a_j||\omega_j|^{\frac{1}{2}} = 0\}$. One defines on E_ω the norm $\|x\| = \sup\limits_{j \in I} |a_j||\omega_j|^{\frac{1}{2}}$ and the scalar product $< x, y >_\omega = \sum\limits_{j \in I} a_j b_j \omega_j$ and one has $| < x, y >_\omega | \leq \|x\|\|y\|$.
In such an ulrametric Hilbert space E_ω, continuous linear endomorphisms may fail to possess an adjoint. We have given in [**4**] the description of the Banach algebra $\mathcal{L}_0(E_\omega) \subset \mathcal{L}(E_\omega)$ of continuous linear operators that admit an adjoint and $\mathcal{C}_0(E_\omega) = \mathcal{C}(E_\omega) \cap \mathcal{L}_0(E_\omega)$ is a closed two-sided ideal of $\mathcal{L}_0(E_\omega)$. One can define a Calkin algebra $\mathfrak{B}_0(E_\omega) = \mathcal{L}_0(E_\omega)/\mathcal{C}_0(E_\omega)$ which with the quotient of the operation of adjunction becomes an involutive ultrametric Banach algebra.

When E_ω is of countable type one obtains by restriction of the above representation of $\mathfrak{B}(E_\omega)$ a linear representation of $\mathfrak{B}_0(E_\omega)$.

Considering the subspace $\ell_{00}^\infty(\mathbb{N}^\star, E_\omega)$ of the elements $x = (x_i)_{i \geq 1} \in \ell^\infty(\mathbb{N}^\star, E_\omega)$ such that with respect to the scalar product one has $\lim\limits_{i \to +\infty} < z, x_i >_\omega = 0$, $\forall z \in E_\omega$, one obtains a representation of $\mathfrak{B}_0(E_\omega)$ in $\ell_{00}^\infty(\mathbb{N}^\star, E_\omega)/\mathcal{U}$.

-(b)- To get a bilinear form on an ultrapower $\ell^\infty(I, E_\omega)/\mathcal{U}$ from the bilinear form of an ulrametric Hilbert space E_ω, it is necessary to consider on $\ell^\infty(I, E_\omega)/\mathcal{U}$ the $\ell^\infty(I, K)/\mathcal{U}$-bilinear form defined for $x = (x_i)_{i \in I}$ and $y = (y_i)_{i \in I}$ two elements of $\ell^\infty(I, E_\omega)/\mathcal{U}$ by setting $< x, y > = < (x_i)_{i \in I}, (y_i)_{i \in I} > = (< x_i, y_i >_\omega)_{i \in I} \in \ell^\infty(I, K)/\mathcal{U}$, with $| < x, y > | \leq \|x\|\|y\|$.

Notice that if K is locally compact, then $\ell^\infty(I,K)/\mathcal{U} = K$. In this case, one has $< x, y >=< (x_i)_{i\in I}, (y_i)_{i\in I} >= \lim_{\mathcal{U}} < x_i, y_i >_\omega \in K$. □

5. Some additional remarks

REMARK 5.1. An ultrametric version of Atkinson theorem.

In what follows, we shall deal with an ultrametric analogue of a theorem of Atkison which states that an operator on a complex Banach space is a Fredholm operator if and only if it is invertible modulo the closed two-sided ideal of compact operators.

DEFINITION 5.2. Let E be an ultrametric Banach space.
A continuous linear endomorphism u of E is said to be a *Fredholm operator* if:
-(1)- ker(u) is finite dimensional
-(2)- the range $Im(u)$ of u is closed in E with finite codimension.

It is clear that any linear automorphism of E is a Fredholm operator. In other words setting Aut(E) equal the group of the linear automorphisms of E and $\mathcal{F}(E)$ the set of Fredholm operators of E, one has Aut$(E) \subset \mathcal{F}(E)$.

For Banach spaces, in the definition (2), $Im(u)$ closed is *implied* by the finite codimension condition.

In [**11**], C. Perez-Garcia has studied the relationship of (semi)-Fredholm operators with some elements of the Calkin algebra $\mathfrak{B}(E) = \mathcal{L}(E)/\mathcal{C}(E)$ (see also [**17**], where the ground field is assumed to be spherically complete).

Let us remind that an ultrametric Banach space is said to be *pseudoreflexive* if the canonical map $j_E : E \longrightarrow E''$ of E in its bidual is isometric. This definition is seen to be equivalent to the fact that the space E is a polar normed space [**12**]. Here, following the ideas of the proof of the classical Atkinson theorem (cf. [**1**] or [**8**]), we shall prove :

THEOREM 5.3. *Let E be an infinite dimensional pseudoreflexive ultrametric K-Banach space and $\mathfrak{B}(E)$ be the Calkin algebra of E.*

Then a continuous linear endomorphism of E is a Fredholm operator of E if and only if its canonical image in $\mathfrak{B}(E)$ is invertible.

PROOF. -(a)- First, assume that u is a Fredholm operator. Since ker(u) is a finite dimensional subspace of the pseudoreflexive Banach space E, it has a topological complement V [**12**]. Let P be a continuous linear projection of E onto ker(u) such that $E = \ker(u) \oplus V$, where $V = \ker(P)$. If $x \in E$, one has $x = x_1 + x_2$, with $x_1 \in\ker(u)$ and $x_2 \in V$. Hence $u(x) = u(x_1) + u(x_2) = u(x_2)$. It follows that $u(E) = u(V)$. The restriction u_0 of u to V is a linear bijection of the Banach V onto the Banach space $u(E)$. Then, by the Banach isomorphism theorem the reciprocal v_0 of u_0 is a continuous linear map of $u(E)$ onto V. On the other hand, since $u(E) = u_0(V)$ is a closed subspace of E with finite codimension, there exists also a continuous linear projection Q of E onto $u(E) = u_0(V)$ such that $E = u(E) \oplus W$, where W =ker(Q). The space W is finite dimensional with dimension equal to the codimension of $u(E)$. Let us define the map v of E into itself such that if $y = y_1+y_2$, with $y_1 \in u(E)$ and $y_2 \in W$, one has $v(y) = v_0(y_1)$, that is $v_0(y_2) = 0$. Let us set $Q_1 = id_E - Q$, then $Im(Q_1) = W$ is finite dimensional, that is Q_1 s of finite rank and as such is a completely continuous operator. One has $u \circ v = id_E - Q_1$ and

$v \circ u = id_E - P$. Since Q_1 and P are of finite rank, one sees that class of u in $\mathfrak{B}(E)$ is invertible, with inverse the class of v.

-(b)- Conversely, assume that $u \in \mathcal{L}(E)$ is such that its class in $\mathfrak{B}(E)$ is invertible. Hence there exists $v \in \mathcal{L}(E)$ such that $u \circ v - id_E = A_1$ and $v \circ u - id_E = A_2$ are completely continuous linear operators.

Let $x \in \ker(u)$, then $v(u(x)) = v(0) = 0 = x + A_2(x)$. Hence x is an eigenvector of A_2 associated with the eigenvalue -1. It is well known (se [**7**] extension of results of [**18**]) that one has a Riesz decomposition $E = \mathrm{N}(-1) \oplus F(-1)$ such that $N(-1)$ is finite dimensional and contains the eigenvectors associated with -1. Then since $\ker(u) \subset \mathrm{N}(-1)$, one sees that $\ker(u)$ is of finite dimension.

For $0 < \alpha < 1$, there exists a continuous linear endomorphism C_2 of finite rank such that $\|A_2 - C_2\| < \alpha$. For $x \in \ker(C_2)$, one has $\|v\|\|u(x)\| \geq \|v \circ u(x)\| = \|x + A_2(x)\| = \|x + C_2(x) + A_2(x) - C_2(x)\| \geq \|x\| - \|(A_2 - C_2)(x)\| \geq \|x\| - \|A_2 - C_2\|\|x\| \geq (1-\alpha)\|x\|$. Then one sees that the restriction of u to $\ker(C_2)$ is injective and if $(u(x_n))_{n\geq 1}$ is a Cauchy sequence in $u(\ker(C_2))$, since $(1-\alpha)\|x_n - x_m\| \leq \|v\|\|u(x_n) - u(x_m)\|$, one sees that $(x_n)_{n\geq 1}$ is a Cauchy sequence in $\ker(C_2)$. It immediately follows that $u(\ker(C_2))$ is norm complete and then is a closed subspace of E.

Since C_2 is of finite rank and $E/\ker(C_2) \simeq Im(C_2)$, then $\ker(C_2)$ is finite codimensional. It follows that the closed subspace $\ker(C_2)$ of E has topological complement. Let W be a topological complement of $\ker(C_2)$, one has $W \simeq E/\ker(C_2) \simeq Im(C_2)$ and then W is of finite dimension.

Since $E = \ker(C_2) \oplus W$, with $u(\ker(C_2))$ a closed subspace and W of finite dimension (hence $u(W)$ is finite dimensional), one obtains that $u(E) = u(\ker(C_2)) + u(W)$ is a closed subspace of E.

Let us remind that if A is an element of $C(E)$, then $A = \sum_{j\geq 1} x'_j \otimes y_j$. But if $x' \in E'$ and $y \in E$, for any $z' \in E'$ and any $x \in E$, one has $< {}^t(x' \otimes y)(z'), x > \; = \; < z', (x' \otimes y)(x) > \; = \; < z', y >< x', x > \; = \; < j_E(y), z' >< x', x > \; = \; < j_E(y) \otimes x'(z'), x >$. Hence ${}^t(x' \otimes y) = j_E(y) \otimes x'$, here j_E is the canonical application of E in its bidual E''.

Then the transpose of A can be written as a convergent sum of rank one operators ${}^tA = \sum_{j\geq 1} j_E(y_j) \otimes x'_j$ and therefore ${}^tA$ belongs to $C(E')$.

Since $u \circ v - id_E = A_1$, with $A_1 \in C(E)$, one has ${}^tv \circ {}^tu - id_{E'} = {}^tA_1$, with ${}^tA_1 \in C(E')$. Hence, as above one obtains that $\ker({}^tu)$ is finite dimensional. Since $\ker({}^tu) = Im(u)^\perp \simeq (E/u(E))'$, the dual space $(E/u(E))'$ of $E/u(E)$ is finite dimensional. It follows that $\dim_K E/u(E) < \infty$, that is $u(E)$ is finite codimensional.

One then concludes that u is a Fredholm operator. □

EXAMPLES. If K is spherically complete, any ultrametric K-Banach space is pseudoreflexive.

If for $0 < \alpha \leq 1$, the space E has an α-orthogonal base, then E is pseudoreflexive. In particular, if E is of countable type, it is pseudoreflexive.

From the ultrametric Atkinson theorem, we deduce the following corollary.

COROLLARY 5.4. *Let E be an ultrametric pseudoreflexive Banach space.*

$-(1)-$ *Let u be a Fredholm operator of E and A be a completely continuous linear operator of E, then $u + A$ is a Fredholm operator.*

$-(2)-$ *The set $\mathcal{F}(E)$ of the Fredholm operators of E is a unitary monoid for the law of composition.*

$-(3)-$ *$\mathcal{F}(E)$ is an open subset of $\mathcal{L}(E)$*

PROOF. Let $\phi_{\aleph_0} = \Phi : \mathcal{L}(E) \longrightarrow \mathfrak{B}(E)$ be the canonical map obtained by reduction modulo $C(E)$. It is a surjective continuous algebra homomorphism.

If we denote by $\mathrm{Inv}(\mathfrak{B}(E))$ the group of the invertible elements of $\mathfrak{B}(E)$, the theorem means that $\Phi(\mathcal{F}(E)) = \mathrm{Inv}(\mathfrak{B}(E))$ and $\mathcal{F}(E) = \Phi^{-1}(\mathrm{Inv}(\mathfrak{B}(E))$.

-(1)- Let $u \in \mathcal{F}(E)$ and $A \in C(E)$. Then $\Phi(u+A) = \Phi(u)$ is invertible in $\mathfrak{B}(E)$; it follows that $u + A$ is a Fredholm operator. In particular $id_E + A$ is a Fredholm operator and for $u \in \mathrm{Aut}(E)$, $u + A$ is a Fredholm operator.

-(2)- Let $u, v \in \mathcal{F}(E)$. Since $\Phi(u)$ and $\Phi(v)$ are invertible in $\mathfrak{B}(E)$, one sees that $\Phi(u \circ v) = \Phi(u)\Phi(v)$ is invertible in $\mathfrak{B}(E)$. It follows that $u \circ v$ belongs to $\mathcal{F}(E)$.

-(3)- It is well known that in a unitary Banach algebra, the group of invertible elements is an open subset. Consequently, here $\mathrm{Inv}(\mathfrak{B}(E))$ is an open subset of the Banach algebra $\mathfrak{B}(E)$.

Since Φ is continuous, it follows that $\mathcal{F}(E) = \Phi^{-1}(\mathrm{Inv}(\mathfrak{B}(E)))$ is an open subset of $\mathcal{L}(E)$. □

QUESTION. Let E_ω be an ultrametric Hilbert space, one can ask if the adjoint of a Fredholm operator $u \in \mathcal{L}_0(E_\omega)$ is a Fredholm operator too?

Equivalently, is any element of $\mathfrak{B}_0(E_\omega) \subset \mathfrak{B}(E_\omega)$, that is invertible in $\mathfrak{B}(E_\omega)$ has its inverse in $\mathfrak{B}_0(E_\omega)$?

REMARK 5.5. It was a long standing question in classical analysis if the automorphisms of the Calkin algebra $\mathfrak{B}(H)$ of a separable infinite dimensional complex Hilbert space are inner ?

Two different answers have been given depending deeply on foundations of set theory.

The first by N. Phillips and N. Weaver [**14**], which assuming the Continuum Hypothesis, gave a partial answer by constructing outer automorphisms of $\mathfrak{B}(H)$.

The second answer by I. Farah [**5**] asserts that if an axiom in set theory said Todorcevic's axiom is assumed then all automorphisms of $\mathfrak{B}(H)$ are inner.

What happens for the ultrametric Calkin algebra $\mathfrak{B}(E)$ when E is an ultrametric Banach space of countable type ?

References

[1] F. V. Atkinson, *The normal solubility of linear equations in normed spaces* (Russian), Mat. Sbornik N.S. **28 (70)** (1951), 3–14. MR0042059

[2] J. W. Calkin, *Two-sided ideals and congruences in the ring of bounded operators in Hilbert space*, Ann. of Math. (2) **42** (1941), 839–873, DOI 10.2307/1968771. MR0005790

[3] B. Diarra, *Ultraproduits ultramétriques de corps values* (French), Ann. Sci. Univ. Clermont-Ferrand II Math. **22** (1984), 1–37. MR789302

[4] B. Diarra, *Bounded linear operators on ultrametric Hilbert spaces*, Afr. Diaspora J. Math. (N.S.) **8** (2009), no. 2, 173–181. MR2538713

[5] I. Farah, *All automorphisms of the Calkin algebra are inner*, Ann. of Math. (2) **173** (2011), no. 2, 619–661, DOI 10.4007/annals.2011.173.2.1. MR2776359

[6] B. Gramsch, *Eine Idealstruktur Banachscher Operatoralgebren* (German), J. Reine Angew. Math. **225** (1967), 97–115, DOI 10.1515/crll.1967.225.97. MR0206727

[7] L. Gruson, *Théorie de Fredholm p-adique* (French), Bull. Soc. Math. France **94** (1966), 67–95. MR0226381

[8] S. Haller, *The Atiyah-Singer Index Theorem*, https://stefanhaller.files.wordpress.com
[9] S. Heinrich, *Ultraproducts in Banach space theory*, J. Reine Angew. Math. **313** (1980), 72–104, DOI 10.1515/crll.1980.313.72. MR552464
[10] E. Luft, *The two-sided closed ideals of the algebra of bounded linear operators of a Hilbert space*, Czechoslovak Math. J. **18 (93)** (1968), 595–605. MR0238086
[11] C. Pérez-García, *Semi-Fredholm operators and the Calkin algebra in p-adic analysis. I, II*, Bull. Soc. Math. Belg. Sér. B **42** (1990), no. 1, 69–101. MR1061430
[12] C. Perez-Garcia and W. H. Schikhof, *Finite-dimensional orthocomplemented subspaces in p-adic normed spaces*, Ultrametric functional analysis (Nijmegen, 2002), Contemp. Math., vol. 319, Amer. Math. Soc., Providence, RI, 2003, pp. 281–298, DOI 10.1090/conm/319/05580. MR1977453
[13] C. Perez-Garcia and W. H. Schikhof, *Locally convex spaces over non-Archimedean valued fields*, Cambridge Studies in Advanced Mathematics, vol. 119, Cambridge University Press, Cambridge, 2010. MR2598517
[14] N. C. Phillips and N. Weaver, *The Calkin algebra has outer automorphisms*, Duke Math. J. **139** (2007), no. 1, 185–202, DOI 10.1215/S0012-7094-07-13915-2. MR2322680
[15] M. van der Put, *The ring of bounded operators on a non-Archimedean normed linear space*, Nederl. Akad. Wetensch. Proc. Ser. A 71=Indag. Math. **30** (1968), 260–264. MR0226437
[16] A. C. M. van Rooij, *Non-Archimedean functional analysis*, Monographs and Textbooks in Pure and Applied Math., vol. 51, Marcel Dekker, Inc., New York, 1978. MR512894
[17] P. Schneider, *Nonarchimedean functional analysis*, Springer Monographs in Mathematics, Springer-Verlag, Berlin, 2002. MR1869547
[18] J.-P. Serre, *Endomorphismes complètement continus des espaces de Banach p-adiques* (French), Inst. Hautes Études Sci. Publ. Math. **12** (1962), 69–85. MR0144186

Université Clermont-Auvergne, Laboratoire de Mathématiques Blaise Pascal, CNRS, BP 10448, 63 000 Clermont-Ferrand

Email address: bertin.diarra@math.univ-bpclermont.fr

Contemporary Mathematics
Volume **704**, 2018
http://dx.doi.org/10.1090/conm/704/14164

On summation of p-adic series

Branko Dragovich

ABSTRACT. Summation of the p-adic functional series $\sum \varepsilon^n \, n! \, P_k^\varepsilon(n;x) \, x^n$, where $P_k^\varepsilon(n;x)$ is a polynomial in x and n with rational coefficients, and $\varepsilon = \pm 1$, is considered. The series is convergent in the domain $|x|_p \leq 1$ for all primes p. It is found the general form of polynomials $P_k^\varepsilon(n;x)$ which provide rational sums when $x \in \mathbb{Z}$. A class of generating polynomials $A_k^\varepsilon(n;x)$ plays a central role in the summation procedure. These generating polynomials are related to many sequences of integers. This is a brief review with some new results.

1. Introduction

In the last thirty years p-adic analysis has had successful applications in p-adic mathematical physics (from strings to complex systems and the universe as a whole) and in some related fields (in particular in bioinformation systems, see, e.g. [**1**]), see [**2, 3**] for an early review and [**4**] for a recent one. The p-adic series, as a part of p-adic analysis, have been also considered as mathematical tools of potential applications in physics. The main reason was in the following. There are series, in particular the power series, of the same form in physics and p-adic analysis. Recall that the results of physical measurements are rational numbers which belong to all p-adic number fields $\mathbb{Q}_p$ as well as to $\mathbb{R}$. Hence, there is a sense to consider p-adic series in rational points, i.e. when argument and sum are some rational numbers, and look for their possible physical content. One of motivations was related to the convergence of series. Namely, the power series $\sum a_n \, x^n$, where $a_n \in \mathbb{Q}$, can be treated simultaneously as p-adic or as real, depending on values of the argument x. Many series which are divergent in the real case are convergent in the p-adic one. There are many examples of divergent series, in particular, in perturbation expansions of quantum field theory and string theory, which diverge due to factorials. This problem of divergences in some real series motivated an investigation of various series with factorials with respect to the p-adic norm. To this end, many p-adic convergent series have been constructed and found their rational sums for some rational arguments.

In this paper we are interested in p-adic invariant summation of a class of infinite functional series which terms contain $n!$, i.e. $\sum \varepsilon^n \, n! \, P_k^\varepsilon(n;x) \, x^n$, where $\varepsilon = \pm 1$ and $P_k^\varepsilon(n;x)$ are polynomials in x and n of degree k. We show that

2010 *Mathematics Subject Classification.* Primary 40A05, 40A30.

Key words and phrases. p-adic series, p-adic numbers, sequences of integers, generating polynomials.

there exist polynomials $P_k^\varepsilon(n;x)$ for any degree k, such that for any $x \in \mathbb{Z}$ the corresponding sums are also integer or rational numbers. Moreover, we have found recurrence relations to calculate all ingredients of such $P_k^\varepsilon(n;x)$.

All necessary general information on p-adic series can be found in standard books on p-adic analysis, see, e.g. [**5**].

2. Series and p-adic invariant summation in integer points

We will give now a brief review of some p-adic series with factorials presented in the papers [**6**–**15**, **20**].

Recall that necessary and sufficient condition for the p-adic power series to be convergent is that the general term vanishes as degree goes to infinity [**3**, **5**], i.e.

$$S(x) = \sum_{n=1}^{+\infty} a_n x^n, \quad a_n \in \mathbb{Q}_p, \quad x \in \mathbb{Q}_p, \quad |a_n x^n|_p \to 0 \text{ as } n \to \infty, \tag{1}$$

where $|\cdot|_p$ denotes the p-adic absolute value (also called p-adic norm). As in the real case, convergent infinite p-adic series can be summed using usual rules as for the finite series.

We are interested here in the series which are simultaneously convergent with respect to p-adic absolute value for every prime number p. One of the simplest such examples is the well known series

$$\sum_{n=0}^{\infty} n! = 0! + 1! + 2! + ... + n! + ..., \qquad |n!|_p = p^{-\frac{n-s_n}{p-1}} \to 0 \ (n \to \infty), \tag{2}$$

where $s_n = n_0 + n_1 + ... + n_r$ is the sum of digits in the canonical expansion of number n in base p, i.e. $n = n_0 + n_1 p + ... + n_r p^r$. It can be shown that if the sum of series (2) has a rational value then it cannot be the same for all p-adic cases. Probably the sum of series (2) is irrational number in all $\mathbb{Q}_p$.

From an application point of view it is more interesting to investigate p-adic series which sum is a rational number, the same for all primes p. As a simple illustrative example one can refer to

$$\sum_{n=1}^{\infty} n!\; n = 1!\; 1 + 2!\; 2 + ... + n!\; n + ... = -1\,. \tag{3}$$

In the proof of (3) one can employ any of the following two properties:

$$(i):\ n!\; n = (n+1)! - n!\,, \qquad (ii):\ \sum_{n=1}^{N-1} n!\, n = -1 + N!\,. \tag{4}$$

In this section we present a review of some convergent p-adic series, where summation is performed using mainly property like (i) in (4). Method similar to (ii) of (4) will be employed in the next section.

In [**6**] was shown that the ground-state energy perturbation series of the anharmonic oscillator

$$E(\lambda) = \frac{1}{2} + \sum_{n=1}^{\infty} (-1)^{n+1} a_n 2^{-2n} \lambda^n\,, \quad a_n \in \mathbb{N}\,, \tag{5}$$

which is divergent in the real case ($a_1 = 3, a_2 = 2 \cdot 3 \cdot 7, a_3 = 2^2 \cdot 3^2 \cdot 37, ...$), is p-adic convergent for $|\lambda|_p < 1$ if $p \neq 2$ and $|\lambda|_2 < \frac{1}{4}$. Paper [**7**] contains two similar examples to (5) and the following one:

$$F(x) = \sum_{n=1}^{\infty} (-1)^{n+1} (n!)^k (n^\ell + w)^m x^n, \quad w \in \mathbb{Z},\, k \in \mathbb{N},\, m \in \mathbb{N} \cup \{0\}, \tag{6}$$

which is convergent for every p in the range $|x|_p \leq 1$ and consequently includes all integers. After suitable rearrangement of the series terms for $F(x)$ and $-F(x)$, and then taking $F(x) + (-F(x)) = 0$, one obtains formula

$$\sum_{n=1}^{\infty} (n!)^k \{(n+1)^k [(n+1)^\ell + w]^m x - (n^\ell + w)^m\} x^{n-1} = -(1+w)^m. \tag{7}$$

One can easily see that (7) contains (3) as a particular case.

In [**7**] was also discussed possible connection between integer sums of p-adic series convergent in $|x|_p \leq 1$ for all p and its real counterpart which is divergent. Recall that the sum of the divergent series depends on the way how summation is performed. Using the same summation procedure as in p-adic convergent cases will give the same integer value in the real divergent counterpart. This kind of summation of divergent series was called adelic summation. For example, series (3) is highly divergent from real point of view, but according to the adelic summation the corresponding real sum should be also -1. Perhaps this adelic summation will find an application in physics and related sciences.

Inspired by divergent series in a zero-dimensional model of quantum field theory, in paper [**9**] the following p-adic series was considered

$$F(x) = \sum_{n=1}^{\infty} (-1)^{n+1} \prod_{i=1}^{\alpha} ((\mu_i n + \nu_i)!)^{k_i} P_\ell(n)\, x^n, \tag{8}$$

where $\alpha, \mu_i \in \mathbb{N}$, $\nu_i \in \mathbb{Z}$, $\mu_i + \nu_i \geq 1$ and $\ell, k_i \in \mathbb{N}_0$, and at least one of $k_i \geq 1$. The domain of convergence contains $|x|_p \leq 1$ and hence includes $x \in \mathbb{Z}$. By suitable rearrangement of terms for $F(x)$ and $-F(x)$, and then taking $F(x) + (-F(x)) = 0$, one obtains summation equality

$$\begin{aligned} &\sum_{n=1}^{\infty} \prod_{i=1}^{\alpha} ((\mu_i n + \nu_i)!)^{k_i} \left[\prod_{i=1}^{\alpha} (\mu_i n + \nu_i + 1)_{\mu_i}^{k_i} P_\ell(n+1)\, x - P_\ell(n) \right] x^{n-1} \\ &\qquad = - \prod_{i=1}^{\alpha} ((\mu_i + \nu_i)!)^{k_i} P_\ell(1), \end{aligned} \tag{9}$$

where $(\mu_i n + \nu_i + 1)_{\mu_i}^{k_i} = (\mu_i n + \nu_i + 1)^{k_i} (\mu_i n + \nu_i + 2)^{k_i} ... (\mu_i n + \nu_i + \mu_i)^{k_i}$. Polynomial $P_\ell(n)$ is

$$P_\ell(n) = c_\ell n^\ell + c_{\ell-1} n^{\ell-1} + ... + c_0, \tag{10}$$

where $c_0, c_1, ..., c_\ell$ are some integers. Taking $x = +1$ or -1 and various values of parameters in (9) one gets sum for plenty of p-adic numerical series.

Paper [**10**] is devoted to rational summation of the wider class of p-adic series

$$S^\varepsilon(x) = \sum_{n=1}^{\infty} \varepsilon^n \prod_{i=1}^{I} ((\mu_i n + \nu_i)!)^{\lambda_i} P_k(n)\, x^{\alpha n + \beta}, \tag{11}$$

where $\varepsilon = \pm 1$, $I, \mu_i, \alpha \in \mathbb{N}$, $\nu_i \in \mathbb{Z}$, $\mu_i + \nu_i \geq 1$, $\lambda_i, \beta \in \mathbb{N}_0$ and at least one of $\lambda_i \in \mathbb{N}$. $P_k(n)$ is a polynomial

$$P_k(n) = C_k n^k + C_{k-1} n^{k-1} + ... + C_0\,, \tag{12}$$

where $C_0, C_1, ..., C_k \in \mathbb{Z}$. The domain of convergence of the series (11) is

$$|x|_p < p^{\frac{1}{(p-1)\alpha} \sum_{i=1}^I \mu_i \lambda_i} \tag{13}$$

which includes $x \in \mathbb{Z}$ for all p-adic norms.

As rational summation one has in mind summation of convergent series, like (11), that for a rational argument x exists a rational sum $S(x)$. For the series (11), rational summation makes a restriction on possible polynomials $P_k(n)$. To find suitable polynomials $P_k(n)$ it is useful to introduce an auxiliary polynomial in (11)

$$P_\ell(n) = a_\ell\, n^\ell + a_{\ell-1}\, n^{\ell-1} + ... + a_0\,, \tag{14}$$

where $0 \leq \ell < k$ and $a_0, a_1, ..., a_\ell$ are some integer numbers. Using the same summation procedure as in the previous cases, and replacing $P_k(n)$ by $A_\ell(n)$ in (11), follows the equality

$$\begin{aligned} &\sum_{n=1}^{\infty} \varepsilon^n \prod_{i=1}^{I} ((\mu_i n + \nu_i)!)^{\lambda_i} \left[\prod_{i=1}^{I} (\mu_i n + \nu_i + 1)_{\mu_i}^{\lambda_i} A_\ell(n+1)\, x^\alpha - \varepsilon A_\ell(n) \right] x^{\alpha n + \beta} \\ &\qquad = -\prod_{i=1}^{I} ((\mu_i + \nu_i)!)^{\lambda_i} A_\ell(1)\, x^{\alpha+\beta}, \end{aligned} \tag{15}$$

where $(\mu_i n + \nu_i + 1)_{\mu_i}^{\lambda_i} = (\mu_i n + \nu_i + 1)^{\lambda_i} (\mu_i n + \nu_i + 2)^{k_i} ... (\mu_i n + \nu_i + \mu_i)^{\lambda_i}$. Summation expression (15) is valid in the domain of convergence of (11), i.e. includes all $x \in \mathbb{Z}$. Taking $x = t \in \mathbb{Q}$ which satisfies the domain of convergence (13) one can construct a polynomial

$$P_k(n;t) = \prod_{i=1}^{I} (\mu_i n + \nu_i + 1)_{\mu_i}^{\lambda_i} A_\ell(n+1)\, t^\alpha - \varepsilon A_\ell(n) \tag{16}$$

where $k = \ell + \sum_{i=1}^I \mu_i \lambda_i$, which depends on t and leads to a rational sum of (11). For a given polynomial $P_k(n)$ the series (11) has a rational sum if there exists an auxiliary polynomial $A_\ell(n)$ such that expression (16) is satisfied. When $t \in \mathbb{Z}$ then the corresponding sum is an integer the same in all $\mathbb{Z}_p$, where $\mathbb{Z}_p$ is the ring of p-adic integers. In the next section we shall see how one can construct all related polynomials $P_k(n;x)$ for any degree k.

In [**11**] is presented extension of summation formula (15) in the form

$$\begin{aligned} &\sum_{n=1}^{\infty} \varepsilon^n \frac{\prod_{i=1}^{I} ((\mu_i n + \nu_i)!)^{\lambda_i}}{\prod_{j=1}^{J} ((\rho_j n + \sigma_j)!)^{\eta_j}} \left[\frac{\prod_{i=1}^{I} (\mu_i n + \nu_i + 1)_{\mu_i}^{\lambda_i}}{\prod_{j=1}^{J} (\rho_j n + \sigma_j + 1)_{\rho_j}^{\eta_j}} A_\ell(n+1)\, x^\alpha - \varepsilon A_\ell(n) \right] x^{\alpha n + \beta} \\ &\qquad = -\frac{\prod_{i=1}^{I} ((\mu_i + \nu_i)!)^{\lambda_i}}{\prod_{j=1}^{J} ((\rho_j + \eta_j)!)^{\eta_j}} A_\ell(1)\, x^{\alpha+\beta}. \end{aligned} \tag{17}$$

Paper [**12**] contains an elaboration of rational summation considered in [**10**].

Papers [**13**–**15**] were presented at conferences on p-adic functional analysis and published in their proceedings. A very simple case of (15) is

$$\sum_{n=0}^{\infty} n!\,[(n+1)A_\ell(n+1) - A_\ell(n)] = -A_\ell(0), \tag{18}$$

where $A_\ell(n)$ is defined by (14). It is obvious that all possible polynomials $A_\ell(n)$ generate the corresponding polynomials $P_k(n) = (n+1)A_\ell(n+1) - A_\ell(n)$, with $k = \ell + 1 \geq 1$, which give integer sum $-A_\ell(0)$, i.e.

$$\sum_{n=0}^{\infty} n!\,P_k(n) = -A_\ell(0). \tag{19}$$

It is useful to investigate the series of the form

$$\sum_{n=0}^{\infty} n!\,[n^k + u_k] = v_k, \tag{20}$$

where u_k, v_k are pairs of integers. The series (20) is analyzed in [**13**], where

$$(n+1)A_{k-1}(n+1) - A_{k-1}(n) = n^k + u_k, \quad k = 1, 2, 3, \ldots. \tag{21}$$

Note that (21) contains a system of $k+1$ linear equations with $k+1$ unknowns $(a_0, a_1, \ldots, a_{k-1}, u_k)$ which has always solution, and that $v_k = -A_{k-1}(0)$. It was also shown that pairs of integers u_k, v_k are unique. If we know integers $u_1, u_2, \ldots, u_k$ and $v_1, v_2, \ldots, v_k$ then one can write general summation formula

$$\sum_{n=0}^{\infty} n!\,P_k(n) = Q_k, \quad P_k(n) = C_k\,n^k + \ldots + C_1\,n + C_0, \quad Q_k = \sum_{j=1}^{k} C_j\,v_j, \tag{22}$$

where $C_0 = \sum_{j=1}^{k} C_j\,u_j$ and $C_0, C_1, \ldots, C_k \in \mathbb{Q}$. Hence only these polynomials $P_k(n)$ which are defined by (22) give p-adic invariant rational summation, where nontrivial role plays term C_0.

In paper [**14**] a modified generalized hypergeometric series was considered, the domain of convergence was found and a summation formula was derived. It was also reconsidered summation formula (20) and related recurrence relation was found (see also [**15**])

$$S_{k+1}(n) = \delta_{0k} - kS_k(n) - \sum_{\ell=0}^{k-1} \binom{k+1}{\ell} S_\ell(n) + n!n^k\,, \tag{23}$$

where $S_k(n) = \sum_{i=0}^{n-1} i!\,i^k$ and δ_{0k} is the Kronecker symbol ($\delta_{0k} = 1$ if $k = 0$ and $\delta_{0k} = 0$ if $k \neq 0$). From (23) follows

$$\sum_{i=0}^{n-1} i!\,(i^k + u_k) = v_k + n!A_{k-1}(n)\,, \tag{24}$$

where u_k and v_k satisfy the following recurrence relations:

$$u_{k+1} = -ku_k - \sum_{\ell=1}^{k-1} \binom{k+1}{\ell} u_\ell + 1\,, \quad u_1 = 0,\ k \geq 1\,, \tag{25}$$

$$v_{k+1} = -kv_k - \sum_{\ell=1}^{k-1} \binom{k+1}{\ell} v_\ell - \delta_{0k}\,, \quad k \geq 0\,. \tag{26}$$

TABLE 1. The first eleven values of u_k and v_k [**14,15**]. Note that in encyclopedia of integer sequences [**23**], sequence u_k overlaps with A000587 and A014182 in absolute values, while v_k coincides with A014619.

k	1	2	3	4	5	6	7	8	9	10	11
u_k	0	1	-1	-2	9	-9	-50	267	-413	-2180	17731
v_k	-1	1	1	-5	5	21	-105	141	777	-5513	13209

Note that $A_k(n)$ is a polynomial in n and in limit $n \to \infty$ formula (24) becomes (20). Using recurrence relations (25) and (26) one can calculate u_k and v_k, which the first eleven values are presented in Table 1.

Note that in [**14**] is introduced the following functional summation formula:

$$\sum_{i=0}^{\infty} n! \left[n^k x^k + U_k(x)\right] x^n = V_k(x) , \tag{27}$$

where $U_k(x)$ and $V_k(x)$ are certain polynomials in x with integer coefficients. When $x = 1$ we have $U_k(1) = u_k$ and $V_k(1) = v_k$.

Paper [**14**] contains a proof that the sum of p-adic series

$$\sum_{n=0}^{\infty} n! \, n^k \, x^n , \quad k \in \mathbb{N}_0 , \; x \in \mathbb{N} \setminus \{1\} \tag{28}$$

cannot be the same rational number in $\mathbb{Z}_p$ for every p. It also contains conjecture that the sum of (28) is a rational number iff $k = x = 1$. In particular, according to this conjecture $u_k \neq 0$ in (20) when $k \neq 1$ and the corresponding sums $\sum n! n^k$ are p-adic irrational numbers. Following this research in [**16**] was shown that $u_k \neq 0$ for $k \equiv 0$ or $2 \pmod 3$. Then has shown [**17**] that u_k can be zero at most twice (see also [**18, 19**]).

Paper [**20**] is devoted to the first-order and second-order differential equations which contain as a solution an analytic function of the form

$$F_k(x) = \sum_{n=0}^{\infty} n! \, P_k(n) \, x^n , \quad k \in \mathbb{N}_0 , \; |x|_p < p^{\frac{1}{p-1}} , \tag{29}$$

where $P_k(n) = n^k + C_{k-1} n^{k-1} + ... + C_0$ is a polynomial in n with $C_i \in \mathbb{Z}$ (in a more general case $C_i \in \mathbb{Q}$ or $C_i \in \mathbb{C}_p$). It is shown the existence of linear differential equations which solution is $F_k(x)$ for any polynomial $P_k(n)$ and some of such equations are constructed. For example, the related first-order and second-order linear differential equations for $F_0(x) = \sum_{n=0}^{\infty} n! \, x^n$ are:

$$x^2 F_0'(x) + (x-1) F_0(x) = -1 , \tag{30}$$

$$x^2 F_0''(x) + (3x-1) F_0'(x) + F_0(x) = 0 . \tag{31}$$

3. Functional summation formula and polynomials $A_k^\varepsilon(n;x)$

This section is devoted to investigation of the finite functional summation formula of the form

$$\sum_{i=0}^{n-1} \varepsilon^i i! \, [i^k x^k + U_k^\varepsilon(x)] \, x^i = V_k^\varepsilon(x) + n! \, A_{k-1}^\varepsilon(n;x) \, , \tag{32}$$

where $\varepsilon = \pm 1$ and $A_k^\varepsilon(n;x)$ are certain polynomials in x, which coefficients are polynomials in n (also of degree k) with integer coefficients. $U_k^\varepsilon(x)$ and $V_k^\varepsilon(x)$ are polynomials related to the polynomials $A_{k-1}^\varepsilon(n;x)$ as follows:

$$U_k^\varepsilon(x) = x \, A_{k-1}^\varepsilon(1;x) - \varepsilon \, A_{k-1}^\varepsilon(0;x) \, , \quad V_k^\varepsilon(x) = -\varepsilon A_{k-1}^\varepsilon(0;x). \tag{33}$$

If $x = 1$ and $\varepsilon = +1$ then summation formula (32) reduces to (24). When $\varepsilon = +1$ and $n \to \infty$ formula (32) evidently becomes (27).

This section is based on papers [**21, 22**] and their elaboration with some new results. Here the main role is plaid by the polynomials $A_k^\varepsilon(n;x)$, because they contain all information on properties of summation formula (32) .

THEOREM 3.1. *Let $S_k^\varepsilon(n;x) = \sum_{i=0}^{n-1} \varepsilon^i \, i! \, i^k \, x^i$, where $\varepsilon = \pm 1$. Then one has the recurrence formula*

$$S_k^\varepsilon(n;x) = \delta_{0k} + \varepsilon \, x \, S_0^\varepsilon(n;x) + \varepsilon \, x \sum_{\ell=1}^{k+1} \binom{k+1}{\ell} S_\ell^\varepsilon(n;x) - \varepsilon^n \, n! \, n^k \, x^n \, . \tag{34}$$

THEOREM 3.2. *The recurrence relation* (34) *has solution in the form*

$$\sum_{i=0}^{n-1} \varepsilon^i \, i! \, [i^k \, x^k \, + U_k^\varepsilon(x)] \, x^i = V_{k-1}^\varepsilon(x) + A_{k-1}^\varepsilon(n;x) \, \varepsilon^{n-1} \, n! \, x^n \, , \tag{35}$$

where polynomials $U_k^\varepsilon(x) \, , V_{k-1}^\varepsilon(x)$ and $A_{k-1}^\varepsilon(n;x)$ satisfy the following recurrence relations:

$$\sum_{\ell=1}^{k+1} \binom{k+1}{\ell} x^{k-\ell+1} \, U_\ell^\varepsilon(x) - \varepsilon \, U_k^\varepsilon(x) - x^{k+1} \, = 0 \, , \quad U_1^\varepsilon(x) = x - \varepsilon \, , \tag{36}$$

$$\sum_{\ell=1}^{k+1} \binom{k+1}{\ell} x^{k-\ell+1} \, V_{\ell-1}^\varepsilon(x) - \varepsilon \, V_{k-1}^\varepsilon(x) \, + \delta_{0k} \, \varepsilon x = 0 \, , \quad V_0^\varepsilon(x) = -\varepsilon \, , \tag{37}$$

$$\sum_{\ell=1}^{k+1} \binom{k+1}{\ell} x^{k-\ell+1} \, A_{\ell-1}^\varepsilon(n;x) - \varepsilon \, A_{k-1}^\varepsilon(n;x) - n^k \, x^k \, = 0 \, , \tag{38}$$

where $A_0^\varepsilon(n;x) = 1$ and $k \in \mathbb{N}$.

Proofs for the above theorems can be found in [**21, 22**]. Recurrence relations (36)-(38) can be taken as definitions of polynomials $U_k^\varepsilon(x), V_k^\varepsilon(x)$ and $A_k^\varepsilon(n;x)$. Connection of $U_k^\varepsilon(x)$ and $V_k^\varepsilon(x)$ to $A_k^\varepsilon(n;x)$ follows from the above relations (36)-(38). Hence, in the sequel it is sufficient to investigate properties of the generating polynomials $A_k^\varepsilon(n;x)$, because properties of $U_k^\varepsilon(x)$ and $V_k^\varepsilon(x)$ can be derived from (33).

THEOREM 3.3. *Let $A_k^\varepsilon(n;x) = \sum_{j=0}^k A_{kj}^\varepsilon(n)\, x^j$. Then there is the following recurrence formula for calculation of $A_{kj}^\varepsilon(n)$:*

$$\begin{aligned} &\binom{k+1}{k+1} A_{kj}^\varepsilon(n) + \binom{k+1}{k} A_{k-1,j-1}^\varepsilon(n) + \binom{k+1}{k-1} A_{k-2,j-2}^\varepsilon(n) + \ldots \\ &+ \binom{k+1}{k-j+1} A_{k-j,0}^\varepsilon(n) = \begin{cases} \varepsilon\, A_{k-1,j}^\varepsilon(n)\,, & j = 0,1,\ldots,k-1, \\ n^k\,, & j = k\,, \end{cases} \end{aligned} \tag{39}$$

where $k = 1,2,3,\ldots$ and $A_{00}^\varepsilon(n) = 1$.

PROOF. Substituting $A_k^\varepsilon(n;x) = \sum_{j=0}^k A_{kj}^\varepsilon(n)\, x^j$ into recurrence relation (38) we have

$$\sum_{j=0}^{\ell-1} x^j \sum_{\ell=1}^{k+1} \binom{k+1}{\ell} A_{\ell-1,j}^\varepsilon(n) x^{k+1-\ell} - \varepsilon \sum_{j=0}^{k-1} A_{k-1,j}^\varepsilon(n) x^j - n^k x^k = 0\,, \tag{40}$$

where $k = 1,2,3,\ldots$ Looking for terms with x^i and x^k we obtain

$$\begin{aligned} \bullet\, x^i :\quad &\binom{k+1}{k+1} A_{ki}^\varepsilon(n) + \binom{k+1}{k} A_{k-1,i-1}^\varepsilon(n) + \binom{k+1}{k-1} A_{k-2,i-2}^\varepsilon(n) \\ &+ \ldots + \binom{k+1}{k-i+1} A_{k-i,0}^\varepsilon(n) - \varepsilon\, A_{k-1,i}^\varepsilon(n) = 0\,,\ \ i = 0,1,\ldots,k-1\,. \end{aligned} \tag{41}$$

$$\begin{aligned} \bullet\, x^k :\quad &\binom{k+1}{k+1} A_{kk}^\varepsilon(n) + \binom{k+1}{k} A_{k-1,k-1}^\varepsilon(n) + \binom{k+1}{k-1} A_{k-2,k-2}^\varepsilon(n) \\ &+ \ldots + \binom{k+1}{1} A_{00}^\varepsilon(n) - n^k = 0. \end{aligned} \tag{42}$$

□

Using recurrence relation (39) we can calculate $A^{\varepsilon}_{kj}(n)$. Performing calculation for $k = 1, 2, ..., 5$, we obtain $A^{\varepsilon}_{k}(n;x)$ presented below.

$$A^{\varepsilon}_{0}(n;x) = A^{\varepsilon}_{00}(n) = 1. \tag{43}$$

$$A^{\varepsilon}_{1}(n;x) = \sum_{j=0}^{1} A^{\varepsilon}_{1j}(n)\, x^j = (n-2)x + \varepsilon. \tag{44}$$

$$\begin{aligned} A^{\varepsilon}_{2}(n;x) = \sum_{j=0}^{2} A^{\varepsilon}_{2j}(n)\, x^j &= (n^2-3n+3)x^2 \\ &+ (n-5)\varepsilon x + 1. \end{aligned} \tag{45}$$

$$\begin{aligned} A^{\varepsilon}_{3}(n;x) = \sum_{j=0}^{3} A^{\varepsilon}_{3j}(n)\, x^j &= (n^3-4n^2+6n-4)x^3 \\ &+ (n^2-7n+17)\varepsilon x^2 + (n-9)x + \varepsilon. \end{aligned} \tag{46}$$

$$\begin{aligned} A^{\varepsilon}_{4}(n;x) = \sum_{j=0}^{4} A^{\varepsilon}_{4j}(n)\, x^j &= (n^4-5n^3+10n^2-10n+5)x^4 \\ &+ (n^3-9n^2+31n-49)\varepsilon x^3 \\ &+ (n^2-12n+52)x^2 + (n-14)\varepsilon x + 1. \end{aligned} \tag{47}$$

$$\begin{aligned} A^{\varepsilon}_{5}(n;x) = \sum_{j=0}^{5} A^{\varepsilon}_{5j}(n)\, x^j &= (n^5-6n^4+15n^3-20n^2+15n-6)x^5 \\ &+ (n^4-11n^3+49n^2-111n+129)\varepsilon x^4 \\ &+ (n^3-15n^2+88n-246)x^3 \\ &+ (n^2-18n+121)\varepsilon x^2 + (n-20)x + \varepsilon. \end{aligned} \tag{48}$$

In the above examples we can note the following properties:

$$\bullet\ A^{\varepsilon}_{kj}(n) = \sum_{i=0}^{j} a_{kj,i}\, n^i, \quad a_{kj,j} = \varepsilon^{k+j}, \quad \varepsilon = \pm 1. \tag{49}$$

$$\bullet\ A^{\varepsilon}_{kk}(n) = \sum_{i=0}^{k} a_{kk,i}\, n^i, \quad a_{kk,i} = (-1)^{k+i}\binom{k+1}{i+1}. \tag{50}$$

$$\bullet\ A^{\varepsilon}_{k1}(n) = \sum_{i=0}^{1} a_{k1,i}\, n^i = \left(n - \frac{k(k+3)}{2}\right)\varepsilon^{k+1}, \tag{51}$$

where $\frac{k(k+3)}{2}$ in (51) is the sum of $2+3+...+(k+1)$.

Using expressions (43)-(48) we can write some useful values of $A^{\varepsilon}_{k}(n;x)$ in Table 2.

TABLE 2. The first six values of $A_k^\varepsilon(n;x)$, where $n = 0, 1$ and $x = \pm 1$.

k	1	2	3	4	5	6
$A_k^\varepsilon(0;1)$	1	-2 + ε	4 - 5ε	-13 + 18ε	58 -63ε	-272 + 251ε
$A_k^\varepsilon(1;1)$	1	-1+ε	2-4ε	-9 +12 ε	43 - 39ε	-192+162ε
$A_k^\varepsilon(0;-1)$	1	2 + ε	4 + 5ε	13 + 18ε	58 +63ε	272 + 251ε
$A_k^\varepsilon(1;-1)$	1	1+ε	2+4ε	9 +12 ε	43 + 39ε	192+162ε

Denoting $A_k^{+1}(n;x) \equiv A_k^+(n;x)$ and $A_k^{-1}(n;x) \equiv A_k^-(n;x)$ from Table 2 we have (cf. [**23**])

$$A_k^+(0;1): \ 1, -1, -1, 5, -5, -21, \ldots \quad cf. \ A014619 \tag{52}$$

$$A_k^-(0;1): \ 1, -3, 9, -31, 121, -523, \ldots \quad cf. \ A040027 \tag{53}$$

$$A_k^+(1;1): \ 1, 0, -2, 3, 4, -30, \ldots \quad cf. \ A007114 \tag{54}$$

$$A_k^-(1;1): \ 1, -2, 6, -21, 82, -354, \ldots \quad cf. \ A032347 \tag{55}$$

$$A_k^+(0;-1): \ 1, 3, 9, 31, 121, 523, \ldots \quad cf. \ A040027 \tag{56}$$

$$A_k^-(0;-1): \ 1, 1, -1, -5, -5, 21, \ldots \quad cf. \ A014619 \tag{57}$$

$$A_k^+(1;-1): \ 1, 2, 6, 21, 82, 354, \ldots \quad cf. \ A032347 \tag{58}$$

$$A_k^-(1;-1): \ 1, 0, -2, -3, 4, 30, \ldots \quad cf. \ A007114\,. \tag{59}$$

$A_k^\varepsilon(n;x)$ in sequences (52)-(59) overlap (up to the sign) with related elements in the on-line encyclopedia of integer sequences [**23**]. Using (52)-(59) it is worth to present also $U_k^\varepsilon(x)$ for $x = \pm 1$ and $\varepsilon = \pm 1$, see below.

$$U_k^+(1): \ 0, 1, -1, -2, 9, -9, \ldots \quad cf. \ A000587 \tag{60}$$

$$U_k^-(1): \ 2, -5, 15, -52, 203, -877, \ldots \quad cf. \ A000110 \tag{61}$$

$$U_k^+(-1): \ -2, -5, -15, -52, -203, -877, \ldots \quad cf. \ A000110 \tag{62}$$

$$U_k^-(-1): \ 0, 1, 1, -2, -9, -9, \ldots \quad cf. \ A000587\,. \tag{63}$$

It is worth pointing out that $-U_k^+(-1) = B_{k+1}$ for $k \in \mathbb{N}$, where B_k are the Bell numbers. Recall that the Bell number B_k is equal to the number of partitions of a set of k elements. These numbers satisfy the recurrence relation

$$B_{k+1} = \sum_{\ell=0}^{k} \binom{k}{\ell} B_\ell\,, \qquad B_0 = 1. \tag{64}$$

Taking limit $n \to \infty$ in (35) and then replacing i by n one obtains

$$\sum_{n=0}^{\infty} n!\,[n^k x^k + U_k^\varepsilon(x)]\,x^n = V_k^\varepsilon(x)\,. \tag{65}$$

Summation formula (65) can be generalized in

$$\sum_{n=0}^{\infty} n!\, P_k^{\varepsilon}(n;x)\, x^n = Q_k^{\varepsilon}(x)\,, \tag{66}$$

where polynomials $P_k^{\varepsilon}(n;x)$ and $Q_k^{\varepsilon}(x)$ are

$$P_k^{\varepsilon}(n;x) = \sum_{j=1}^{k} C_j\,[n^j x^j + U_j^{\varepsilon}(x)]\,, \quad Q_k^{\varepsilon}(x) = \sum_{j=1}^{k} C_j\, V_j^{\varepsilon}(x) \tag{67}$$

and $C_j \in \mathbb{Q}$. Summation formula (67) is the general one, p-adic invariant, valid for all $x \in \mathbb{Z}$ and $Q_k^{\varepsilon}(x) \in \mathbb{Q}$.

Concluding remarks

Summarizing, in this paper we presented a brief review of previously obtained particular results on summation of p-adic functional series of the form $\sum_{n=1}^{\infty} n! P_k^{\varepsilon}(n;x)x^n$. Some new results are contained in section 3, particularly in formulas (65)-(67). Significant role of the polynomials $A_k^{\varepsilon}(n;x)$ is emphasized, because all information on summability of (65) is coded in these polynomials. Moreover, the polynomials $A_k^{\varepsilon}(n;x)$ for $x = \pm 1$ and $\varepsilon = \pm 1$ are related to some well known sequences of integer numbers, including the Bell numbers. In [**22**] some aspects of the more general series of the type $\sum_{n=1}^{\infty}(n+\nu)! P_k^{\varepsilon}(n;x)x^{\alpha n+\beta}$ are considered, which can be easily connected to results obtained in section 3.

It is worth mentioning that power series everywhere convergent on $\mathbb{R}$ and all $\mathbb{Q}_p$, as well as their adelic aspects, are considered in [**8, 11**]. Kurepa's hypothesis on the left factorial is investigated in [**15**] and infinitely many its equivalents are found. Some concrete examples can be found in author's papers, in particular in [**13**].

Acknowledgements

This work was supported in part by Ministry of Education, Science and Technological Development of the Republic of Serbia, project OI 174012. I would like to thank the referee for pointing out misprints.

References

[1] B. Dragovich and A. Yu. Dragovich, *A p-adic model of DNA sequence and genetic code*, p-Adic Numbers Ultrametric Anal. Appl. **1** (2009), no. 1, 34–41, DOI 10.1134/S2070046609010038. MR2566118

[2] L. Brekke and P. G. O. Freund, *p-adic numbers in physics*, Phys. Rep. **233** (1993), no. 1, 1–66, DOI 10.1016/0370-1573(93)90043-D. MR1238475

[3] V. S. Vladimirov, I. V. Volovich, and E. I. Zelenov, *p-adic analysis and mathematical physics*, Series on Soviet and East European Mathematics, vol. 1, World Scientific Publishing Co., Inc., River Edge, NJ, 1994. MR1288093

[4] B. Dragovich, A. Yu. Khrennikov, S. V. Kozyrev, and I. V. Volovich, *On p-adic mathematical physics*, p-Adic Numbers Ultrametric Anal. Appl. **1** (2009), no. 1, 1–17, DOI 10.1134/S2070046609010014. MR2566116

[5] W. H. Schikhof, *Ultrametric calculus: An introduction to p-adic analysis*, Cambridge Studies in Advanced Mathematics, vol. 4, Cambridge University Press, Cambridge, 1984. MR791759

[6] I. Ya. Aref′eva, B. G. Dragović, and I. V. Volovich, *On the p-adic summability of the anharmonic oscillator*, Phys. Lett. B **200** (1988), no. 4, 512–514, DOI 10.1016/0370-2693(88)90161-X. MR926670

[7] B. Dragović, *p-adic perturbation series and adelic summability*, Phys. Lett. B **256** (1991), no. 3-4, 392–396, DOI 10.1016/0370-2693(91)91780-Y. MR1100250

[8] B. G. Dragović, *Power series everywhere convergent on* **R** *and all* Q_p, J. Math. Phys. **34** (1993), no. 3, 1143–1148, DOI 10.1063/1.530189. MR1207974

[9] B. G. Dragović, *On p-adic aspects of some perturbation series* (English, with English and Russian summaries), Teoret. Mat. Fiz. **93** (1992), no. 2, 211–218, DOI 10.1007/BF01083520; English transl., Theoret. and Math. Phys. **93** (1992), no. 2, 1225–1231 (1993). MR1233542

[10] B. G. Dragović, *Rational summation of p-adic series* (English, with English and Russian summaries), Teoret. Mat. Fiz. **100** (1994), no. 3, 342–353, DOI 10.1007/BF01018570; English transl., Theoret. and Math. Phys. **100** (1994), no. 3, 1055–1064 (1995). MR1311893

[11] B. G. Dragović, *On p-adic series in mathematical physics*, Trudy Mat. Inst. Steklov. **203** (1994), no. Izbran. Voprosy Mat. Fiz. i Anal., 255–270; English transl., Proc. Steklov Inst. Math. **3(203)** (1995), 221–232. MR1382609

[12] B. Dragović, *On p-adic series with rational sums*, Sci. Rev. Ser. Sci. Eng. **19-20** (1996), 97–104. Symposium Dedicated to the Memory of Đuro Kurepa (Belgrade, 1996). MR1437391

[13] B. Dragovich, *On some p-adic series with factorials*, p-adic functional analysis (Nijmegen, 1996), Lecture Notes in Pure and Appl. Math., vol. 192, Dekker, New York, 1997, pp. 95–105. MR1459205

[14] B. Dragovich, *On p-adic power series*, p-adic functional analysis (Poznań, 1998), Lecture Notes in Pure and Appl. Math., vol. 207, Dekker, New York, 1999, pp. 65–75. MR1702048

[15] B. Dragović, *On some finite sums with factorials*, Facta Univ. Ser. Math. Inform. **14** (1999), 1–10. MR2015580

[16] M. R. Murty and S. Sumner, *On the p-adic series* $\sum_{n=1}^{\infty} n^k \cdot n!$, Number Theory, CRM Proc. Lecture Notes, vol. 36, Amer. Math. Soc., Providence, RI, 2004, pp. 219–227. MR2076598

[17] N. C. Alexander, Non-vanishing of Uppuluri-Carpenter numbers, *Preprint*; URL: http://tinyurl.com/oo36das.

[18] P. K. Saikia and D. Subedi, *Bell numbers, determinants and series*, Proc. Indian Acad. Sci. Math. Sci. **123** (2013), no. 2, 151–166, DOI 10.1007/s12044-013-0128-5. MR3055434

[19] D. Subedi, *Complementary Bell numbers and p-adic series*, J. Integer Seq. **17** (2014), no. 3, Article 14.3.1, 14 pp. MR3168683

[20] M. de Gosson, B. Dragovich, and A. Khrennikov, *Some p-adic differential equations*, p-adic functional analysis (Ioannina, 2000), Lecture Notes in Pure and Appl. Math., vol. 222, Dekker, New York, 2001, pp. 91–102. MR1838284

[21] B. Dragovich and N. Z. Misic, *p-adic invariant summation of some p-adic functional series*, p-Adic Numbers Ultrametric Anal. Appl. **6** (2014), no. 4, 275–283, DOI 10.1134/S2070046614040025. MR3286072

[22] B. Dragovich, A. Yu. Khrennikov, and N. Ž. Mišić, *Summation of p-adic functional series in integer points*, Filomat **31** (2017), no. 5, 1339–1347. MR3630051

[23] N. J. A. Sloane, The on-line encyclopedia of integer sequences, https://oeis.org/.

INSTITUTE OF PHYSICS, UNIVERSITY OF BELGRADE, BELGRADE, SERBIA

MATHEMATICAL INSTITUTE, SERBIAN ACADEMY OF SCIENCES AND ARTS, BELGRADE, SERBIA
Email address: dragovich@ipb.ac.rs

Contemporary Mathematics
Volume **704**, 2018
http://dx.doi.org/10.1090/conm/704/14165

Spectrum of ultrametric Banach algebras of strictly differentiable functions

Alain Escassut and Nicolas Maïnetti

ABSTRACT. Let $\mathbb{K}$ be an ultrametric complete field and let E be an open subset of $\mathbb{K}$ of strictly positive codiameter. Let $\mathcal{D}(E)$ be the Banach $\mathbb{K}$-algebra of bounded strictly differentiable functions from E to $\mathbb{K}$, a notion whose definition is detailed. It is shown that all elements of $\mathcal{D}(E)$ have a derivative that is continuous in E. Given a positive number $r > 0$, all functions that are bounded and are analytic in all open disks of diameter r are strictly differentiable. Maximal ideals and continuous multiplicative semi-norms on $\mathcal{D}(E)$ are studied by recalling the relation of contiguity on ultrafilters: an equivalence relation. So, the maximal spectrum of $\mathcal{D}(E)$ is in bijection with the set of equivalence classes with respect to contiguity. Every prime ideal of $\mathcal{D}(E)$ is included in a unique maximal ideal and every prime closed ideal of $\mathcal{D}(E)$ is a maximal ideal, hence every continuous multiplicative semi-norm on $\mathcal{D}(E)$ has a kernel that is a maximal ideal. If $\mathbb{K}$ is locally compact, every maximal ideal of $\mathcal{D}(E)$ is of codimension 1. Every maximal ideal of $\mathcal{D}(E)$ is the kernel of a unique continuous multiplicative semi-norm and every continuous multiplicative semi-norm is defined as the limit along an ultrafilter on E. Consequently, the set of continuous multiplicative semi-norms defined by points of E is dense in the whole set of all continuous multiplicative semi-norms. The Shilov boundary of $\mathcal{D}(E)$ is equal to the whole set of continuous multiplicative semi-norms. Many results are similar to those concerning algebras of uniformly continuous functions but some specific proofs are required.

Introduction and preliminaries

Let $\mathbb{K}$ be a field which is complete with respect to an ultrametric absolute value that will be denoted by $| \,.\, |$.

Consider a Banach $\mathbb{K}$-algebra T. It is well known that the set of maximal ideals is not sufficient to describe spectral properties of T: we have to consider the set of continuous multiplicative semi-norms [**4**], [**5**], [**7**], [**8**], [**9**], [**10**], [**11**], [**12**], [**13**], [**14**]. Many studies were made on continuous multiplicative semi-norms on algebras of analytic functions, analytic elements and their applications to holomorphic functional calculus [**1**], [**3**], [**4**]. Continuous multiplicative semi-norms of the Banach algebras of bounded continuous functions and those of bounded uniformly continuous functions were studied in [**10**], [**11**]. After these studies, it would be natural to look for Banach algebras of derivable functions. Unfortunately, such Banach

2010 *Mathematics Subject Classification.* Primary 46S10; Secondary 12J25.
Key words and phrases. Strictly differentiable functions, ultrametric fields.

algebras do not seem to exist, as shows the remark below. This is why it seems interesting to consider strictly differentiable functions.

Definitions and notation: Throughout the paper, we denote by E an open subset of $\mathbb{K}$. Given $a \in \mathbb{K}$ and $r > 0$, we set $d(a, r) = \{x \in \mathbb{K} \mid |x - a| \leq r\}$, $d(a, r^-) = \{x \in \mathbb{K} \mid |x - a| < r\}$ $d_E(a, r) = \{x \in E \mid |x - a| \leq r\}$ and $d_E(a, r^-) = \{x \in E \mid |x - a| < r\}$.

We denote by δ the distance between two subsets of $\mathbb{K}$: given two subsets B_1, B_2 of $\mathbb{K}$, we set $\delta(B_1, B_2) = \inf\{|x - y| \mid x \in B_1,\ y \in B_2\}$. We denote by diam the diameter of a subset B of $\mathbb{K}$ and we set $\text{codiam}(B) = \delta(B, \mathbb{K} \setminus B)$. Similarly, given a subset B of E, we set $\text{codiam}_E(B) = \delta(B, E \setminus B)$. A subset B of E will be said to be *uniformly open in* E or *uniformly open subset of* E if $\text{codiam}_E(B) > 0$.

Given a bounded function f from E to $\mathbb{K}$, we put $\|f\|_0 = \sup_{x \in E} |f(x)|$.

Given a disk $d(a, R^-)$, we denote by $\mathcal{A}(d(a, R^-))$ the $\mathbb{K}$-algebra of analytic functions in $d(a, R^-)$ i.e. the set of power series $\sum_{n=0}^{\infty} a_n(x - a)^n$ of radius of convergence $r \geq R$. For every $r \in]0, s]$, we call r-analytic function a function from E to $\mathbb{K}$ whose restriction in all disks $d(a, r^-) \subset E$ belongs to $\mathcal{A}(d(a, r^-))$ and we denote by $\mathcal{A}_b(E, r)$ the set of r-analytic functions bounded in E.

Let $D = \{(x, x) \mid x \in E\}$ and let $\mathcal{D}(E)$ be the $\mathbb{K}$-vector space of bounded functions f from E to $\mathbb{K}$ such that the mapping ϕ defined in $(E \times E) \setminus D$ into $\mathbb{K}$ as $\phi(x, y) = \dfrac{f(x) - f(y)}{x - y}$ is bounded by a bound M_f and expands to a continuous function from $E \times E$ to $\mathbb{K}$. The functions $f \in \mathcal{D}(E)$ will be called *the strictly differentiable functions from E to* $\mathbb{K}$. Given $f \in \mathcal{D}(E)$, we put $\|f\|_1 = \sup_{(x,y) \in (E \times E) \setminus D} \phi(x, y)$ and we check that $\|\ .\ \|_1$ is another $\mathbb{K}$-vector space norm on $\mathcal{D}(E)$. Finally we put $\|f\| = \max(\|f\|_0, \|f\|_1)$.

REMARK 1. A uniformly open subset in $\mathbb{K}$ is open and closed.

REMARK 2. Suppose $\mathbb{K}$ is algebraically closed and let $E = d(0, 1)$. For every $r \in [0, 1] \cap |\mathbb{K}|$, we denote by $\xi(r)$ an element b of E such that $|b|^2 = r$. The set $E \setminus \{0\}$ obviously admits a partition of the form $\{d(a_j, |a_j|^-)_{j \in I}\}$.

Now, let f be the function defined in E in the following way. Given $x \in d(a_j, |a_j|^-)$, we put $f(x) = \xi(|a_j|)$ and $f(0) = 0$. In this way, f is constant in each disk $d(a_j, |a_j|^-)$, $a \in E$ and therefore f has a derivative equal to 0 at each point $a \in E \setminus \{0\}$ but f has no derivative at 0 because

$$\lim_{x \to 0} \left| \frac{f(x)}{x} \right| = \lim_{x \to 0} \frac{\sqrt{|x|}}{|x|} = +\infty.$$

Now, let us take a decreasing sequence $(r_n)_{n \in \mathbb{N}}$ in $|\mathbb{K}|$, of limit 0 and for each $n \in \mathbb{N}$, let f_n be the function defined in E by $f_n(x) = f(x)$ for every $x \in E \setminus d(0, r_n)$ and $f_n(x) = \xi(a_j)$ for every $x \in d(0, r_n)$, with $a_j \in d(0, r_n)$. Thus, we can check that f_n has a derivative equal to 0 in all E. Therefore, the sequence $(f_n')_{n \in \mathbb{N}}$ trivially is uniformly convergent to the function that is identically zero in all E.

On the other hand, consider $\|f_n - f\|_0$. By construction, we check $|f(x)| \leq \sqrt{r_n}$ for every $x \in d(0, r_n)$ and hence $|f(x) - f_n(x)| \leq \sqrt{r_n}$ for every $x \in d(0, r_n)$. But since $f_n(x) = f(x)$ for every $x \in E \setminus d(0, r_n)$, we derive $\|f_n - f\|_0 \leq \sqrt{r_n}$. Consequently, the sequence $(f_n)_{n \in \mathbb{N}}$ is uniformly convergent in E to f. And f is not derivable at 0, although the sequence $(f'_n)_{n\in\mathbb{N}}$ is also uniformly convergent in E.

Theorems 1 and 2 are designed to recall properties that we can also find in [**16**], with some differences on hypotheses or definitions.

THEOREM 1. *Every function $f \in \mathcal{D}(E)$ is uniformly continuous, derivable in E and f' is bounded and continuous in E. Moreover, if E is compact, then a function from E to $\mathbb{K}$ belongs to $\mathcal{D}(E)$ if and only if for every $a \in E$, $\dfrac{f(x)-f(y)}{x-y}$ has a limit when x and y tend to a separately, while being distinct.*

PROOF. Let $f \in \mathcal{D}(E)$. Then f is Lipschitzian and hence uniformly continuous in E. Next, for every $a \in E$, $\dfrac{f(x)-f(y)}{x-y}$ has a limit $l(a)$ when x and y tend to a separately, while being distinct. For each $\epsilon > 0$ and let $r(\epsilon) > 0$ be such that $\left|\dfrac{f(x)-f(y)}{x-y} - l\right| \leq \epsilon$ for all $x,\ y \in d_E(a, r(\epsilon))$, $x \neq a$, $y \neq a$, $x \neq y$. Without loss of generality, we can assume $|x-a| = r(\epsilon) \geq |y-a|$. Now, if y tends to a, the inequality remains and since f is continuous, we obtain $\left|\dfrac{f(x)-f(a)}{x-a} - l(a)\right| \leq \epsilon$. Consequently, $\lim\limits_{x\to a} \dfrac{f(x)-f(a)}{x-a} = l(a)$. And hence $\dfrac{f(x)-f(y)}{x-y}$ has a limit when x and y tend to a separately, while being distinct, even if x or y is equal to a. Particularly, f has a derivative equal to $l(a)$ at each point $a \in E$.

By definition $\left|\dfrac{f(x)-f(y)}{x-y}\right|$ is bounded in E, hence, so is $\left|\dfrac{f(x)-f(a)}{x-a}\right|$ and hence so is $|f'(x)|$. Finally, let us take $b \in d_E(a, r(\epsilon))$. We have $\left|\dfrac{f(b)-f(a)}{b-a} - f'(a)\right| \leq \epsilon$ and $\left|\dfrac{f(y)-f(x)}{y-x} - f'(a)\right| \leq \epsilon$ for all $x,\ y \in d_E(a, r(\epsilon))$, hence $\left|\dfrac{f(y)-f(b)}{y-b} - f'(a)\right| \leq \epsilon$ for every $y \in d_E(a, r(\epsilon))$ hence $|f'(b) - f'(a)| \leq \epsilon$, which proves that f' is continuous.

Now, suppose that E is compact and let f be a function from E to $\mathbb{K}$ such that for every $a \in E$, $\dfrac{f(x)-f(y)}{x-y}$ has a limit $l(a)$ when x and y tend to a separately, while being distinct.

We will check that $\left|\dfrac{f(x)-f(y)}{x-y}\right|$ is bounded by a number M_f. Suppose that it is not bounded. Then, there exists a sequence of pairs $(x_n, y_n)_{n\in\mathbb{N}}$ in $E \times E$ such that $\lim\limits_{n\to+\infty} \left|\dfrac{f(x_n)-f(y_n)}{x_n-y_n}\right| = +\infty$. Since E is compact we can extract from the sequence $(x_n, y_n)_{n\in\mathbb{N}}$ a subsequence converging to a pair $(a, b) \in E \times E$. So, without loss of generality, we can assume that the sequence $(x_n, y_n)_{n\in\mathbb{N}}$ converges to (a, b). If $a \neq b$, then we have a contradiction due to the continuity of f. And if $a = b$, then we have a contradiction due to the very hypothesis. In the same way, by fixing y, this proves that f is bounded in E. □

REMARK 3. A function f from E to $\mathbb{K}$ which is derivable with a continuous derivative is not automatically strictly differentiable. Indeed, suppose E has an infinite residue class field. We can find a sequence $(a_n)_{n\in\mathbb{N}}$ such that $|a_n| = |a_n -$

$a_m| = 1$ for all $n \neq m$. Now take a sequence $(r_n)_{n\in\mathbb{N}}$ of $\mathbb{R}^+$ with $\lim_{n\to+\infty} r_n = 0$. Then the set $E \setminus \bigcup_{n=0}^{\infty} d(a_n, r_n^-)$ is open. Now, take $b_n \in d(a_n, r_n)$ such that $|a_n - b_n| = r_n$ for every $n \in \mathbb{N}$.

We can define a function f from E to $\mathbb{K}$ such that $f(x) = 0$ for all $x \in \bigcup_{n=0}^{\infty} d(a_n, r_n^-)$ and $f(x) = 1$ for every $x \in E \setminus \bigcup_{n=0}^{\infty} d(a_n, r_n^-)$. Of course, f is derivable and f' is continuous in E. However, $\left|\frac{f(a_n) - f(b_n)}{a_n - b_n}\right| = \frac{1}{r_n}$ and therefore $\left|\frac{f(x) - f(y)}{x - y}\right|$ is not bounded in E.

THEOREM 2. *$\| \, . \, \|$ is a norm of $\mathbb{K}$-vector space on $\mathcal{D}(E)$ and $\mathcal{D}(E)$ is complete for that norm.*

PROOF. Both $\| \, . \, \|_0$ and $\| \, . \, \|_1$ are norms of $\mathbb{K}$-vector space, hence so is $\| \, . \, \|$. Let us show that $\mathcal{D}(E)$ is complete for that norm.

Let g be a function from E to $\mathbb{K}$ that belongs to the completion of $\mathcal{D}(E)$ with respect to the norm $\| \, . \, \|$. For all $\epsilon > 0$, there exists $f_\epsilon \in \mathcal{D}(E)$ such that $\|f_\epsilon - g\| \leq \epsilon$. Since f_ϵ is bounded, so is g. Next, we check that $\|f_\epsilon - g\|_1 \leq \epsilon$, hence

$$\left|\frac{(f_\epsilon - g)(x) - (f_\epsilon - g)(y)}{x - y}\right| \leq \epsilon$$

for all $x,\ y \in (E \times E) \setminus D$. Therefore since $\left|\frac{f_\epsilon(x) - f_\epsilon(y)}{x - y}\right|$ is bounded by $\|f_\epsilon\|_1$ in $(E \times E) \setminus D$, so is $\left|\frac{g(x) - g(y)}{x - y}\right|$. Now we have to check that $\left|\frac{g(x) - g(y)}{x - y}\right|$ expands to a continuous function from $E \times E$ to $\mathbb{K}$. Let $(a, a) \in D$ and let us fix $\epsilon > 0$. By definition, $\left|\frac{f_\epsilon(x) - f_\epsilon(y)}{x - y}\right|$ has a limit $(f_\epsilon)'(a)$ when x and y tend to a. Consequently, there exists a disk $d_E(a, r)$ in E such that for all $x,\ y \in d_E(a, r),\ x \neq y$,

$$\left|\frac{f_\epsilon(x) - f_\epsilon(y)}{x - y} - (f_\epsilon)'(a)\right| \leq \epsilon. \tag{1}$$

But since $\|f_\epsilon - g\|_1 \leq \epsilon$, we have

$$\left|\frac{(f_\epsilon(x) - g(x)) - (f_\epsilon(y) - g(y))}{x - y}\right| \leq \epsilon$$

for all $x,\ y \in d_E(a, r)$. Consequently, by (1) we derive (2) for all $x, y \in d_E(a, r),\ x \neq y$:

$$\left|\frac{g(x) - g(y)}{x - y} - (f_\epsilon)'(a)\right| \leq \epsilon. \tag{2}$$

Now, let us take $\omega \in]0, \epsilon[$ and take $f_\omega \in \mathcal{D}(E)$ such that $\|f_\omega - g\| \leq \omega$. There exists a disk $d_E(a, s)$ in E, with $0 < s \leq r$ such that

$$\left|\frac{f_\omega(x) - f_\omega(y)}{x - y} - (f_\omega)'(a)\right| \leq \omega$$

for all $x,\ y \in d_E(a, s),\ x \neq y$. Consequently, when $x,\ y \in d_E(a, s)$, we can see that $|(f_\omega)'(a) - (f_\epsilon)'(a)| \leq \epsilon$. Therefore each interval $[|(f_\epsilon)'(a)| - \epsilon, |(f_\epsilon)'(a)| + \epsilon]$ contains $\frac{g(x) - g(y)}{x - y}$ whenever $x,\ y \in d_E(a, s),\ x \neq y$. Therefore the family of intervals $[|(f_\epsilon)'(a)|-\epsilon, |(f_\epsilon)'(a)|+\epsilon]$ is a basis of a Cauchy filter whose limit $L(a)$, by (2), is the limit of $\dfrac{g(x)-g(y)}{x-y}$ when x and y tend to a, which finishes the proof that g belongs to $\mathcal{D}(E)$. □

THEOREM 3. *$\mathcal{D}(E)$ is a $\mathbb{K}$-algebra and $\| \ . \ \|$ is a norm of $\mathbb{K}$-algebra.*

PROOF. Let $f,\ g\in\mathcal{D}(E)$. Given $x,\ y\in E$, we have

$$\frac{f(x)g(x)-f(y)g(y)}{x-y}=f(x)\frac{g(x)-g(y)}{x-y}+g(y)\frac{f(x)-f(y)}{x-y}$$

hence the limit at a of $\dfrac{f(x)g(x)-f(y)g(y)}{x-y}$ when $x,\ y$ tend to a with $x\neq y$ is $f(a)g'(a)+g(a)f'(a)$. Next, we can easily check that $\|fg\|=\max\Big(\|fg\|_0,\|fg\|_1\Big)\leq\max\Big(\|f\|_0\|g\|_0,\|fg\|_1\Big)$. Now $\|fg\|_1$

$$=\sup_{x,y}\Big|\frac{f(x)g(x)-f(y)g(y)}{x-y}\Big|=\max\Big(\sup_{x,y}\Big|\frac{f(x)(g(x)-g(y)}{x-y}\Big|,\sup_{x,y}\Big|\frac{g(y)(f(x)-f(y)}{x-y}\Big|\Big).$$

Consequently, $\|fg\|_1\leq\|f\|\|g\|$ and therefore $\|fg\|\leq\|f\|\|g\|$. □

COROLLARY 3.1. *$\mathcal{D}(E)$ is a Banach $\mathbb{K}$-algebra.*

THEOREM 4. *Let $r\in]0,s]$. Suppose E uniformly open in $\mathbb{K}$, of codiameter s and let $r\in]0,s]$. Then $\mathcal{A}_b(E,r)$ provided with the norm $\| \ . \ \|_E$ is a Banach $\mathbb{K}$-algebra included in $\mathcal{D}(E)$.*

PROOF. By definition, E admits a partition by a family of disks $(E_i)_{i\in I}$ of diameter r. Now, it is easily seen that $\mathcal{A}(E,r)$ is a $\mathbb{K}$-algebra algebraically and topologically isomorphic to the direct product $\prod_{i\in I}\mathcal{A}(E_i,r)$. Let us show that $\mathcal{A}_b(E,r)$ is included in $\mathcal{D}(E)$. Let $f\in\mathcal{A}_b(E,r)$, let $M=\|f\|_E$ and take $x,\ y\in E$. We will prove that $\Big|\dfrac{f(x)-f(y)}{x-y}\Big|$ is bounded in E.

Suppose first that $|x-y|\geq r$. Then $\Big|\dfrac{f(x)-f(y)|}{x-y}\Big|\leq\dfrac{M}{r}$. Suppose now that $|x-y|<r$. Then both x and y belong to a same disk E_i. Clearly, we can assume that E_i is the disk $d(0,r^-)$ without loss of generality and then, inside that disk, $f(t)$ is a power series $\sum_{n=0}^{+\infty}a_nt^n$, with $\sup_{n\in\mathbb{N}}|a_n|r^n\leq M$. Consequently,

$$\frac{f(x)-f(y)}{x-y}=\sum_{n=1}^{+\infty}a_n\sum_{j=0}^{n-1}x^jy^{n-1-j}$$

and therefore,

$$\Big|\frac{f(x)-f(y)}{x-y}\Big|\leq\sup_{n\geq1}|a_n|r^{n-1}\leq\lim_{s\to r^-}\frac{|f|(s)}{s}\leq\frac{M}{r}.$$

Consequently, we have proven that

$$\Big|\frac{f(x)-f(y)}{x-y}\Big|\leq\frac{M}{r}\tag{1}$$

for all $x,y\in E$.

Now when x and y approach independently to a same point $a \in E$, both x and y belong to a same disk E_i and we can again assume that $E_i = d(0, r^-)$. Moreover, without loss of generality, we can also assume that $a = 0$. Then we have again

$$\frac{f(x) - f(y)}{x - y} = \sum_{n=1}^{+\infty} a_n \sum_{j=0}^{n-1} x^j y^{n-1-j}.$$

When x and y go separately to 0, clearly the sum goes to a_1 which is $f'(0)$. Therefore $\mathcal{A}_b(E, r)$ is a sub-algebra of $\mathcal{D}(E)$.

On the other hand, by (1), we can see that $\|f\|_1 \leq \frac{\|f\|_E}{r}$ therefore the norms $\| \,.\, \|_E$ and $\| \,.\, \|$ are equivalent on $\mathcal{A}_b(E, r)$.

Let us show that $\mathcal{A}_b(E, r)$ is closed with respect to the norm $\| \,.\, \|_E$ and hence with respect to the norm $\| \,.\, \|$. Let $(f_n)_{n \in \mathbb{N}}$ be a Cauchy sequence of $\mathcal{A}_b(E, r)$. For every disk $d(a, r^-)$, it is a Cauchy sequence of $\mathcal{A}_b(d(a, r^-))$ with respect to the norm $\| \,.\, \|_{d(a,r^-)}$, hence the sequence converges to a function $g_a \in \mathcal{A}_b(d(a, r^-))$. Then the function f defined in E as $f(x) = g_a(x)$ for every $x \in d(a, r^-)$ is clearly the limit of the sequence $(f_n)_{n \in \mathbb{N}}$ with respect to the norm $\| \,.\|_E$. On the other hand, since the sequence (f_n) is a Cauchy sequence of bounded functions, its limit is obviously bounded. Thus, $\mathcal{A}_b(E, r)$ is closed in $\mathcal{D}(E)$. □

The role of ultrafilters here is essential as in a few previous works [**10**]. They were deeply studied in [**15**].

Notation and definitions. Let $\mathcal{F}$ be a filter on E. Given a function f from E to $\mathbb{K}$ admitting a limit along $\mathcal{F}$, we will denote by $\lim_{\mathcal{F}} f(x)$ that limit.

Let $Ul(E)$ be the set of ultrafilters on E. Two filters $\mathcal{F}, \mathcal{G}$ on E will be said to be *contiguous* if for every $H \in \mathcal{F}$, $L \in \mathcal{G}$, we have $\delta(H, L) = 0$. We shall denote by $(\mathcal{T})$ the relation defined on $Ul(E)$ as $\mathcal{U}(T)\mathcal{V}$ if $\mathcal{U}$ and $\mathcal{V}$ are contiguous.

An ultrafilter $\mathcal{U}$ on the set E is said to be *principal* if it converges to a point $a \in E$.

REMARK 4. Let $\mathcal{U}, V$ be contiguous ultrafilters on E and assume $\mathcal{U}$ is convergent. Then $\mathcal{V}$ is convergent and has the same limit as $\mathcal{U}$.

The following Lemma L is immediate:

LEMMA L. *Let $X_1, \ldots, X_q$ be uniformly open subsets of E. Then if $\bigcap_{j=1}^{q} X_j$ is not empty, it is uniformly open in E.*

THEOREM 5. *Let $X \subset E$ be uniformly open in E and let u be defined as $u(x) = 1$ for every $x \in X$ and $u(x) = 0$ for every $x \notin X$. Then u belongs to $\mathcal{D}(E)$.*

PROOF. Let $r = \text{codiam}(X)$. Of course, we have $\frac{u(x) - u(y)}{x - y} = 0$ for all $(x,\ y) \in X \times X,\ x \neq y$. Finally, let $x \in X,\ y \in (E \setminus X)$. Then $\left|\frac{u(x) - u(y)}{x - y}\right| \leq \frac{1}{r}$. Moreover, if x and y tend to a point $a \in E$, then we can assume that either both belong to X or both belong to $E \setminus X$, hence $\frac{u(x) - u(y)}{x - y} = 0$. □

NOTATION. Let $f_1, ..., f_q \in \mathcal{D}(E)$ and let $\epsilon > 0$. We set $W(f_1, ..., f_q, \epsilon) = \{x \in E \mid |f(x)| \le \epsilon\}$.

THEOREM 6. *Let $f_1, ..., f_q \in \mathcal{D}(E)$, let $\epsilon > 0$. Then if $W(f_1, ..., f_q, \epsilon)$ is not empty, it is uniformly open in E.*

PROOF. By Lemma L, we can assume $q = 1$. So, consider a set $W(f, \epsilon)$. Since f is uniformly continuous, there exists $r(\epsilon) > 0$ such that $|f(x) - f(y)| \le \epsilon$ for all $x, y \in E$ such that $|x - y| \le r(\epsilon)$. Let $a \in E$ be such that $|f(a)| > \epsilon$. Then $|f(x)| > \epsilon$ for every $x \in d_E(a, r(\epsilon))$, which proves that $W(f, \epsilon)$ is uniformly open in E. □

Theorems 7 and 8 were proven in [**10**].

THEOREM 7. *Let $\mathcal{U}, \mathcal{V}$ be two ultrafilters on E that are not contiguous. There exist uniformly open subsets $H \in \mathcal{U}$, $L \in \mathcal{V}$ of E and $f \in \mathcal{D}(E)$ such that $f(x) = 1$ for every $x \in H$, $f(x) = 0$ for every $x \in L$.*

PROOF. Let $F \in \mathcal{U}$, $G \in \mathcal{V}$ be such that $\delta(F, G) = r > 0$. Let $H = \bigcup_{b \in F} d_E(b, r^-)$. Then $F \cap G = \emptyset$.

Let f be defined as $f(x) = 1$ for every $x \in H$ and $f(x) = 0$ for every $x \notin H$. Then $f(x) - f(y) = 0$ for every $x, y \in H$ and $f(x) - f(y) = 0$ for all $x, y \notin H$. Now, take $x \in H$ and $y \notin H$. We have $\frac{f(x) - f(y)}{x - y} = \frac{1}{x - y}$ and hence $\left|\frac{f(x) - f(y)}{x - y}\right| \le \frac{1}{r}$. Consequently, $\frac{f(x) - f(y)}{x - y}$ is bounded in E. Moreover, when x and y tend to a point $a \in E$, either both x, y belong to H or both x, y do not belong to H, therefore $\frac{f(x) - f(y)}{x - y} = 0$, so f belongs to $\mathcal{D}(E)$. On the other hand, by construction, H belongs to $\mathcal{U}$, and putting $L = E \setminus H$, we see that $G \subset L$, hence L belongs to $\mathcal{V}$. Then, f satisfies $f(x) = 1$ for every $x \in H$, $f(x) = 0$ for every $x \in L$. Moreover, by construction, we check that both H and L are uniformly open in E, which ends the proof. □

NOTATION. We denote by E' another open subset of $\mathbb{K}$. Let f be a mapping from E to E' and let $\mathcal{U}$ be an ultrafilter on E. We denote by $\overline{f}(\mathcal{U})$ the ultrafilter admitting the basis $f(\mathcal{U})$.

LEMMA M. *Let f be a uniformly continuous mapping from E to E'. Let $\mathcal{U}$, $\mathcal{V}$ be contiguous. Then $\overline{f}(\mathcal{U})$ and $\overline{f}(\mathcal{V})$ are contiguous.*

PROOF. Set $\mathcal{U}' = \overline{f}(\mathcal{U})$ and $\mathcal{V}' = \overline{f}(\mathcal{V})$ and let $X \in \mathcal{U}'$, $Y \in \mathcal{V}'$. The family of subsets $f^{-1}(f(\mathcal{U}))$ is obviously included in $\mathcal{U}$, hence the set $P = f^{-1}(X)$ belongs to $\mathcal{U}$ and similarly, the set $Q = f^{-1}(Y)$ belongs to $\mathcal{V}$. Now, we have $\delta(P, Q) = 0$, so there exist a sequence $(a_n)_{n \in \mathbb{N}}$ of P and a sequence $(b_n)_{n \in \mathbb{N}}$ of Q such that $\lim_{n \to +\infty} (a_n - b_n) = 0$. But since f is uniformly continuous, in E' we have $\lim_{n \to +\infty} \delta'(f(a_n), f(b_n)) = 0$ and hence $\delta'(X, Y) = 0$ i.e. $\mathcal{U}'(\mathcal{T})\mathcal{V}'$. □

NOTATION. Given a uniformly continuous mapping f from E to E' and a class of contiguity H on E, we will denote by $\overline{f}(H)$ the class of contiguity on E': $\{\overline{f}(\mathcal{U}) \mid \mathcal{U} \in H\}$.

Given a filter $\mathcal{F}$ on E, we will denote by $\mathcal{I}(\mathcal{F})$ the ideal of the $f \in \mathcal{D}(E)$ such that $\lim\limits_{\mathcal{F}} f(x) = 0$. We will denote by $\mathcal{I}^*(\mathcal{F})$ the ideal of the $f \in \mathcal{D}(E)$ such that there exists a subset $L \in \mathcal{F}$ such that $f(x) = 0$ for every $x \in L$. Given $a \in E$ we will denote by $\mathcal{I}(a)$ the ideal of the $f \in \mathcal{D}(E)$ such that $f(a) = 0$.

We will denote by $Max(\mathcal{D}(E))$ the set of maximal ideals of $\mathcal{D}(E)$ and by $Max_E(\mathcal{D}(E))$ the set of maximal ideals of the form $\mathcal{I}(a)$, $a \in E$.

The proof of Theorem 8 is easy and is not specific to the algebra $\mathcal{D}(E)$ [**10**].

THEOREM 8. *Given an ultrafilter $\mathcal{U}$ on E, $\mathcal{I}(\mathcal{U})$, $\mathcal{I}^*(\mathcal{U})$ are prime ideals of $\mathcal{D}(E)$.*

NOTATION. We will denote by $|\cdot|_\infty$ the Archimedean absolute value of $\mathbb{R}$.

THEOREM 9. *Let $\mathcal{U}$, $\mathcal{V}$ be two ultrafilters on E. Then $\mathcal{I}(\mathcal{U}) = \mathcal{I}(\mathcal{V})$ if and only if $\mathcal{U}$ and $\mathcal{V}$ are contiguous.*

PROOF. Suppose that $\mathcal{U}$, $\mathcal{V}$ are not contiguous. By Theorem 7, there exist $H \in \mathcal{U}$, $L \in \mathcal{V}$ and $f \in \mathcal{D}(E)$ such that $f(x) = 1$ for every $x \in H$, $f(x) = 0$ for every $x \in L$. Consequently, f belongs to $\mathcal{I}(\mathcal{U})$ but does not belong to $\mathcal{I}(\mathcal{V})$. Thus, $\mathcal{I}(\mathcal{F}) \neq \mathcal{I}(\mathcal{G})$.

Now, suppose that $\mathcal{U}$, $\mathcal{V}$ are contiguous. Let $f \in \mathcal{I}(\mathcal{U})$. Since $\mathcal{U}, \mathcal{V}$ are ultrafilters and since f is bounded, $|f(x)|$ admits limits on both filters. Let $l = \lim\limits_{\mathcal{V}} |f(x)|$, suppose $l > 0$ and let $L \in \mathcal{V}$ be such that $\Big| |f(x)| - l \Big|_\infty \leq \frac{l}{3}$ for every $x \in L$, hence $|f(x)| \geq \frac{2l}{3}$ for every $x \in L$. Let $H \in \mathcal{U}$ be such that $|f(x)| \leq \frac{l}{3}$ for all $x, y \in d_E(a, r)$, $x \neq y$. $x \in H$. Since $f \in \mathcal{D}(E)$, f is uniformly continuous, hence there exists $\rho > 0$ such that $\delta(x, y) \leq \rho$ implies $|f(x) - f(y)| \leq \frac{l}{4}$. And there exist $a \in H$, $b \in L$ such that $|a - b| \leq \rho$, hence $|f(a) - f(b)| \leq \frac{l}{4}$, a contradiction because $|f(a)| \leq \frac{l}{3}$ and $|f(b)| \geq \frac{2l}{3}$. □

REMARK 5. As noticed in [**10**], Relation $(\mathcal{T})$ is not transitive in the case of the set of all filters on E. However, given a topological space X satisfying the normality axiom, particularly, given a metric space X, then $(\mathcal{T})$ is transitive for ultrafilters and therefore is an equivalence relation on $Ul(X)$ [**10**].

NOTATION. We will denote by $Y_{(\mathcal{T})}(E)$ the set of equivalence classes on $Ul(E)$ with respect to Relation $(\mathcal{T})$. Given $H \in Y_{(\mathcal{T})}(E)$, we will denote by $\mathcal{I}(H)$ the ideal $\mathcal{I}(\mathcal{U})$, $\mathcal{U} \in H$.

Let $f \in \mathcal{D}(E)$ and let ϵ be > 0. We set $B(f, \epsilon) = \{x \in E \mid |f(x)| \leq \epsilon\}$.

Theorem 10 looks like certain Bezout-Corona statements [**8**], [**17**], [**9**], [**10**]. The proof is close to that given in [**10**] but here the functions in $\mathcal{D}(E)$ have more properties, allowing for a more specific proof.

THEOREM 10. *Let $f_1, ..., f_q \in \mathcal{D}(E)$ satisfying $\inf\limits_{x \in E}(\max\limits_{1 \leq j \leq q} |f_j(x)|) > 0$. Then there exist*

$g_1, ..., g_q \in \mathcal{D}(E)$ such that $\sum\limits_{j=1}^{q} f_j(x) g_j(x) = 1$ for all $x \in E$.

PROOF. Let $M = \inf_{x\in E}(\max_{1\leq j\leq q}|f_j(x)|)$. Let $H_j = \{x \in E \mid |f_j(x)| \geq M\}$, $j = 1, ..., q$ and let $F_j = \bigcup_{m=1}^{j} H_m$, $j = 1, ..., q$. Let $g_1(x) = \frac{1}{f_1(x)}$ for all $x \in H_1$ and $g_1(x) = 0$ for all $x \in E\setminus H_1$. In H_1, we have $\dfrac{g_1(x) - g_1(y)}{x-y} = \dfrac{f_1(y) - f_1(x)}{f_1(x)f_1(y)(x-y)}$, hence $\dfrac{g_1(x) - g_1(y)}{x-y}$ converges at each point $a \in H_1$ to $g_1'(a)$.

Let us show that g_1 belongs to $\mathcal{D}(E)$. Let us fix $i \in I$ and let $h = \|f_1\|_1$. Let x, $y \in E_i$. If both x, y belong to $E_i \setminus H_1$, we have $g_1(x) - g_1(y) = 0$. If both x, y belong to H_1, we have $g_1(x) - g_1(y) = \dfrac{f_1(y) - f_1(x)}{f_1(x)f_1(y)(x-y)}$ hence $|g_1(x) - g_1(y)| \leq \dfrac{\|f_1\|_1}{M^2}$.

Finally, suppose $x \in E_1 \setminus H_1$ and $y \in H_1$. Now, we have $|g_1(x) - g_1(y)| = \dfrac{1}{|f_1(y)}| \leq \dfrac{1}{M}$. On the other hand, by hypothesis, $|f_1(x)| < M$, $|f_1(y)| \geq M$, hence $|f_1(x) - f_1(y)| \geq M$, therefore $\dfrac{1}{|x-y|} \leq \dfrac{\|f_1\|_1}{|f_1(x) - f_1(y)|}$, hence

$$\Big|\frac{g_1(x) - g_1(y)}{x-y}\Big| \leq \frac{\|f_1\|_1}{|f_1(y)(f_1(x) - f_1(y))|} \leq \frac{\|f_1\|_1}{M^2}.$$

Consequently, g_1 belongs to $\mathcal{D}(E)$.

Suppose now we have constructed $g_1, ..., g_k \in \mathcal{D}(E)$ satisfying $\sum_{j=1}^{k} f_j g_j(x) = 1$ for all $x \in F_k$ and $\sum_{j=1}^{k} f_j g_j(x) = 0$ for all $x \in E \setminus F_k$. Let g_{k+1} be defined on E by $g_{k+1}(x) = \dfrac{1}{f_{k+1}(x)}$ for every $x \in F_{k+1} \setminus F_k$ and $g_{k+1}(x) = 0$ for every $x \in E \setminus (F_{k+1} \setminus F_k)$. In the same way as we did for g_1, we can check that g_{k+1} belongs to $\mathcal{D}(E)$.

Next, we have $\sum_{j=1}^{k+1} f_j g_j(x) = 1$ for every $x \in F_{k+1}$ and $\sum_{j=1}^{k} f_j g_j(x) = 0$ for every $x \in E\setminus F_{k+1}$. So, by an immediate recurrence, we can get functions $g_1, ..., g_q \in \mathcal{D}(E)$ such that
$\sum_{j=1}^{q} f_j g_j(x) = 1$ for every $x \in E$, which ends the proof. □

COROLLARY 10.1. *Let I be an ideal of $\mathcal{D}(E)$ different from $\mathcal{D}(E)$. The family $B(f, \epsilon)$, $f \in I$, $\epsilon > 0$, generates a filter $\mathcal{F}_I$ on E such that $I \subset \mathcal{I}(\mathcal{F}_I)$.*

Main results

THEOREM 11. *Let $\mathcal{M}$ be a maximal ideal of $\mathcal{D}(E)$. There exists an ultrafilter $\mathcal{U}$ on E such that $\mathcal{M} = \mathcal{I}(\mathcal{U})$.*

PROOF. By Corollary 10.1, we have $\mathcal{M} \subset \mathcal{I}(\mathcal{F}_{\mathcal{M}})$. Let $\mathcal{U}$ be an ultrafilter thinner than $\mathcal{F}_{\mathcal{M}}$. Then $\mathcal{I}(\mathcal{F}_{\mathcal{M}}) \subset \mathcal{I}(\mathcal{U})$, hence $\mathcal{M} \subset \mathcal{I}(\mathcal{U})$. But since $\mathcal{M}$ is a maximal ideal of $\mathcal{D}(E)$, $\mathcal{M}$ is equal to $\mathcal{I}(\mathcal{U})$. □

COROLLARY 11.1. *For every maximal ideal $\mathcal{M}$ of $\mathcal{D}(E)$ there exists a unique $\mathcal{H} \in Y_{(\mathcal{T})}(E)$ such that $\mathcal{M} = \mathcal{I}(\mathcal{U})$ for every $\mathcal{U} \in \mathcal{H}$.*

Moreover, the mapping Ψ that associates to each $\mathcal{M} \in Max(\mathcal{D}(E))$ the unique $\mathcal{H} \in Y_{(\mathcal{T})}(E)$ such that $\mathcal{M} = \mathcal{I}(\mathcal{U})$ for every $\mathcal{U} \in \mathcal{H}$, is a bijection from $Max(\mathcal{D}(E))$ onto $Y_{(\mathcal{T})}(E)$.

THEOREM 12. *Let $\mathcal{U}$ be an ultrafilter on E such that, for every $f \in \mathcal{D}(E)$, $f(x)$ converges on $\mathcal{U}$ in $\mathbb{K}$. Then $\mathcal{I}(\mathcal{U})$ is of codimension 1.*

PROOF. For each $f \in \mathcal{D}(E)$ we set $\chi(f) = \lim_{\mathcal{U}} f(x)$. Then $\mathcal{I}(\mathcal{U})$ is the ideal $Ker(\chi)$ of $\mathcal{D}(E)$ and this is a maximal ideal because clearly $\chi(\mathcal{D}(E)) = \mathbb{K}$. So, $\mathcal{I}(\mathcal{U})$ is a maximal ideal of $\mathcal{D}(E)$ of codimension 1. □

COROLLARY 12.1. *Let $\mathcal{U}$ be a Cauchy ultrafilter on E. Then $\mathcal{I}(\mathcal{U})$ is of codimension 1.*

PROOF. If $\mathcal{U}$ is convergent, there is nothing to prove. Suppose now that $\mathcal{U}$ does not converge. Since the functions of $\mathcal{D}(E)$ are uniformly continuous, they have a continuation to the completion $\widetilde{E}$ of E and $\mathcal{U}$ defines an ultrafilter that converges in $\widetilde{E}$ to a point a. Given $f \in \mathcal{D}(E)$, let $\widetilde{f}$ be the continuation of f in $\widetilde{E}$: we have $\lim_{\mathcal{U}} f(x) = \widetilde{f}(a)$. So by Theorem 12, $\mathcal{M}$ is of codimension 1. □

COROLLARY 12.2. *Let $\mathbb{K}$ be a locally compact field. Then every maximal ideal of $\mathcal{D}(E)$ is of codimension 1.*

PROOF. Let $\mathcal{M}$ be a maximal ideal of $\mathcal{D}(E)$. By Corollary 9.1 there exists an ultrafilter $\mathcal{U}$ such that $\mathcal{M} = \mathcal{I}(\mathcal{U})$. Let $f \in \mathcal{D}(E)$. Since f is bounded, $f(E)$ is included in a compact subset B of $\mathbb{K}$. Consequently, f has a limit $\chi(f)$ along $\mathcal{U}$. Thus, the mapping χ from $\mathcal{D}(E)$ to $\mathbb{K}$ is a $\mathbb{K}$-algebra homomorphism and therefore $\mathcal{M}$ is of codimension 1. □

We will now examine prime closed ideals of $\mathcal{D}(E)$.

THEOREM 13. *Let $\mathcal{U}$ be an ultrafilter on E and let $\mathcal{P}$ be a prime ideal included in $\mathcal{I}(\mathcal{U})$. Let $L \in \mathcal{U}$ be uniformly open in E and let $H = E \setminus L$. Let u be the function defined on E by $u(x) = 1$ for every $x \in H$, $u(x) = 0$ for every $x \in L$. Then u belongs to $\mathcal{P}$.*

PROOF. By Theorem 7 we know that $u \in \mathcal{D}(E)$. By construction, $1 - u$ does not belong to $\mathcal{I}(\mathcal{U})$ because $\lim_{\mathcal{U}} u(x) = 1$. But $u(1-u) = 0$, hence u belongs to $\mathcal{P}$ because $\mathcal{P}$ is prime. □

COROLLARY 13.1. *Let $\mathcal{U}$ be an ultrafilter on E. The ideal of the $f \in \mathcal{D}(E)$ such that there exists a uniformly open subset $H \in \mathcal{U}$ of E such that $f(x) = 0$ for every $x \in H$ is included in every prime ideal $\mathcal{P} \subset \mathcal{I}(\mathcal{U})$.*

THEOREM 14. *The closure of a prime ideal of $\mathcal{D}(E)$ is a maximal ideal.*

PROOF. The proof is similar to Theorem 12 in [**10**]. Let $\mathcal{P}$ be a prime ideal of $\mathcal{D}(E)$ included in a maximal ideal $\mathcal{M} = \mathcal{I}(\mathcal{U})$. Let f belong to $\mathcal{I}(\mathcal{U})$. Let us take $\epsilon > 0$ and let us find $h \in \mathcal{P}$ such that $\|f - h\| \leq \epsilon$. By Theorem 7, we can find a uniformly open subset $L \in \mathcal{U}$ of E such that $|f(x)| \leq \epsilon$ for every $x \in L$. Let u be the characteristic function of $E \setminus L$. By Theorem 13, u belongs to $\mathcal{P}$ and hence so does uf. We then check that $\|f - uf\| \leq \epsilon$. Thus, $\mathcal{P}$ is dense in $\mathcal{M}$. □

COROLLARY 14.1. *Let $\mathcal{P}$ be a prime ideal of $\mathcal{D}(E)$. There exists a unique maximal ideal $\mathcal{M}$ of $\mathcal{D}(E)$ containing $\mathcal{P}$.*

COROLLARY 14.2. *Every prime closed ideal of $\mathcal{D}(E)$ is a maximal ideal.*

Let us recall the main definitions concerning multiplicative semi-norms [**3**], [**4**], [**5**], [**8**], [**10**], [**13**], [**14**].

Notation and definition: We denote by $Mult(\mathcal{D}(E), \| \, . \, \|)$ the set of multiplicative semi-norms of $\mathcal{D}(E)$ provided with the topology of pointwise convergence. Given $\phi \in Mult(\mathcal{D}(E), \| \, . \, \|)$, the set of the $f \in \mathcal{D}(E)$ such that $\phi(f) = 0$ is a closed prime ideal called *the kernel of* ϕ. It is denoted by $Ker(\phi)$.

We denote by $Mult_m(\mathcal{D}(E), \| \, . \, \|)$ the set of multiplicative semi-norms of $\mathcal{D}(E)$ whose kernel is a maximal ideal and by $Mult_1(\mathcal{D}(E), \| \, . \, \|)$ the set of multiplicative semi-norms of $\mathcal{D}(E)$ whose kernel is a maximal ideal of codimension 1.

Let $a \in E$. The mapping φ_a from $\mathcal{D}(E)$ to $\mathbb{R}$ defined by $\varphi_a(f) = |f(a)|$ belongs to $Mult(\mathcal{D}(E), \| \, . \, \|)$. Let $\mathcal{U}$ be an ultrafilter on E. By Urysohn's Theorem, given $f \in \mathcal{D}(E)$, the mapping from E to $\mathbb{R}$ that sends x to $|f(x)|$ admits a limit along $\mathcal{U}$. We set $\varphi_{\mathcal{U}}(f) = \lim_{\mathcal{U}} |f(x)|$. Moreover, we denote by $Mult_E(\mathcal{D}(E), \| \, . \, \|)$ the set of multiplicative semi-norms of $\mathcal{D}(E)$ of the form φ_a, $a \in E$. Consequently, by definition, $Mult_E(\mathcal{D}(E), \| \, . \, \|)$ is a subset of $Mult_1(\mathcal{D}(E), \| \, . \, \|)$.

The following Theorems 15 and 16 are immediate and well known:

THEOREM 15. *Let $a \in E$. Then $\mathcal{I}(a)$ is a maximal ideal of $\mathcal{D}(E)$ of codimension 1 and φ_u belongs to $Mult_1(\mathcal{D}(E), \| \, . \, \|)$.*

COROLLARY 15.1. *$Mult_E(\mathcal{D}(E), \| \, . \, \|)$ is included in $Mult_1(\mathcal{D}(E), \| \, . \, \|)$.*

THEOREM 16. *Let $\mathcal{U}$ be an ultrafilter on E. Then $\varphi_{\mathcal{U}}$ belongs to the closure of $Mult_E(\mathcal{D}(E), \| \, . \, \|)$.*

Now, Corollary 16.1 is an immediate consequence of Theorems 14, 15 and 16 and Corollary 12.1:

COROLLARY 16.1. *$Mult(\mathcal{D}(E), \| \, . \, \|) = Mult_m(\mathcal{D}(E), \| \, . \, \|)$. Furthermore, if $\mathbb{K}$ is locally compact then $Mult(\mathcal{D}(E), \| \, . \, \|) = Mult_1(\mathcal{D}(E), \| \, . \, \|)$.*

REMARK 6. Suppose $\mathbb{K}$ is locally compact and E is a disk in an algebraically closed complete ultrametric field. There do exist ultrafilters on E that do not converge. Let $\mathcal{U}$ be such an ultrafilter. Then $\varphi_{\mathcal{U}}$ belongs to $Mult_1(\mathcal{D}(E), \| \, . \, \|)$ but does not belong to $Mult_E(\mathcal{D}(E), \| \, . \, \|)$.

REMARK 7. In $\mathcal{H} \in Y_{(\mathcal{T})}(E)$ the various ultrafilters $\mathcal{U} \in \mathcal{H} \in Y_{(\mathcal{T})}(E)$ define various prime ideals of $\mathcal{D}(E)$ and it is not clear whether these ideals are minimal among the set of prime ideals of $\mathcal{D}(E)$.

THEOREM 17. *The topology induced on E by the one of $Mult_E(\mathcal{D}(E), \| \, . \, \|)$ is equivalent to the metric topology induced on E by the field $\mathbb{K}$.*

PROOF. The proof made for Theorem 15 in [**10**] holds with no change. Let us recall it. Let $a \in E$. The filter of neighborhoods of a admits for basis the family of disks $d_E(a, r^-)$, $r > 0$. But we can check that such a disk is induced by a neighborhood of φ_a with respect to the topology of $Mult_E(\mathcal{D}(E), \| \,.\, \|)$ and to the metric topology of E. Given $\varphi_a \in Mult_E(\mathcal{D}(E), \| \,.\, \|)$, we set $W'(\varphi_a, f_1, ..., f_q, \epsilon) = W(\varphi_a, f_1, ..., f_q, \epsilon) \cap Mult_E(\mathcal{D}(E), \| \,.\, \|)$. Let $r \in]0, 1[$. There exist $u \in \mathcal{D}(E)$ such that $u(x) = 0$ for every $x \in d_E(a, r^-)$ and $u(x) = 1$ for every $x \in E \setminus d_E(a, r^-)$. Consequently, $W'(\varphi_a, u, r)$ is the set of the φ_b such that $|b - a| \leq r$. Hence the topology of $Mult_E(\mathcal{D}(E), \| \,.\, \|)$ is thinner or equal to the metric topology of E. Now, since each f_j is continuous, the set of the $x \in E$ such that $|\ |f_j(x)| - |f_j(a)|\ |_\infty \leq \epsilon$ for every $j = 1, ..., q$ is open and hence contains a ball $d_E(a, r^-)$ of E. Consequently, the topology of E is thinner or equal to this of $Mult_E(\mathcal{D}(E), \| \,.\, \|)$, which finishes the proof that the topology induced on E by $Mult(\mathcal{D}(E), \| \,.\, \|)$ coincides with the metric topology of E. □

Theorem 18 was proven in [19] for the algebra of bounded continuous functions. In [**10**] we proved it again for algebras of bounded continuous functions and for algebras of bounded uniformly continuous functions, in a different way. Here we adapt that proof to the algebra $\mathcal{D}(E)$.

THEOREM 18. *Let $\mathcal{M}$ be a maximal ideal of $\mathcal{D}(E)$. Let T be the field $\dfrac{\mathcal{D}(E)}{\mathcal{M}}$ and let θ be the canonical surjection from $\mathcal{D}(E)$ onto T. Given any ultrafilter $\mathcal{U}$ such that $\mathcal{I}(\mathcal{U}) = \mathcal{M}$, the quotient norm $\| \,.\, \|'$ on T is defined by $\|\theta(f)\|' = \lim_{\mathcal{U}} |f(s)|$ and hence is multiplicative.*

PROOF. Let $t \in T$ and let $f \in \mathcal{D}(E)$ be such that $\theta(f) = t$. Let $\mathcal{U}$ be an ultrafilter such that $\mathcal{I}(\mathcal{U}) = \mathcal{M}$. So, on T we have an absolute value ψ defined as $\psi(t) = \lim_{\mathcal{U}} |f(s)|$. That absolute value is continuous with respect to the quotient norm $\| \,.\, \|'$ of T. Indeed, by construction, we have $\|f\| \geq \lim_{\mathcal{U}} |f(s)| = \psi(t)$. But $\|t\|' = \inf_{h \in \mathcal{M}} \|f + h\|$ and of course, $\psi(\theta(t + h)) = \psi(t) = \lim_{\mathcal{U}} |f(s)|$. Consequently $\|t\|' \geq \psi(t)$.

Conversely, let $V \in \mathcal{U}$ be such that $|f(x)| \leq \lim_{\mathcal{U}} |f(s)| + \epsilon$ for every $x \in V$. There exist $f_1, ..., f_q \in \mathcal{M}$ and $\epsilon > 0$ sucht that $\bigcap_{j=1}^{q} B(f_j, \epsilon) \subset V$. Let $X = \bigcap_{j=1}^{q} B(f_j, \epsilon)$. By Theorem 6, since $f_j \in \mathcal{D}(E)$ for every $j = 1, ..., q$, X is uniformly open in E. Consequently, by Theorem 5 there exists $u \in \mathcal{D}(E)$ such that $u(x) = 0$ for every $x \in X$, $u(x) = 1$ for every $x \in E \setminus X$. Then $u(1 - u) = 0$. But $1 - u \notin \mathcal{M}$. Hence, u belongs to $\mathcal{M}$ and then $\theta(f - uf) = \theta(f) = t$. But by construction, $(f - uf)(x) = 0$ for every $x \in E \setminus X$ and $(f - uf)(x) = f(x)$ for every $x \in X$. Consequently, $\|f - uf\| \leq \lim_{\mathcal{U}} |f(s)| + \epsilon$ and therefore $\|t\|' = \|\theta(f - uh)\| \leq \lim_{\mathcal{U}} |f(s)| + \epsilon$. This finishes the proof that the equality $\|\theta(f)\|' = \lim_{\mathcal{U}} |f(s)|$. Now, such a norm defined as $\|\theta(f)\|' = \lim_{\mathcal{U}} |f(s)|$ is obviously multiplicative. □

REMARK 8. A similar proof is given in [**10**] for Theorem 16 where we forgot to mention the use of a lemma such as Theorem 6 in the case of algebra B.

DEFINITION. Recall that given a commutative Banach $\mathbb{K}$-algebra T with unity, every maximal ideal of T is the kernel of at least one continuous multiplicative semi-norm [**2**], [**3**]. The algebra T is said to be *multbijective* if every maximal ideal is the kernel of only one continuous multiplicative semi-norm.

REMARK 9. There exist ultrametric Banach $\mathbb{K}$-algebras that are not multbijective [**1**], [**2**], [**3**].

THEOREM 19. *$\mathcal{D}(E)$ is multbijective.*

PROOF. Let $\mathcal{M}$ be a maximal ideal of $\mathcal{D}(E)$ and let T be the field $\frac{\mathcal{D}(E)}{\mathcal{M}}$. By Theorem 18, the quotient norm of T is multiplicative. But then, T admits only one continuous multiplicative semi-norm: its quotient $\mathbb{K}$-algebra norm. Consequently, $\mathcal{D}(E)$ admits only one continuous multiplicative semi-norm whose kernel is $\mathcal{M}$, which proves that $\mathcal{D}(E)$ is multbijective. $\square$

By Theorem 19 and Corollary 14.2, we can now state Corollary 19.1:

COROLLARY 19.1. *The mapping that associates to each $\phi \in Mult(\mathcal{D}(E), \| \, . \, \|)$ its kernel $Ker(\phi)$ is a bijection from $Mult(\mathcal{D}(E), \| \, . \, \|)$ onto $Max(\mathcal{D}(E))$.*

By Theorem 11, Corollary 11.1 and Theorem 19, we have Corollary 19.2:

COROLLARY 19.2. *For every $\phi \in Mult(\mathcal{D}(E), \| \, . \, \|)$ there exists a unique $\mathcal{H} \in Y_{(\mathcal{T})}(E)$ such that $\phi(f) = \lim_{\mathcal{U}} |f(x)|$ for every $f \in \mathcal{D}(E)$, for every $\mathcal{U} \in \mathcal{H}$.*

Moreover, the mapping $\widetilde{\Psi}$ that associates to each $\phi \in Mult(\mathcal{D}(E), \| \, . \, \|)$ the unique $\mathcal{H} \in Y_{(\mathcal{T})}(E)$ such that $\phi(f) = \lim_{\mathcal{U}} |f(x)|$ for every $f \in \mathcal{D}(E)$, for every $\mathcal{U} \in \mathcal{H}$, is a bijection from $Mult(\mathcal{D}(E), \| \, . \, \|)$ onto $Y_{(\mathcal{T})}(E)$.

Now, by Theorems 15 and 16, we have Corollary 19.3:

COROLLARY 19.3. *$Mult_E(\mathcal{D}(E), \| \, . \, \|)$ is dense in $Mult(\mathcal{D}(E), \| \, . \, \|)$.*

PROOF. Indeed, let $\psi \in Mult(\mathcal{D}(E), \| \, . \, \|)$. By Theorem 16, $Ker(\psi)$ is a maximal ideal $\mathcal{I}(\mathcal{U})$ with $\mathcal{U}$ an ultrafilter on E; but then by Theorem 19, $\psi = \varphi_{\mathcal{U}}$ and hence ψ belongs to the closure of $Mult_E(\mathcal{D}(E), \| \, . \, \|)$. $\square$

THEOREM 20. *For every $\phi \in Mult(\mathcal{D}(E), \| \, . \, \|)$, ϕ satisfies $\phi(f) \leq \|f\|_0$ for every $f \in \mathcal{D}(E)$.*

PROOF. By Theorem 19, $\mathcal{D}(E)$ is multbijective, then by Theorem 11, every element of $Mult_m(\mathcal{D}(E), \| \, . \, \|)$ is of the form $\varphi_{\mathcal{U}}$ where $\mathcal{U}$ is an ultrafilter on E. Of course, we have $\varphi_{\mathcal{U}}(f) \leq \|f\|_0$ for every $f \in \mathcal{D}(E)$ and consequently, $\phi(f) \leq \|f\|_0$ for every $f \in \mathcal{D}(E)$, for every $\phi \in Mult_m(\mathcal{D}(E), \| \, . \, \|)$. But by Corollary 16.1, $Mult_m(\mathcal{D}(E), \| \, . \, \|) = Mult(\mathcal{D}(E), \| \, . \, \|)$. Therefore $\phi(f) \leq \|f\|_0$ for every $f \in \mathcal{D}(E)$, for every $\phi \in Mult(\mathcal{D}(E), \| \, . \, \|)$. $\square$

NOTATION. On $\mathcal{D}(E)$ we denote by $\| \, . \, \|_{si}$ the semi-norm of $\mathcal{D}(E)$ defined as $\|f\|_{si} = \lim_{n \to +\infty} \sqrt[n]{\|f^n\|}$.

Let us recall Proposition F (See for instance Theorem 6.19 in [**3**]):

PROPOSITION F. *Let T be a normed $\mathbb{K}$-algebra whose norm is $\| \, . \, \|$. Then $\| \, . \, \|_{si}$ is a power multiplicative semi-norm on T such that $\|f\|_{si} = \sup\{\varphi(f) \mid \varphi \in Mult(\mathcal{D}(E), \| \, . \, \|\}$.*

Then we can state Theorem 21:

THEOREM 21. *$\|f\|_{si} = \|f\|_0$ for every $f \in \mathcal{D}(E)$.*

PROOF. We have $\|f\|_{si} = \sup\{\varphi(f) \mid \varphi \in Mult(\mathcal{D}(E), \|\ .\ \|\}$. But by Corollary 16.1, we have $Mult(\mathcal{D}(E), \|\ .\ \|) = Mult_m(\mathcal{D}(E), \|\ .\ \|)$, hence $\|f\|_{si} = \sup\{\varphi(f) \mid \varphi \in Mult_m(\mathcal{D}(E), \|\ .\ \|\}$. Consequently, $\|f\|_{si} \geq |f(x)|$ for every $x \in E$, for every $f \in \mathcal{D}(E)$ and hence $\|f\|_{si} \geq \|f\|_0$ for every $f \in \mathcal{D}(E)$. On the other hand, by Theorem 20, we have $\phi(f) \leq \|f\|_0$ for every $f \in \mathcal{D}(E)$, for every $\phi \in Mult(\mathcal{D}(E), \|\ .\ \|)$, hence $\|f\|_{si} \leq \|f\|_0$ for every $f \in \mathcal{D}(E)$ and finally $\|f\|_{si} = \|f\|_0$ for every $f \in \mathcal{D}(E)$. □

THEOREM 22. *Suppose $\mathbb{K}$ is algebraically closed. Let $\mathcal{U}$ be an ultrafilter on $\mathbb{K}$ and suppose there exists $P \in \mathbb{K}[x]$, $P \neq 0$ satisfying $\lim_{\mathcal{U}} P(x) = 0$. Then $\mathcal{U}$ is a principal ultrafilter.*

PROOF. Let $P(x) = \prod_{j=1}^{q}(x - a_j)$. Let $\mathcal{F}$ be the filter admitting for basis the family of sets $\Lambda(r) = \bigcup_{j=1}^{q} d(a_j, r)$, $r > 0$. Suppose first that $\mathcal{U}$ is not secant with $\mathcal{F}$. There exist $\rho > 0$ and $H \in \mathcal{U}$ such that $\Lambda(\rho) \cap H = \emptyset$. Then $|P(x)| \geq \rho^q$ for every $x \in H$, a contradiction to the hypothesis $\lim_{\mathcal{U}} P(x) = 0$. Consequently, $\mathcal{U}$ is secant with $\mathcal{F}$. Hence it is obviously secant with the filter of neighborhoods of one of the points a_j and therefore, it converges to this point. □

As a consequence, we have Theorem 23:

THEOREM 23. *Suppose that $\mathbb{K}$ is algebraically closed and that E is a closed bounded subset of $\mathbb{K}$ with infinitely many points and let $\mathcal{M}$ be a maximal ideal of $\mathcal{D}(E)$ of the form $\mathcal{I}(\mathcal{U})$ where $\mathcal{U}$ is not principal. Then $\mathcal{M}$ is of infinite codimension.*

PROOF. Indeed, by Theorem 22 the ideal $\mathcal{M}$ contains no polynomials different from 0, hence $\dfrac{\mathcal{D}(E)}{\mathcal{M}}$ contains a subfield isomorphic to $\mathbb{K}(x)$ and therefore is not of finite codimension. □

REMARK 10. Suppose $\mathbb{K}$ is algebraically closed and let $E = \mathbb{K}$. Then the algebra $\mathcal{D}(E)$ contains no polynomial. In such a case, it is not clear whether all maximal ideals not defined by points of $\mathbb{K}$ are of infinite codimension.

REMARK 11. Concerning uniformly continuous functions, it has been shown in **[10]** that two ultrafilters that are not contiguous define two distinct continuous multiplicative semi-norms. By Corollary 11.1 we know that the same holds in $\mathcal{D}(E)$.

Now, concerning bounded analytic functions inside the disk $F = \{x \in \mathbb{K}\ |x| < 1\}$, in **[7]**, it was shown that the same property holds for a large set of ultrafilters on F. However, the question remains whether it holds for all ultrafilters on F.

The following lemma is well known and comes from the fact that on a vector space of finite dimension over a complete field, two norms are equivalent.

LEMMA U. *Let $\mathbb{L}$ be an algebraic extension of $\mathbb{K}$ of degree t of the form $\mathbb{K}[a]$ provided with the absolute value $|\ .\ |$ which expands that of $\mathbb{K}$. Given $x \in \mathbb{L}$, let*

$x = \sum_{j=0}^{t-1} a^j x_j$ *and let* $\|x\| = \max_{0\le j\le t-1}|x_j|$. *Then the norm* $\| \ . \ \|$ *is equivalent to the absolute value* $| \ . \ |$.

THEOREM 24. *Let* $\mathbb{L}$ *be an algebraic extension of* $\mathbb{K}$ *of degree* t *of the form* $\mathbb{K}[a]$ *provided with the absolute value which expands that of* $\mathbb{K}$. *Let* f *be a strictly differentiable function from* E *to* $\mathbb{L}$. *There exists* $f_0, ..., f_{t-1} \in \mathcal{D}(E)$ *such that* $f = \sum_{j=0}^{t-1} a^j f_j$.

PROOF. For every $x \in E$, we can write $f(x) = \sum_{j=0}^{t-1} a^j f_j(x)$ with $f_j(x) \in \mathbb{K}$, $0 \le j \le t-1$.We will show that each function f_j belongs to $\mathcal{D}(E)$.

By definition, given any $a \in E$, $\dfrac{f(x)-f(y)}{x-y}$ has a limit $f'(a) \in \mathbb{L}$ when x and y tend to a while $x \ne y$. Let us write $f'(a) = \sum_{j=0}^{t-1} a^j l_j$, hence

$$\frac{f(x)-f(y)}{x-y} - f'(a) = \sum_{j=0}^{t-1} a^j \Big(\frac{f_j(x)-f_j(y)}{x-y} - l_j\Big).$$

Since $\displaystyle\lim_{\substack{x\to a,\ y\to a\\ x\ne y}} \frac{f(x)-f(y)}{x-y} - f'(a) = 0$, by Lemma U, we can see that for each $j = 0, ..., t-1$, we have $\displaystyle\lim_{\substack{x\to a,\ y\to a\\ x\ne y}} \frac{f_j(x)-f_j(y)}{x-y} - l_j = 0$. Consequently, f_j does belong to $\mathcal{D}(E)$ and that ends the proof of Theorem 24. □

We can now prove Theorem 25 that will let us show Theorem 26:

THEOREM 25. *Let* $\mathbb{L}$ *be a finite algebraic extension of* $\mathbb{K}$ *provided with the absolute value which expands that of* $\mathbb{K}$. *Suppose there exists a morphism of* $\mathbb{K}$-*algebra*, χ, *from* $\mathcal{D}(E)$ *onto* $\mathbb{L}$. *Let* $\widehat{\mathcal{D}}(E)$ *be the* $\mathbb{L}$-*algebra of strictly differentiable functions from* E *to* $\mathbb{L}$. *Then* χ *has continuation to a morphism of* $\mathbb{L}$-*algebra* $\widehat{\chi}$ *from* $\widehat{\mathcal{D}}(E)$ *to* $\mathbb{L}$.

PROOF. The proof is similar to the proof of Proposition 3.5 in [**10**]. Let $t = [\mathbb{L} : \mathbb{K}]$. By the primitive element theorem, there exists $a \in \mathbb{L}$ such that $\mathbb{L} = \mathbb{K}[a]$. Let $f,\ g \in \widehat{\mathcal{D}}(E)$. Then f is of the form $\sum_{j=0}^{t-1} a^j f_j$, $f_j \in \mathcal{D}(E)$, $j = 0, \dots, d-1$ and g is of the form $\sum_{j=0}^{t-1} a^j g_j$, $g_j \in \mathcal{D}(E)$, $j = 0, ..., t-1$. By Theorem 24, all the f_j and the g_j belong to $\mathcal{D}(E)$.

Now, we can define $\widehat{\chi}$ on $\widehat{\mathcal{D}}(E)$ as $\widehat{\chi}(f) = \sum_{j=0}^{t-1} a^j \chi(f_j)$. Then obviously, $\widehat{\chi}$ is $\mathbb{K}$-linear. On the other hand, for each $q \in \mathbb{N}$, a^q is of the form $P_q(a)$ where

$P_q \in \mathbb{K}[x]$, $\deg(P_q) \leq t-1$. Then $\widehat{\chi}(a^q) = \widehat{\chi}(P_q(a)) = P_q(\widehat{\chi}(a)) = P_q(a) = a^q$. Next,

$$\widehat{\chi}(fg) = \widehat{\chi}\Big((\sum_{j=0}^{t-1} a^j f_j)(\sum_{j=0}^{t-1} a^j g_j)\Big) = \widehat{\chi}\Big(\sum_{\substack{0\leq m\leq t-1\\ 0\leq n\leq t-1}} a^{m+n} f_m g_n\Big)$$

$$= \sum_{\substack{0\leq m\leq t-1\\ 0\leq n\leq t-1}} a^{m+n}\chi(f_m)\chi(g_n) = (\sum_{j=0}^{t-1} a^j\chi(f_j))(\sum_{j=0}^{t-1} a^j\chi(g_j)) = \chi(f)\chi(g).$$

That finishes the proof that χ is a $\mathbb{L}$-algebra morphism from $\widehat{\mathcal{D}}(E)$ to $\mathbb{L}$. □

THEOREM 26. *Every maximal ideal of finite codimension of $\mathcal{D}(E)$ is of codimension* 1.

PROOF. The proof follows from Theorem 3.7 in [**10**]. Let $\mathcal{M}$ be a maximal ideal of finite codimension of $\mathcal{D}(E)$, let $\mathbb{L}$ be the field $\dfrac{\mathcal{D}(E)}{\mathcal{M}}$ and let χ be the canonical surjective morphism from $\mathcal{D}(E)$ over $\mathbb{L}$. Since $\mathcal{M}$ is of finite codimension, $\mathbb{L}$ is a finite extension of $\mathbb{K}$ and hence has a unique absolute value expanding that of $\mathbb{K}$. Consequently, $\mathbb{L}$ is complete for this absolute value. Let $\widehat{\mathcal{D}}(E)$ be the $\mathbb{L}$-algebra of strictly differentiable functions from E to $\mathbb{L}$. By Theorem 25, χ has continuation to a $\mathbb{L}$-algebra morphism $\widehat{\chi}$ from $\widehat{\mathcal{D}}(E)$ to $\mathbb{L}$.

Let $f \in \mathcal{D}(E)$. Then $\chi(f)$ is an element b of $\mathbb{L}$ and of course, $f - b \in \widehat{\mathcal{D}}(E)$. Now, by Theorem 25, $\widehat{\chi}$ is surjective on $\mathbb{L}$, hence $Ker(\widehat{\chi})$ is a maximal ideal $\widehat{\mathcal{M}}$ of $\widehat{\mathcal{D}}(E)$. Then by Theorem 11, there exists an ultrafilter $\mathcal{U}$ on E such that $\widehat{\mathcal{M}}$ is the ideal of the $g \in \widehat{A}$ such that $\lim_{\mathcal{U}} g(x) = 0$. So, we have $\lim_{\mathcal{U}}(f(x) - b) = 0$, therefore $\lim_{\mathcal{U}}(f(x)) = b$. But since $\mathbb{K}$ is complete, this proves that b belongs to $\mathbb{K}$ and therefore $\chi(\mathcal{D}(E)) = \mathbb{K}$ i.e. $\mathbb{L} = \mathbb{K}$. □

By Theorem 26 and Corollary 12.1, we can state this corollary:

COROLLARY 26.1. *Let $\mathcal{U}$ be an ultrafilter on E. The following* 3 *statements are equivalent:*

i) $\mathcal{I}(\mathcal{U})$ is of codimension 1,

ii) $\mathcal{I}(\mathcal{U})$ is of finite codimension,

iii) for every $f \in \mathcal{D}(E)$, the filter generated by $f(\mathcal{U})$ converges in $\mathbb{K}$.

COROLLARY 26.2. *Every maximal ideal of $\mathcal{D}(E)$ of finite codimension is of codimension 1.*

The following Theorem is almost classical [**2**], [**3**]:

THEOREM 27. *Suppose $\mathbb{K}$ is not locally compact. There exists a sequence $(b_n)_{n\in\mathbb{N}}$ in $\mathbb{K}$ such that either the sequence $|b_n - b_{n+1}|$ is a strictly monotonous, of limit $r \in]0, +\infty[$, or $|b_n - b_m|$ is a constant whenever $m \neq n \in \mathbb{N}$.*

PROOF. Indeed, since $\mathbb{K}$ is not locally compact, either the set $|\mathbb{K}| = \{|x| \mid x \in \mathbb{K}\}$ is not discrete and hence is dense in $[0, +\infty[$ or the residue field of $\mathbb{K}$ is infinite. If $|\mathbb{K}|$ is dense in $[0, +\infty[$, there exists a sequence $(b_n)_{n\in\mathbb{N}}$ such that $|b_n - b_{n+1}|$ is strictly monotonous, of limit $r \in]0, +\infty[$. And if the residue field of $\mathbb{K}$ is infinite, then there exists sequences $(b_n)_{n\in\mathbb{N}}$ such that $|b_n - b_m| = c$ for every $n \neq m$. □

THEOREM 28. *Suppose that $\mathbb{K}$ is not locally compact and that E contains a sequence $(a_n)_{n\in\mathbb{N}}$ such that $|a_n - a_m| \geq r >$ for all $m \neq n$. Let $(b_n)_{n\in\mathbb{N}}$ be a sequence of $\mathbb{K}$ such that either the sequence $|b_n - b_{n+1}|$ is a strictly monotonous, of limit $r \in]0, +\infty[$, or $|b_n - b_m|$ is a constant whenever $m \neq n \in \mathbb{N}$.*

Then, there exists an ultrafilter $\mathcal{U}$ on E thinner than the sequence $(a_n)_{n\in\mathbb{N}}$ and $f \in \mathcal{D}(E)$ such that the filter generated by $f(\mathcal{U})$ is thinner than the sequence $(b_n)_{n\in\mathbb{N}}$ and does not converge.

PROOF. By hypothesis, there exists $\lambda > 0$ such that $|b_m - b_n| \geq \lambda$ for all $m \neq n$. Now, we can define a function f from E to $\mathbb{K}$ as $f(x) = b_n$ for every $x \in d_E(a_n, r^-)$ and (for instance) $f(x) = 0$ for every $x \in E \setminus \left(\bigcup_{n\in\mathbb{N}} d_E(a_n, r^-)\right)$. We can check that f belongs to $\mathcal{D}(E)$.

Now, let $\mathcal{U}$ be an ultrafilter on E thinner than the sequence $(a_n)_{n\in\mathbb{N}}$. Then, the filter $\mathcal{W}$ generated by $f(\mathcal{U})$ on $\mathbb{K}$ is thinner than the sequence $(b_n)_{n\in\mathbb{N}}$. Every element X of $\mathcal{W}$ contains several terms of the sequence $(b_n)_{n\in\mathbb{N}}$, for instance b_m, b_n with $m \neq n$, hence $\operatorname{diam}(X) \geq \lambda$ and therefore $\mathcal{W}$ does not converge. □

THEOREM 29. *The algebra $\mathcal{D}(E)$ admits maximal ideals of infinite codimension if and only if $\mathbb{K}$ is not locally compact.*

PROOF. If $\mathbb{K}$ is locally compact, then by Theorem 12, every maximal ideal of $\mathcal{D}(E)$ is of codimension 1.

Now, suppose that $\mathbb{K}$ is not locally compact. By Theorem 27 there exists a sequence $(b_n)_{n\in\mathbb{N}}$ in $\mathbb{K}$ such that either the sequence $|b_n - b_{n+1}|$ is a strictly monotonous, of limit $r \in]0, +\infty[$, or $|b_n - b_m|$ is a constant whenever $m \neq n \in \mathbb{N}$. So, by Theorem 28 in both cases there exists an ultrafilter $\mathcal{U}$ on E and $f \in B$ such that the filter generated by $f(\mathcal{U})$ is thinner than the sequence $(b_n)_{n\in\mathbb{N}}$ and does not converge. Consequently, $\mathcal{I}(\mathcal{U})$ is a maximal ideal of $\mathcal{D}(E)$ of infinite codimension. □

NOTATION. Given $r \in]0, +\infty[$, and $a \in E$, we denote by $d_E(a, r^-)$ the ball of E $\{x \in E : |x - a| < r\}$ and we denote by $\operatorname{card}(E, r)$ the cardinal of the set of disks $d_E(a, r^-)$, $a \in E$. We denote by U the disk $d(0, 1)$ of $\mathbb{K}$.

We can now add some precisions on maximal ideals of infinite codimension by making certain hypotheses on E and $\mathbb{K}$.

THEOREM 30. *Suppose that we have $\operatorname{card}(E, r) \leq \operatorname{card}(U, r)$ for every $r \in]0, 1]$. Let $\mathcal{U}$ be an ultrafilter on E. Either $\mathcal{U}$ is convergent and then $\mathcal{I}(\mathcal{U})$ is of codimension 1, or $\mathcal{U}$ is not convergent and then $\mathcal{I}(\mathcal{U})$ is of infinite codimension.*

PROOF. If $\mathcal{U}$ is convergent, by Corollary 12.1 $\mathcal{I}(\mathcal{U})$ is of codimension 1. Now, suppose that $\mathcal{U}$ is not convergent. There exists $\rho > 0$ such that $\operatorname{diam}(X) \geq \rho$ for every $X \in \mathcal{U}$. Let $r \in]0, \rho[$. Consider the covering of E by the family of disks $(d_E(a_i, r^-))_{i\in J}$ that are pairwise disjoint. Since $\operatorname{card}(E, r) \leq \operatorname{card}(U, r)$, there exists a covering of U by a family of disks $(d(b_j, r^-)_{i\in I})$ pairwise disjoint, with an injection θ from I to J. So, we have $|b_{\theta(i)} - b_{\theta(j)}| \leq r$ for every $i \neq j$.

Consider now the function f from E to $\mathbb{K}$ defined by $f(x) = b_{\theta(i)}$ for every $x \in d_E(a_i, r^-)$. Let $X \in \mathcal{U}$. Since $\operatorname{diam}(X) \leq r$, there exists a_i, $a_j \in X$ with $a_i \neq a_j$ and we have $f(a_i) = b_i$, $f(a_j) = b_j$, hence $|f(a_i) - f(a_j)| \geq r$ which shows that f does not converge on $\mathcal{U}$ and therefore, by Corollary 26.1, $\mathcal{I}(\mathcal{U})$ is not of finite codimension. □

COROLARY 30.1. *Suppose that E is separable and that $\mathbb{K}$ is not locally compact. Let $\mathcal{U}$ be a non-convergent ultrafilter on E. Then $\mathcal{I}(\mathcal{U})$ is of infinite codimension.*

REMARK 12. Particularly Theorem 30 applies when $\mathbb{K}$ is algebraically closed and E is separable.

Let us recall some results on the Shilov boundary of an ultrametric normed algebra [**6**].

Given a norm of $\mathbb{K}$-algebra, we call *Shilov boundary of T* a closed subset S of $Mult(T, \| \,.\, \|)$ that is minimum with respect to inclusion, such that, for every $x \in T$, there exists $\phi \in S$ such that $\phi(x) = \|x\|_{si}$.

PROPOSITION G ([**3**], [**5**]). *Every normed $\mathbb{K}$-algebra admits a Shilov boundary.*

THEOREM 31. *The Shilov boundary S of $\mathcal{D}(E)$ is equal to $Mult(\mathcal{D}(E), \| \,.\, \|)$.*

PROOF. Henceforth we take $\epsilon \in]0, \frac{1}{2}[$. Suppose that the Shilov boundary S of $\mathcal{D}(E)$ is not equal to $Mult(\mathcal{D}(E), \| \,.\, \|)$ and let $\psi \in Mult(\mathcal{D}(E), \| \,.\, \|) \setminus S$. Since S is a closed subset of $Mult(\mathcal{D}(E), \| \,.\, \|)$, the set $Mult(\mathcal{D}(E), \| \,.\, \|) \setminus S$ is an open subset of $Mult(\mathcal{D}(E), \| \,.\, \|)$ and hence, there exist $f_1, ..., f_q$ such that $W(\psi, f_1, ..., f_q, \epsilon) \subset (Mult(\mathcal{D}(E), \| \,.\, \|) \setminus S)$. Let $L = \{x \in E \mid |\psi(f_j) - |f(_j(x)|\,|_\infty \leq \dfrac{\epsilon}{2}\}$. By Theorem 6, L is uniformly open in E. Consequently, by Theorem 5 the characteristic function u of L belongs to $\mathcal{D}(E)$ and obviously satisfies $\psi(u) = 1$. On the other hand, we have $u(x) = 0$ for every $x \notin L$.

Now, there exists $\theta \in S$ such that $\theta(u) = \|u\| = 1$. Consider the neighborhood $W(\theta, f_1, ..., f_q, u, \frac{\epsilon}{2})$. Since $Mult_E(\mathcal{D}(E), \| \,.\, \|)$ is dense in $Mult(\mathcal{D}(E), \| \,.\, \|)$, we can take $\varphi_a \in W(\theta, f_1, ..., f_q, u, \frac{\epsilon}{2})$ we have $|\ |u(a)| - \theta(u)|_\infty \leq \frac{\epsilon}{2}$, hence $|u(a)| \geq 1 - \frac{\epsilon}{2} > 0$. But since $\theta \in S$, we have $W(\theta, f_1, ..., f_q, \frac{\epsilon}{2}) \cap W(\psi, f_1, ..., f_q, \frac{\epsilon}{2}) = \emptyset$ and so much the more: $W(\theta, f_1, ..., f_q, u, \frac{\epsilon}{2}) \cap W(\psi, f_1, ..., f_q, u, \frac{\epsilon}{2}) = \emptyset$. Let $H = \{a \in E \mid \varphi_a \in W(\theta, f_1, ..., f_q, u, \frac{\epsilon}{2})\}$. Then $H \cap L = \emptyset$. But by definition of u, we have $u(a) = 0$ for every $a \in H$, a contradiction. $\square$

Morphisms between algebras of strictly differentiable functions

NOTATIONS. Henceforth, we denote by E' another open subset of $\mathbb{K}$ and by $\mathcal{D}(E')$ the $\mathbb{K}$-algebra of strictly differentiable functions from E' to $\mathbb{K}$ whose norm is denoted by $\| \,.\, \|'$. We denote by γ an injective strictly differentiable function from E' to E and for every $f \in \mathcal{D}(E')$, we set $\widetilde{\gamma}(f) = f \circ \gamma$.

THEOREM 32. *Let $f \in \mathcal{D}(E)$. Then $f \circ \gamma$ belongs to $\mathcal{D}(E')$ and $\widetilde{\gamma}$ is a morphism from $\mathcal{D}(E)$ to $\mathcal{D}(E')$ such that $\|\widetilde{\gamma}(f) - \widetilde{\gamma}(g)\|' \leq \|f - g\|$ for every $f,\ g \in \mathcal{D}(E)$. Moreover, if γ is a bijection from E' onto E, then $\widetilde{\gamma}$ is injective and particularly and if γ^{-1} is also strictly differentiable, then $\widetilde{\gamma}$ is an isometric isomorphism.*

PROOF. Let $f \in \mathcal{D}(E)$. Let us first show that $\widetilde{\gamma}(f)$ belongs to $\mathcal{D}(E)'$. Set $h = f \circ \widetilde{\gamma}$. Given $u,\ v \in E',\ u \neq v$, since γ is injective, we can write

$$\frac{h(u) - h(v)}{u - v} = \Big(\frac{f(\gamma(u)) - f(\gamma(v))}{\gamma(u) - \gamma(v)}\Big)\Big(\frac{\gamma(u) - \gamma(v)}{u - v}\Big).$$

By definition, the mapping ϕ from $(E \times E) \setminus D$ to $\mathbb{K}$ given by $\phi(x, y) = \dfrac{f(x) - f(y)}{x - y}$ is bounded and expands to a continuous function from $E \times E$ to $\mathbb{K}$. Similarly, denoting by D' the set $\{(u,u) \mid u \in E'\}$, γ is bounded in $(E' \times E') \setminus D'$ and expands to a continuous function from $E' \times E'$ to $\mathbb{K}$. Consequently, h is bounded in $(E' \setminus E') \setminus D'$ and expands to a continuous function from $E' \times E'$ to $\mathbb{K}$. Therefore h belongs to $\mathcal{D}(E)'$. It is then obvious that $\|\widetilde{\gamma}(f) - \widetilde{\gamma}(g)\|' \leq \|f-g\|$ for every $f,\ g \in \mathcal{D}(E)$. Suppose now that γ is surjective. Let $f,\ g \in \mathcal{D}(E)$ be such that $\widetilde{\gamma}(f) = \widetilde{\gamma}(g)$. Then for all $u \in E'$, we have $f(\gamma(u)) = g(\gamma(u))$. But since γ is surjective, for each $u \in E'$, there exists $x \in E$ such that $x = \gamma(u)$, hence $f(u) = g(u)$ and hence $f = g$, which shows that $\widetilde{\gamma}$ is injective. Particularly, if γ^{-1} is also strictly differentiable, then $\widetilde{\gamma}$ is an isomorphism from $\mathcal{D}(E)$ onto $\mathcal{D}(E')$ such that $\|\widetilde{\gamma}(f) - \widetilde{\gamma}(g)\|' = \|f-g\|$ for every $f,\ g \in \mathcal{D}(E)$. □

NOTATION. Let $\mathcal{H}$ be a class of contiguity on $Ul(E')$. By Lemma M, we know that two contiguous ultrafilters on E' have images by γ that are also contiguous. We denote by $\overline{\gamma}(\mathcal{H})$ the class of contiguity of the $\overline{f}(\mathcal{U})$, $\mathcal{U} \in \mathcal{H}$.

THEOREM 33. *Let ψ be a continuous morphism from $\mathcal{D}(E)$ to $\mathcal{D}(E')$ and let $\mathcal{M}'$ be a maximal ideal of $\mathcal{D}(E')$. Then, $\psi^{-1}(\mathcal{M}')$ is a maximal ideal of $\mathcal{D}(E)$.*

PROOF. We know that $\psi^{-1}(\mathcal{M}')$ is a prime ideal of $\mathcal{D}(E)$. Since ψ is continuous and since $\mathcal{M}'$ is closed, $\psi^{-1}(\mathcal{M}')$ is closed, hence it is a closed prime ideal of $\mathcal{D}(E)$. Then by Corollary 14.2, $\psi^{-1}(\mathcal{M}')$ is a maximal ideal of $\mathcal{D}(E)$. □

COROLLARY 33.1. *For every class of contiguity $\mathcal{H}'$ on E', the maximal ideal $\mathcal{M}' = \mathcal{I}(\mathcal{H}')$ satisfies $(\widetilde{\gamma})^{-1}(\mathcal{M}') = \mathcal{I}(\overline{\gamma}(\mathcal{H}'))$.*

THEOREM 34. *γ is supposed to be a strictly differentiable bijection from E' onto E. Then $\overline{\gamma}$ defines a surjective mapping from the set of the classes of contiguity on E' onto the set classes of contiguity on E and $(\widetilde{\gamma})^{-1}$ defines a surjection from $Max(\mathcal{D}(E'))$ onto $Max(\mathcal{D}(E))$ as $(\widetilde{\gamma})^{-1}(\mathcal{I}(\mathcal{H}')) = \mathcal{I}(\overline{\gamma}(\mathcal{H}'))$.*

PROOF. Let $\mathcal{H}'$ be a class of contiguity on E' and let $\mathcal{M}' = \mathcal{I}(\mathcal{H}')$. By Lemma M $\overline{\gamma}(\mathcal{H}')$ is a class of contiguity on E, hence $\mathcal{I}(\overline{\gamma}(\mathcal{H}'))$ is a maximal ideal of $\mathcal{D}(E)$. And by Corollary 14.2 , $\widetilde{\gamma}^{-1}(\mathcal{M}')$ is a maximal ideal of $\mathcal{D}(E)$, hence $\widetilde{\gamma}^{-1}(\mathcal{M}')$ is of the form $\mathcal{I}(\mathcal{H})$ where $\mathcal{H}$ is a class of contiguity of E. Consequently, $\mathcal{I}(\overline{\gamma}(H')) = \mathcal{I}(H)$. But since the mapping $\mathcal{I}$ defines a bijection from the set of classes of contiguity on E onto $Max(\mathcal{D}(E))$, we have $\overline{\gamma}(\mathcal{H}') = \mathcal{H}$ and hence $(\widetilde{\gamma})^{-1}(\mathcal{M}') = \mathcal{I}(\overline{\gamma}(\mathcal{H}'))$. □

NOTATION. Let $(T, \|\cdot\|)$ be a $\mathbb{K}$-normed algebra and let $\varphi \in Mult(T, \|\cdot\|)$. Let $f_1, ..., f_q \in T$ and let $\epsilon \in]0, +\infty[$. We will denote by $\mathcal{N}(\varphi, f_1, ..., f_q, \epsilon)$ the set $\{\phi \in Mult(T, \|\cdot\|) \mid |\varphi(f_k) - \phi(f_k)|_\infty \leq \epsilon\}$.

Recall that given a class of contiguity $\mathcal{H}$ on E we denote by $\mathcal{I}(\mathcal{H})$ the maximal ideal $\mathcal{I}(\mathcal{U})$, $\mathcal{U} \in \mathcal{H}$. Similarly, given a class of contiguity $\mathcal{H}'$ on E' we will denote by $\mathcal{I}'(\mathcal{H}')$ the maximal ideal $\mathcal{I}(\mathcal{U}')$, $\mathcal{U}' \in \mathcal{H}'$.

THEOREM 35. *Let γ_* be the mapping from $Mult(\mathcal{D}(E'), \|\cdot\|')$ to $Mult(\mathcal{D}(E), \|\cdot\|)$ defined by $(\gamma_*(\varphi'))(f) = \varphi'(f \circ \gamma)$, $(f \in \mathcal{D}(E))$. Then γ_* is continuous and surjective.*

PROOF. By Corollary 19.1 the mapping that associates to each $\varphi \in Mult(\mathcal{D}(E), \|\cdot\|)$ the ideal $Ker(\varphi)$ is a bijection from $Mult(\mathcal{D}(E), \|\cdot\|)$ onto $Max(\mathcal{D}(E))$. Next, given $\varphi' \in Mult(\mathcal{D}(E'), \|\cdot\|')$, by Corollary 19.2 φ' is defined by a class of contiguity $\mathcal{H}'$ on E' such that

$$\varphi'(f \circ \gamma) = \lim_{\mathcal{H}'} f(\gamma(u)) = \lim_{\widetilde{\gamma}(\mathcal{H}')} (f(x)).$$

Consequently, $Ker(\varphi') = \mathcal{I}'(\mathcal{H}')$ and $Ker(\gamma_*(\varphi')) = \mathcal{I}(\mathcal{H})$.

Let us now show the continuity of γ_*. Consider again $\varphi' \in Mult(\mathcal{D}(E'), \|\cdot\|')$ and a neighborhood $W = \mathcal{N}(\gamma_*(\varphi'), f_1, ..., f_q, \epsilon)$ of $\gamma_*(\varphi')$ in $Mult(\mathcal{D}(E), , \|\cdot\|)$ with $f_j \in \mathcal{D}(E)$, $j = 1, ..., q$. Set $g_k = f_k \circ \gamma, k = 1, ..., q$ and let $V = \mathcal{N}(\varphi', f_1, ..., f_q, \epsilon)$ be a neighborhood of φ' in $Mult(\mathcal{D}(E'), \|\cdot\|')$. Then

$$W = \{\psi \in Mult(\mathcal{D}(E), \|\cdot\|) \mid |\psi(f_k) - (\gamma_*(\varphi'))(f_k)|_\infty \leq \epsilon,\ k = 1, ..., q\}$$

and $= \{\phi \in Mult(\mathcal{D}(E'), \|\cdot\|') \mid |\phi(g_k) - (\varphi')(g_k)|_\infty \leq \epsilon,\ k = 1, ..., q\}$
$= \{\phi \in Mult(\mathcal{D}(E'), \|\cdot\|') \mid |\phi(f_k \circ \gamma) - (\varphi')(f_k \circ \gamma)|_\infty \leq \epsilon,\ k = 1, ..., q\}$
$= \{\phi \in Mult(\mathcal{D}(E'), \|\cdot\|') \mid |(\gamma_*(\phi))(f_k) - (\gamma_*(\varphi'))(f_k)|_\infty \leq \epsilon,\ k = 1, ..., q\}$
$= \{\phi \in Mult(\mathcal{D}(E'), \|\cdot\|') \mid (\gamma_*(\phi)) \in W\}$, hence $\gamma_*(V) \subset W$. That shows the continuity of γ_*. □

REMARK 13. We can derive the following interpretation: let $\varphi' \in Mult(\mathcal{D}(E'), \|\cdot\|')$ and let $\mathcal{M}' = Ker(\varphi')$. Then $(\widetilde{\gamma})^{-1}(\mathcal{M}')$ is a maximal ideal $\mathcal{M}$ of $\mathcal{D}(E)$ which is the kernel of a unique $\varphi \in Mult(\mathcal{D}(E), \|\cdot\|)$. Consequently, φ is defined as $\varphi(f) = \varphi'(f \circ \gamma)$ described by Theorem 34.

Acknowledgements

We are very grateful to the referee for pointing out to us many misprints.

References

[1] A. Escassut, *Spectre maximal d'une algèbre de Krasner* (French), Colloq. Math. **38** (1977/78), no. 2, 339–357. MR0508094

[2] A. Escassut, *The ultrametric spectral theory*, Period. Math. Hungar. **11** (1980), no. 1, 7–60, DOI 10.1007/BF02019983. MR571136

[3] A. Escassut, *Analytic elements in p-adic analysis*, World Scientific Publishing Co., Inc., River Edge, NJ, 1995. MR1370442

[4] A. Escassut, *Ultrametric Banach algebras*, World Scientific Publishing Co., Inc., River Edge, NJ, 2003. MR1978961

[5] A. Escassut and N. Mainetti, *Spectral semi-norm of a p-adic Banach algebra*, Bull. Belg. Math. Soc. Simon Stevin **5** (1998), no. 1, 79–91. MR1610739

[6] A. Escassut and N. Maïnetti, *Shilov boundary for ultrametric algebras*, p-adic Numbers in Number Theory, Analytic Geometry and Functional Analysis, Belgian Mathematical Society, pp. 81–89 (2002).

[7] A. Escassut and N. Maïnetti, *On ideals of the algebra of p-adic bounded analytic functions on a disk*, Bull. Belg. Math. Soc. Simon Stevin **14** (2007), no. 5, 871–876. MR2378994

[8] A. Escassut and N. Maïnetti, *About the ultrametric corona problem*, Bull. Sci. Math. **132** (2008), no. 5, 382–394, DOI 10.1016/j.bulsci.2007.09.002. MR2426642

[9] A. Escassut, *The ultrametric corona problem and spherically complete fields*, Proc. Edinb. Math. Soc. (2) **53** (2010), no. 2, 353–371, DOI 10.1017/S0013091508000837. MR2653237

[10] A. Escassut and N. Maïnetti, *Multiplicative spectrum of ultrametric Banach algebras of continuous functions*, Topology Appl. **157** (2010), no. 16, 2505–2515, DOI 10.1016/j.topol.2010.08.003. MR2719394

[11] A. Escassut and N. Maïnetti, *Morphisms between ultrametric Banach algebras and maximal ideals of finite codimension*, Advances in ultrametric analysis, Contemp. Math., vol. 596, Amer. Math. Soc., Providence, RI, 2013, pp. 63–71, DOI 10.1090/conm/596/11923. MR3156537

[12] A. Escassut, *Value Distribution in p-adic Analysis.* WSCP Singapore, (2015).

[13] G. Garandel, *Les semi-normes multiplicatives sur les algèbres d'éléments analytiques au sens de Krasner* (French), Nederl. Akad. Wetensch. Proc. Ser. A **78**=Indag. Math. **37** (1975), no. 4, 327–341. MR0390286

[14] B. Guennebaud, *Sur une notion de spectre pour les algèbres normées ultramétriques*, thèse d'Etat, Université de Poitiers, (1973).

[15] L. Haddad, *Sur quelques points de topologie générale. Théorie des nasses et des tramails* (French), Ann. Fac. Sci. Univ. Clermont-Ferrand **44** (1970), 5–80. MR0301685

[16] W. H. Schikhof, *Ultrametric calculus. An introduction to p-adic analysis*, Cambridge Studies in Advanced Mathematics, vol. 4, Cambridge University Press, Cambridge, 1984. MR791759

[17] M. van der Put, *The non-archimedean corona problem*, Table Ronde d'Analyse Non Archimédienne (Paris, 1972), Soc. Math. France, Paris, 1974, pp. 287–317. Bull. Soc. Math. France, Mém. No. 39-40. MR0364685

Université Clermont Auvergne, Laboratoire de Mathématiques Blaise Pascal, UMR-CNRS, BP 10448F-63000 Clermont-Ferrand, France

Email address: `alain.escassut@math.univ-bpclermont.fr`

Département GEA IUT d'Aurillac, Université Clermont Auvergne, 100 Rue de l'Égalité, 15000 Aurillac

Email address: `Nicolas.Mainetti@iut.u-clermont1.fr`

Contemporary Mathematics
Volume **704**, 2018
http://dx.doi.org/10.1090/conm/704/14166

Classical p-adic Nevanlinna theory and Nevalinna Theory out of a hole

Alain Escassut and Ta Thi Hoai An

ABSTRACT. In the first section called Classical theory, we recall basic properties of the analytic and meromorphic functions, Motzkin's factoraization of analytic elements and the classical p-adic Nevanlinna theory. The second section is devoted to meromorphic functions in the complement of an "open" disk with the use of Motzkin's factorization and we show the existence of a Nevanlinna theory in that field of meromorphic functions. Applications are examined: Nevanlinna theory on 3 small functions, parametrization of algebraic curves, qusai-exceptional small functions and branched small functions.

I. Classical theory

I.1 Basic definitions. We denote by $\mathbb{K}$ an algebraically closed field of characteristic 0, complete with respect to an ultrametric absolute value $|\ .\ |$. Analytic functions inside a disk or in the whole field $\mathbb{K}$ were introduced and studied in many books. Given $\alpha \in \mathbb{K}$ and $R \in \mathbb{R}_+^*$, we denote by $d(\alpha, R)$ the disk $\{x \in \mathbb{K} \mid |x-\alpha| \leq R\}$, by $d(\alpha, R^-)$ the disk $\{x \in \mathbb{K} \mid |x-\alpha| < R\}$ and by $C(\alpha, r)$ the circle $\{x \in \mathbb{K} \mid |x-\alpha| = r\}$. We denote by $|\mathbb{K}|$ the set $\{|x| \mid x \in \mathbb{K}\}$.

During the forties, Mark Krasner had the idea of introducing a kind of analytic functions based upon the following property in complex analysis. Let D be an open bounded subset of $\mathbb{C}$. Then, by Runge's Theorem, every holomorphic functions in D which is continuous on the closure of D is the uniform limit on $\overline{D}$ of sequences of rational functions with no pole in $\overline{D}$. This property suggested Marc Krasner defining analytic elements in a certain kind of sets called quasi-connected sets, as the uniform limit of a sequence of rational functions with no pole in such a set. Later, it appeared that such analytic elements may also be defined in a more general class of sets, particularly *infraconnected sets*. Recall that a subset of $\mathbb{K}$ is said to be *infraconnected* if for every $a \in D$, the closure of the image of the function I_a defined in D as $I_a(x) = |x-a|$, is an interval.

Given a closed bounded subset D of $\mathbb{K}$, we denote by $\widetilde{D}$ the smallest closed disk containing D i.e. if $R = \text{diam(D)}$, then $\widetilde{D} = d(a, R)$ with $a \in D$.

2010 *Mathematics Subject Classification.* Primary 12J25; 46S10.

The second named author was supported by Vietnam's National Foundation for Science and Technology Development (NAFOSTED) under the Grant No. 101.04-2014.41.

Lemma I.1.1 ([12], Lemma 1.10). *let D be a closed bounded subset of $\mathbb{K}$. Then $\widetilde{D}\setminus D$ admits a partition of the form $\{d(a_i, r_i^-)_{i\in I}\}$ where each disk $d(a_i, r_i^-)$ is maximal.*

Definitions and notation. Let D be a closed bounded subset of $\mathbb{K}$. The disks $d(a_i, r_i^-)$ lying in the partition of $\widetilde{D}\setminus D$ shown above are called *the holes of D.*

Let $R(D)$ be the $\mathbb{K}$-algebra of rational functions without poles in D provided with the norm of uniform convergence on D and let $H(D)$ be the completion of $R(D)$ with respect to that norm. So, $H(D)$ is a $\mathbb{K}$-Banach algebra.

Now consider a closed unbounded subset D of $\mathbb{K}$ and let $R_b(D)$ be the algebra of bounded rational functions having no pole in D, provided with the norm of uniform convergence on D. Then the completion of $R_b(D)$ is a $\mathbb{K}$-Banach algebra $H_b(D)$ again whose elements are called *bounded analytic elements in D.* Particularly this applies to sets of the form $\mathbb{K}\setminus d(a, r^-)$. Moreover we denote by $H_0(D)$ the $\mathbb{K}$-Banach algebra of elements f such that $\lim\limits_{|x|\to+\infty,\ x\in D} f(x) = 0$.

Let us recall that a filter $\mathcal{F}$ on a set E is said to be *secant with a subset B of E* if the family of sets $H \cap B$, $H \in \mathcal{F}$ makes a filter on B.

Here we will not recall the theory of analytic functions due to Krasner, soon generalized by Philippe Robba and next by Kamal Boussaf. But we have to describe some properties of analytic elements on infraconnected sets. Properties of these functions are nice because they derive from those of rational functions.

Particularly, the absolute value of an analytic element admits a limit along a *circular filter* and so, multiplicative semi-norms on the $\mathbb{K}$-algebra of the analytic elements on a bounded closed subset D of $\mathbb{K}$ are characterized by circular filters on D. These properties explain the behavior of analytic and meromorphic functions in the whole field $\mathbb{K}$ and inside a disk [**9**], [**10**], [**12**].

Notation. Throughout this chapter, D is a closed bounded infraconnected subset of $\mathbb{K}$. We will denote by $Mult(H(D), \|\ .\ \|_D)$ the set of continuous multiplicative semi-norms ψ of the $\mathbb{K}$-algebra $H(D)$.

Let $a \in \mathbb{K}$ and let r', r'' be such that $0 < r' < r''$. We denote by $\Gamma(a, r', r'')$ the annulus $\{x \in K \mid r' < |x - a| < r''\}$ and by $\Delta(a, r', r'')$ the annulus

$\{x \in K \mid r' \leq |x - a| \leq r''\}$. Now, let $a \in \mathbb{K}$, $r > 0$ and consider the filter admitting for basis the annuli $\Gamma(b, r', r'')$ with $b \in d(a, r)$ and $r' < r < r''$. This filter is called *circular filter of center a, of diameter r.* Moreover if $\mathbb{K}$ is not spherically complete, each decreasing sequence of disks (D_n) whose intersection is empty also defines a filter that is called *circular filter of basis (D_n).* Finally, for every $a \in \mathbb{K}$, the filter of neighborhoods of a is called *circular filter of neighborhoods of a* and such a circular filter is said to be *punctual.*

Theorem I.1.2 (B. Guennebaud, G. Garandel [E0], Theorem 13.1). *Each circular filter $\mathcal{F}$ on $\mathbb{K}$ defines a multiplicative semi-norm on $\mathbb{K}[x]$ which is a norm if and only if it is not punctual and the semi-norm is continuous with respect to the norm $\|\ .\ \|_D$ if and only if the filter is secant with D. Each circular filter $\mathcal{F}$ secant with D defines on $(R(D),\ \|\ .\ \|_D)$ a continuous multiplicative semi-norm $\varphi_{\mathcal{F}}$ that has continuation to $H(D)$ and the mapping associating to each circular filter secant with D, its multiplicative semi-norm $\varphi_{\mathcal{F}}$ is a bijection from the set of circular filters secant with D onto the set of continuous multiplicative semi-norms on $H(D)$ and* *on $R(D)$. $Mult(H(D), \| \,.\, \|_D)$ is compact with respect to the topology of pointwise convergence.*

The following theorem is easily proven and is given as Corollary 14.7 in [**10**]:

Theorem I.1.3. *A set D is infraconnected if and only if $Mult(R(D), \| \,.\, \|_D)$ is connected.*

I.2. Power and Laurent series. All analytic elements in $d(0, R^-)$ are power series converging in $d(0, R^-)$. The converse is false. However, concerning a "closed " disk $d(0, R)$, we will see that the analytic elements in $d(0, R)$ are exactly the power series converging in this disk.

Definitions. Let $f(x) = \sum_{n=0}^{\infty} a_n x^n$ be a power series with coefficients in $\mathbb{K}$.

As usual, when $\limsup_{n\to\infty} \sqrt[n]{|a_n|} \neq 0$, we call *radius of convergence of f* the number $r = \dfrac{1}{\limsup_{n\to\infty} \sqrt[n]{|a_n|}}$ (with $r = 0$ when $\limsup_{n\to\infty} \sqrt[n]{|a_n|} = +\infty$).

When $\limsup_{n\to\infty} \sqrt[n]{|a_n|} = 0$, we define the radius of convergence of f as $+\infty$

Example. Let $f(x) = \sum_{n=1}^{\infty} nx^n$. The radius of convergence of this series is 1. This function obviously defines the rational function $\dfrac{x}{(1-x)^2}$ in $d(1, 1^-)$.

Notations and definitions. Power series whose radius of convergence is ∞ are called *entire functions on* $\mathbb{K}$ and the set of entire functions is denoted by $\mathcal{A}(\mathbb{K})$.

For every $a \in \mathbb{K}$, $r \in \mathbb{R}_+^*$, similarly we denote by $\mathcal{A}(d(a, r^-))$ the set of power series in $x - a$ whose radius of convergence is superior or equal to r and by $\mathcal{A}_b(d(a, r^-))$ the set of functions $f \in \mathcal{A}(d(a, r^-))$ that are bounded in $d(a, r^-)$. The set $\mathcal{A}(d(a, r^-)) \setminus \mathcal{A}_b(d(a, r^-))$ will be denoted by $\mathcal{A}_u(d(a, r^-))$.

Similarly, we denote by $\mathcal{A}(\mathbb{K} \setminus d(a, r^-))$ the set of Laurent series converging whenever $|x - a| \geq r$ and by $\mathcal{A}_b(\mathbb{K} \setminus d(a, r^-))$ the $\mathbb{K}$-Banach algebra of bounded Laurent series converging whenever $|x - a| \geq r$.

Let $R \in \mathbb{R}_+^*$ and let $f \in \mathcal{A}(d(a, R^-))$. Given $r \in]0, R[$, f belongs to $H(d(a, r))$, hence for every circular filter $\mathcal{F}$ secant with $d(a, r)$, $\varphi_{\mathcal{F}}(f)$ is defined. Given $b \in d(a, R)$, if $s < R$, and if $\mathcal{F}$ is the circular filter of center b and diameter s, we put $\varphi_{b,s}(f) = \varphi_{\mathcal{F}}(f) = \|f\|_{d(b,s)}$. Particularly, if $a = 0$ we put $|f|(s) = \lim_{\mathcal{F}} |f(x)| = \lim_{|x|\to s,\ |x|\neq r} |f(x)|$.

The following theorem is important in the Nevanlinna Theory.

Theorem I.2.1 ([12], Theorem 14.7). *Let $f \in \mathcal{A}(K)$. The following three statements are equivalent:*

i) $\lim_{r\to+\infty} \dfrac{|f|(r)}{r^q} = +\infty \ \forall q \in \mathbb{N}$,

ii) there exists no $q \in \mathbb{N}$ such that $\lim_{r\to+\infty} \dfrac{|f|(r)}{r^q} = 0$,

iii) f is not a polynomial.

Theorem I.2.2 ([12], Theorem 14.6) . *Let $r \in \mathbb{R}_+^*$ and let $D = d(0,r)$. Then $H(D)$ is the set of power series $f(x) = \sum_{n=0}^{\infty} a_n x^n$ such that $\lim_{|n|_\infty \to \infty} |a_n| r^n = 0$ and we have*

$$\|f\|_D = \max_{n \in \mathbb{N}} |a_n| r^n = \varphi_{\mathcal{F}}(f).$$

Moreover, the norms $\| \, . \, \|_{d(0,r)}$ and $\| \, . \, \|_{C(0,r)}$ are multiplicative and coincide on $H(C(0,r))$.

Theorem I.2.3 (Corollary 14.15). *Let $f \in \mathcal{A}(d(0,r^-))$ be not identically zero. For every $\alpha \in d(0,r^-)$, $f(x)$ is equal to a power series*

$$\sum_{n=0}^{\infty} b_n(\alpha)(x-\alpha)^n.$$

If f is not identically zero and if α is a zero of f in $d(0,r^-)$, α is an isolated zero and f factorizes in $\mathcal{A}(d(0,r^-))$ in the form $(x-\alpha)^q g(x)$, with $g \in \mathcal{A}(d(0,r^-))$, $q \in \mathbb{N}^*$ $g(\alpha) \neq 0$.

Definition. Let D be an infraconnected subset of $\mathbb{K}$, let $f \in H(D)$, let $\alpha \in \overset{\circ}{D}$, let $r > 0$ be such that $d(\alpha, r) \subset D$ and suppose $f(x) = \sum_{n=q}^{\infty} b_n(x-\alpha)^n$ whenever $x \in d(\alpha, r)$, with $b_q(\alpha) \neq 0$ and $q > 0$. Then α is called *a zero of multiplicity order* q, or more simply, *a zero of order* q. In the same way, q is called *the multiplicity order of* α.

Theorem I.2.4 ([12], Theorem 14.19). *Let $R \in \mathbb{R}_+^*$ and let $f \in \mathcal{A}(d(a,R^-))$. Then f is invertible in $\mathcal{A}(d(a,R^-))$ if and only if f has no zero in $d(a,R^-)$.*

Theorem I.2.5 ([12], Theorem 14.20). *Let $R \in \mathbb{R}_+^*$. The $\mathbb{K}$-subalgebra $\mathcal{A}_b(d(0,R^-))$ of $\mathcal{A}(d(0,R^-))$ is a Banach $\mathbb{K}$-algebra with respect to the norm $\| \, . \, \|_{d(0,R^-)}$. Further, this norm is multiplicative and satisfies $\|f\|_{d(0,R^-)} = \lim_{r \to R} |f|(r) = \sup_{n \in \mathbb{N}} |a_n| R^n$.*

Let $f(x) = \sum_{n=0}^{\infty} a_n x^n \in \mathcal{A}(d(0,R^-))$. Then f is bounded in $d(0,R^-)$ if and only if so is the sequence $(|a_n| R^n)_{n \in \mathbb{N}}$. Moreover, if f is bounded, then $\|f\|_{d(0,R^-)} = \sup_{n \in \mathbb{N}} |a_n| R^n$.

Corollary I.2.5.a. *Let $R \in \mathbb{R}_+^*$ and let $f, \ g \in \mathcal{A}(d(a,R^-))$. Then fg belongs to $\mathcal{A}_b(d(a,R^-))$ if and only if so do both f and g and $\mathcal{A}_b(d(a,R^-))$ is $\mathbb{K}$-subalgebra of $\mathcal{A}(d(a,R^-))$.*

Theorem I.2.6 ([12], Theorem 14.22). *Suppose that $\mathbb{K}$ has characteristic different from 2. Let $f, \ g \in \mathcal{A}(\mathbb{K}) \setminus \mathbb{K}$ (resp. $f, \ g \in \mathcal{A}_u(d(0,r^-))$) be distinct. Then $f^2 - g^2$ belongs to $\mathcal{A}(\mathbb{K}) \setminus \mathbb{K}$ (resp. $f, \ g \in \mathcal{A}_u(d(0,r^-))$).*

Theorem I.2.7 ([12], Theorem 14.23). *For every $r \in \mathbb{R}_+^*$, $H(d(0,r^-))$ is included in $\mathcal{A}_b(d(0,r^-))$.*

Remarks. Actually, $H(d(0,r^-))$ is much smaller than $\mathcal{A}_b(d(0,r^-))$. In particular, $\sqrt[q]{1+x}$ belongs to $\mathcal{A}_b(d(0,1^-)$, but does not belong to $H(d(0,1^-))$.

Let $\sum_{0}^{+\infty} a_n x^n$ be a power series whose radius of convergence is r. Suppose first that $r \in |\mathbb{K}|$. If there is at least one point $\alpha \in C(0,r)$ such that the series converges at α, then this implies that $\lim_{n\to+\infty} |a_n|r^n = 0$ and hence the series converges in all $C(0,r)$ and defines an element of $H(d(0,r))$. If r does not belong to $|\mathbb{K}|$, the power series converging in $d(0,r)$ are just the power series converging in $d(0,r^-)$ and hence, this is why we don't have to consider *analytic functions* inside a disk $d(a,r)$, whenever $r > 0$.

I.3. The Mittag-Leffler Theorem. Throughout the paragraph I.3, D is a closed infraconnected subset of $\mathbb{K}$. The wonderful Mittag-Leffler Theorem for analytic elements is due to Marc Krasner who showed it on quasi-connected sets. The same proof holds on infraconnected sets as it was shown by Philippe Robba. The theorem shows that a Banach space $H(D)$ is a direct topological sum of elementary subspaces and is indispensable to have a clear image of the space $H(D)$.

Theorem I.3.1. (M.Krasner, [19], [12], Theorem 15.1) *Let $f \in H(D)$. There exists a unique sequence of holes $(T_n)_{n\in\mathbb{N}^*}$ of D and a unique sequence $(f_n)_{n\in\mathbb{N}}$ in $H(\mathbb{K}\setminus T_n)$ such that $f_0 \in H(\widetilde{D})(n>0)$, $\lim_{n\to\infty} f_n = 0$ satisfying further*

(1) $f = \sum_{n=0}^{\infty} f_n$ *and* $\|f\|_D = \sup_{n\in\mathbb{N}} \|f_n\|_D$.

For every hole $T_n = d(a_n, r_n^-)$, we have

(2) $\|f_n\|_D = \|f_n\|_{\mathbb{K}\setminus T_n} = \varphi_{a_n,r_n}(f_n) \le \varphi_{a_n,r_n}(f) \le \|f\|_D$.

If D is bounded and if $\widetilde{D} = d(a,r)$ we have

(3) $\|f_0\|_D = \|f_0\|_{\widetilde{D}} = \varphi_{a,r}(f_0) \le \varphi_{a,r}(f) \le \|f\|_D$.

Let $D' = \widetilde{D} \setminus \left(\bigcup_{n=1}^{\infty} T_n\right)$. Then f belongs to $H(D')$ and its decomposition in $H(D')$ is given again by (1) *and f satisfies $\|f\|_{D'} = \|f\|_D$.*

The Mittag-Leffler Theorem suggests some new definitions

Definitions and notations. Let $f \in H(D)$. We consider the series $\sum_{n=0}^{\infty} f_n$ obtained in the last Theorem, whose sum is equal to f in $H(D)$, with $f_0 \in H(\widetilde{D})$, $f_n \in H(\mathbb{K}\setminus T_n)\setminus\{0\}$ and with the T_n holes of D. Each T_n will be called a *f-hole* and f_n will be called the *Mittag-Leffler term of f associated to T_n*, whereas f_0 will be called the *principal term* of f. For each f-hole T of D, the Mittag-Leffler term of f associated to T will be denoted by $\overline{\overline{f_T}}$ whereas the principal term of f will be denoted by $\overline{\overline{f_0}}$. The series $\sum_{n=0}^{\infty} f_n$ will be called *the Mittag-Leffler series of f on the infraconnected set D.*

Example 1. *Let $f \in H(d(0,1^-))$ and let $(d(\alpha_m,1^-))_{m\in\mathbb{N}^*}$ be the family of f-holes. Then f is of the form*

$$(1) \quad \sum_{n=0}^{\infty} a_{n,0}x^n + \sum_{m,n\in\mathbb{N}^*} \frac{a_{n,m}}{(x-\alpha_m)^n}$$

with $\lim_{n\to\infty} a_{n,0} = 0$, $\lim_{n\to\infty} |a_{n,m}| = 0$ *whenever* $m \in \mathbb{N}^*$ *and* $\lim_{m\to\infty} \big(\sup_{n\in\mathbb{N}^*} |a_{n,m}|\big) = 0$. *On the other hand, f satisfies*

$$\|f\|_{d(0,1^-)} = \sup_{m\in\mathbb{N},\ n\in\mathbb{N}^*} |a_{n,m}|.$$

Conversely, every function of the form (1), *with the α_m satisfying $|\alpha_m| = |\alpha_j - \alpha_m| = 1$ whenever $m \neq j$, belongs to $H(d(0,1^-))$. The norm $\|\ .\ \|_{d(0,1^-)}$ is multiplicative and equal to $\varphi_{0,1}$.*

Example 2. *Let $r_1, r_2 \in \mathbb{R}_+$ satisfy $0 < r_1 < r_2$. Then $H(\Delta(0,r_1,r_2))$ is equal to the set of the Laurent series $\sum_{-\infty}^{+\infty} a_n x^n$ with $\lim_{n\to-\infty} |a_n|r_1^n = \lim_{n\to\infty} |a_n|r_2^n = 0$ and we have*

$$\Big\|\sum_{-\infty}^{+\infty} a_n x^n\Big\|_{\Delta(0,r_1,r_2)} = \max\Big(\sup_{n\geq 0} |a_n|r_1^n,\ \sup_{n<0} |a_n|r_2^n\Big).$$

Example 3. *Let $r \in \mathbb{R}_+^*$. Then $H(C(0,r))$ is equal to the set of the Laurent series $\sum_{-\infty}^{+\infty} a_n x^n$ with $\lim_{|n|_\infty\to\infty} |a_n|r^n = 0$ and we have $\|\sum a_n x^n\|_{C(0,r)} = \sup_{n\in\mathbb{Z}} |a_n|r^n$. Next, the norm $\|\ .\ \|_{C(0,r)}$ is multiplicative and equal to $\varphi_{0,r}$.*

Theorem I.3.2 ([12], Theorem 15.12). *Let D_1, D_2 be infraconnected subsets of $\mathbb{K}$ having finitely many holes, such that $D_1 \cap D_2 \neq \emptyset$ and let $f_j \in H(D_j)$, $j = 1,2$ be such that $f_1(x) = f_2(x)\ \forall x \in D_1 \cap D_2$. Then the function f defined in $D_1 \cup D_2$ as $f(x) = f_j(x)\ \forall x \in D_j$, $j = 1,2$, belongs to $H(D_1 \cup D_2)$.*

Notation. Let E be a $\mathbb{K}$-Banach space. We will denote by E^* the $\mathbb{K}$-Banach space of continuous linear forms of E provided with its usual norm. The dual of a Banach space $H(D)$ was thoroughly studied by Yvette Amice (Theorem 15.13 in [**12**]).

Theorem I.3.3 (Y. Amice, [12], Theorem 15.13). *Let $r \in \mathbb{R}_+$. Given $h(t) = \sum_{n=0}^{\infty} \frac{b_n}{t^n} \in \mathcal{A}_b(\mathbb{K} \setminus d(0,r))$ there exists a unique $\phi_h \in H(d(0,r))^*$ satisfying $\phi_h(x^q) = b_q$, $(q \in \mathbb{N})$. Moreover, on the space $\mathcal{A}_b(\mathbb{K} \setminus d(0,r))$ provided with the norm $\|\ .\ \|_{\mathbb{K}\setminus d(0,r)}$, the mapping $h \to \phi_h$ is an isometric isomorphism from $\mathcal{A}_b(\mathbb{K} \setminus d(0,r))$ onto $H(d(0,r))^*$.*

Theorem I.3.4 ([12], Theorem 15.17). *Let $f \in H_b(\mathbb{K} \setminus d(a,r^-))$, and for each $\alpha \in d(a,r^-)$, let $f(x) = \sum_{n=0}^{\infty} \frac{b_n(\alpha)}{(x-\alpha)^n}$. Then $b_1(\alpha)$ does not depend on α in $d(a,r^-)$.*

Definition and notation. Let $f \in H_b(D)$, let T be a hole of D and let $a \in T$. Let $f_T(x) = \sum_{n=1}^{\infty} \frac{b_n(a)}{(x-a)^n}$. Since $b_1(a)$ actually does not depend on a in T, we set $\text{res}(f,T) = b_1(a)$ and this number $\text{res}(f,T)$ will be called *the residue of f on the hole T.*

I.4. Divisors of analytic functions and Lazard's problem. In this paragraph we shall define divisors in $\mathbb{K}$ or in a disk $d(a, R^-)$. We then shall define the divisor of an analytic function and of an ideal. Given a divisor T on $\mathbb{K}$, there is no problem to construct an entire function whose divisor is T. But given a divisor T on a disk $d(a, r^-)$, it is not always possible to find an analytic function (in that disk) whose divisor is T. This is Lazard's problem that we will recall.

Definition. We call *a divisor in* $\mathbb{K}$ (resp. *a divisor in a disk* $d(a, R^-)$) a mapping T from $\mathbb{K}$ (resp. from $d(a, R^-)$) to $\mathbb{N}$ whose support is countable and has a finite intersection with each disk $d(a, r)$, $\forall r > 0$ (resp. $\forall r \in]0, R[$). Thus, a divisor on $\mathbb{K}$ (resp. of $d(a, R^-)$) is characterized by a sequence $(a_n, q_n)_{n \in \mathbb{N}}$ with $a_n \in \mathbb{K}$, $\lim_{n\to\infty} |a_n| = \infty$, (resp. $a_n \in d(a, R^-)$, $\lim_{n\to\infty} |a_n - a| = R$), $|a_n| \leq |a_{n+1}|$ and $q_n \in \mathbb{N}^*$ $\forall n \in \mathbb{N}$. So, we will frequently denote a divisor by the sequence $(a_n, q_n)_{n\in\mathbb{N}}$ which characterizes it.

The set of divisors on $\mathbb{K}$ (resp. on $d(a, R^-)$) is provided with a natural multiplicative law that makes it a semi-group. It is also provided with a natural order relation: given two divisors T and T', we can set $T \leq T'$ when $T(\alpha) \leq T'(\alpha)$ $\forall \alpha \in d(a, R^-)$. Moreover, if T, T' are two divisors such that $T(\alpha) \geq T'(\alpha)$ $\forall \alpha \in d(0, R^-)$, we can define the divisor $\dfrac{T}{T'}$.

Given $f \in \mathcal{A}(\mathbb{K})$ (resp. $f \in \mathcal{A}(d(a, R^-))$), we can define *the divisor of* f, denoted by $\mathcal{D}(f)$ on K (resp. on $d(a, R^-)$) as $\mathcal{D}(f)(\alpha) = 0$ whenever $f(\alpha) \neq 0$ and $\mathcal{D}(f)(\alpha) = s$ when f has a zero of order s at α.

Similarly, given an ideal I of $\mathcal{A}(\mathbb{K})$ (resp. of $\mathcal{A}(d(a, R^-))$) we will denote by $\mathcal{D}(I)$ the lower bound of the the $\mathcal{D}(f)$ $f \in I$ and $\mathcal{D}(I)$ will be called *the divisor of* I.

Finally, given a divisor $T = (a_n, q_n)_{n\in\mathbb{N}}$, we shall denote by $\overline{T}$ the divisor $(a_n, 1)_{n\in\mathbb{N}}$. Let $T = (a_n, q_n)_{n\in\mathbb{N}}$ be a divisor on $\mathbb{K}$ (resp. of $d(a, R^-)$). For every $r > 0$ (resp. $r \in]0, R[$) we set $|T|(r) = \prod\limits_{|a_j| \leq r} \left(\dfrac{r}{|a_j|}\right)^{q_j}$. The divisor T on $d(a, R^-)$ is said to be *bounded* if $\lim\limits_{r\to R} |T|(r) < \infty$ and then we put $\|T\| = \lim\limits_{r\to R} |T|(r)$.

Remark. Let $f \in \mathcal{A}(d(a, R^-))$ and let $(a_n, q_n)_{n\in\mathbb{N}} = \mathcal{D}(f)$. Then $\omega_{a_n}(f) = q_n$ $\forall n \in \mathbb{N}$ and $\omega_\alpha(f) = 0$ $\forall \alpha \in d(a, R^-) \setminus \{a_n \mid n \in \mathbb{N}\}$.

Theorem I.4.1 ([12], Theorem 28.1) *Let $a \in \mathbb{K}$, $R > 0$. Let f, $g \in \mathcal{A}(\mathbb{K})$ (resp. f, $g \in \mathcal{A}(\delta(a, R^-))$) be such that $\mathcal{D}(f) \geq \mathcal{D}(g)$. Then there exists $h \in \mathcal{A}(\mathbb{K})$ (resp. $h \in \mathcal{A}(d(a, R^-))$) such that $f = gh$.*

Corollary I.4.1.a. Let $a \in \mathbb{K}$, $R > 0$. Let I be an ideal of $\mathcal{A}(\mathbb{K})$ (resp. an ideal of $\mathcal{A}(d(a, R^-))$) and suppose that there exists $g \in I$ such that $\mathcal{D}(g) = \mathcal{D}(I)$. Then $I = g\mathcal{A}(\mathbb{K})$ (resp. $I = g\mathcal{A}(d(a, R^-))$).

In the whole field $\mathbb{K}$, given a divisor T, it is always possible to find an entire function admitting T for divisor.

Theorem I.4.2 ([12], Theorem 28.4). *Let $T = (\alpha_n, q_n)_{n\in\mathbb{N}}$ be divisor of $\mathbb{K}$ The infinite product $\prod\limits_{n=1}^{\infty}(1 - \dfrac{x}{\alpha_n})^{q_n}$ is uniformly convergent in all bounded subsets of $\mathbb{K}$ and defines an entire function $f \in \mathcal{A}(\mathbb{K})$ such that $\mathcal{D}(f) = T$. Moreover, given $g \in \mathcal{A}(\mathbb{K})$ such that $\mathcal{D}(g) = T$, then g is of the form λf.*

Corollary I.4.2.a. *For every divisor T on $\mathbb{K}$, there exists $f \in \mathcal{A}(\mathbb{K})$ such that $\mathcal{D}(f) = T$. Moreover, if $f(0) = 1$, f satisfies $|f|(r) = |T|(r)\ \forall r > 0$.*

So, given a divisor T on $\mathbb{K}$, we can find an entire function whose divisor is just T. It is natural to consider the same problem inside a disk $d(a, r^-)$. Indeed, in $\mathbb{C}$, it is known that the similar problem always admits a solution, in the whole field $\mathbb{C}$ as well as inside an open disk. Actually, in the general context of a complete ultrametric algebraically closed field $\mathbb{K}$, given a divisor T in $d(a, r^-)$, the problem of finding an analytic function f in $d(a, r^-)$, admitting T for divisor, has no solution when $\mathbb{K}$ is not spherically complete, in the general case, but it has a solution when $\mathbb{K}$ is spherically complete, has proven by M. Lazard [**20**]. On the other hand, in the general case, we can construct an analytic function f whose divisor is a little bit bigger than the given divisor T but narrows it.

Theorem I.4.3 ([12], Theorem 28.14). *Let $T = (a_n, q_n)_{n\in\mathbb{N}}$ be a divisor on the disk $d(a, R^-)$ with $a_n \neq 0\ \forall n \in \mathbb{N}$ and let $\epsilon > 0$. There exists $f \in \mathcal{A}(d(a, R^-))$ such that $\mathcal{D}(f) \geq T$, $f(0) = 1$ and $|f|(r) \leq |T|(r)(1 + \epsilon)\ \forall r \in]0, R[$.*

Remark. Here we may notice that $H(d(0, R^-))$ is smaller than $\mathcal{A}_b(d(0, R^-))$. Indeed, there exist functions $f \in \mathcal{A}_b(d(0, R^-))$ having infinitely many zeros in $d(0, R^-)$. But any element of $H(d(0, R^-))$ is quasi-invertible and hence has finitely many zeros.

Theorem I.4.4 (M. Lazard, [20], [12], Theorem 29.1). *Let $\mathbb{K}$ be not spherically complete and let $(D_n)_{n\in\mathbb{N}}$ be a decreasing sequence of disks $d(u_n, \rho_n)$ such that $\bigcap_{n=0}^{\infty} D_n = \emptyset$. Let $R = \dfrac{1}{\lim_{n\to\infty} \rho_n}$. There exists sequences $(c_n)_{n\in\mathbb{N}}$ of $d(0, R^-)$ such that $\lim_{n\to\infty} |c_n| = R$ and such that no function $f \in \mathcal{A}(d(0, R^-))$ admits for divisor the divisor $T = (c_n, 1)_{n\in\mathbb{N}}$.*

Theorem I.4.5 (M. Lazard [20], [12], Theorem 29.4). *Suppose $\mathbb{K}$ is spherically complete. Assume that $|Q_m|(\rho_m) \leq |T|(\rho_m)\ \forall m \in \mathbb{N}$. Let $R \in]0, +\infty[$. There exists $f \in \mathcal{A}(d(0, R^-))$ satisfying*

i) $f(0) = 1$,
ii) $|f|(r) \leq |T|(r)\ \forall r < R$,
iii) P_m divides $f - Q_m$ in $\mathcal{A}(d(0, R^-))$.

Corollary I.4.5.a. *Suppose $\mathbb{K}$ is spherically complete. Let T be a divisor on $d(a, R^-)$. There exists $f \in \mathcal{A}(d(a, R^-))$ such that $\mathcal{D}(f) = T$.*

I.5. Motzkin factorization. The idea of factorizing quasi-invertible analytic elements into a product of singular factors is a remarkable idea due to E. Motzkin [**22**]. This factorization has tight links with the Mittag-Leffler series. In the paragraph, we denote by D an infraconnected open closed bounded subset of $\mathbb{K}$.

According to the properties of power series [**12**], we have first the following lemma.

Lemma I.5.1. *Let $E = \mathbb{K} \setminus d(a, r^-)$ with $a \in \mathbb{K}$ and $r > 0$. Let $f \in H(E)$ be invertible in $H(E)$. Then $f(x)$ is a Laurent series of the form $\sum_{-\infty}^{q} a_n(x - a)^n$ with $|a_q|r^q > |a_n|r^n\ \forall n < q$.*

Theorem I.5.2 derives from Lemma I.5.1:

Theorem I.5.2 ([12], Lemma 31.1). *Let $T = d(a, r^-)$, with $a \in \mathbb{K}$ and $r > 0$, let $E = \mathbb{K} \setminus T$ and take $b \in T$. Let $g \in H(E)$ be invertible in $H(E)$. Then there exist $\lambda \in \mathbb{K}$, $q \in \mathbb{Z}$ and $h \in H(E)$ invertible in $H(E)$, satisfying $\|h - 1\|_E < 1$, $\lim_{|x| \to +\infty} h(x) = 1$ and $g(x) = \lambda(x - b)^q h(x)$. Moreover, λ, q are respectively unique, satisfying those relations. Further, both λ, q do not depend on b in T.*

Definitions. Let $E = \mathbb{K} \setminus d(a, r^-)$ with $a \in \mathbb{K}$ and $r > 0$. Let $f \in H(E)$ be invertible in $H(E)$ and let $\lambda(x - a)^q h(x)$ be the factorization given in the previous theorem. The integer q will be named *the index of* f associated to $d(a, r^-)$ and will be denoted by $m(f, d(a, r^-))$. If $\lambda = 1$, the element f will be called *a pure factor associated to* $d(a, r^-)$. Let $\mathcal{G}^T$ be the group of invertible elements of $H(\mathbb{K} \setminus T)$.

The following Corollary is then immediate:

Corollary I.5.2.a. *Let $T = d(a, r^-)$. The set of pure factors associated to T is a sub-multiplicative group of the group $\mathcal{G}^T$. Further, every element of $\mathcal{G}^T$ is of the form λh with h a pure factor associated to T and $\lambda \in \mathbb{K}^*$.*

Definition. Let f belong to $H(D)$. Let T be a hole of D and let h be a pure factor associated to T. If $\frac{f}{h}$ belongs to $H(D \cup T)$ and has no zero inside T, h is called *Motzkin factor of f in the hole T* and the number $m(h, T)$ is called *Motzkin index of f* associated to T.

Theorem I.5.3 ([12], Theorem 31.16). *Let $f \in H(D)$. For each hole T of D, f admits a unique Motzkin factor f^T. Moreover, the set of holes T such that $f^T \neq 1$ is a sequence $(f^{T_n})_{n \in \mathbb{N}}$ converging to 1 in $H(D)$, the set of indices n such that $m(f, T_n) \neq 0$ is finite and f is equal to the infinite product $f^0 \prod_{n \in I} f^{T_n}$, that converges in $H(D)$, with f^0 a power series converging in $\widetilde{D}$ and I a countable set.*

I.6. Meromorphic functions. In this chapter, we will define and examine the basic properties of meromorphic functions: relations with poles of analytic elements, absolute values on fields of meromorphic functions defined by circular filters, values of the derivative on a circular filter, developement in a Laurent series in an annulus, existence of primitives.

Definitions and notation. We denote by $\mathcal{M}(\mathbb{K})$ the field of fractions of $\mathcal{A}(\mathbb{K})$. The elements of $\mathcal{M}(\mathbb{K})$ are called *meromorphic functions in* $\mathbb{K}$.

In the same way, given $a \in \mathbb{K}$ and $r > 0$, we denote by $\mathcal{M}(d(a, r^-))$ (resp. $\mathcal{M}_b(d(a, r^-))$, resp. $\mathcal{M}_u(d(a, r^-)))$ the field of fractions of $\mathcal{A}(d(a, r^-))$ (resp. the field of fractions of $\mathcal{A}_b(d(a, r^-))$, resp. the set $\mathcal{M}(d(a, r^-)) \setminus \mathcal{M}_b(d(a, r^-)))$. The elements of $\mathcal{M}(d(a, r^-))$ are called *meromorphic functions in* $d(a, r^-)$.

Let $b \in \mathbb{K}$ (resp. $b \in d(a, R^-)$) and let $r \in \mathbb{R}_+^*$ (resp. $r \in]0, R[$). The absolute value $\varphi_{b,r}$ defined on $\mathcal{A}(\mathbb{K})$ (resp. on $\mathcal{A}(d(a, R^-)))$ has an immediate continuation to $\mathcal{M}(\mathbb{K})$ (resp. to $\mathcal{M}(d(a, R^-)))$ that we shall denote again by $\varphi_{b,r}$. In the same way, $\varphi_{0,r}$ will be denoted by $| \,.\, |(r)$ on $\mathcal{M}(\mathbb{K})$ and on $\mathcal{M}(d(0, R^-))$.

Let $f = \frac{h}{l} \in \mathcal{M}(\mathbb{K})$ (resp. $f = \frac{h}{l} \in \mathcal{M}(d(a, R^-)))$. For each $\alpha \in \mathbb{K}$ (resp. $\alpha \in d(a, R^-)$) the number $\omega_\alpha(h) - \omega_\alpha(l)$ does not depend on the functions h, l choosed to make $f = \frac{h}{l}$. Thus, we can generalize the notation by setting $\omega_\alpha(f) = \omega_\alpha(h) - \omega_\alpha(l)$.

If $\omega_\alpha(f)$ is an integer $q > 0$, α is called *a zero of f of order q*.
If $\omega_\alpha(f)$ is an integer $q < 0$, α is called *a pole of f of order $-q$*.
If $\omega_\alpha(f) \geq 0$, f will be said to be *holomorphic* at α.

Similarly as for $\mathcal{A}(\mathbb{K})$, given $f \in \mathcal{M}(\mathbb{K})$ (resp. $f \in \mathcal{M}(d(a, R^-))$), we can define the divisor $\mathcal{D}(f)$ on $\mathbb{K}$ (resp. of $d(a, R^-)$) as $\mathcal{D}(f)(\alpha) = 0$ whenever $f(\alpha) \neq 0$ and $\mathcal{D}(f)(\alpha) = s$ when f has a zero of order s at α.

Lemma I.6.1 ([12], Lemma 32.3). *Let $f \in \mathcal{M}(\mathbb{K})$. There exists $h \in \mathcal{A}(\mathbb{K})$ such that $\mathcal{D}(h) = \mathcal{D}(f)$ and then the function $l = \dfrac{h}{f}$ belongs to $\mathcal{A}(\mathbb{K})$. Then $\mathcal{D}(\frac{1}{f}) = \mathcal{D}(l)$ and we can write f in the form $\dfrac{h}{l}$ with $h,\ l \in \mathcal{A}(\mathbb{K})$, having no common zero.*

Theorem I.6.2 ([12], Theorem 32.4). *Let $f \in \mathcal{M}(\mathbb{K})$ (resp. $f \in \mathcal{M}(d(a, R^-))$) have no pole in $\mathbb{K}$ (resp. in $d(a, R^-)$). Then f belongs to $\mathcal{A}(\mathbb{K})$ (resp. to $\mathcal{A}(d(a, R^-))$).*

Corollary I.6.2.a. *Let $f,\ g \in \mathcal{A}(\mathbb{K})$ (resp. $f,\ g \in \mathcal{A}(d(a, R^-))$) be such that $\mathcal{D}(g) \leq \mathcal{D}(f)$. There exists $h \in \mathcal{A}(\mathbb{K})$ (resp. $h \in \mathcal{A}(d(a, R^-))$) such that $f = gh$.*

Corollary I.6.2.b. *Let $f \in \mathcal{M}(\mathbb{K})$ (resp. $f \in \mathcal{M}(d(a, R^-))$) have no zero and no pole in $\mathbb{K}$ (resp. in $d(a, R^-)$). Then it is a constant (resp. an invertible element of $\mathcal{A}_b(d(a, R^-))$).*

Theorem I.6.3 ([12], Corollary 32.10). *Let $\mathbb{K}$ be spherically complete, let $a \in \mathbb{K}$, $r \in \mathbb{R}_+^*$ and let $f \in \mathcal{M}(d(a, R^-))$. There exist $g,\ h \in \mathcal{A}(d(a, R^-))$, having no common zero, such that $f = \dfrac{g}{h}$.*

Remark. If $\mathbb{K}$ is not spherically complete, in the general case, we cannot find an analytic function $h \in \mathcal{A}(d(a, R^-))$ such that $\mathcal{D}(h) = \mathcal{D}(f)$. Consequently, in a field such as $\mathbb{C}_p$, we can't write f in the form $f = \dfrac{h}{l}$ with $h,\ l \in \mathcal{A}(d(a, R^-))$, having no common zero (this gap was forgotten in several works).

However, we can take an algebraically closed spherically complete extension $\widehat{\mathbb{K}}$ of $\mathbb{K}$ and consider f as an analytic function on the disk $\widehat{d}(a, R^-)$ in the field $\widehat{\mathbb{K}}$: then f may be written in the form $f = \dfrac{\widehat{h}}{\widehat{l}}$ with $\widehat{h},\ \widehat{l} \in \mathcal{A}(\widehat{d}(a, R^-))$, with $\widehat{h},\ \widehat{l}$ having no common zero.

Definitions. Let $f \in \mathcal{M}(\mathbb{K})$ (resp. $f \in \mathcal{M}_u(d(a, R^-))$) and let $b \in \mathbb{K}$. Then b will be said to be *an exceptional value for f* if $f - b$ has no zero in $\mathbb{K}$ (resp. in $d(a, R^-)$) and b will be said to be *a pseudo-exceptional value for f* if $\lim\limits_{r \to \infty} |f - b|(r) = 0$ (resp. $\lim\limits_{r \to R^-} |f - b|(r) = 0$).

Moreover, if $f \in \mathcal{M}(\mathbb{K}) \setminus \mathbb{K}(x)$ (resp. if $f \in \mathcal{M}_u(d(a, R^-))$), b will be said to be *a quasi-exceptional value for f* if $f - b$ has finitely many zeros in $\mathbb{K}$ (resp. in $d(a, R^-)$).

Theorem I.6.4 ([12], Theorem 32.21). *Let $f \in \mathcal{M}(\mathbb{K}) \setminus \mathbb{K}$, (resp. $f \in \mathcal{M}_u(d(a, R^-))$). If b is an exceptional value for f then it is a pseudo-exceptional value for f. Let $f \in \mathcal{M}(\mathbb{K}) \setminus \mathbb{K}(x)$, (resp. $f \in \mathcal{M}_u(d(a, R^-))$). If b is a quasi-exceptional value for f then it is a pseudo-exceptional value for f.*

Theorem I.6.5 ([12], Theorem 32.22). *Let $f \in \mathcal{M}(\mathbb{K}) \setminus \mathbb{K}$, (resp. $f \in \mathcal{M}_u(d(a, R^-))$). Then f admits at most one pseudo-exceptional value. Moreover, if $f \in \mathcal{A}(\mathbb{K}) \setminus \mathbb{K}$, (resp. $f \in \mathcal{A}(d(a, R^-))$), then f has no pseudo-exceptional value.*

Corollary I.6.5.a. *Let $f \in \mathcal{M}(\mathbb{K}) \setminus \mathbb{K}$, (resp. $f \in \mathcal{M}_u(d(a, R^-))$). Then f admits at most one exceptional value. Moreover, if $f \in \mathcal{M}(\mathbb{K}) \setminus \mathbb{K}(x)$, (resp. $f \in \mathcal{M}_u(d(a, R^-))$), then f admits at most one quasi-exceptional value. Further, if $f \in \mathcal{A}(\mathbb{K}) \setminus \mathbb{K}$ (resp. if $f \in \mathcal{A}_u(d(a, R^-))$) then f admits no exceptional value. And if $f \in \mathcal{A}(\mathbb{K}) \setminus \mathbb{K}[x]$ (resp. if $f \in \mathcal{A}_u(d(a, R^-))$) then f admits no quasi-exceptional value.*

Definition and notation. Let $f \in \mathcal{M}(\mathbb{K})$ (resp. $f \in \mathcal{M}(d(0, R^-))$ have a pole α of order q and let $f(x) = \sum_{k=-q}^{-1} a_k(x-\alpha)^k + h(x)$ with $a_{-q} \neq 0$ and $h \in \mathcal{M}(\mathbb{K})$ (resp. $f \in \mathcal{M}(d(0, R^-))$ and h holomorphic at α. Accordingly to usual notations the coefficient a_{-1} is called *residue of f at α* and denoted by $\text{res}(f, \alpha)$.

We can now compare residues on a hole defined for analytic elements and residues at a point, we just defined for a meromorphic function:

Theorem I.6.6 ([12], Theorem 33.1). *Let $a \in \mathbb{K}$, let $R \in \mathbb{R}_+^*$, let $f \in \mathcal{M}(d(a, R^-))$ and let $r \in]0, R[$. Let α_j, $1 \leq j \leq q$ be the poles of f in $d(a, r)$, let $\rho \in]0, \min_{i \neq j} |\alpha_i - \alpha_j|[$ and for each $j = 1, ..., q$, let $\rho_j \in]0, \rho[$, let $T_j = d(\alpha_j, \rho_j^-)$. Let $D = d(\alpha, r) \setminus \left(\bigcup_{j=1}^q T_j\right)$. Then f belongs to $H(D)$ and $\text{res}(f, \alpha_j) = \text{res}(f, T_j)$, $j = 1, ..., q$.*

Corollary I.6.6.a. *Let $f \in H_b(D)$ be meromorphic in $T = d(b, r^-)$ and admit only one pole b inside T. Let q be the multiplicity order of b. Then the Mittag-Leffler term of f associated to T is of the form $\sum_{j=1}^{q} \frac{a_j}{(x-b)^j}$, with $a_q \neq 0$ and also is of the form $\frac{P}{(x-a_j)^q}$ where P is a polynomial of degree $s < q$. Moreover, it does not depend on r when r tends to 0.*

It seems obvious that the condition for a meromorphic function to admit primitives is that all residues are null. Actually, the proof is not this immediate.

Theorem I.6.7 ([12], Theorem 33.12). *A function $f \in \mathcal{M}((\mathbb{K})$ (resp. $f \in \mathcal{M}((d(a, R^-))$, $a \in \mathbb{K}$, $R > 0$) admits primitives in $\mathcal{M}(\mathbb{K})$ (resp. in $\mathcal{M}((d(a, R^-)))$ if and only if all residues of f are nul.*

Corollary I.6.7.a. *Let $f \in \mathcal{M}((\mathbb{K})$ (resp. $f \in \mathcal{M}((d(a, R^-))$, $a \in \mathbb{K}$, $R > 0$). Then f' belongs to $\mathbb{K}(x)$ if and only if so does f.*

The following Theorem is a strong improvement of the classical upper bound of $|f'|(r)$ in function of $|f|(r)$ that is due to J.P. Bézivin [**2**]. The theorem is particularly useful for examining the derivative admitting no zero in $\mathbb{K}$.

Theorem I.6.8 ([12], Theorem 33.10). *Let $f \in \mathcal{M}(d(0, R^-))$. For each $n \in \mathbb{N}$ and for all $r \in]0, R[$, we have*

$$|f^{(n)}|(r) \leq |n!| \frac{|f|(r)}{r^n}.$$

Here, we can recall this theorem obtained in whose proof is mainly due to Jean-Paul Bézivin [**2**]].

Theorem I.6.9 ([**12**], **Theorem 44.16**). *Let $f \in \mathcal{M}(\mathbb{K})$ and for each $r > 0$, let $n(f, r)$ be the number of multiple poles of f in $d(0, r)$. If there exists $c > 0$ and $s \in \mathbb{N}$ such that $n(r, f) \le cr^s$ $\forall r > 0$, then f' admits no quasi-exceptional value.*

I.7. Nevanlinna Theory in the classical context. The Nevanlinna Theory was made by Rolf Nevanlinna on complex functions in the 1920th [**23**]. It consists of defining counting functions of zeros and poles of a meromorphic function f and giving an upper bound for multiple zeros and poles of various functions $f-b$, $b \in \mathbb{C}$.

A similar theory for functions in a p-adic field was constructed by A. Boutabaa [**4**], after some previous works by Ha Huy Khoai [**16**]. The p-adic Nevanlinna Theory was first stated and correctly proved by Abedelbaki Boutabaa in $\mathcal{M}(\mathbb{K})$ in 1988. The theory was extended to functions in $\mathcal{M}(d(0, R^-))$ by taking into account Lazard's problem in 1999 [**5**].

Throughout the next paragraphs, we assume that, we denote by I the interval $[t, +\infty[$ and by J an interval of the form $[t, R[$ with $t > 0$.

We have to introduce the counting function of zeros and poles of f, counting or not multiplicity. Here we will choose a presentation that avoids assuming that all functions we consider admit no zero and no pole at the origin.

Definitions. We denote by $Z(r, f)$ the counting function of zeros of f in $d(0, r)$ in the following way.

Let (a_n), $1 \le n \le \sigma(r)$ be the finite sequence of zeros of f such that $0 < |a_n| \le r$, of respective order s_n.

We set $Z(r, f) = \max(\omega_0(f), 0) \log r + \sum_{n=1}^{\sigma(r)} s_n(\log r - \log |a_n|)$ and so, $Z(r, f)$ is called *the counting function of zeros of f in $d(0, r)$, counting multiplicity.*

In order to define the counting function of zeros of f without multiplicity, we put $\overline{\omega_0}(f) = 0$ if $\omega_0(f) \le 0$ and $\overline{\omega_0}(f) = 1$ if $\omega_0(f) \ge 1$.

Now, we denote by $\overline{Z}(r, f)$ the counting function of zeros of f without multiplicity:

$\overline{Z}(r, f) = \overline{\omega_0}(f) \log r + \sum_{n=1}^{\sigma(r)} (\log r - \log |a_n|)$ and so, $\overline{Z}(r, f)$ is called *the counting function of zeros of f in $d(0, r)$ ignoring multiplicity.*

In the same way, considering the finite sequence (b_n), $1 \le n \le \tau(r)$ of poles of f such that $0 < |b_n| \le r$, with respective multiplicity order t_n, we put

$N(r, f) = \max(-\omega_0(f), 0) \log r + \sum_{n=1}^{\tau(r)} t_n(\log r - \log |b_n|)$ and then $N(r, f)$ is called *the counting function of the poles of f, counting multiplicity.*

Next, in order to define the counting function of poles of f without multiplicity, we put $\overline{\overline{\omega_0}}(f) = 0$ if $\omega_0(f) \ge 0$ and $\overline{\overline{\omega_0}}(f) = 1$ if $\omega_0(f) \le -1$ and we set

$\overline{N}(r, f) = \overline{\overline{\omega_0}}(f) \log r + \sum_{n=1}^{\tau(r)} (\log r - \log |b_n|)$ and then $\overline{N}(r, f)$ is called *the counting function of the poles of f, ignoring multiplicity.*

Now, we can define the Nevanlinna function $T(r, f)$ in I or J as $T(r, f) = \max(Z(r, f), N(r, f))$ and the function $T(r, f)$ is called *characteristic function of f or Nevanlinna function of f*.

Finally, if S is a subset of $\mathbb{K}$ we will denote by $Z^S(r, f')$ the counting function of zeros of f', excluding those which are zeros of $f - a$ for any $a \in S$.

Remark. If we change the origin, the functions Z, N, T are not changed, up to an additive constant.

Lemma I.7.1 ([12], Lemma 40.1). *Let $\widehat{\mathbb{K}}$ be a complete algebraically closed extension of $\mathbb{K}$ whose absolute value extends that of $\mathbb{K}$ and let $f \in \mathcal{M}(\mathbb{K})$ (resp. let $f \in \mathcal{M}(d(0, R^-))$). Let $\widehat{d}(0, R) = \{x \in \widehat{\mathbb{K}} \mid |x| < R\}$. The meromorphic function $\widehat{f}$ defined by f in $\widehat{d}(0, R)$ has the same Nevanlinna functions as f.*

Theorem I.7.2 ([12], Theorem 40.2). *Let $f \in \mathcal{M}(\mathbb{K})$ (resp. $f \in \mathcal{M}(d(0, R^-))$) have no zero and no pole at 0. Then*

$$\log(|f|(r)) = \log(|f(0)|) + Z(r, f) - N(r, f).$$

Theorem I.7.3 ([12], Theorem 40.3). *Let $f,\ g \in \mathcal{M}(\mathbb{K})$ (resp. $f,\ g \in \mathcal{M}(d(0, R^-))$). Then*

$$Z(r, fg) \le Z(r, f) + Z(r, g),\ N(r, fg) \le N(r, f) + N(r, g),$$
$$T(r, fg) \le T(r, f) + T(r, g),$$
$$T(r, f + g) \le T(r, f) + T(r, g) + O(1)$$
$$T(r, cf) = T(r, f)\ \forall c \in \mathbb{K}^*,\ T(r, \frac{1}{f}) = T(r, f)),$$
$$T(r, \frac{f}{g}) \le T(r, f)) + T(r, g).$$

Suppose now $f, g \in \mathcal{A}(\mathbb{K})$ (resp. $f,\ g \in \mathcal{A}(d(0, R^-))$). Then $Z(r, fg) = Z(r, f) + Z(r, g)$, $T(r, f) = Z(r, f))$ $T(r, fg) = T(r, f) + T(r, g) + O(1)$ and $T(r, f + g) \le \max(T(r, f), T(r, g))$.

Moreover, if $\lim\limits_{r \to +\infty} T(r, f) - T(r, g) = +\infty$ then $T(r, f + g) = T(r, f)$ when r is big enough.

Lemma I.7.4 ([12], Lemm 40.4). *Let $\alpha_1, \cdots, \alpha_n \in \mathbb{K}$ be pairwise distinct, let $P(u) = \prod_{i=1}^{n}(u - \alpha_i)$ and let $f \in \mathcal{M}(d(0, R^-))$. Then $Z(r, P(f)) = \sum_{i=1}^{n} Z(r, f - \alpha_i)$ and $\overline{Z}(r, P(f)) = \sum_{i=1}^{n} \overline{Z}(r, f - \alpha_i)$.*

Now we can derive the following lemma:

Lemma I.7.5 ([12], Lemm 40.5). *Let $f \in \mathcal{M}(\mathbb{K})$. Then f belongs to $\mathbb{K}(x)$ if and only if $T(r, f) = O(\log r)$.*

Corollary I.7.5.a. *Let $f \in \mathcal{M}(\mathbb{K})$ (resp. $f \in \mathcal{M}(d(0, R^-))$). Then*

$$Z(r, \frac{f'}{f}) - N(r, \frac{f'}{f}) \le -\log r + O(1).$$

Theorem I.7.6 ([12], Theorem 40.7). *Let $f \in \mathcal{A}(\mathbb{K})$ (resp. $f \in \mathcal{A}(d(0,R^-))$) and let $b \in \mathbb{K}$. Then $Z(r,f) = Z(r,f-b) + O(1)$ $r \in I$ (resp. $r \in J$).*

Theorem I.7.7 (First Main Theorem) ([12], Theorem 40.8). *Let $f,\ g \in \mathcal{M}(\mathbb{K})$ (resp. let $f,\ g \in \mathcal{M}(d(0,R^-))$). Then $T(r,f+b) = T(r,f)+O(1)$. Let h be a Moebius function. Then $T(r,f) = T(r,h \circ f) + O(1)$. Let $P(X) \in \mathbb{K}[X]$. Then $T(r,P(f)) = \deg(P)T(r,f) + O(1)$ and $T(r,f'P(f) \geq T(r,P(f))$.*

*Suppose now $f, g \in \mathcal{A}(\mathbb{K})$ (resp. $f,\ g \in \mathcal{A}(d(0,R^-))$). Then $Z(r,fg) = Z(r,f) + Z(r,g)$,
$T(r,f) = Z(r,f))$, $T(r,fg) = T(r,f) + T(r,g) + O(1)$ and
$T(r,f+g) \leq \max(T(r,f),T(r,g))$.*

Moreover, if $\lim_{r \to +\infty} T(r,f) - T(r,g) = +\infty$ then $T(r,f+g) = T(r,f)$ when r is big enough.

Theorem I.7.8 ([12], Theorem 40.9). *Let $f \in \mathcal{M}(\mathbb{K})$ (resp. $f \in \mathcal{M}(d(0,R^-))$). There exists $\phi,\ \psi \in \mathcal{A}(\mathbb{K})$ (resp. $\phi,\ \psi \in \mathcal{A}(d(0,R^-))$) such that $f = \dfrac{\phi}{\psi}$ and $\max(T(r,\phi),T(r,\psi)) \leq T(r,f) + O(1)$, $r \in I$ (resp. ($r \in J$)).*

Theorem I.7.9 ([12], Theorem 40.10). *Let $f \in \mathcal{M}(d(0,R^-))$. Then f belongs to $\mathcal{M}_b(d(0,R^-))$ if and only if $T(r,f)$ is bounded in $[0,R[$.*

Corollary I.7.9.a. *Let $f \in \mathcal{M}_u(d(a,R^-))$, and let $h \in \mathcal{M}_b(d(a,R^-))$, $h \neq 0$. Then fh belongs to $\mathcal{M}_u(d(a,R^-))$.*

We can also derive this Corollary :

Corollary I.7.9.b. *Let $f \in \mathcal{M}(d(a,R^-))$ and let $P \in \mathbb{K}[x]$. Then $P(f)$ belongs to $\mathcal{M}_b(d(a,R^-))$ if and only if so does f.*

The following theorem has proven to be very useful in certain problems of value sharing:

Theorem I.7.10 ([12], Theorem 40.14). *Let $f \in \mathcal{M}(\mathbb{K})$ (resp. $f \in \mathcal{M}(d(0,R^-))$). Then $Z(r,f') - N(r,f') \leq Z(r,f) - N(r,f) - \log r + O(1)$, $r \in I$, (resp. $r \in J$). Moreover, $N(r,f^{(k)}) = N(r,f) + k\overline{N}(r,f) + O(1)$, $r \in I$ (resp. $r \in J$) and $Z(r,f^{(k)}) \leq Z(r,f) + k\overline{N}(r,f) - k\log r + O(1)$, $r \in I$ (resp. $r \in J$).*

Corollary: I.7.10.a *Let $f \in \mathcal{M}(\mathbb{K})$ (resp. $f \in \mathcal{M}(d(0,R^-))$). Then $T(r,f^{(k)}) \leq (k+1)T(r,f) + O(1)$ ($r \in I$) (resp. $r \in J$).*

Theorem I.7.11. *Let $f \in \mathcal{M}(\mathbb{K})$ (resp. $f \in \mathcal{M}(d(0,R^-))$). Then,*

$$T(r,f) - Z(r,f) \leq T(r,f') - Z(r,f') + O(1).$$

Lemma I.7.12 ([12], Lemma 40.17). *Let $f \in \mathcal{M}(\mathbb{K})$ (resp. $f \in \mathcal{M}(d(0,R^-))$) and let $G = \dfrac{f'}{f}$. Then, G satisfies $Z(r,G) \leq N(r,G) - \log r + O(1)$ $r \in I$ (resp. ($r \in J$).*

We can now prove the Second Main Theorem under different forms. The following Theorem is essential and directly leads to the theorems. That theorem was first given in **[21]** in a different form.

Theorem I.7.13 ([12], Theorem 40.19). *Let $f \in \mathcal{M}(\mathbb{K})$ and let $a_1, ..., a_q \in \mathbb{K}$ be distinct. Then*

$$(q-1)T(r,f) \leq \max_{1\leq k\leq q}\Big(\sum_{j=1,j\neq k}^{q} Z(r,f-a_j)\Big) + O(1).$$

Corollary I.7.13.a. *Let $f \in \mathcal{M}(\mathbb{K})$ and let $a_1, ..., a_q \in \mathbb{K}$ be distinct. Then*

$$(q-1)T(r,f) \leq \sum_{j=1}^{q} Z(r,f-a_j) + O(1).$$

Theorem I.7.14 ([12], Theorem 40.21). *Let $f \in \mathcal{M}(d(0,R^-))$ and let $\tau_1, ..., \tau_q \in \mathcal{M}_b(d(0,R^-))$ be distinct. Then*

$$(q-1)T(r,f) \leq \max_{1\leq k\leq q}\Big(\sum_{j=1,j\neq k}^{q} Z(r,f-\tau_j)\Big) + O(1).$$

Corollary I.7.14.a. *Let $f \in \mathcal{M}(d(0,R^-))$ and let $\tau_1, ..., \tau_q \in \mathcal{M}_b(d(0,R^-))$ be distinct. Then*

$$(q-1)T(r,f) \leq \sum_{j=1}^{q} Z(r,f-\tau_j) + O(1).$$

Theorem I.7.15 ([12], Theorem 40.23). *Let $\alpha_1, ..., \alpha_q \in \mathbb{K}$, with $q \geq 2$, let $S = \{\alpha_1, ..., \alpha_q\}$ and let $f \in \mathcal{M}(\mathbb{K})$ (resp. $f \in \mathcal{M}(d(0,R^-))$). Then*

$$(q-1)T(r,f) \leq \sum_{j=1}^{q} \overline{Z}(r,f-\alpha_j) + Z(r,f') - Z_0^S(r,f') + O(1)\ \forall r \in I\ \textit{(resp. } \forall r \in J).$$

Moreover, if f belongs to $f \in \mathcal{A}(\mathbb{K})$ (resp. $\mathcal{A}(d(0,R^-))$), then

$$qT(r,f) \leq \sum_{j=1}^{q} \overline{Z}(r,f-\alpha_j) + Z(r,f') - Z_0^S(r,f') + O(1) \quad \forall r \in I\ \textit{(resp. } \forall r \in J).$$

Remark. The last Theorem does not hold in complex analysis. Indeed, let f be a meromorphic function in $\mathbb{C}$ omitting two values a and b, such as $f(x) = \dfrac{e^x}{e^x - 1}$. Then $Z(r,f-a) + Z(r,f-b) = 0$.

Theorem I.7.16 (Second Main Theorem, [12] Theorem 40.24). *Let $\alpha_1, ..., \alpha_q \in \mathbb{K}$, with $q \geq 2$, let $S = \{\alpha_1, ..., \alpha_q\}$ and let $f \in \mathcal{M}(\mathbb{K})$ (resp. $f \in \mathcal{M}_u(d(0,R^-))$). Then*

$$(q-1)T(r,f) \leq \sum_{j=1}^{q} \overline{Z}(r,f-\alpha_j) + \overline{N}(r,f) - Z_0^S(r,f') - \log r + O(1) \quad \forall r \in I\ \textit{(resp. } \forall r \in J).$$

Moreover, if f belongs to $f \in \mathcal{A}(\mathbb{K})$ (resp. $\mathcal{A}(d(0,R^-))$), then

$$(q-1)T(r,f) \leq \sum_{j=1}^{q} \overline{Z}(r,f-\alpha_j) - Z_0^S(r,f') + O(1) \quad \forall r \in I\ \textit{(resp. } \forall r \in J).$$

I.8. Classical applications of the Nevanlinna Theory. Theorem I.8.1 ([12], Theorem 41.1). *Let $a_1,\ a_2 \in \mathbb{K}$ and let $f,\ g \in \mathcal{A}(\mathbb{K})$ (with $a_1 \neq a_2$) satisfy $f^{-1}(\{a_i\}) = g^{-1}(\{a_i\})$ ($i = 1,\ 2$). Then $f = g$.*

Theorem I.8.2 ([12], Theorem 41.2). *Let $a_1,\ a_2,\ a_3 \in \mathbb{K}$ (with $a_i \neq a_j\ \forall i \neq j$) and let $f,\ g \in \mathcal{A}_u(d(a, R^-))$ satisfy $f^{-1}(\{a_i\}) = g^{-1}(\{a_i\})$ ($i = 1,\ 2,\ 3$). Then $f = g$.*

Theorem I.8.3 ([12], Theorem 41.3). *Let $a_1,\ a_2,\ a_3,\ a_4 \in \mathbb{K}$ (with $a_i \neq a_j\ \forall i \neq j$) and let $f,\ g \in \mathcal{M}(\mathbb{K})$ satisfy $f^{-1}(\{a_i\}) = g^{-1}(\{a_i\})$ ($i = 1,\ 2,\ 3,\ 4$). Then $f = g$.*

Theorem I.8.4 ([12], Theorem 41.4). *Let $a_1,\ a_2,\ a_3,\ a_4,\ a_5 \in \mathbb{K}$ (with $a_i \neq a_j\ \forall i \neq j$) and let $f,\ g \in \mathcal{M}_u(d(a, R^-))$ satisfy $f^{-1}(\{a_i\}) = g^{-1}(\{a_i\})$ ($i = 1,\ 2,\ 3,\ 4,\ 5$). Then $f = g$.*

Definition. A polynomial $P \in \mathbb{K}[x]$ is called *a polynomial of uniqueness for a family of functions $\mathcal{F}$* if $P(f) = P(g)$ implies $f = g$ for all $f,\ g \in \mathcal{F}$.

Theorem I.8.5 ([12], Theorem 41.9). *Let*

$$Q(x) = (n+2)(n+3)x^{n+3} - 2(n+3)(n+1)x^{n+2} + (n+3)(n+2)x^{n+1}.$$

Then Q is a polynomial of uniqueness for $\mathcal{M}(\mathbb{K})$ whenever $n \geq 2$ and is a polynomial of uniqueness for $\mathcal{M}_u(d(a, R^-))$ whenever $n \geq 3$, $a \in \mathbb{K}$, $R > 0$.

(This remarkable polynomial was presented by G. Frank and M. Reinders [**14**].)

We can now apply the Second Main Theorem to obtain results concerning certain algebraic curves:

Theorem I.8.6 ([12], Theorem 42.2). *Let Λ be a non-degenerate elliptic curve of equation $y^2 = (x - a_1)(x - a_2)(x - a_3)$. There do not exist $g,\ f \in \mathcal{M}(\mathbb{K})$ such that $g(t) = y,\ f(t) = x,\ t \in \mathbb{K}$. Moreover, given $a \in \mathbb{K},\ R > 0$ there do not exist $g,\ f \in \mathcal{A}_u(d(a, R^-))$ such that $g(t) = y,\ f(t) = x,\ t \in \mathbb{K}$.*

And similarly:

Theorem I.8.7. *Let $a_i \in \mathbb{K},\ i = 1, ..., q,\ q \geq 4$ be pairwise distinct, let $c \in \mathbb{K}$ and let R be > 0. Let $m \in \mathbb{N}$ be ≥ 2. There exist no $f,\ g \in \mathcal{M}(\mathbb{K}) \setminus \mathbb{K}$ such that $g(t)^m = \prod_{i=1}^{q}(f(t) - a_i)$.*

Theorem I.8.8 ([12], Theorem 42.4). *Let $a_i \in \mathbb{K},\ i = 1, ..., q,\ q \geq 5$ be pairwise distinct, let $c \in \mathbb{K}$ and let R be > 0. Let $m \in \mathbb{N}$ be ≥ 2 and prime to q. There exist neither $f,\ g \in \mathcal{M}(\mathbb{K}) \setminus \mathbb{K}$ nor $f,\ g \in \mathcal{M}_u(d(c, R^-))$ such that $g(t)^m = \prod_{i=1}^{q}(f(t) - a_i)$.*

Example. Let Λ be a curve of equation $y^2 = P(x)$ with $\deg(P) = 5$, P admitting five distinct zeros. If two functions $f,\ g \in \mathcal{M}(K)$ satisfy $g(t)^2 = P(f(t))$, then the two both functions $f,\ g$ are constant. And if $f,\ g \in \mathcal{M}(d(a, R^-))$ satisfy $g(t)^2 = P(f(t))$, then the two both functions $f,\ g$ belong to $\mathcal{M}_b(d(a, R^-))$.

Theorem I.8.9 ([12], Theorem 42.10). *Let $f,\ g \in \mathcal{M}(\mathbb{K})$ satisfy $g^m + f^n = 1$, with $\min(m, n) \geq 2$ and $\max(m, n) \geq 3$. Then f and g are constant. Moreover, if $f,\ g \in \mathcal{A}(\mathbb{K})$ and satisfy $g^m + f^n = 1$, with $\min(m, n) \geq 2$, then f and g are constant.*

Let us now recall the famous Hayman conjecture concerning complex meromorphic functions. Let f be a function meromorphic on $\mathbb{C}$. W. Hayman showed that for every $n \geq 3$, $f'f^n$ takes every value infinitely many times and he conjectured that this was true for all $n \geq 1$. That was proven for $n = 2$ by E. Mues and for $n = 1$ by W. Bergweiler and A. Eremenko [**1**] and separately by H.Chen and M. Fang [**7**].

On the field $\mathbb{K}$, we can consider the same problem. Jacqueline Ojeda [**24**] (and separately y H. Jurvanen) solved the problem for $n \geq 3$ also. Moreover, applying the Nevanlinna theory inside a disk, Jacqueline Ojeda also proved the same for unbounded meromorphic functions inside a disk $d(a, R^-)$. In 2013, Jacqueline Ojeda and I proved the conjecture for meromorphic functions on $\mathbb{K}$ when $n = 2$ [**13**]. The conjecture remains for $n = 1$.

Theorem I.8.10 ([24], [13], [12], Corollary 45.10 and Theorem 45.15,) *Let $a \in \mathbb{K}$ and $R > 0$. Let $f \in \mathcal{M}(\mathbb{K}) \setminus \mathbb{K}(x)$ (resp. $f \in \mathcal{M}_u(a, R^-)$). For every $n \geq 3$, $f'f^n$ takes every value infinitely many times. Given $f \in \mathcal{M}(\mathbb{K}) \setminus \mathbb{K}(x)$, $f'f^2$ takes every value infinitely many times.*

I.9. Small functions. Small functions with respect to a meromorphic function are well known in the general theory of complex functions. Particularly, one knows the Nevanlinna Theorem on 3 small functions. Here we will construct a similar theory.

Definitions and notation. Throughout the chapter we set $a \in K$ and $R \in]0, +\infty[$. For each $f \in \mathcal{M}(\mathbb{K})$ (resp. $f \in \mathcal{M}(d(a, R^-))$) we denote by $\mathcal{M}_f(\mathbb{K})$, (resp. $\mathcal{M}_f(d(a, R^-))$) the set of functions $h \in \mathcal{M}(\mathbb{K})$, (resp. $h \in \mathcal{M}(d(a, R^-))$) such that $T(r, h) = o(T(r, f))$ when r tends to $+\infty$ (resp. when r tends to R). Similarly, if $f \in \mathcal{A}(\mathbb{K})$ (resp. $f \in \mathcal{A}(d(a, R^-))$) we shall denote by $\mathcal{A}_f(\mathbb{K})$ (resp. $\mathcal{A}_f(d(a, R^-))$) the set $\mathcal{M}_f(\mathbb{K}) \cap \mathcal{A}(\mathbb{K})$, (resp. $\mathcal{M}_f(d(a, R^-)) \cap \mathcal{A}(d(a, R^-))$).

The elements of $\mathcal{M}_f(\mathbb{K})$ (resp. $\mathcal{M}_f(d(a, R^-))$) are called *small meromorphic functions with respect to f, small functions* in brief. Similarly, if $f \in \mathcal{A}(\mathbb{K})$ (resp. $f \in \mathcal{A}(d(a, R^-))$) the elements of $\mathcal{A}_f(\mathbb{K})$ (resp. $\mathcal{A}_f(d(a, R^-))$) are called *small analytic functions with respect to f small functions in brief.*

Remark. In [**25**], Yamanoi proved a Nevanlinna theorem on n small complex meromorphic functions. Unfortunately, no similar theorem exists for functions in $\mathbb{K}$.

Theorem I.9.1 ([12], Theorem 43.1). *Let $a \in \mathbb{K}$ and $r > 0$. $\mathcal{A}_f(\mathbb{K})$ is a $\mathbb{K}$-subalgebra of $\mathcal{A}(\mathbb{K})$, $\mathcal{A}_f(d(a, R^-))$ is a $\mathbb{K}$-subalgebra of $\mathcal{A}(d(a, R^-))$, $\mathcal{M}_f(\mathbb{K})$ is a subfield of $\mathcal{M}(\mathbb{K})$ and $\mathcal{M}_f(d(a, R^-))$ is a subfield of field of $\mathcal{M}(a, R^-)$). Moreover, $\mathcal{A}_b(d(a, R^-)$ is a sub-algebra of $\mathcal{A}_f(d(a, R^-)$ and $\mathcal{M}_b(d(a, R^-)$ is a subfield of $\mathcal{M}_f(d(a, R^-)$.*

Theorem I.9.2 ([12], Theorem 43.2). *Let $f \in \mathcal{M}(\mathbb{K})$, (resp.$f \in \mathcal{M}(d(0, R^-))$) and let $g \in \mathcal{M}_f(\mathbb{K})$, (resp.$g \in \mathcal{M}_f(d(0, R^-))$). Then $T(r, fg) = T(r, f) + o(T(r, f))$ and $T(r, \frac{f}{g}) = T(r, f) + o(T(r, f))$.*

Theorem I.9.3 ([12], Theorem 43.3). *Let $f \in \mathcal{A}(\mathbb{K})$ (resp. let $a \in \mathbb{K}$ and $r > 0$ and let $f \in \mathcal{A}_u(d(a, r^-))$). Let g, $h \in \mathcal{A}_f(\mathbb{K})$ (resp. let g, $h \in \mathcal{A}_f(d(a, r^-))$ with g and h not identically zero. If gh belongs to $\mathcal{A}_f(\mathbb{K})$ (resp. to $\mathcal{A}_f(d(a, r^-))$), then so do g and h.*

The following theorem is known as the Second Main Theorem on three small functions and is proven in the same way as in $\mathbb{C}$:

Theorem I.9.4 ([12], Theorem 43.10). *Let $f \in \mathcal{M}(\mathbb{K})$ (resp. $f \in \mathcal{M}_u(d(0,R^-))$) and let $w_1,\ w_2,\ w_3 \in \mathcal{M}_f(\mathbb{K})$ (resp. $w_1,w_2,w_3 \in \mathcal{M}_f(d(0,R^-))$) be pairwise distinct. Then $T(r,f) \le \sum_{j=1}^{3} \overline{Z}(r,f-w_j) + o(T(r,f))$.*

Theorem 19.5 closely derives from Theorem 19.4.

Theorem I.9.5 ([12], Theorem 43.11). *Let $f \in \mathcal{M}(\mathbb{K})$ (resp. $f \in \mathcal{M}_u(d(0,R^-))$) and let $w_1,\ w_2 \in \mathcal{M}_f(\mathbb{K})$ (resp. $w_1,w_2 \in \mathcal{M}_f(d(0,R^-))$) be distinct. Then $T(r,f) \le \overline{Z}(r,f-w_1) + \overline{Z}(r,f-w_2) + \overline{N}(r,f) + o(T(r,f)))$.*

Next, by setting $g = f - w_1$ and $w = w_1 + w_2$, we can write this Corollary.

Corollary I.9.5.a. *Let $g \in \mathcal{M}(\mathbb{K})$ (resp. $g \in \mathcal{M}_u(d(0,R^-))$) and let $w \in \mathcal{M}_g(\mathbb{K})$ (resp. $w \in \mathcal{M}_g(d(0,R^-))$). Then $T(r,g) \le \overline{Z}(r,g) + \overline{Z}(r,g-w) + \overline{N}(r,g) + o(T(r,g)))$.*

Corollary I.9.5.b. *Let $f \in \mathcal{A}(\mathbb{K})$ (resp. $f \in \mathcal{A}_u(d(0,R^-))$) and let $w_1,\ w_2 \in \mathcal{A}_f(\mathbb{K})$ (resp. $w_1,w_2 \in \mathcal{A}_f(d(0,R^-))$) be distinct. Then*

$$T(r,f) \le \overline{Z}(r,f-w_1) + \overline{Z}(r,f-w_2) + o(T(r,f)).$$

Corollary I.9.5.c. *Let $f \in \mathcal{A}(\mathbb{K})$ (resp. $f \in \mathcal{A}_u(d(0,R^-))$) and let $w \in \mathcal{A}_f(\mathbb{K})$ (resp. $u \in \mathcal{A}_f(d(0,R^-))$.) Then $T(r,f) \le \overline{Z}(r,f) + \overline{Z}(r,f-w) + o(T(r,f)))$.*

Here is now an application of that theory.

Theorem I.9.6 ([12], Theorem 43.15). *Let $h,\ w \in \mathcal{A}_b(d(a,R^-))$ and let $m,\ n \in \mathbb{N}^*$ be such that $\min(m,n) \ge 2$, $\max(m,n) \ge 3$. Then the functional equation*

$$(\mathcal{E}) \quad (g(x))^n = h(x)(f(x))^m + w(x)$$

has no solution in $\mathcal{A}_u(d(a,R^-))$.

II. Nevanlinna Theory out of a hole

II.1. Definitions. We will now show that another Nevanlinna Theory exists for meromorphic functions defined not in an "open" disk, but on the contrary in the complement of such a disk.

Recently, M.O. Hanyak and A.A. Kondratyuk constructed a Nevanlinna theory for meromorphic functions in a *punctured complex plane*, i.e. in the set of the form $\mathbb{C} \setminus \{a_1,...,a_m\}$, where we understand that the meromorphic functions can admit essential singularities at $a_1,...,a_m$ [**17**]. Here we consider the situation in our field $\mathbb{K}$ of characteristic 0.

So, we mean to construct a Nevanlinna theory for meromorphic functions in the complement of an open disk thanks to the use of specific properties of the Analytic Elements on infraconnected subsets of $\mathbb{K}$ and particularly the Motzkin Factorization [**22**].

Once the Nevanlinna Theory is established for such functions, we can apply it to obtain results on uniqueness, Picard's values and branched values as it was done in similar problems.

Throughout that study, we fix a number $R > 0$ and denote by I the interval $[R,+\infty[$. Next, we denote by S the disk $d(0,R^-)$ and put $D = \mathbb{K} \setminus S$.

Recall that we denote by $\mathcal{A}(D)$ the $\mathbb{K}$-algebra of Laurent series converging in D. Similarly, we will denote by $\mathcal{M}(D)$ the field of fractions of $\mathcal{A}(D)$ that we will call *field of meromorphic functions in D*.

Given $f \in \mathcal{M}(D)$, for $r > R$, we will denote by $Z_R(r, f)$ the counting function of zeros of f between R and r, i.e. if $\alpha_1, ..., \alpha_m$ are the distinct zeros of f in $\Delta(0, R, r)$, with respective multiplicity u_j, $1 \leq j \leq m$, then $Z_R(r, f) = \sum_{j=1}^{m} u_j(\log(r) - \log(|\alpha_j|))$. Similarly, we denote by $N_R(r, f)$ the counting function of poles of f between R and r, i.e. if $\beta_1, ..., \beta_n$ are the distinct poles of f in $\Delta(0, R, r)$, with respective multiplicity v_j, $1 \leq j \leq m$, then $N_R(r, f) = \sum_{j=1}^{n} v_j(\log(r) - \log(|\beta_j|))$. Finally we put $T_R(r, f) = \max\big(Z_R(r, f), N_R(r, f)\big)$.

Next, we denote by $\overline{Z}_R(r, f)$ the counting function of zeros without counting multiplicity. if $\alpha_1, ..., \alpha_m$ are the distinct zeros of f in $\Delta(0, R, r)$, then we put $\overline{Z}_R(r, f) = \sum_{j=1}^{m} \log(r) - \log(|\alpha_j|)$.

Similarly, we denote by $\overline{N}_R(r, f)$ the counting function of poles without counting multiplicity. if $\beta_1, ..., \beta_n$ are the distinct poles of f in $\Delta(0, R, r)$, then we put $\overline{N}_R(r, f) = \sum_{j=1}^{n} \log(r) - \log(|\beta_j|)$.

Finally, putting $W = \{a_1, ..., a_q\}$, we denote by $Z_R^W(r, f')$ the counting function of zeros of f' on points where $f(x) \notin W$.

By Theorem I.5.3 we can derive Theorem II.1.1.

Theorem II.1.1. *Let $f \in \mathcal{M}(D)$. We can write f in a unique way in the form $f^S f^0$ with $f^S \in H(D)$ a Motzkin factor associated to S and $f^0 \in \mathcal{M}(\mathbb{K})$, having no zero and no pole in S.*

PROOF. Suppose first $f \in \mathcal{A}(D)$ and take $V > R$. Then as a quasi-invertible element of $H(\Delta(0, R, V))$ [**9**], [**12**], by Theorem I.5.3, f admits a factorization in the form $f^S f^0$ where f^S is a Motzkin factor and f^0 belongs to $H(d(0, V))$ and has no zero in S. Moreover by Theorem I.5.2, f^S does not depend on V. Consequently, since f^S is obviously invertible in $\mathcal{A}(D)$, we can factorize $f \in \mathcal{A}(D)$ in the form $f^S f^0$ where f^0 belongs to $\mathcal{A}(\mathbb{K})$ and has no zero in S.

Consider now the general case. $f = \dfrac{g}{h}$ with g, $h \in \mathcal{A}(D)$. Then we can write $g = g^S g^0$, $h = h^S h^0$, hence $f = \Big(\dfrac{g^S}{h^S}\Big)\Big(\dfrac{g^0}{h^0}\Big)$. Then we check that this is the factorization announced in the statement. $f^S = \dfrac{g^S}{h^S}$ and $f^0 = \dfrac{g^0}{h^0}$. The uniqueness also comes from Theorem I.5.3. □

The following lemmas are immediate.

Lemma II.1.2. Let f, $g \in \mathcal{A}(D)$. If $|f|(r) > |g|(r)$, then $|f + g|(r) = |f|(r)$.

Lemma II.1.3. The set of Motzkin factors associated to S makes a multiplicative group. Let f, $g \in \mathcal{M}(D)$. Then $(fg)^S = (f^S)(g^S)$, $\Big(\dfrac{1}{f}\Big)^S = \dfrac{1}{f^S}$, $(fg)^0 = (f^0)(g^0)$, $\Big(\dfrac{1}{f}\Big)^0 = \dfrac{1}{f^0}$ and $m(fg, S) = m(f, S) + m(g, S)$, $m(\dfrac{1}{f}, S) = -m(f, S)$.

PROOF. We have $f = f^S f^0$. Since f^S has no zero and no pole in D, by Lemma I.1 it satisfies $|f^S|(r)) = r^{m(f,S)}$ *for all* $r \in I$, hence $\log(|f^S|(r)) - \log(|f^S|(R)) = m(f,S)(\log r - \log R)$ $(r \in I)$. Next, since f^0 has no zero and no pole in S, we have $\log(|f^0|(r)) - \log(|f^0|(R)) = Z_R(r, f^0) - N_R(r, f^0)$ $(r \in I)$. Therefore the statement is clear. □

Notation. We will denote by $\mathcal{M}^*(D)$ the set of $f \in \mathcal{M}(D)$ such that $f^0 \notin \mathbb{K}(x)$ i.e. the set of $f \in \mathcal{M}(D)$ admitting at least infinitely many zeros in D or infinitely many poles in D. Similarly, we will denote by $\mathcal{A}^*(D)$ the set of $f \in \mathcal{A}(D)$ such that $f^0 \notin \mathbb{K}[x]$ i.e. the set of f admitting infinitely many zeros in D. Next, we set $\mathcal{M}^0(D) = \mathcal{M}(D) \setminus \mathcal{M}^*(D)$ and $\mathcal{A}^0(D) = \mathcal{A}(D) \setminus \mathcal{A}^*(D)$.

Theorem II.1.4 is similar to Theorem I.7.2.

Theorem II.1.4. *Let* $f \in \mathcal{M}(D)$. *Then*

$$\log(|f|(r)) - \log(|f|(R)) = Z_R(r, f) - N_R(r, f) + m(f, S)(\log r - \log R), \quad (r \in I).$$

PROOF. By Theorem T.I.1, we have $f = f^S f^0$. Since f^S has no zero and no pole in D, by Lemma I.1 it satisfies $|f^S|(r)) = r^{m(f,S)}$ *for all* $r \in I$, hence $\log(|f^S|(r)) - \log(|f^S|(R)) = m(f,S)(\log r - \log R)$ $(r \in I)$. Next, since f^0 has no zero and no pole in S, we have $\log(|f^0|(r)) - \log(|f^0|(R)) = Z_R(r, f^0) - N_R(r, f^0)$ $(r \in I)$. Therefore the statement is clear. □

Corollary II.1.4.a. *Let* $f \in \mathcal{M}(D)$. *Then* $T_R(r, f)$ *is identically zero if and only if* f f^0 *is a constant.*

Corollary II.1.4.b. *Let* $f \in \mathcal{A}(D)$ *and let* $\phi \in H_0(D)$. *Then* $Z_R(r, f + \phi) = Z_R(r, f) + O(\log(r))$ $(r \in I)$.

Corollary II.1.4.c. *Let* $f,\ g \in \mathcal{A}(D)$ *satisfy* $\log(|f|(r)) \leq \log(|g|(r))$ $\forall r \geq R$ $(r \in I)$. *Then* $Z_R(r, f) \leq Z_R(r, g) + (m(g, S) - m(f, S))(\log(r) - \log(R))$, $(r \in I)$.

We can now characterize the set $\mathcal{M}^*(D)$.

Theorem II.1.5. *Let* $f \in \mathcal{A}(D)$. *Then* $Z_R(r, f') \leq Z_R(r, f) + O(\log(r))$ $(r \in I)$.

PROOF. Indeed, by Theorem 18.1 in [**12**], we have $|f'|(r) \leq |f|(r)$ and hence the conclusion comes from Theorem II.1.4. □

Corollary II.1.5.1. *Let* $f \in \mathcal{A}(D)$. *Then* $Z_R(\frac{f'}{f}) - N_R(r, \frac{f'}{f}) \leq O(\log(r))$.

Theorem II.1.6. *Let* $f \in \mathcal{M}(D)$. *The three following statements are equivalent:*

i) $\displaystyle\lim_{r\to+\infty} \frac{T_R(r, f)}{\log(r)} = +\infty$ $(r \in I)$,

ii) $\dfrac{T_R(r, f)}{\log(r)}$ *is unbounded,*

iii) f *belongs to* $\mathcal{M}^*(D)$.

PROOF. Consider an increasing sequence $(u_n)_{n\in\mathbb{N}}$ in $\mathbb{R}_+$ such that $\displaystyle\lim_{n\to+\infty} u_n = +\infty$ and let $(k_n)_{n\in\mathbb{N}}$ be a sequence of $\mathbb{N}^*$. Clearly, we have

$$\lim_{r\to+\infty} \frac{\sum_{u_n \leq r} k_n(\log(r) - \log(u_n))}{\log(r)} = +\infty.$$

Consequently, if a function $f \in \mathcal{M}^*(D)$ has infinitely many zeros (resp. infinitely many poles in D) then $\lim\limits_{n\to+\infty} \dfrac{Z_R(r,f)}{\log(r)} = +\infty$ (resp. $\lim\limits_{n\to+\infty} \dfrac{N_R(r,f)}{\log(r)} = +\infty$) hence in both cases, $\lim\limits_{n\to+\infty} \dfrac{T_R(r,f)}{\log(r)} = +\infty$. Conversely, if f has finitely many zeros and finitely many poles in D, then we check that $\lim\limits_{n\to+\infty} \dfrac{T_R(r,f)}{\log(r)} < +\infty$. Thus the equivalence of the three statements is clear. $\square$

Operations on $\mathcal{M}(D)$ work almost like for meromorphic functions in the whole field.

Theorem II.1.7. *Let $f,\ g \in \mathcal{M}(D)$. Then for any $b \in \mathbb{K}$ we have*
$T_R(r, f+b) = T_R(r,f) + O(\log(r)), (r \in I)$
$T_R(r, f.g) \le T_R(r,f) + T_R(r,g) + O(\log(r)),\ (r \in I),$
$T_R(r, \frac{1}{f}) = T_R(r,f),\ (r \in I),$
$T_R(r, f+g) \le T_R(r,f) + T_R(r,g) + O(\log(r)),\ (r \in I)$
and $T_R(r, f^n) = nT_R(r,f),\ (r \in I)$.
Let h be a Moebius function. Then
$T_R(r, h\circ f) = T_R(r,f) + O(\log(r)),\ (r \in I).$

Moreover, if both f and g belong to $\mathcal{A}(D)$, then

$$T_R(r, f+g) \le \max(T_R(r,f), T_R(r,g)) + O(\log(r)),\ (r \in I)$$

and $T_R(r, fg) = T_R(r,f) + T_R(r,g),\ (r \in I)$.

Particularly, if $f \in \mathcal{A}^*(D)$, then $T_R(r, f+b) = T_R(r,f) + O(1),\ (r \in I)$.

Given a polynomial P of degree q, then $T_R(r, P\circ f) = qT_R(r,f)$.

Corollary II.1.7.a. *Let $f,\ g \in \mathcal{M}^0(D)$. Then*
$T_R(r, \frac{f}{g}) \ge T_R(r,f) - T_R(r,g)$ $(r \in I)$ *and* $T_R(r, \frac{f}{g}) \ge T_R(r,g) - T_R(r,f)$ $(r \in I)$.

By Theorems II.1.6 and II.1.7, now we have this immediate corollary:

Corollary II.1.7.b. *$\mathcal{M}^0(D)$ is a subfield of $\mathcal{M}(D)$.*

Theorem II.1.8. *Every $f \in \mathcal{M}^*(D)$ is transcendental over $\mathcal{M}^0(D)$.*

Proof. Consider a polynomial $P(Y) = \sum\limits_{j=0}^{n} a_j Y^j \in \mathcal{M}^0(D)[Y]$ with $a_n = 1$. Let $f \in \mathcal{M}^*(D)$ and suppose that $P(f) = 0$. Then $f^n = -\sum\limits_{j=0}^{n-1} a_j f^j$.

Set $\Xi = \sum\limits_{j=0}^{n-1} a_j f^j$ and $f = f^0 \dfrac{g}{h}$ with $g,\ h \in \mathcal{A}(D)$ having no zero in S. Then
$\Xi = \dfrac{\sum_{j=0}^{n-1} a_j g^j h^{n-1-j}}{h^{n-1}}.$

Since $\sum_{j=0}^{n-1} a_j g^j h^{n-1-j}$ belongs to $\mathcal{A}(D)$, by Theorem II.3 we have

$$T_R(r, \sum_{j=0}^{n-1} a_j g^j h^{n-1-j}) \leq (n-1)T_R(r,f) + O(\log(r)),\ (r \in I)$$

and of course $T_R(r, h^{n-1}) \leq (n-1)T_R(r,f)$, $(r \in I)$. Consequently

$$T_R(r, \Xi) \leq (n-1)T_R(r,f) + O(\log(r),\ (r \in I).$$

But on the other hand, by Theorem II.3, $T_R(r, f^n) = nT_R(r,f)$. Therefore we should have $nT_R(r,f) \leq (n-1)T_R(r,f) + O(\log(r),\ (r \in I)$, which is impossible by Theorem II.2 because f belongs to $\mathcal{M}^*(D)$. Consequently, the equality $P(f) = 0$ is impossible, which proves that f is transcendental over $\mathcal{M}^0(D)$. □

In Theorem II.1.9 we find results already stated about derivatives, for functions in the whole field or in an open disk. The difference only holds in the rest:

Theorem II.1.9. *Let $f \in \mathcal{M}(D)$. Then*

$$N_R(r, f^{(k)}) = N_R(r,f) + k\overline{N}_R(r,f),\ (r \in I)$$

and

$$Z_R(r, f^{(k)}) \leq Z_R(r,f) + k\overline{N}_R(r,f) + O(\log(r)),\ (r \in I).$$

Proof. The inequality $N_R(r, f^{(k)}) = N_R(r,f) + k\overline{N}_R(r,f) + O(1)$ is obvious. Next consider f in the form $\frac{g}{h}$ with $g,\ h \in \mathcal{A}(\mathbb{K})$. Recall that we can write h in the form $\overline{h}\widetilde{h}$ with $\overline{h}$ and $\widetilde{h}$ in $\mathcal{A}(\mathbb{K})$, each zero of $\overline{h}$ being of order one and all zeros of h being a zero of $\overline{h}$. So, h' is of the form $\widetilde{h}\widehat{h}$ where $\widehat{h}$ belong to $\mathcal{A}(\mathbb{K})$ and none of the zeros of $\widehat{h}$ is a zero of h. Then f' is of the form $\dfrac{g'\overline{h} - g\widehat{h}}{h\overline{h}}$. So,
$Z_R(r, f') \leq Z_R(r, g'\overline{h} - g\widehat{h})$ and hence,
$Z_R(r, f') \leq \max(Z_R(r, g'\overline{h}), Z_R(r, g\widehat{h})$. (1)

On the other hand, by Theorem II.1.5, $Z_R(r, g') \leq Z_R(r, g) + O(\log r)$. Obviously, $Z_R(r, \overline{h}) \leq \overline{Z}_R(r,h) = \overline{N}_R(r, \ell) = \overline{N}_R(r,f)$ hence $Z_R(r, (r, g'\overline{h}) \leq Z_R(r,f) + \overline{N}_R(r,f)$.

Now, let us estimate $Z_R(r, \widehat{h})$. Since $\log(|h'|(r)) \leq \log(|h|(r)) - \log r$, we have

$$Z_R(r, h') \leq Z_R(r,h) + O(\log(r)).$$

But since $h' = \widetilde{h}\widehat{h}$, we have

$$Z_R(r, \widehat{h}) = Z_R(r, h') - Z_R(r, \widetilde{h}) \leq Z_R(r,h) - Z_R(r, \widetilde{h}) + O(\log(r))$$
$$= Z_R(r, \overline{h}) + O(\log(r)) = \overline{N}_R(r,f) + O(\log(r)).$$

Consequently,

$$Z_R(r, g\widehat{h}) \leq Z_R(r,g) + \overline{N}_R(r,f) + O(\log(r)) = Z_R(r,f) + \overline{N}_R(r,f) + O(\log(r)).$$

Thus, by (1), we have proven the claim when $k = 1$ and then it is immediately derived by induction on k. □

We can now define small functions in the same way as on the full field $\mathbb{K}$ or as on the disk $d(0, R^-)$

Definitions and notation. For each $f \in \mathcal{M}(D)$ we denote by $\mathcal{M}_f(D)$ the set of functions $h \in \mathcal{M}(D)$ such that $T_R(r, h) = o(T_R(r, f))$ when r tends to $+\infty$. Similarly, if $f \in \mathcal{A}(D)$ we denote by $\mathcal{A}_f(D)$ the set $\mathcal{M}_f(D) \cap \mathcal{A}(D)$.

The elements of $\mathcal{M}_f(D)$ are called *small meromorphic functions with respect to f, small functions* in brief. Similarly, if $f \in \mathcal{A}(D)$ the elements of $\mathcal{A}_f(D)$ are called *small analytic functions with respect to f, small functions* in brief.

A function $w \in \mathcal{M}_f(D)$ will be called *an exceptional small function with respect to f* if $f - w$ has no zero in D and it will be called *a quasi-exceptional small function with respect to f* if $f - w$ has finitely many zeros in D.

Theorem II.1.10. *$\mathcal{A}_f(D)$ is a $\mathbb{K}$-subalgebra of $\mathcal{A}(D)$, $\mathcal{M}_f(D)$ is a subfield of $\mathcal{M}(D)$.*

Let $f \in \mathcal{M}(D)$ and let $g \in \mathcal{M}_f(D)$. Then $T_R(r, fg) = T_R(r, f) + o(T_R(r, f))$ $(r \in I)$ and $T_R(r, \frac{f}{g}) = T_R(r, f) + o(T_R(r, f))$ $(r \in I)$.

Let g, $h \in \mathcal{A}(D)$ with g and h not identically zero. If gh belongs to $\mathcal{A}_f(D)$ then so do g and h.

Theorem II.1.11. *Let $f \in \mathcal{M}^*(D)$. There exists at most one function $w \in \mathcal{M}_f(D)$, such that $f - w$ have finitely many zeros in D. Moreover, if f has finitely many poles, then there exists no function $w \in \mathcal{M}_f(D)$, such that $f - w$ have finitely many zeros in D.*

PROOF. Suppose that f admits two distinct quasi-exceptional small functions. Without loss of generality we may assume that these functions are 0 and $w \in \mathcal{M}_f(D)$. Let $g = f - w$, let f^S be the Motzkin factor of f associated to S and let g^S be the Motzkin factor of g associated to S. Then f is of the form $f^S \frac{P}{h}$ with $h \in \mathcal{A}^*(D)$ and $P \in \mathbb{K}[x]$ having all its zeros in D and g is of the form $g^S \frac{Q}{l}$ with $l \in \mathcal{A}^*(D)$ and $Q \in \mathbb{K}[x]$ having all its zeros in D. Consequently, we have $hg^S Q - lf^S P = hlw$. Now, by Theorem II.1.7, we have

$$T_R(r, hg^S Q - lf^S P) \leq \max(T_R(r, hg^S Q), T_R(r, lf^S P)) + O(\log(r)),$$
$$\leq (T_R(r, h), T_R(r, l) + O(\log(r)), \ (r \in I),$$

hence

$$(1) \qquad T_R(r, hlw) \leq \max((T_R(r, h), T_R(r, l)) + O(\log(r)), \ (r \in I).$$

Next, by Theorem II.1.7 and Corollary II.1.7.a we have

$$(2) \qquad T_R(r, wlh) \geq T_R(r, h) + T_R(r, l) - T_R(r, w).$$

Now, since $f = f^S \frac{P}{h}$, clearly $T_R(r, f) = T_R(r, h) + O(\log(r))$ and similarly, $T_R(r, g) = T_R(r, l) + O(\log(r))$. But since $T_R(r, w) = o(T_R(r, f)$, by Theorem II.1.10 we can check that $T_R(r, g) = T_R(r, f) + o(T_R(r, f))$. Consequently, $T_R(r, l) = T_R(r, h) + o(T_R(r, h))$. Therefore by (1) we have $T_R(r, hlw) \leq (T_R(r, h) + o(T_R(r, h), \ (r \in I)$ and by (2) we obtain $T_R(r, wlh) \geq 2T_R(r, h) + o(T_R(r, f))$ $(r \in I)$, a contradiction. This proves that f cannot have two small functions $w \in \mathcal{M}_f(D)$ such that $f - w$ have finitely many zeros.

Suppose now that $f \in \mathcal{M}^*(D)$ has finitely many poles and admits a quasi-exceptional small function w. Set $g = f - w$. Then g is of the form $g^S\left(\frac{P}{h}\right)$ where P is a polynomial whose zeros lie in D and h belongs to $\mathcal{A}(D)$ and it admits for zeros the poles of f and those of w. Consequently, h belongs to $\mathcal{A}_f(D)$. Therefore we can check that $T_R(r,g) = o(T_R(r,f)$. But by Corollary II.1.7.a we have $T_R(r,g) \geq T_R(r,f) - T_R(r,w) = T_R(r,f) + o(T_R(r,f))$, a contradiction. □

Corollary II.1.11.a. *Let $f \in \mathcal{A}^*(D)$. Then f has no quasi-exceptional small function.*

Lemma II.1.12 will be necessary in the proof of Theorem II.1.13. It looks like Corollary 2.15 in [**5**].

Lemma II.1.12. *Let $f \in \mathcal{M}(D)$. Suppose that there exists $\xi \in \mathbb{K}$ and a sequence of intervals $J_n = [u_n, v_n]$ such that $u_n < v_n < u_{n+1}$, $\lim_{n\to+\infty} u_n = +\infty$ and* $\lim_{n\to+\infty}\left[\inf_{r\in J_n} \frac{T_R(r,f) - Z_R(r,f-\xi)}{\log(r)}\right] = +\infty.$

Let $\tau \in \mathbb{K}$ $\tau \neq \xi$. Then $Z_R(r,f-\tau) = T_R(r,f) + O(\log(r)))$ $\forall r \in J_n$ when n is big enough.

PROOF. Without loss of generality, we can obviously suppose that $\xi = 0$. By Theorem II.1.1, f is of the form $f^S f^0$ and f^0 is of the form $\frac{g}{h}$ with $g,\ h \in \mathcal{A}(D)$, having no zero in S. Set $w = f^S$. Thus we have

$$\lim_{n\to+\infty}\left[\inf_{r\in J_n} \frac{Z_R(r,h) - Z_R(r,g)}{\log(r)}\right] = +\infty.$$

Consequently, by Theorem II.1.4, we have

$$\lim_{n\to+\infty}\left[\inf_{r\in J_n} \frac{\log(|h|(r) - \log(|g|(r)}{\log(r)}\right] = +\infty. \tag{1}$$

Consider now $f - \tau$. We have $f - \tau = \frac{wg - \tau h}{h}$, hence

$$\log(|f|(r)) = \log\left(|wg - \tau h|(r) - \log(|h|(r)\right).$$

But by (A), we have $\log(|\tau h|(r)) > \log(|wg|(r))$ because $\log(|w|(r) = O(\log(r))$, therefore by Lemma II.1.2 $\log\left(|wg - \tau h|(r)\right) = \log(|\tau h|(r))$ $\forall r \in J_n$ when n is big enough and hence

$$\lim_{n\to+\infty}\left[\sup_{r\in J_n} \frac{\log(|\tau h - wg|(r) - \log(|h|(r)}{\log(r)}\right] = 0. \tag{2}$$

Consequently, by (2) and by Theorem II.1.4,

$$\lim_{n\to+\infty}\left[\sup_{r\in J_n} \frac{Z_R(r,\tau h - wg) - Z_R(r,h)}{\log(r)}\right] = 0$$

i.e.

$$\lim_{n\to+\infty}\left[\sup_{r\in J_n} \frac{Z_R(r,f-\tau) - T_R(r,f)}{\log(r)}\right] = 0$$

which proves the claim. □

Similarly to the reasoning when we considered meromorphic functions in $\mathbb{K}$ or inside a disk $d(a, R^-)$, the Nevanlinna second Main Theorem is based on the following theorem:

Theorem II.1.13. *Let $f \in \mathcal{M}(D)$ and let $a_1, ..., a_q \in \mathbb{K}$ be distinct. Then*

$$(q-1)T_R(r,f) \leq \max_{1 \leq k \leq q} \Big(\sum_{j=1, j \neq k}^{q} Z_R(r, f - a_j) \Big) + O(\log(r)) \ (r \in I).$$

PROOF. Suppose Theorem II.1.13 is wrong. Thus, there exists $f \in \mathcal{M}(D)$ and $a_1, ..., a_q \in \mathbb{K}$ such that

$$(q-1)T_R(r,f) - \max_{1 \leq k \leq q} \Big(\sum_{j=1, j \neq k}^{q} Z_R(r, f - a_j) \Big)$$

admits no superior bound in $]0, +\infty[$. So, there exists a sequence of intervals $J_s = [w_s, y_s]$ such that $w_s < y_s < w_{s+1}$, $\lim_{s \to +\infty} w_s = +\infty$ and two distinct indices $m \leq q$ and $t \leq q$ such that

$$\lim_{s \to +\infty} \Big[\inf_{r \in J_s} \frac{(T_R(r,f) - Z_R(r, f - a_m))}{\log(r)} \Big] = +\infty$$

and

$$\lim_{s \to +\infty} \Big[\inf_{r \in J_s} \frac{(T_R(r,f) - Z_R(r, f - a_t))}{\log(r)} \Big] = +\infty$$

But by Lemma II.1, that is impossible. This ends the proof of Theorem II.1.13. $\square$

Corollary II.1.13.a. *Let $f \in \mathcal{M}(\mathbb{K})$ and let $a_1, ..., a_q \in \mathbb{K}$ be distinct. Then $(q-1)T_R(r,f) \leq \sum_{j=1}^{q} Z_R(r, f - a_j) + O(\log(r))$ $(r \in I)$.*

We can now state and prove the Second Main Theorem for $\mathcal{M}(D)$.

Theorem II.1.14. *Let $f \subset \mathcal{M}(D)$, let $\alpha_1, ..., \alpha_q \in \mathbb{K}$, with $q \geq 2$ and let $W = \{\alpha_1, ..., \alpha_q\}$. Then $(q-1)T_R(r,f) \leq$*

$$\sum_{j=1}^{q} \overline{Z}_R(r, f - \alpha_j) + Z_R(r, f') - Z_R^W(r, f') + O(\log(r)) \quad (r \in I).$$

Moreover, if f belongs to $\mathcal{A}(D)$ then $qT_R(r,f) \leq$

$$\leq \sum_{j=1}^{q} \overline{Z}_R(r, f - \alpha_j) + Z_R(r, f') - Z_R^W(r, f') + O(\log(r)) \ (r \in I).$$

Theorem II.1.15 (Second Main Theorem).
Let $f \in \mathcal{M}(D)$, let $\alpha_1, ..., \alpha_q \in \mathbb{K}$, with $q \geq 2$ and let $W = \{\alpha_1, ..., \alpha_q\}$. Then $(q-1)T_R(r,f) \leq$

$$\sum_{j=1}^{q} \overline{Z}_R(r, f - \alpha_j) + \overline{N}_R(r, f) - Z_R^W(r, f') + O(\log(r)) \ (r \in I).$$

Corollary II.1.15.a. *Let $f \in \mathcal{M}(D)$, let $\alpha_1, ..., \alpha_q \in \mathbb{K}$, with $q \geq 2$ and let $W = \{\alpha_1, ..., \alpha_q\}$. Then*

$$\sum_{j=1}^{q}\left(Z_R(r,f-\alpha_j)-\overline{Z}_R(r,f-\alpha_j)\right)$$
$$\leq T_R(r,f)+\overline{N}_R(r,f)-Z_R^W(r,f')+O(\log(r))\quad (r\in I).$$

PROOF. of Theorems II.1.14 and II.1.15. By Theorem II.1.13 there exists a constant $B>0$ and for each $r>R$ there exists $k(r)\in\mathbb{N}$, $k(r)\leq q$, such that

$$(q-1)T_R(r,f)\leq \sum_{j=1,j\neq k(r)}^{q} Z_R(r,f-a_j)+B\log(r)$$

i.e. $(q-1)T_R(r,f)\leq\sum_{j=1}^{q}Z_R(r,f-a_j)-Z_R(r,a_{k(r)}+O(\log(r))$. Now,

$$\sum_{j=1}^{q}Z_R(r,f-a_j)=\sum_{j=1}^{q}\overline{Z}_R(r,f-a_j)+Z_R(r,f')-Z_R^W(r,f')+B\log(r).$$

Consequently,
(1)
$$(q-1)T_R(r,f)\leq\sum_{j=1}^{q}\overline{Z}_R(r,f-a_j)+Z_R(r,f')-Z_R^W(r,f')-Z_R(r,f-a_{k(r)})+O(\log(r))$$

and this proves the first claim of Theorem II.1.14.

Particularly, if $f\in\mathcal{A}(D)$ then we have $Z_R(r,f-a_j)=T_R(r,f-a_j)=T_R(r,f)+O(\log(r))$ $\forall j=1,...,q$, hence $Z_R(r,f-a_{k(r)})=T_R(r,f)+O(\log(r))$ and therefore

$$qT_R(r,f)\leq\sum_{j=1}^{q}\overline{Z}_R(r,f-a_j)+Z_R(r,f')-Z_R^W(r,f')+O(\log(r)),$$

which ends the proof of Theorem II.1.14.

Consider now the situation in Theorem II.1.15. By Theorem II.1.13, for each $j=1,...,q$, there exists a constant $B_j>0$ such that $Z_R(r,f')\leq Z_R(r,f-a_j)+\overline{N}_R(r,f-a_j)+B_j\log(r))$. Consequently, there exists a constant $C>0$ such that $Z_R(r,f')\leq Z_R(r,f-a_{k(r)})+\overline{N}_R(r,f-a_{k(r)})+C\log(r)$ $\forall r>R$.

Therefore, by Relation (1) that remains true in Theorem II.1.15, we can derive

$$(q-1)T_R(r,f)\leq\sum_{j=1}^{q}\overline{Z}_R(r,f-\alpha_j)+\overline{N}_R(r,f)-Z_R^W(r,f')+O(\log(r))\quad\forall r\in I.$$

□

We can now find again a Second Main Theorem on Three Small Functions.

Theorem II.1.16. *Let $f\in\mathcal{M}^*(D)$ and let $w_1,\ w_2,w_3\in\mathcal{M}_f(D)$ be pairwise distinct. Then*

$$T_R(r,f)\leq\sum_{j=1}^{3}\overline{Z}_R(r,f-w_j)+\sum_{j=1}^{3}T(r,w_j)+O(\log(r))\ (r\in I).$$

PROOF. of Theorem II.1.16. Let $\phi(x) = \dfrac{(f(x) - w_1(x))(w_2(x) - w_3(x))}{(f(x) - w_3(x))(w_2(x) - w_1(x))}$. By Theorem II.12 we have

(1) $T_R(r, \phi) \leq \overline{Z}_R(r, \phi) + \overline{Z}_R(r, \phi - 1) + \overline{N}_R(r, \phi) + O(\log(r))$.

On the other hand, we have $T_R(r, f) \leq T_R(r, f - w_j) + T_R(r, w_j)$ $(j = 1, 2, 3)$, hence $T_R(r, f) \leq T_R(r, \dfrac{w_3 - w_1}{f - w_3}) + o(T_R(r, f))$, thereby

$T_R(r, f) \leq T_R(r, \dfrac{w_3 - w_1}{f - w_3} + 1) + o(T_R(r, f)) = T_R(r, \dfrac{f - w_1}{f - w_3}) + o(T_R(r, f))$.

Now, $T_R(r, \dfrac{w_2 - w_1}{w_2 - w_3}) = o(T_R(r, f))$. Consequently, by writing $\dfrac{f - w_1}{f - w_3} = \Big(\dfrac{w_2 - w_1}{w_2 - w_3}\Big)\phi$ we have

$$T_R(r, \frac{f - w_1}{f - w_3}) \leq T_R(r, \phi) + T_R(r, \frac{w_2 - w_1}{w_2 - w_3}) \leq T_R(r, \phi) + o(T_R(r, f))$$

and finally $T_R(r, f) \leq T_R(r, \phi) + o(T_R(r, f))$. Thus, by (1) we obtain

(2) $T_R(r, f) \leq \overline{Z}_R(r, \phi) + \overline{Z}_R(r, \phi - 1) + \overline{N}_R(r, \phi) + o(T_R(r, f))$.

Now, we can check that

$\overline{Z}_R(r, \phi) + \overline{Z}_R(r, \phi - 1) + \overline{N}_R(r, \phi) \leq \sum_{j=1}^{3} \overline{Z}_R(r, f - w_j) + \sum_{1 \leq j < k \leq 3} \overline{Z}_R(r, w_k - w_j)$

$\leq \sum_{j=1}^{3} \overline{Z}_R(r, f - w_j) + o(T_R(r, f))$ which, by (2), completes the proof. □

Corollary II.1.16.a. *Let $f \in \mathcal{M}^*(D)$ and let w_1, $w_2, w_3 \in \mathcal{M}^0(D)$ be pairwise distinct. Then*

$$T_R(r, f) \leq \sum_{j=1}^{3} \overline{Z}_R(r, f - w_j) + O(\log(r)) \ (r \in I).$$

As an easy consequence, we have

Theorem II.1.17. *Let $f \in \mathcal{M}^*(D)$ and let w_1, $w_2 \in \mathcal{M}_f(D)$ be distinct. Then*

$$T_R(r, f) \leq \overline{Z}_R(r, f - w_1) + \overline{Z}_R(r, f - w_2) + \overline{N}_R(r, f)$$
$$+ T_R(r, w_1) + T_R(r, w_2) + O(\log(r)) \ (r \in I).$$

Corollary II.1.17.a. *Let $f \in \mathcal{M}^*(D)$ and let w_1, $w_2 \in \mathcal{M}^0(D)$ be distinct. Then*

$$T_R(r, f) \leq \overline{Z}_R(r, f - w_1) + \overline{Z}_R(r, f - w_2) + \overline{N}_R(r, f) + O(\log(r)) \ (r \in I).$$

Corollary II.1.17.b. *Let $f \in \mathcal{A}^*(D)$ and let w_1, $w_2 \in \mathcal{A}^0(D)$ be distinct. Then*

$$T_R(r, f) \leq \overline{Z}_R(r, f - w_1) + \overline{Z}_R(r, f - w_2) + O(\log(r)) \ (r \in I).$$

PROOF. *of Theorem II.1.17* Let $g = \frac{1}{f}$, $h_j = \frac{1}{w_j}$, $j = 1, 2$, $h_3 = 0$. Clearly,

$$T_R(r, g) = T_R(r, f),\ T_R(r, h_j) = T_R(r, w_j),\ j = 1, 2,$$

so we can apply Theorem II.13 to g, h_1, h_2, h_3. Thus we have:

$$T_R(r, g) \leq \overline{Z}_R(r, g - h_1) + \overline{Z}_R(r, g - h_2) + \overline{Z}_R(r, g) + o(T_R(r, g)).$$

But we notice that $\overline{Z}_R(r, g - h_j) = \overline{Z}_R(r, f - w_j)$ for $j = 1, 2$ and $\overline{Z}_R(r, g) = \overline{N}_R(r, f)$. Moreover, we know that $o(T_R(r, g)) = o(T_R(r, f))$. Consequently, the claim is proven when $w_1 w_2$ is not identically zero.

Now, suppose that $w_1 = 0$. Let $\lambda \in \mathbb{K}^*$, let $l = f + \lambda$ and $\tau_j = u_j + \lambda$, ($j = 1,\ 2,\ 3$). Thus, we have $T_R(r, l) = T_R(r, f) + O(\log(r))$, $T_R(r, \tau_j) = T_R(r, w_j) + O(\log(r))$, ($j = 1,\ 2$), $\overline{N}_R(r, l) = \overline{N}_R(r, f)$. By the claim already proven whenever $w_1 w_2 \neq 0$ we may write $T_R(r, l) \leq \overline{Z}_R(r, l - \tau_1) + \overline{Z}_R(r, l - \tau_2) + \overline{N}_R(r, l) + o(T_R(r, l)))$ hence $T_R(r, f) \leq \overline{Z}_R(r, f - w_1) + \overline{Z}_R(r, f - w_2) + \overline{N}_R(r, l) + o(T_R(r, f)))$. □

II.2. Applications of the Nevanlinna Theory out of a hole. Theorems II.2.1 and II.2.2 are similar to Theorems I.8.2 and I.8.4.

Theorem II.2.1. *Let a_1, a_2, $a_3 \in \mathbb{K}$ (with $a_i \neq a_j\ \forall i \neq j$) and let f, $g \in \mathcal{A}^*(D)$ satisfy $f^{-1}(\{a_i\}) = g^{-1}(\{a_i\})$ ($i = 1,\ 2,\ 3$). Then $f = g$.*

Theorem II.2.2. *Let a_1, a_2, a_3, a_4, $a_5 \in \mathbb{K}$ (with $a_i \neq a_j\ \forall i \neq j$) and let $f, g \in \mathcal{M}^*(D)$ satisfy $f^{-1}(\{a_i\}) = g^{-1}(\{a_i\})$ ($i = 1,\ 2,\ 3,\ 4,\ 5$). Then $f = g$.*

PROOF. of Theorems II.2.1 and II.2.2. In both Theorems, we suppose that f and g are two different functions. Let us remark that since f, g belong to $\mathcal{M}^*(D)$, by Theorem II.1.6 we have $\lim\limits_{r\to+\infty} \frac{T_R(r, f)}{\log(r)} = +\infty$ and $\lim\limits_{r\to+\infty} \frac{T_R(r, g)}{\log(r)} = +\infty$.

Now, for each $j = 1, ..., n$, let Y_j be the set of all zeros of $f - a_j$ (without taking multiplicities into account). Since $a_i \neq a_j\ \forall i \neq j$, we have $Y_i \cap Y_j = \emptyset\ \forall i \neq j$. Next, we notice that $f(x) = a_j$ implies $f(x) - g(x) = 0$. Consequently, we check that

$$\sum_{j=1}^{n} \overline{Z}_R(r, f - a_j) \leq \overline{Z}_R(r, f - g). \tag{1}$$

By applying Theorem II.1.15 to f we obtain

$$(n-1)T_R(r, f) \leq \sum_{j=1}^{n} \overline{Z}_R(r, f - a_j) + \overline{N}_R(r, f) + O(\log(r))$$

$$\leq n\overline{Z}_R(r, f - g) + \overline{N}_R(r, f) + O(\log(r))\ (r \in I)$$

hence by (1),

$$(n-1)T_R(r, f) \leq T_R(r, f - g) + \overline{N}_R(r, f) + O(\log(r))\ (r \in I)$$

and finally

$$(n-1)T_R(r, f) \leq T_R(r, f - g) + N_R(r, f) + O(\log(r))\ (r \in I).$$

Similarly,

$$(n-1)T_R(r, g) \leq T_R(r, f - g) + N_R(r, g) + O(\log(r))\ (r \in I),$$

therefore we obtain

$$(2) \qquad \begin{aligned} &(n-1)\max(T_R(r,f),T_R(r,g)) \\ &\leq T_R(r,f-g)) + \max(N_R(r,f),N_R(r,g)) + O(\log(r)) \ (r \in I). \end{aligned}$$

Assume first we are in the hypothesis of Theorem II.2.1. We have $N_R(r,f) = N_R(r,g) = 0$, and by Theorem II.1.7, $T_R(r,f-g) \leq \max(T_R(r,f),T_R(r,g)) + O(\log(r))$. Consequently,

$$(n-1)\max(T_R(r,f),T_R(r,g)) \leq \max(T_R(r,f),T_R(r,g)) + O(\log(r)) \ (r \in I).$$

Then by the remark at the beginning of the proof, we can see that the inequality does not hold with $n = 3$, when r goes to $+\infty$. Consequently, $f = g$.

Assume now the hypothesis of Theorem II.2.2. Since

$$\max(N_R(r,f),N_R(r,g)) \leq \max(T_R(r,f),T_R(r,g)),$$

by (2) and by Theorem II.1.7 we have

$$(n-1)\max(T_R(r,f),T_R(r,g)) \leq 3\max(T_R(r,f),T_R(r,g)) + O(\log(r)) \ (r \in I).$$

The inequality does not hold with $n = 5$, when r goes to $+\infty$, hence $f = g$. □

We can now apply the Second Main Theorem to obtain results concerning certain algebraic curves.

Theorem II.2.3. *Let Λ be a curve of equation $y^q = P(x)$ with $P \in \mathbb{K}[x]$. If Λ admits a parametrization of the form $y = g(t)$, $x = f(t)$ with $f,\ g \in \mathcal{M}(D)$ and if f (resp. g) belongs to $\mathcal{M}^0(D)$, then g (resp. f) also belongs to $\mathcal{M}^0(D)$.*

PROOF. Let $a_1, ..., a_n$ be n distinct zeros of P. Suppose that two functions $f,\ g \in \mathcal{M}(D)$ satisfy the equation $g(t)^q = P(f(t))$. If f belongs to $\mathcal{M}^0(D)$, then by Theorem II.1.6, g also belongs to $\mathcal{M}^0(D)$. Conversely, if g belongs to $\mathcal{M}^0(D)$, then f satisfies $P(f) - g^q = 0$, hence by Theorem II.1.6, f belongs to $\mathcal{M}^0(D)$. □

Theorem II.2.4. *Let Λ be a non-degenerate elliptic curve of equation $y^2 = (x-a_1)(x-a_2)(x-a_3)$. There do not exist $g,\ f \in \mathcal{A}^*(D)$ such that $g(t) = y,\ f(t) = x,\ t \in \mathbb{K}$.*

PROOF. Indeed, suppose that there exist $g,\ f \in \mathcal{A}^*(D)$ such that $g(t) = y,\ f(t) = x,\ t \in \mathbb{K}$. Since for each $j = 1,\ 2,\ 3,\ f - a_i$ must have infinitely many zeros in D, each one is a zero of g^2 and hence it is a zero of even order, hence $\overline{Z}_R(r, f-a_j) = \dfrac{Z_R(r,f-a_j)}{2}$. Consequently, by Theorem II.1.15 , we have

$$2T_R(r,f) \leq \sum_{j=1}^{3} \frac{Z_R(r,f-a_j)}{2} + O(\log(r)) \leq \frac{3T_R(r,f)}{2} + O(\log(r)),$$

a contradiction when $f \in \mathcal{A}^*(D)$. □

Theorem II.2.5. *Let Λ be a curve of equation $y^q = P(x)$ with $P \in \mathbb{K}[x]$ admitting n distinct zeros of order 1. If Λ admits a parametrization of the form $y = g(t)$, $x = f(t)$ with $f,\ g \in \mathcal{M}^*(D)$ and $t \in D$, then $n \leq \dfrac{2q}{q-1}$. Moreover, if $\deg(P) = n$ and if n and q are relatively prime, then $n \leq \dfrac{q+1}{q-1}$.*

PROOF. Let $a_1, ..., a_n$ be n distinct zeros of P. Suppose that two functions $f,\ g \in \mathcal{M}^*(D)$ satisfy the equation $(g(t))^q = P(f(t))$. Let $\alpha \in D$ be a zero of $f - a_j$ of order s. It is a zero of order l of $g - a_j$, hence $lq = s$ therefore q divides s.

Consequently, for each $j = 1, ..., n$, we have $\overline{Z}_R(r, f - a_j) \leq \frac{1}{q} Z_R(r, f - a_j)$, therefore, by Theorems II.1.7 and II.1.15 we have

$$(n-1)T_R(r,f) \leq \frac{n}{q} T_R(r,f) + \overline{N}_R(r,f) + O(\log(r)). \tag{1}$$

Since $f,\ g$ belong to $\mathcal{M}^*(D)$, that implies $n \leq \frac{2q}{q-1}$.

Suppose now that $\deg(P) = n$ and that n and q are relatively prime. Let β be a pole of f of order l in D. Since $\deg(P) = n$, all zeros of P are of order 1 and β is a pole of g^q of order ln. But since n and q are relatively prime, q must divide l. Consequently, by (1) now we have

$$(n-1)T_R(r,f) \leq \frac{n}{q} T_R(r,f) + \frac{1}{q} T_R(r,f) + O(\log(r))$$

hence $(n-1)T_R(r,f) \leq (\frac{n+1}{q})T_R(r,f) + O(\log(r))$. And since $f,\ g$ belong to $\mathcal{M}^*(D)$, that implies $n \leq \frac{q+1}{q-1}$. □

By Theorems II.2.3 and II.2.5, we can get these corollaries:

Corollary II.2.5.a. *Let Λ be a curve of equation $y^q = P(x)$ with $q \geq 2$, $P \in \mathbb{K}[x]$ admitting n distinct zeros of order 1. If Λ admits a parametrization of the form $y = g(t)$, $x = f(t)$ with $f,\ g \in \mathcal{M}(D)$ and $t \in D$ and if $n > \dfrac{2q}{q-1}$ then the two both functions f and g belong to $\mathcal{M}^0(D)$.*

Corollary II.2.5.b. *Let Λ be a curve of equation $y^q = P(x)$ with $q \geq 2$ relatively prime to n and $P \in \mathbb{K}[x]$ of degree n admitting n distinct zeros. If Λ admits a parametrization of the form $y = g(t)$, $x = f(t)$ with $f,\ g \in \mathcal{M}(D)$ and if $n > \dfrac{q+1}{q-1}$, then the two both functions f and g belong to $\mathcal{M}^0(D)$.*

Particularly Corollary II.2.5.b applies to hyper-elliptic curves.

Example. Let Λ be a curve of equation $y^2 = P(x)$ with $\deg(P) = 5$, P admitting five distinct zeros. If two functions $f,\ g \in \mathcal{M}(D)$ satisfy $g(t)^2 = P(f(t))$, then the two both functions belong to $\mathcal{M}^0(D)$.

Another application concerns analytic functions:

Theorem II.2.6. *Let $f,\ g \in \mathcal{M}(\mathbb{K})$ satisfy $g^m + f^n = 1$, with $\min(m,n) \geq 3$ with $(m,n) \neq (3,3)$. Then f and g belong to $\mathcal{M}^0(D)$. Moreover, if $f,\ g \in \mathcal{A}(D)$ and satisfy $g^m + f^n = 1$, with $\min(m,n) \geq 2$ and $(m,n) \neq (2,2)$, then f and g belong to, $\mathcal{A}^0(D)$.*

Concerning the Hayman conjecture we can easily have a theorem that looks like this one for a disk, by following a similar way as in [**24**]..

Theorem II.2.7. *Let $f \in \mathcal{M}(D)$. For every $n \geq 3$, $f'f^n$ takes every value infinitely many times.*

However, so far, here we don't have a theorem concerning $\mathcal{M}(D)$ in case $n = 2$.

We can now consider the problem of branched functions.

Definition. Let $f \in \mathcal{M}^*(D)$ and let $w \in \mathcal{M}_f(D)$. Then w is called *a perfectly branched function, with respect to* f if all zeros of $f - w$ are multiple except maybe finitely many. Particularly, the definition applies to constants that are then called *branched values.*

Theorem II.2.8. *Let $f \in \mathcal{M}^*(D)$. Then f admits at most 4 perfectly branched values.*

PROOF. Suppose f has q perfectly branched values b_j with $j = 1, ..., q$. For each j, let s_j be the number of simple zeros of $f - b_j$ and let $s = \sum_{j=1}^{q} s_j$. Applying Theorem II.1.15 , we have

$$(q-1)T_R(r,f) \leq \sum_{j=1}^{q} \overline{Z}_R(r, f-b_j) + \overline{N}_R(r,f) + O(\log r). \tag{1}$$

But since $f - b_j$ has s_j simple zeros, we have

$$\begin{aligned}\overline{Z}_R(r, f-b_j) &\leq \frac{Z_R(r, f-b_j) + s_j(\log r - \log R)}{2} + O(1)\\ &\leq \frac{T_R(r,f) + s_j(\log r - \log R)}{2} + O(1)\ \forall j = 1, ..., q\end{aligned}$$

hence, by (1), we have

$$(q-1)T_R(r,f) \leq \frac{qT_R(r,f)}{2} + T_R(r,f) + O(\log(r)). \tag{2}$$

By (2) clearly we have $q \leq 4$ in all cases, which shows the statement of Theorem II.2.8 whenever $f \in \mathcal{M}(D)$. $\square$

In the proof of Theorem II.2.10, we will use the following lemma:

Lemma II.2.9. *Let $\Theta(x) = \sum_{-\infty}^{0} a_n x^n \in \mathcal{A}(D)$, with $a_0 = 1$, have no zero in D. Take $R' = 4R$ and set $D' = \mathbb{K} \setminus d(0, R'^-)$. Then there exists a function $f \in \mathcal{A}(D')$ such that $(f(x))^2 = \Theta(x)$.*

PROOF. Since Θ belongs to $\mathcal{A}(D)$ and has no zero in D, while $a_0 = 1$, we have $|a_n|R^{-n} < 1\ \forall n > 0$ and $\lim_{n\to-\infty} |a_n|R^{-n} = 0$. Then, there exists a unique function $\ell \in \mathcal{A}(d(1, (\frac{1}{4})^-))$ with value in $d(1, 1^-)$ such that $(\ell(u))^2 = u\ \ \forall u \in d(1, (\frac{1}{4})^-)$ (see for instance Theorem 31.23 in [**12**]). Here, we put $u = \sum_{-\infty}^{0} a_n x^n$. Since Θ has no zero in D, we have $\frac{|a_n|}{R^n} < 1\ \forall n > 0$, hence $\frac{|a_n|}{(4R)^n} < \frac{1}{4}\ \forall n > 0$, therefore $\Theta(x) \in d(1, \frac{1}{4})^-)$. Consequently, we can apply ℓ to $\Theta(x)$ in D'. Then, putting $f(x) = \ell(\Theta(x))$, we have $f(x)^2 = \Theta(x)\ \forall x \in D'$. $\square$

The following Theorem II.2.10 generalizes Theorem 50.12 in [**12**].

Theorem II.2.10. *Let $f \in \mathcal{M}^*(D)$ have finitely many poles. Then f admits at most one perfectly branched function in $\mathcal{M}^0(D)$.*

Proof. Suppose that f admits two perfectly branched rational functions w_1 and w_2. If we consider the function $g = f - w_1$, we can see that g has two perfectly branched rational functions 0 and $w_1 - w_2$. So, without loss of generality, we may assume that f admits two perfectly branched rational functions that are 0 and $w(x) \neq 0$.

Suppose first that f has infinitely many zeros of order ≥ 3. Then $Z_R(r, f) - 2\overline{Z}_R(r, f)$ is a function $\zeta(r)$ such that

$$\lim_{r\to+\infty} \frac{\zeta(r)}{\log r} = +\infty. \tag{1}$$

therefore

$$\overline{Z}_R(r, f) \leq \frac{T_R(r, f) - \zeta(r)}{2}.$$

On the other hand, by Theorem II.1.17, we have

$$T_R(r, f) \leq \overline{Z}_R(r, f) + \overline{Z}_R(r, f - w) + \overline{N}_R(r, f) + o(T_R(r, f)).$$

Consequently by (1), we can see a contradiction proving that f cannot admit 0 and w as branched rational functions.

Suppose now that all zeros of both f and $f - w$ are of order 2 except finitely many. Since f has infinitely many zeros in D and since $w \in \mathcal{M}^0(D)$, there exists $R' > 4R$ satisfying the following properties.
i) all poles of f in D lie in $\Delta(0, R, R')$
ii) $|f|(r) > |w|(r)\ \forall r \geq R'$,
iii) all zeros of f and of $f - w$ in $D \setminus \Delta(0, R, R')$ are of order 2 exactly.

Let $S' = d(0, R'^-)$ and let $D' = \mathbb{K} \setminus S'$. Then f obviously belongs to $\mathcal{M}(D')$. Therefore, by Theorem II.1.1, f admits a Motzkin factor of the form $x^s\theta$ with $s = m(f, S')$ and then we can write f in the form $x^s\theta g^2$ and $g \in \mathcal{A}(\mathbb{K})$, having no zero in S'. Similarly, $f - w$ admits a Motzkin factor of the form $x^t\tau$ with $t = m(f - w, S')$ and we can then write $f - w$ in the form $x^t\tau h^2$ and $h \in \mathcal{A}(\mathbb{K})$, having no zero in S'.

Since $|f|(r) > |w|(r)\ \forall r \geq R'$, we can check that f and $f - w$ have the same number of zeros on each circle $C(0, r)$ $(r \geq R')$ and $|f|(r) = |f - w|(r)\ \forall r \geq R'$. Consequently, $s = t$. We have the equality

$$g^2 - \frac{\tau}{\theta}h^2 = \frac{b}{x^s\theta}. \tag{2}$$

Now, $\dfrac{\tau}{\theta}$ is a Motzkin factor of index zero, hence by Lemma II.2.9, it admits a square root $\Xi \in H(D')$ of the form $\sum_{-\infty}^{0} a_n x^n$ with $|a_n|S^n < |a_0|\ \forall n < 0$. Then by (2) we have

$$(g - \Xi h)(g + \Xi h) = \frac{w}{x^s\theta}. \tag{3}$$

We will check that this equality is impossible. Indeed, both functions $g - \Xi h$ and $g + \Xi h$ belong to $\mathcal{A}(D')$. Suppose $Z_{R'}(r, g - \Xi h) = O(\log(r))$. Then $g + \Xi h = (g - \Xi h) + 2\Xi h$ satisfies $Z_{R'}(r, g - \Xi h) + Z_S(r, \Xi h) = Z_{R'}(r, \Xi h) + O(\log(r))$ and consequently,

$$\lim_{r\to+\infty} \Big[\frac{Z_{R'}(r, (g - \Xi h)(g + \Xi h))}{\log(r)}\Big] = +\infty$$

because

$$\lim_{r\to+\infty}\left[\frac{Z_{R'}(r,\Xi h)}{\log(r)}\right]=+\infty.$$

But on the other hand, by construction, $Z_{R'}(r,\dfrac{w}{x^s\theta})$ is of the form $O(\log(r))$, which shows that (3) is impossible. That ends the proof of Theorem II.2.10. $\square$

Corollary II.2.10.a. *Let $f\in\mathcal{A}^*(D)$. Then f admits at most one perfectly branched function in $\mathcal{M}^0(D)$.*

Notation Let $h\in\mathcal{M}(\mathbb{K})\setminus\mathbb{K}$ (resp. $h\in E(x)\setminus E$) and let $\Lambda(h)$ be the set of zeros c of h' such that $h(c)\neq h(d)$ for every zero d of h' other than c. If $\Lambda(h)$ is finite, we denote by $\Upsilon(h)$ its cardinal and if $\Lambda(h)$ is not finite, we put $\Upsilon(h)=+\infty$.

Given a property $\mathcal{Q}$ satisfied by h, we will denote by $Z_R(r,h'\mid\mathcal{Q})$ the counting function of zeros of h' whenever $\mathcal{Q}$ is satisfied.

In the proof of Theorems II.2.13 and II.2.14 we will need the following lemma which is similar to Lemma 10 in [**11**] (see also Lemma 53.1 in [**12**]).

Lemma II.2.11. *Let $P(x)\in\mathbb{K}[x]\setminus\mathbb{K}$ and let $f,\ g\in\mathcal{M}(D)$ satisfy $P\circ f=P\circ g$. Let $W=\{c_1,...,c_l\}$ be the set of zeros of P'.*

For each $j=1,...,k$ $(k\leq l)$ let $q_j=\omega_{c_j}(P')$. We assume that $P(c_j)\neq P(c_n)$ $\forall j=1,...,k,\ \forall n\leq l$. Then $f,\ g$ satisfy

$$\overline{N}_R(r,f)+\sum_{j=1}^{k}\overline{Z}_R(r,f-c_j)\leq\overline{Z}_R(r,\frac{1}{f}-\frac{1}{g})+\sum_{j=1}^{k}\frac{1}{q_j}Z_R(r,g'\mid f(x)=c_j,\ g(x)\notin W).$$

Furthermore, if $f,\ g\in\mathcal{A}(D)$, then

$$\sum_{j=1}^{k}\overline{Z}_R(r,f-c_j)\leq\overline{Z}_R(r,f-g)+\sum_{j=1}^{k}\frac{1}{q_j}Z_R(r,g'\mid f(x)=c_j,\ g(x)\notin W).$$

PROOF. Without loss of generality, we may assume that $0\notin S$ and that $c_n\neq 0$ $\forall n\leq j$. Let $\mathcal{P}(f)$ be the set of poles of f and let $\phi=\dfrac{1}{f}-\dfrac{1}{g}$. Since $P\circ f=P\circ g$, each pole α of f is a pole of same order of g and hence, we have $\phi(\alpha)=0$. Therefore

$$\overline{N}_R(r,f)\leq\overline{Z}_R(r,\phi\mid x\in\mathcal{P}(f)).\tag{1}$$

Let us fix $j\in\{1,...,k\}$ and let $\alpha\in D$ satisfy $f(\alpha)=c_j$. Suppose first that $g(\alpha)$ lies in W. Thanks to the hypothesis $P(c_n)\neq P(c_j)$ $\forall n\neq j$, if $g(\alpha)\neq c_j$ then $P(g(\alpha))\neq P(c_j)$, a contradiction to $P(g(\alpha))=P(f(\alpha))$. So we have $g(\alpha)=f(\alpha)=c_j$ and since $c_j\neq 0$, then $\phi(\alpha)=0$. Consequently,

$$\overline{Z}_R(r,f-c_j\mid g(x)\in W)\leq\overline{Z}_R(r,\phi\mid f(x)=c_j)\tag{2}$$

and similarly if $f,\ g\in\mathcal{A}(D)$

$$\overline{Z}_R(r,f-c_j\mid g(x)\in W)\leq\overline{Z}_R(r,f-g\mid f(x)=c_j).\tag{3}$$

Consequently, by (1) and (2) we can derive

$$\overline{N}_R(r,f)+\sum_{j=1}^{k}\overline{Z}_R(r,f-c_j\mid g(x)\in W)\le\overline{Z}_R(r,\phi). \tag{4}$$

Similarly, if $f,\ g\in\mathcal{A}(D)$, by (3) we have

$$\sum_{j=1}^{k}\overline{Z}_R(r,f-c_j\mid g(x)\in W)\le\overline{Z}_R(r,f-g). \tag{5}$$

In order to complete the proof, we will show

$$\overline{Z}_R(r,f-c_j\mid g(x)\notin W)\le\frac{1}{q_j}Z_R(r,g'\mid f(x)=c_j,\ g(x)\notin W). \tag{6}$$

Indeed, consider α such that $g(\alpha)\notin W$.

Since $P'(f(\alpha))=P'(c_j)=0$, we notice that $f'(\alpha)P'(f(\alpha))=g'(\alpha)P'(g(\alpha))=0$. But since $g(\alpha)\notin W$, we have $P'(g(\alpha))\neq 0$, hence $g'(\alpha)=0$. Consequently, we obtain

$$\overline{Z}_R(r,f-c_j\mid g(x)\notin W)\le\overline{Z}_R(r,g'\mid f(x)=c_j,\ g(x)\notin W). \tag{7}$$

On the other hand, since $f(\alpha)=c_j$, we see that $\omega_\alpha(f'(x)P'(f(x)))\ge q_j$ hence

$$\omega_\alpha(g'(x)P'(g(x)))\ge q_j. \tag{8}$$

But since $g(\alpha)\notin W$, we have $P'(g(\alpha))\neq 0$, hence by (8), $\omega_\alpha(g')\ge q_j$, and consequently

$$\overline{Z}_R(r,g'\mid f(x)=c_j,\ g(x)\notin W)\le\frac{1}{q_j}Z_R(r,g'\mid f(x)=c_j\ ,\ g(x)\notin W). \tag{9}$$

Thus, by (24) and (26) we obtain (23) which, by (4) proves

$$\overline{N}_R(r,f)+\sum_{j=1}^{k}\overline{Z}_R(r,f-c_j)\le\overline{Z}_R(r,\frac{1}{f}-\frac{1}{g})+\sum_{j=1}^{k}\frac{1}{q_j}Z_R(r,g'\mid f(x)=c_j,\ g(x)\notin W).$$

Similarly, if $f,\ g\in\mathcal{A}(D)$, by (5) and (6) we have

$$\sum_{j=1}^{k}\overline{Z}_R(r,f-c_j)\le\overline{Z}_R(r,f-g)+\sum_{j=1}^{k}\frac{1}{q_j}Z_R(r,g'\mid f(x)=c_j,\ g(x)\notin W).$$

□

Lemma II.2.12. *Let $P\in\mathbb{K}[x]$ and let $f,\ g\in\mathcal{A}^*(D)$ satisfy $P\circ f=P\circ g$. Then $T_R(r,f)=T_R(r,g)+O(\log(r))$ $r>R)$.*

PROOF. Let $q=\deg(P)$. By Theorem T.II.5, we have $T_R(r,P\circ f)=qT_R(r,f)+O(\log(r))$ and $T_R(r,P\circ g)=qT_R(r,g)+O(\log(r))$. But since, $P\circ f=P\circ g$ that means $T_R(r,f)=T_R(r,g)+O(\log(r))$. □

Theorem II.2.13. *Let $P\in\mathbb{K}[x]$ be such that $\Upsilon(P)\ge 2$. Then P is a polynomial of uniqueness for $\mathcal{A}^*(D)$.*

Remark. One can show that when a polynomial is of degree 3 and such that P' has 2 distinct zeros, then the condition $\Upsilon(P)=2$ is automatically satisfied.

Example. Without loss of generality we can assume that P' is of the form $3x^2-b$, hence $P(x)=x^3-bx+c$. Then $P(b)=b^3-b^2+c,\ P(-b)=-b^3+b^2+c\neq P(b)$.

Theorem II.2.14. *Let $P \in \mathbb{K}[x]$ be such that $\Upsilon(P) \geq 4$. Then P is a polynomial of uniqueness for $\mathcal{M}^*(D)$.*

Example. 1) Let $P(x) = \dfrac{x^4}{4} + \dfrac{x^3}{3} - x^2$. Then
$P'(x) = x(x-1)(x+2)$.
And $P(0) = 0,\ P(1) = \dfrac{1}{4} + \dfrac{1}{3} - 1,\ P(2) = \dfrac{8}{3}$.
Thus the three zeros a_j of P' satisfy $P(a_i) \neq P(a_j)\ \forall i \neq j$. Consequently, $\Upsilon(P) = 3$ and hence P is a polynomial of uniqueness for $\mathcal{A}^*(D)$.
2) Let $P(x) = \dfrac{x^5}{5} - \dfrac{5x^3}{3} + 4x$. Then
$P'(x) = (x-1)(x+1)(x-2)(x+2)$. $P(1) = \dfrac{1}{5} - \dfrac{5}{3} + 4$,
$P(-1) = -P(-1),\ P(2) = \dfrac{32}{5} - \dfrac{24}{3} + 8 = -P(-2)$.
Thus the four zeros a_j of P' satisfy $P(a_i) \neq P(a_j)\ \forall i \neq j$. Consequently, $\Upsilon(P) = 4$ and hence P is a polynomial of uniqueness for $\mathcal{M}^*(D)$.

PROOF. *of Theorems II.2.13 and II.2.14* Suppose that f and g are not identical. Let W be the set of zeros of P' and let $c_1, ..., c_k$ lie in W. Clearly by applying Theorem II.1.15 we obtain respectively in Theorems II.2.13 and II.2.14

$$(1)\ \ (k-1)T_R(r,f) \leq \sum_{j=1}^{k} \overline{Z}_R(r, f-c_j) + \overline{N}_R(r,f) - Z_R^W(r,f') + O(\log(r)),\ (r > R),$$

$$(2)\ \ (k-1)T_R(r,g) \leq \sum_{j=1}^{k} \overline{Z}_R(r, g-c_j) + \overline{N}_R(r,g) - Z_R^W(r,g') + O(\log(r)),\ (r > R).$$

Now, let $\phi = \dfrac{1}{f} - \dfrac{1}{g}$ and for each $j = 1, ..., k$, let $q_j = \omega_{c_j}(P')$. By (1) and (2) and by Lemma II.2.11 we obtain

$$(k-1)T_R(r,f) \leq \overline{Z}_R(r,\phi) + \sum_{j=1}^{k} \frac{1}{q_j} Z_R(r, g' \mid f(x) = c_j,\ g(x) \notin W)$$

$$(3) \qquad -Z_R(r, f' \mid f(x) \notin W + O(\log(r))$$

and similarly

$$(k-1)T_R(r,g) \leq \overline{Z}_R(r,\phi) + \sum_{j=1}^{k} \frac{1}{q_j} Z_R(r, f' \mid g(x) = c_j,\ f(x) \notin W)$$

$$(4) \qquad -Z_R(r, g' \mid g(x) \notin W + O(\log(r)).$$

By adding in each case the two inequalities by (3) and (4) in Theorems II.2.13 and II.2.14, we obtain:

$$(k-1)(T_R(r,f) + T_R(r,g)) \leq 2\overline{Z}_R(r,\phi)$$

$$+\sum_{j=1}^{k}\frac{1}{q_j}[Z_R(r,f'\mid g(x)=c_j,\ f(x)\notin S+Z_R(r,g'\mid f(x)=c_j,\ g(x)\notin W)]$$

$$-Z_R(r,f'\mid f(x)\notin W)-Z_R(r,g'\mid g(x)\notin W)+O(\log(r)). \tag{5}$$

Now, in each inequality (5), we notice that in the left side member we have the term:

$$\sum_{j=1}^{k}\frac{1}{q_j}[Z_R(r,f'\mid g(x)=c_j,\ f(x)\notin S)]-Z_R(r,f'\mid f(x)\notin W)$$

which is clearly less than or equal to zero and similarly

$$\sum_{j=1}^{k}\frac{1}{q_j}[Z_R(r,g'\mid f(x)=c_j,\ g(x)\notin S)]-Z_R(r,g'\mid g(x)\notin W)\le 0.$$

Consequently, by Lemma II.2.11, in Theorems II.2.14 we obtain

$$(k-1)(T_R(r,f)+T_R(r,g))\le 2\overline{Z}_R(r,\phi)+O(\log(r))$$

and hence

$$(k-1)(T_R(r,f)+T_R(r,g))\le 2\overline{Z}_R(r,f-g)+O(\log(r)). \tag{6}$$

Now, by Theorem II.1.7 we have $Z_R(r,\phi)\le T_R(r,f)+T_R(r,g)+O(\log(r))$ therefore $k\le 3$ and hence, if $\Upsilon(P)\ge 4$, we have $f=g$.

Now assume the hypotheses of Theorem II.2.13. By Lemma II.2.11 we can replace $Z_R(r,\phi)$ by $Z_R(r,f-g)$. Next, by Lemma II.2.12, $T_R(r,f)=T_R(r,g)+O(\log(r))$, hence, by Theorem II.1.7, we can derive $T_R(r,f-g)\le T_R(r,f)+O(\log(r))=T_R(r,g)+O(\log(r))$. Consequently in place of (6), in Theorem II.2.13 we obtain

$$(k-1)(T_R(r,f)+T_R(r,g))\le 2\overline{Z}_R(r,f-g)+O(\log(r))\le T_R(r,f)+T_R(r,g)+O(\log(r)).$$

Thus we can conclude that $k\le 2$ in Theorem II.2.13 and hence, if $\Upsilon(P)\ge 3$, we have $f=g$. □

Remark. The hypothesis $\Upsilon(P)\ge 4$ however is not necessary to prove that a polynomial P is a polynomial of uniqueness for $\mathcal{M}^*(D)$, as shows Theorem II.2.16 below. The proof is similar to that of Theorem 41.9 in [**12**] and first came from [**14**].

In the proof of Theorem II.2.16 we will need the following basic Lemma II.2.15 (stated in [**12**]).

Lemma II.2.15. *Let E be an algebraically closed field of characteristic 0 and let $P(x)=(n-1)^2(x^n-1)-n(n-2)(x^{n-1}-1)^2\in E[x]$. Then P admits 1 as a zero of order 4 and all other zeros u_j $(1\le j\le 2n-6)$ are simple.*

Theorem II.2.16. *Let*

$$Q(x)=\Big((n+2)(n+1)x^{n+3}-2(n+3)(n+1)x^{n+2}+(n+3)(n+2)x^{n+1}\Big)$$

with $b\in\mathbb{K}^*$. Then Q is a polynomial of uniqueness for $\mathcal{M}^*(D)$ for every $n\ge 3$.

PROOF. Suppose $f,\ g \in \mathcal{M}(D)$ and suppose that $Q(f) = Q(g)$. Let $h = \frac{f}{g}$. We can derive

$(n+2)(n+1)(h^{n+3}-1)g^2 - 2(n+3)(n+1)(h^{n+2}-1)g + (n+3)(n+2)(h^{n+1}-1) = 0.$

If h is a constant, it is 1, a contradiction. So, we suppose h is not constant.

Let $P(x) = (n+2)^2(x^{n+3}) - 1) - (n+3)(n+1)(x^{n+2} - 1)^2 \in \mathbb{K}[x]$. By Lemma II.2.15 P admits 1 as a zero of order 4 and all other zeros u_j $(1 \leq j \leq 2n)$ are simple. By change of variable, we can obviously assume that $h - u_j$ has no zero and no pole at 0. Consequently, we check that

$$\Big(g - \big(\frac{n+3}{n+2}\big)\big(\frac{h^{n+2}-1}{h^{n+3}-1}\big)\Big)^2 = \frac{(n+3)(h-1)^4\prod_{j=1}^{2n}(h-u_j)}{(n+2)^2(n+1)(h^{n+3}-1)^2}.$$

Since $\dfrac{(n+3)(h-1)^4\prod_{j=1}^{2n}(h-u_j)}{(n+2)^2(n-+1)(h^{n+3}-1)^2}$ is equal to a square, clearly each zero of $h - u_j, (1 \leq j \leq 2n)$ has order at least 2. Consequently

$$\sum_{j=1}^{2n}\overline{Z}_R(r, h-u_j) \leq \frac{1}{2}\sum_{j=1}^{2n} Z_R(r, h-u_j) \leq \frac{1}{2}(2n)T_R(r,h) + O(1)\ (r > R).$$

Then, applying Theorem II.1.15 to h at the points u_j $(1 \leq j \leq 2n)$, we obtain

$$(2n-1)T_R(r,h) \leq \sum_{j=1}^{2n}\overline{Z}_R(r, h-u_j) + \overline{N}_R(r,h) + O(\log(r)) \leq$$

$$\leq \frac{1}{2}\sum_{j=1}^{2n} Z_R(r, h-u_j) + \overline{N}_R(r,h) + O(\log(r))$$

$$\leq \frac{1}{2}(2n)T_R(r,h) + \overline{N}_R(r,h) + O(\log(r))\ (r > R)$$

and therefore $(2n-1)T_R(r,h) \leq nT_R(r,h) + T_R(r,h) + O(\log(r))$ $(r > R)$. Now, since $f,\ g$ belong to $\mathcal{M}^*(D)$, by Theorem II.1.6 we have $\lim\limits_{r\to+\infty}\dfrac{T_R(r,h)}{\log(r)} = +\infty$ and hence we can conclude that $n \leq 2$. That finishes the proof of Theorem II.2.16. $\square$

Notation. Following Theorem II.2.16, given $n \in \mathbb{N}$ and let $c \in \mathbb{K}$) we denote by $P_{n,c}$ the polynomial introduced in [**12**] and also used in [**11**]:
$P_{n,c}(x) = (n-1)(n-2)x^n - 2n(n-2)x^{n-1} + n(n-1)x^{n-2} + c$
and by $L(n,c)$ be the set of zeros of $P_{n,c}$ in $\mathbb{K}$.

In the proof of Theorems II.2.19 and II.2.20, we will also need the following lemmas given in [**5**]. and [**14**].

Lemma II.2.17. *Let $F,\ G \in \mathcal{M}(D)$ have the same poles, ignoring multiplicitiy, and let $H = \frac{F''}{F'} - \frac{G''}{G'}$. Every pole of H has multiplicity order 1. Let α be a pole of F and G. If α has same multiplicity for F and G, then H has no pole at α. Moreover, if α has a multiplicity order 1 for both F and G, then α is a zero of H.*

Lemma II.2.18. *Let $f,\ g \in \mathcal{M}(D))$ be two different non-constant functions satisfying $\frac{f''}{f'} = \frac{g''}{g'}$. Then f and g are linked by a relation of the form $f = ag + b$.*

Theorem II.2.19. *Let $f,\ g \in \mathcal{M}^*(D)$ be two different non-constant functions satisfying $f^{-1}(L(n,c)) = g^{-1}(L(n,c))$. Then $n \leq 16$. Moreover, if $f,\ g \in \mathcal{A}^*(D)$, then $n \leq 9$.*

Corollary II.2.19.a. *Let $n \geq 17$. Then $L(n,c)$ is an ursim for $\mathcal{M}^*(D)$. Let $n \geq 10$. Then $L(n,c)$ is an ursim for $\mathcal{A}^*(D)$.*

In order to state Theorem II.2.20, we need to recall the notation $\mathcal{E}$ used with URSCM. Given a subset B of $\mathbb{K}$ and $f \in \mathcal{M}(D)$ we denote by $\mathcal{E}(f,B)$ the set in $\mathbb{K}\times\mathbb{N}^*$:
$\bigcup_{a\in B}\{(z,q) \in \mathbb{K}\times\mathbb{N}^* |\ z \text{ a zero of order } q \text{ of } f(x)-a\}$.

And given a subset B the projective space of dimension 1 on $\mathbb{K}$, $\mathbb{P}^1(\mathbb{K})$, containing $\{\infty\}$ and $f \in \mathcal{M}(D)$, we denote by $\mathcal{E}(f,B)$ the subset of $\mathbb{K}\times\mathbb{N}^*$. $\mathcal{E}(f,S\cap \mathbb{K}) \cup \{(z,q)\ z \text{ a pole of order } q \text{ of } f\}$.

Theorem II.2.20. *Let $f,\ g \in \mathcal{M}^*(D)$ be two distinct non-constant functions satisfying $\mathcal{E}(f,L(n,c)) = \mathcal{E}(g,L(n,c))$. Then $n \leq 10$. Moreover, if $f,\ g \in \mathcal{A}^*(D)$ then $n \leq 6$.*

Corollary II.2.20.a. *For every $n \geq 11$, $L(n,c)$ is an urscm for $\mathcal{M}^*(D)$ and for every $n \geq 7$, $L(n,c)$ is an urscm for $\mathcal{A}^*(D)$.*

PROOF. of Theorems II.2.19 and II.2.20. In order to simplify the notation, we will just denote by P the polynomial $P_{n,c}$ and by L the set $L(n,c)$.

We then consider two distinct non-constant meromorphic functions $f,\ g \in \mathcal{M}(D)$ and we assume $f^{-1}(L) = g^{-1}(L)$ in Theorem II.2.19 and $\mathcal{E}(f,S) = \mathcal{E}(g,S)$ in Theorem II.2.20.

Let $F = \dfrac{1}{P(f)}$ $G = \dfrac{1}{P(g)}$ and $H = \dfrac{F''}{F'} - \dfrac{G''}{G'}$.

We first suppose that H is identically 0. By Lemma II.2.18, it is immediate to derive that there exist $A \in \mathbb{K}^*$ and $B \in \mathbb{K}$ such that $P(f) = \dfrac{P(g)}{BP(g)+A}$. Therefore, by Theorems II.1.7 we have:

(1) $T_R(r,f) = T_R(r,g) + O(\log(r))$ $(r \in]0,R[)$.

We can check that $P(X)+c$ is of the form $bX^{n-2}(X-e_1)(X-e_2)+c$, with $b, e_1, e_2 \in \mathbb{K}$, (resp. $b, e_1, e_2 \in \mathbb{K}$) and $e_1e_2 \neq 0,\ e_1 \neq e_2$.

We have to distinct three cases. $i)$ $B \neq 0$, $ii)$ $B = 0,\ A = 1$, $iii)$ $B = 0,\ A \neq 1$.

$i)$ Since $AB \neq 0$, every zero of $P(g) + \dfrac{A}{B}$ is a pole of $P(f)$ and therefore is a zero of order at least n of $P(g) + \dfrac{A}{B}$. On the other hand, we check that whenever $A,\ B \in \mathbb{K}^*$, the polynomial $P(X) + \dfrac{A}{B}$ admits at least two distinct zeros b_1 and b_2 of order 1 and therefore it admits another zero l of order at most $n-2$. Consequently, every zero of $g-b_1$ or $g-b_2$ has order at least n and every zero of $g-l$ has order at least 2. By a change of variable, we can obviously assume that $g,\ g-l,\ g-b_1, g-b_2$ have neither any zeros nor any poles at 0. Then, by applying Theorem II.1.15, we obtain:

$$\begin{aligned}2T_R(r,g) - \overline{N}_R(r,g) &\leq \overline{Z}_R(r,g-l) + \overline{Z}_R(r,g-b_1) + \overline{Z}_R(r,g-b_2) + O(\log(r))\\ &\leq \frac{1}{2}Z_R(r,g-l) + \frac{1}{n}(Z_R(r,g-b_1) + Z_R(r,g-b_2)) + O(\log(r))\ (r \in J)\end{aligned}$$

which leads to $n \leq 4$.

ii) In this case we have $P(f) = P(g)$. But by Theorem II.2.15, P is a polynomial of uniqueness, for $\mathcal{M}_u(d(0, R^-))$, hence $f = g$.

iii) Let $\lambda = \frac{1}{A}$. We will check that at least one of the two polynomials
$Q_1(X) = (n-1)(n-2)X^n - 2n(n-2)X^{n-1} + n(n-1)X^{n-2} + c(\lambda - 1)$ and
$Q_2(X) = (n-1)(n-2)X^n - 2n(n-2)X^{n-1} + n(n-1)X^{n-2} + c(\frac{1}{\lambda} - 1)$ admits n distinct zeros. Indeed, suppose this not true. We notice that
$Q_1'(X) = Q_2'(X) = P'(X) = n(n-1)(n-2)X^{n-3}(X-1)^2$. Consequently, if both Q_1 and Q_2 admit at least one zero of order > 1, then this must be 1, because it cannot be 0. But then, we have $c(\lambda - 1) = c(\frac{1}{\lambda} - 1) = -2$. Since $c \neq 0$, and since $\lambda \neq 1$, this implies $\lambda = -1$, and $c = 1$, which is excluded by hypothesis.

Now, since λ and $\frac{1}{\lambda}$ play symmetric roles, without loss of generality we can assume that Q_1 admits n distinct zeros $b_1, ..., b_n$. Thus, putting
$\gamma(X) = \lambda X^{n-2}(n-1)\Big((n-2)X^2 - 2n(n-2)X + n(n-1)\Big)$, we have
(2) $Q_1(f) = \gamma(g)$.
So, applying Theorem II.1.15, we obtain

$$(n-1)T_R(r,f) \leq \sum_{j=1}^{n} \overline{Z}_R(r, f - b_j) + \overline{N}_R r, f) + O(\log(r)) \ (r \in J).$$

On the other hand, we have

$$\sum_{j=1}^{n} \overline{Z}_R(r, f - b_j) \leq \overline{Z}_R(r, Q_1(f)) = \overline{Z}_R(r, g) + \overline{Z}_R(r, g - e_1) + \overline{Z}_R(r, g - e_2)$$

$$\leq 3T_R(r, g) + O(\log(r)).$$

But by (2) it is seen that $T_R(r, g) \leq T_R(r, f) + O(\log(r))$ $(r \in J)$, hence finally

$$(n-1)T_R(r,f) \leq 3T_R(r,f) + \overline{N}_R(r,f) + O(\log(r)) \ (r \in J). \tag{3}$$

Consequently, we have $n \leq 5$. Moreover, if $f,\ g \in \mathcal{A}(d(0, R^-))$, then $N_R(r, f) = 0$, and then (3) leads to $n \leq 4$. That finishes proving the claims of the Theorems when $H = 0$.

Henceforth, we suppose that H is not identically 0 and we will apply Fujimoto's method [**15**]. Let α be a zero of $f - a_j$, for some j. Then both $P(f)$ and $P(g)$ vanish at α. So, we have $\sum_{j=1}^{n} \overline{Z}_R(r, f - a_j) = \overline{Z}_R(r, P(f)) = \overline{Z}_R(r, P(g))$. According to Lemma II.2.17, if both $P(f), P(g)$ have a zero of order one, then H also has a zero. Else, at least one of the two functions $P(f)$ and $P(g)$ admits α as a zero of order strictly greater than 1, and then, (since all zeros of P are simple), at least one of the two functions f', g' has a zero at α, while α is a zero $f - a_i$ for some i and a zero of $g - a_j$ for some j. Consequently, we obtain

$$\sum_{j=1}^{n} \overline{Z}_R(r, f - a_j) \leq \overline{Z}_R(r, H) + \overline{Z}_R(r, f') + \overline{Z}_R(r, g') - \overline{Z}^S_R(r, f') - \overline{Z}^S_R(r, g'). \tag{3A}$$

Moreover, in the hypothesis of Theorems II.2.20, since any zero α of $f - a_i$ is a zero of certain $g - a_j$ with the same multiplicity order, we have $\omega_\alpha(f') = \omega_\alpha(g')$. Consequently,we obtain this improvement of (3A).

(3B)
$$\sum_{j=1}^{n}\overline{Z}_R(r,f-a_j)\le \overline{Z}_R(r,H)+\frac{1}{2}[\overline{Z}_R(r,f')+\overline{Z}_R(r,g')-\overline{Z}_R^S(r,f')-\overline{Z}_R^S(r,g')].$$
By Corollary II.1.5.1 we know that $\overline{Z}_R(r,H)\le N_R(r,H)+O(\log(r))$, $(r\in J)$, hence
$$\sum_{j=1}^{n}\overline{Z}_R(r,f-a_j)\le N_R(r,H)+\overline{Z}_R(r,f')+\overline{Z}_R(r,g')-\overline{Z}_R^S(r,f')-\overline{Z}_R^S(r,g')+O(\log(r))$$
$(r\in I)$, and similarly
$$\sum_{j=1}^{n}\overline{Z}_R(r,g-a_j)\le N_R(r,H)+\overline{Z}_R(r,f')+\overline{Z}_R(r,g')-\overline{Z}_R^S(r,f')-\overline{Z}_R^S(r,g')+O(\log(r))$$
$(r\in I)$, hence
$$\sum_{j=1}^{n}\overline{Z}_R(r,f-a_j)+\overline{Z}_R(r,g-a_j)$$
$$(4A)\quad \le 2N_R(r,H)+2[\overline{Z}_R(r,f')+\overline{Z}_R(r,g')-\overline{Z}_R^S(r,f')-\overline{Z}_R^S(r,g')]+O(\log(r)).$$
And in the hypothesis of Theorem II.2.20, by (3B) we obtain
$$\sum_{j=1}^{n}\overline{Z}_R(r,f-a_j)+\overline{Z}_R(r,g-a_j)$$
$$(4B)\quad \le 2N_R(r,H)+[\overline{Z}_R(r,f')+\overline{Z}_R(r,g')-\overline{Z}_R^S(r,f')-\overline{Z}_R^S(r,g')]+O(\log(r)).$$
Consider now $N_R(r,H)$, and let η be a pole of H. Either η is a zero of $F'G'$ or it is a pole of FG. Let ζ be the counting function of the poles of H occuring when FG has a pole. So, we have
$$(5)\qquad N(r,H)=\overline{Z}_R(r,F'G')+\zeta(r)$$
Suppose first η is a zero of F' but is not a pole of FG. Either η is a zero of f', or it is a zero of $P'(f)$, or it is a pole of f. Consequently, η is not a zero of $P(f)$ and we have $\overline{Z}_R(r,F')\le\overline{Z}_R^S(r,f')+\overline{Z}_R(r,P'(f))+\overline{N}_R(r,f)$. Since
$$P'(X)=n(n-1)(n-2)(X-1)^2X^{n-3},$$
we have $\overline{Z}_R(r,P'(f))\le 2T_R(r,f)$, $\quad\overline{Z}_R(r,P'(g))\le 2T_R(r,g)$, and therefore
$$(6)\qquad \overline{Z}_R(r,F')\le\overline{Z}_R^S(r,f')+2T(r,f)+\overline{N}_R(r,f).$$
Similarly, if η is a zero of G' but is not a pole of FG, we have
$$(7)\qquad \overline{Z}_R(r,G')\le\overline{Z}_R^S(r,g')+2T(r,g)+\overline{N}_R(r,g).$$
Suppose now η is a pole of FG. Then η is a zero of $P(f)$ and $P(g)$ (we notice that when η is a zero of $P(f)$, it is a zero of $P(g)$, and vice-versa), and then by Lemma II.2.17, it may not be a pole of H when it is a zero of same order of $P(f)$ and $P(g)$. Consequently, in the hypothesis of Theorem II.2.19 $\zeta(r)$ satisfies
$$\zeta(r)\le\overline{Z}_R(r,f')-\overline{Z}_R^S(r,f')+\overline{Z}_R(r,g')-\overline{Z}_R^S(r,g'),$$

and therefore by (5), (6) and (7) we obtain

$$N_R(r,H) \le \overline{Z}^S_R(r,f') + \overline{Z}^S_R(r,g') + 2T(r,f) + 2T(r,g) + \overline{N}_R(r,f) + \overline{N}(r,g)$$
$$+\overline{Z}_R(r,f') - \overline{Z}^S_R(r,f') + \overline{Z}_R(r,g') - \overline{Z}^S_0(r,g'),$$

hence

$$N_R(r,H) \le 2T(r,f) + 2T(r,g) + \overline{N}_R(r,f) + \overline{N}_R(r,g) + \overline{Z}_R(r,f') + \overline{Z}_R(r,g').$$

Thus by (4A), in the hypothesis of Theorems II.2.19, we obtain

$$(8A) \qquad \sum_{j=1}^{n} \overline{Z}_R(r,f-a_j) + \overline{Z}_R(r,g-a_j) \le$$

$$4(T(r,f)+T(r,g))+2(\overline{N}_R(r,f)+2\overline{N}_R(r,g))+4(\overline{Z}_R(r,f')+\overline{Z}_R(r,g'))+O(\log(r)).$$

Now, in the hypotheses of Theorem II.2.20 since the order of a zero is the same for $P(f)$ and $P(g)$, the counting function ζ is identically 0, so by, (5) (6), (7) we have

$$N_R(r,H) \le 2T_R(r,f) + 2T_R(r,g) + \overline{N}_R(r,f) + \overline{N}(r,g) + \overline{Z}^S_R(r,f') + \overline{Z}^S_R(r,g'),$$

and therefore by (4B) we obtain

$$\sum_{j=1}^{n} \overline{Z}_R(r,f-a_j) + \overline{Z}_R(r,g-a_j) \le 4(T_R(r,f) + T_R(r,g)) + 2(\overline{N}(r,f)$$

$$(8B) \quad +2\overline{N}(r,g)) + (\overline{Z}_R(r,f') + \overline{Z}_R(r,g')) + (\overline{Z}^S_R(r,f') + \overline{Z}^S_R(r,g')) + O(\log(r)).$$

Now, applying Theorem II.1.15 to f and g at the points a_j $(1 \le j \le n)$, we have:

$$(n-1)T_R(r,f) \le \overline{N}(r,f) + \sum_{j=1}^{n} \overline{Z}_R(r,f-a_j) - Z^S_R(r,f') + O(\log(r)) \ (r \in I),$$

$$(n-1)T_R(r,g) \le \overline{N}_R(r,g) + \sum_{j=1}^{n} \overline{Z}_R(r,g-a_j) - Z^S_R(r,g') - \log r + O(\log(r)), \ (r \in I)$$

hence,

$$(n-1)(T_R(r,f) + T_R(r,g)) \le \overline{N}_R(r,f) + \overline{N}_R(r,g)$$
$$+\sum_{j=1}^{n} \overline{Z}_R(r,f-a_j) + \overline{Z}_R(r,g-a_j) - Z^S_R(r,f') - Z^S_R(r,g') + O(\log(r)) \ (r \in I).$$

hence by (8A), in the hypotheses of Theorems II.2.19, we obtain

$$(n-1)(T_R(r,f) + T_R(r,g)) \le 3(\overline{N}_R(r,f) + \overline{N}(r,g))$$
$$(9A) \qquad +4(T(r,f) + T(r,g)) + 4(\overline{Z}_R(r,f') + \overline{Z}_R(r,g')) + O(\log(r)) \ (r \in I)$$

So, by Theorem II.1.9 we can derive: $\quad (n-1)(T_R(r,f) + T_R(r,g))$

$$\le 3(\overline{N}_R(r,f) + \overline{N}_R(r,g)) + 4(T_R(r,f) + T_R(r,g)) + 4(Z_R(r,f) + Z_R(r,g))$$
$$+4(\overline{N}_R(r,f) + \overline{N}_R(r,g)) + O(\log(r)) \ (r \in I)$$

(resp. $r \in J$.)

Therefore we obtain $n \le 16$ and if f, g belong to $\mathcal{A}^*(D)$, then $n \le 9$.

Next, by (8B), in the hypotheses of Theorem II.2.20, we obtain

$$(n-1)(T_R(r,f)+T_R(r,g)) \le 3(\overline{N}_R(r,f)+\overline{N}_R(r,g))+4(T_R(r,f)+T_R(r,g))$$

$+\overline{Z}_R(r,f')+\overline{Z}_R(r,g')+O(\log(r))$ $(r \in I)$ (resp. $r \in J$)

Hence by Theorem II.1.9 we obtain:
$(n-1)(T_R(r,f)+T_R(r,g)) \le 3(\overline{N}_R(r,f)+\overline{N}_R(r,g))+4(T_R(r,f)+T_R(r,g))$
$+(Z_R(r,f)+Z_R(r,g))+\overline{N}_R(r,f)+\overline{N}_R(r,g)+O(\log(r))$ $(r \in I)$ (resp. $r \in J$)
Consequently, $n \le 10$ in Theorem II.2.20. Moreover, if f, g belong to $\mathcal{A}(\mathbb{K})$ or to $\mathcal{A}^*(D)$, then we obtain $n \le 6$. □

Acknowledgement

We are grateful to the referee for many good remarks on notations.

References

[1] W. Bergweiler and A. Eremenko, *On the singularities of the inverse to a meromorphic function of finite order*, Rev. Mat. Iberoamericana **11** (1995), no. 2, 355–373, DOI 10.4171/RMI/176. MR1344897

[2] J.-P. Bézivin, K. Boussaf, and A. Escassut, *Zeros of the derivative of a p-adic meromorphic function*, Bull. Sci. Math. **136** (2012), no. 8, 839–847, DOI 10.1016/j.bulsci.2012.07.003. MR2995004

[3] K. Boussaf, *Motzkin factorization in algebras of analytic elements*, Proceedings of the 3rd International Conference on p-adic Functional Analysis (Aubière, 1994), Ann. Math. Blaise Pascal **2** (1995), no. 1, 73–91. MR1342806

[4] A. Boutabaa, *Théorie de Nevanlinna p-adique* (French, with English summary), Manuscripta Math. **67** (1990), no. 3, 251–269, DOI 10.1007/BF02568432. MR1046988

[5] A. Boutabaa and A. Escassut, *Urs and ursims for p-adic meromorphic functions inside a disc*, Proc. Edinb. Math. Soc. (2) **44** (2001), no. 3, 485–504, DOI 10.1017/S0013091599000759. MR1875765

[6] K. S. Charak, *Value distribution theory of meromorphic functions*, Mathematics Newsletter, Vol. 18, no. 4, pp. 1-35 (2009).

[7] H. H. Chen and M. L. Fang, *The value distribution of $f^n f'$*, Sci. China Ser. A **38** (1995), no. 7, 789–798. MR1360682

[8] W. Cherry and J. T.-Y. Wang, *Non-Archimedean analytic maps to algebraic curves*, Value distribution theory and complex dynamics (Hong Kong, 2000), Contemp. Math., vol. 303, Amer. Math. Soc., Providence, RI, 2002, pp. 7–35, DOI 10.1090/conm/303/05235. MR1943525

[9] A. Escassut, *Analytic elements in p-adic analysis*, World Scientific Publishing Co., Inc., River Edge, NJ, 1995. MR1370442

[10] A. Escassut, *Ultrametric Banach algebras*, World Scientific Publishing Co., Inc., River Edge, NJ, 2003. MR1978961

[11] A. Escassut, *Meromorphic functions of uniqueness*, Bull. Sci. Math. **131** (2007), no. 3, 219–241, DOI 10.1016/j.bulsci.2006.05.004. MR2309863

[12] A. Escassut, *Value Distribution in p-adic Analysis.* WSCP Singapore, (2015)

[13] A. Escassut and J. Ojeda, *The p-adic Hayman conjecture when $n = 2$*, Complex Var. Elliptic Equ. **59** (2014), no. 10, 1451–1455, DOI 10.1080/17476933.2013.854347. MR3219689

[14] G. Frank and M. Reinders, *A unique range set for meromorphic functions with* 11 *elements*, Complex Variables Theory Appl. **37** (1998), no. 1-4, 185–193. MR1687880

[15] H. Fujimoto, *On uniqueness of meromorphic functions sharing finite sets*, Amer. J. Math. **122** (2000), no. 6, 1175–1203. MR1797660

[16] H. H. Khoái, *On p-adic meromorphic functions*, Duke Math. J. **50** (1983), no. 3, 695–711, DOI 10.1215/S0012-7094-83-05033-0. MR714825

[17] M. O. Hanyak and A. A. Kondratyuk, *Meromorphic functions in m-punctured complex planes* (English, with English and Russian summaries), Mat. Stud. **27** (2007), no. 1, 53–69. MR2341474

[18] P.-C. Hu and C.-C. Yang, *Meromorphic functions over non-Archimedean fields*, Mathematics and its Applications, vol. 522, Kluwer Academic Publishers, Dordrecht, 2000. MR1794326

[19] M. Krasner, *Prolongement analytique uniforme et multiforme dans les corps valués complets* (French), Les Tendances Géom. en Algèbre et Théorie des Nombres, Éditions du Centre National de la Recherche Scientifique, Paris, 1966, pp. 97–141. MR0204404

[20] M. Lazard, *Les zéros des fonctions analytiques d'une variable sur un corps valué complet* (French), Inst. Hautes Études Sci. Publ. Math. **14** (1962), 47–75. MR0152519

[21] M. Ru, *A note on p-adic Nevanlinna theory*, Proc. Amer. Math. Soc. **129** (2001), no. 5, 1263–1269. MR1712881

[22] E. Motzkin, *La décomposition d'un élément analytique en facteurs singuliers* (French, with English summary), Ann. Inst. Fourier (Grenoble) **27** (1977), no. 1, ix, 67–82. MR0470247

[23] R. Nevanlinna, *Le théorème de Picard-Borel et la théorie des fonctions méromorphes* (French), Chelsea Publishing Co., New York, 1974. Reprinting of the 1929 original. MR0417418

[24] J. Ojeda, *Hayman's conjecture in a p-adic field*, Taiwanese J. Math. **12** (2008), no. 9, 2295–2313. MR2479056

[25] K. Yamanoi, *The second main theorem for small functions and related problems*, Acta Math. **192** (2004), no. 2, 225–294, DOI 10.1007/BF02392741. MR2096455

Université Clermont Auvergnel, Laboratoire de Mathématiques Blaise Pascal, UMR CNRS, BP 10448, 63000 Clermont-Ferrand, France

Email address: `alain.escassut@math.univ-bpclermont.fr`

Institute of Mathematics, Vietnam Academy of Science and Technology 18, Hoang Quoc Viet Hanoi, Vietnam — and — Institute of Mathematics and Applied Sciences (TIMAS), Thang Long University, Hanoi, Vietnam

Email address: `tthan@math.ac.vn`

Contemporary Mathematics
Volume **704**, 2018
http://dx.doi.org/10.1090/conm/704/14167

The distance preserving mappings and isometrics defined on non-Archimedean Banach spaces

Albert Kubzdela

ABSTRACT. The paper deals with the problem under what conditions is a map of a non-Archimedean Banach space E into itself preserving unit distance an isometry. We prove that every surjective, non-expansive map $T : E \to E$ with the SDOPP is isometric if and only if E is locally compact and we show that the assumptions for T cannot be weakened.

1. Introduction

Let $T : E \to F$ be a map between normed spaces E and F. We say that T has a *conservative distance* $r > 0$ if $||x - y|| = r$ implies $||T(x) - T(y)|| = r$ for all $x, y \in E$. In particular, we say that T has *the distance one preserving property* (DOPP) if for all $x, y \in E$ with $||x - y|| = 1$ we get $||T(x) - T(y)|| = 1$. If additionally $||x - y|| = 1$ follows $||T(x) - T(y)|| = 1$ we will say that T has *the strong distance one preserving property* (SDOPP). T is a *non-expansive map* if $||T(x) - T(y)|| \leq ||x - y||$ for all $x, y \in E$ and T is an *isometry* (T is *isometric*) if $||T(x) - T(y)|| = ||x - y||$ for all $x, y \in E$.

The problem, under what conditions is a mapping of a Banach space into itself preserving unit distance an isometry, known as Aleksandrov problem, has been intensively studied by many specialists in the real and complex case (see for instance [**2**], [**7**], [**8**], [**9**], [**10**], [**12**] and references). Among others, it has been shown (see [**1**] and [**4**]) that every homeomorphism $f : E^n \longrightarrow E^n$ $(2 < n \leq \infty)$, where E^n denotes an euclidean space, with a non-trivial conservative distance $r > 0$ is an isometry and that for any integer $n \geq 1$, there exists an integer n_0 such that for each $m \geq n_0$ there is a map $f : E^n \longrightarrow E^m$ preserving distance one but, which is not an isometry (see [**7**]). Rassias and Semrl (see [**7**, Theorem 5]) demonstrated that every non-expansive with SDOPP, surjective mapping $T : X \to Y$ between real normed spaces X, Y such that one of them has a dimension greater than one is isometric. However, the question if a continuous mapping $f : E^\infty \to E^\infty$, where E^∞ denotes an infinite-dimensional Hilbert space, satisfying the DOPP must be an isometry is still open (see [**10**, Conjecture 1.2]).

2010 *Mathematics Subject Classification.* 46S10.
Key words and phrases. Non-Archimedean normed spaces, conservative distances, maps preserving distances, isometrics.

From now on by $\mathbb{K}$ we will denote a non-Archimedean complete non-trivially valued field, i.e. its valuation satisfies the strong triangle inequality what means $|\lambda+\mu| \leq \max\{|\lambda|, |\mu|\}$ for all $\lambda, \mu \in \mathbb{K}$. Recall that $|\mathbb{K}^*| = \{|\lambda| : \lambda \in \mathbb{K}\backslash\{0\}\}$ is the *value group* of $\mathbb{K}$ and $\mathbb{k} = B_{\mathbb{K}}/B_{\mathbb{K}}^-$ is the *residue class field* of $\mathbb{K}$, where $B_{\mathbb{K}}$ and $B_{\mathbb{K}}^-$ denote the closed and open unit ball in $\mathbb{K}$ centered at zero, respectively. $\mathbb{K}$ is said to be *discretely valued* if 0 is the only accumulation point of $|\mathbb{K}^*|$; then, there exists an *uniformizing element* $\rho \in \mathbb{K}$ with $0 < |\rho| < 1$ such that $|\mathbb{K}^*| = \{|\rho|^n : n \in \mathbb{Z}\}$. Otherwise, we say that $\mathbb{K}$ is *densely valued,* then, $|\mathbb{K}^*|$ is a dense subset of $[0, \infty)$.

By E we will denote a non-Archimedean Banach space, i.e. a Banach space over $\mathbb{K}$ equipped with the norm satisfying the strong triangle inequality, i.e. $\|x+y\| \leq \max\{\|x\|, \|y\|\}$ for all $x, y \in E$. In addition, we will assume $|\mathbb{K}^*| \subset \{||x|| : x \in E\}$, so there is $x_0 \in E$ with $||x_0|| = 1$.

Let $B_r(x) := \{y \in E : ||x-y|| \leq r\}$ and $B_r^-(x) := \{y \in E : ||x-y|| < r\}$ ($r > 0$, $x \in E$) denote a *closed ball* and an *open ball* in E, respectively. Note that both balls are closed and open (clopen) and the topology induced on E by a non-Archimedean norm is zero-dimensional. It follows directly from the strong triangle inequality that every point of any ball is its center and any two balls in E are either disjoint, or one is contained in the other. We will write shortly B_r (B_r^-) instead of $B_r(0)$ ($B_r^-(0)$).

By [**11**, 1.B], $\mathbb{K}$ is locally compact if and only if $\mathbb{K}$ is discretely valued and $\mathbb{k}$ is finite. If $\mathbb{K}$ is locally compact, then every finite-dimensional E is locally compact and each ball in E is compact (see [**11**, Corollary 3.17]).

Let I be a non-empty set and let $s : I \to (0, \infty)$ be a map. By $c_0(I : s)$ we will denote a linear space over $\mathbb{K}$ equipped with the non-Archimedean norm

$$\left|\left|(\lambda_i)_{i \in I}\right|\right| := \sup_{i \in I} |\lambda_i| \cdot s(i)$$

which consists of all $\left(\lambda^i\right)_{i \in I} \in \mathbb{K}^I$ such that for every $\varepsilon > 0$ there exists a finite $J \subset I$ for which $\left|\lambda^i\right| \cdot s(i) < \varepsilon$ for every $i \in I \backslash J$. If $s(i) = 1$ for all $i \in I$ we will write shortly $c_0(I)$. Recall that if $\mathbb{K}$ is locally compact then E is isometrically isomorphic with the space $c_0(I : s)$ for some set I and some map $s : I \to (0, \infty)$, see [**11**, Example 3.H].

For more background on normed spaces over non-Archimedean valued fields we refer the reader to [**6**] and [**11**].

In non-Archimedean setting, maps having conservative distances were considered among others in [**3**] and [**5**]. The paper [**3**] shown (see [**3**, Theorem 5 and Corollary 11]) that for finite-dimensional E every surjective, non-expansive map with the SDOPP $T : E \to E$ such that $T(0) = 0$ is isometric if and only if $\mathbb{K}$ is locally compact. In this paper we extend this result demonstrating

THEOREM 1. *Every surjective, non-expansive map $T : E \to E$ with the SDOPP is isometric if and only if E is locally compact.*

We also show (see Proposition 3) that the assumption in Theorem 1 about the SDOPP cannot be replaced by the DOPP.

2. Results

We start considering maps having conservative distances defined on infinite-dimensional E.

PROPOSITION 2. *Let $\mathbb{K}$ be locally compact. Then for every infinite-dimensional E there exists a surjective, non-expansive map $E \to E$ with SDOPP which is not isometric.*

PROOF. As we remarked in Introduction, without loss of generality we can assume that $E = c_0(I:s)$ for some set I and some map $s : I \to (0, \infty)$ with the norm defined as $||x|| = \max\{|x_i| \cdot s(i) : i \in I\}$. Since $\mathbb{K}$ is discretely valued, we can also suppose that $s : I \to (|\rho|, 1]$, where $\rho \in \mathbb{K}$, $|\rho| < 1$ is an uniformizing element of $|\mathbb{K}^*|$. The proof requires examination of two cases.

First, assume that there is an infinite countable subset $J \subset I$, say $J = \{i_1, i_2, ...\}$, that the sequence $(s(i_n))_n$ is non-decreasing. Denote $s_n := s(i_n), n \in \mathbb{N}$. Then, define the map $T_3 : E \to E$ setting

$$T_3(x) = \begin{cases} \rho x & \text{if } x \in B^-_{s_1|\rho^n|}(\rho^n e_{i_1}),\ n \in \mathbb{N} \\ x + \rho(e_{i_n} - e_{i_{n+1}}) & \text{if } x \in B^-_{s_{n+1}|\rho|}(\rho e_{i_{n+1}}),\ n \in \mathbb{N} \\ x & \text{elsewhere.} \end{cases}$$

Clearly $T_3(0) = 0$. We see that the map T_3 transfers the ball $B^-_{s_1|\rho^n|}(\rho^n e_{i_1})$ onto $B^-_{s_1|\rho^{n+1}|}(\rho^{n+1} e_{i_1})$ and the ball $B^-_{s_{n+1}|\rho|}(\rho e_{i_{n+1}})$ onto $B^-_{s_n|\rho|}(\rho e_{i_n})$ $(n \in \mathbb{N})$, respectively.

Let

$$X_1 := \bigcup_{n\in\mathbb{N}} B^-_{s_1|\rho^n|}(\rho^n e_{i_1}), \qquad X_2 := \bigcup_{n\in\mathbb{N}} B^-_{s_n|\rho|}(\rho e_{i_n}).$$

Since, for every $x \in E\backslash(X_1 \cup X_2)$ we have $T_3(x) = x$ we imply that T_3 is surjective.

Clearly $||T_3(x)|| \le ||x||$ for all $x \in E$. $T_3(x) \ne x$ only if $x \in X_1 \cup X_2$, thus, only if $||x|| < 1$. Hence, T_3 has SDOPP.

To check that T is non-expansive we should consider the following three cases:

a). $x \in X_1, y \in E\backslash(X_1 \cup X_2)$: Then, $||x - \rho^k e_{i_1}|| < s_1 |\rho^k|$ for some $k \in \mathbb{N}$. Hence, $||x|| = s_1|\rho^k|$ since $||\rho^k e_{i_1}|| = s_1|\rho^k|$. But $y \notin X_1$, thus $||y - \rho^k e_{i_1}|| \ge s_1|\rho^k|$. Therefore

$$||x - y|| = ||x - \rho^k e_{i_1} + \rho^k e_{i_1} - y|| = ||y - \rho^k e_{i_1}|| = \max\{||x||, ||y||\}$$

and

$$||T_3(x) - T_3(y)|| = ||\rho x - y|| \le ||x - y||.$$

b). $x \in X_2, y \in E\backslash(X_1 \cup X_2)$: Then, $||x - \rho e_{i_{n+1}}|| < s_{n+1}|\rho|$ for some $n \in \mathbb{N}$ and

$$||x - y|| = ||x - \rho e_{i_{n+1}} + \rho e_{i_{n+1}} - y|| = ||y - \rho e_{i_{n+1}}|| \ge s_{n+1}|\rho|$$

as $y \notin X_2$. Hence,

$$\begin{aligned} ||T_3(x) - T_3(y)|| &= ||x + \rho(e_{i_n} - e_{i_{n+1}}) - y|| \\ &\le \max\{||x - y||, ||\rho(e_{i_n} - e_{i_{n+1}})||\} \le ||x - y||, \end{aligned}$$

since $||\rho(e_{i_n} - e_{i_{n+1}})|| \le s_{n+1}|\rho|$.

c). $x, y \in X_1 \cup X_2$: If x, y are in the same ball $B^-_{s_1|\rho^n|}(\rho^n e_{i_1})$ or $B^-_{s_n|\rho|}(\rho e_{i_n})$ $(n \in \mathbb{N})$, the conclusion is straightforward. Otherwise, $||x|| \ne ||y||$, thus $||x - y|| =$

$\max\{||x||,||y||\}$ and $||T_3(x)-T_3(y)||\leq||x-y||$ since $||T_3(x)||\leq||x||$ and $||T_3(y)||\leq||y||$, respectively.

Also note that if $x,y\in B^-_{s_1|\rho^m|}(\rho^m e_{i_1})$ for some $m\in\mathbb{N}$, then

$$||T_3(x)-T_3(y)||=|\rho|\cdot||x-y||<||x-y||.$$

Thus, T_3 is not isometric.

Now consider the second case, assuming that there is no infinite countable subset $J\subset I$, written as $J=\{i_1,i_2,...\}$, such that the sequence $(s(i_n))_n$ is non-decreasing. Then, we can choose a countable subset $\{k_1,k_2,...\}\subset I$ such that $s(k_n)>s(k_{n+1})$ for every $n\in\mathbb{N}$. Additionally, we can assume that $s(k_1)=1$. Let $E_0:=\overline{[(e_{k_n})_n]}$. Then, $E=E_0\oplus E_1$, where $E_1\simeq c_0(I\backslash\{k_1,k_2,...\}:s)$. First, we construct a surjective, non-expansive map $E_0\to E_0$ with SDOPP which is not isometric.

As we assumed $\mathbb{K}$ is locally compact, hence the residue class field $\Bbbk$ is finite. Therefore, we can find $\lambda_1,...,\lambda_p\in B_{\mathbb{K}}\backslash B^-_{\mathbb{K}}$, where $p=card(\Bbbk)-1$, such that $|\lambda_i-\lambda_j|=1$ if $i\neq j$. Let $\lambda_0:=0$. Note that $||e_{k_1}||>||e_{k_2}||>....$ Hence, for every nonzero $\gamma\in\mathbb{K}$ every ball $B^-_{|\gamma|s(k_n)}(\gamma e_{k_n})$ $(n\in\mathbb{N})$ of E_0 can be covered by finitely many disjoint balls $B^-_{|\gamma|s(k_{n+1})}(\gamma e_{k_n}+\lambda_i\gamma e_{k_{n+1}})$, $i=0,1,...,p$. Indeed, take $x\in B^-_{|\gamma|s(k_n)}(\gamma e_{k_n})$; note that $B^-_{|\gamma|s(k_n)}(\gamma e_{k_n})=B_{|\gamma|s(k_{n+1})}(\gamma e_{k_n})$. Then, $x=\sum_m a_m e_{k_m}$ for some $a_m\in\mathbb{K}$ $(m\in\mathbb{N})$ and

$$\begin{aligned}||x-\gamma e_{k_n}||&=\left|\left|\sum_{m\neq n}a_m e_{k_m}+(a_n-\gamma)e_{k_n}\right|\right|\\&=\max\left\{\max_{m\neq n}\{s(k_m)\cdot|a_m|\},s(k_n)\cdot|a_n-\gamma|\right\}\leq|\gamma|\cdot s(k_{n+1}).\end{aligned}$$

Hence, $|a_n-\gamma|<|\gamma|$, $|a_{n+1}|\leq|\gamma|$, $|a_m|<|\gamma|$ if $m<n$ and $|a_m|\leq|\gamma|$ if $m>n$. Therefore, there is $j\in\{0,...,m\}$ such that $\left|\frac{a_{n+1}}{\gamma}-\lambda_j\right|<1$. Thus,

$$\begin{aligned}&||x-(\gamma e_{k_n}+\lambda_j\gamma e_{k_{n+1}})||\\&\quad=\left|\left|\sum_{m\notin\{n,n+1\}}a_m e_{k_m}+(a_n-\gamma)e_{k_n}+(a_{n+1}-\gamma\lambda_j)e_{k_{n+1}}\right|\right|\\&=\max\left\{\max_{m\notin\{n,n+1\}}\{s(k_m)\cdot|a_m|\},s(k_n)\cdot|a_n-\gamma|,s(k_{n+1})\cdot|\frac{a_{n+1}}{\gamma}-\lambda_j|\cdot|\gamma|\right\}\\&\qquad<|\gamma|\cdot s(k_{n+1}).\end{aligned}$$

So, $x\in B^-_{|\gamma|s(k_{n+1})}(\gamma e_{k_n}+\lambda_j\gamma e_{k_{n+1}})$. Consequently, every ball $B^-_{|\gamma|s(k_n)}(\gamma e_{k_n})$, $n\in\mathbb{N}$, has an infinite countable partition consisting of closed balls with radius equal to $|\rho\gamma|$ and centres in $\gamma e_{k_n}+\sum_{m=n+1}^{\infty}\alpha_m\gamma e_{k_m}$, where $\alpha_m\in\{\lambda_0,...,\lambda_p\}$ $(m=n+1,n+2,...)$ and only finitely many α_m are different from zero.

Now, define the map $T_4:E_0\to E_0$ using two sequences of balls

$$\{B_{|\rho^{-1}|s(k_{n+1})}(\rho^{-1}e_{k_n})\}_{n\in\mathbb{N}},\qquad\{B_{|\rho^{-l}|s(k_2)}(\rho^{-l}e_{k_1})\}_{n\in\mathbb{N}\backslash\{1\}}$$

as follows.

A). If $x\in B_{|\rho^{-1}|s(k_{n+1})}(\rho^{-1}e_{k_n})$ for some $n\in\mathbb{N}$ then there exist $\{\alpha_{n+1},\alpha_{n+2},...\}$, $\alpha_m\in\{\lambda_0,...,\lambda_p\}$ $(m=n+1,n+2,...)$ and only finitely many α_m are different from zero, such that

$$x \in B_1(\rho^{-1} e_{k_n} + \sum_{m=n+1}^{\infty} \alpha_m \rho^{-1} e_{k_m}).$$

Hence, we can define the map $\phi_1 : x \mapsto (\alpha_{n+1}, \alpha_{n+2}, ...)$ and set

$$T_4(x) := \rho^{-1}(e_{k_{n+1}} + \sum_{m=n+1}^{\infty} \alpha_m e_{k_{m+1}}) + x - \rho^{-1}(e_{k_n} + \sum_{m=n+1}^{\infty} \alpha_m e_{k_m}).$$

Observe that

$$(1) \qquad T_4 : B_{|\rho^{-1}|s(k_{n+1})}(\rho^{-1} e_{k_n}) \xrightarrow{onto} B_{|\rho^{-1}|s(k_{n+2})}(\rho^{-1} e_{k_{n+1}}), n \in \mathbb{N}$$

and every closed ball with radius equal to unity which is contained in the ball $B_{|\rho^{-1}|s(k_{n+1})}(\rho^{-1} e_{k_n})$ is transferred isometrically by T_4 onto a ball with a radius of a unit which is contained in $B_{|\rho^{-1}|s(k_{n+2})}(\rho^{-1} e_{k_{n+1}})$.

B). If $x \in B_{|\rho^{-l}|s(k_2)}(\rho^{-l} e_{k_1})$ for some $l \in \{2, 3, ...\}$, then there exist $\{\delta_2, \delta_3, ...\}$ and $\{\beta_1, \beta_2, ...\}$, where $\delta_m, \beta_m \in \{\lambda_0, ..., \lambda_p\}$ and only finitely many δ_m and β_m are different from zero, such that

$$x \in B_{|\rho^{-l+1}|}(\rho^{-l} e_{k_1} + \sum_{m=2}^{\infty} \delta_m \rho^{-l} e_{k_m} + \sum_{n=1}^{\infty} \beta_n \rho^{-l+1} e_{k_n}).$$

Thus, similarly as in the previous case, we can define the map

$$\phi_2 : x \mapsto (\delta_2, \beta_1, \delta_3, \beta_2, \delta_4, ...)$$

and set

$$T_4(x) := \rho^{-l+1}(e_{k_1} + \sum_{m=2}^{\infty} c_{m-1} e_{k_m}) + x$$
$$- (\rho^{-l} e_{k_1} + \sum_{m=2}^{\infty} \delta_m \rho^{-l} e_{k_m} + \sum_{n=1}^{\infty} \beta_n \rho^{-l+1} e_{k_n}),$$

where $(c_1, c_2, c_3, ...) = (\delta_2, \beta_1, \delta_3, \beta_2, \delta_4, ...)$. Then,

$$(2) \qquad T_4 : B_{|\rho^{-l}|s(k_2)}(\rho^{-l} e_{k_1}) \xrightarrow{onto} B_{|\rho^{-l+1}|s(k_2)}(\rho^{-l+1} e_{k_1}), l = 2, 3, ...$$

and every closed ball with radius equal to $|\rho^{-l+2}|$ which is contained in the ball $B_{|\rho^{-l}|s(k_2)}(\rho^{-l} e_{k_1})$ is transferred isometrically by T_4 onto a ball (with radius equal to $|\rho^{-l+2}|$) which is contained in $B_{|\rho^{-l+1}|s(k_2)}(\rho^{-l+1} e_{k_1})$.

C). Otherwise, we set

$$T_4(x) := x.$$

Obviously, $T_4(0) = 0$. From (1) and (2) we imply that T_4 is surjective. Let

$$V_1 = \bigcup_{n \in \mathbb{N}} B_{|\rho^{-1}|s(k_{n+1})}(\rho^{-1} e_{k_n}), \quad V_2 = \bigcup_{l \in \mathbb{N} \setminus \{1\}} B_{|\rho^{-l}|s(k_2)}(\rho^{-l} e_{k_1}).$$

To check that T_4 is non-expansive we shall consider the following few cases, where we also prove that T_4 is not isometric and it has the SDOPP:

i). $x, y \in V_1$: If $x \in B_{|\rho^{-1}|s(k_{n+1})}(\rho^{-1} e_{k_n})$, $y \in B_{|\rho^{-1}|s(k_{m+1})}(\rho^{-1} e_{k_m})$ and $n > m$ then $||x - y|| > 1$ and $||x - y|| = ||y|| = ||\rho^{-1} e_{k_m}||$. But $||T_4(x) - T_4(y)|| =$

$||T_4(y)|| = \left||\rho^{-1}e_{k_{m+1}}\right|| > 1$. Hence, $||T_4(x) - T_4(y)|| < ||x - y||$. If $n = m$, then if $\phi_1(x) = \phi_1(y)$,

$$x, y \in B_1(\rho^{-1}e_{k_n} + \sum_{l=n+1}^{\infty} \alpha_l \rho^{-1}e_{k_l}),$$

where $(\alpha_{n+1}, \alpha_{n+2}, ...) = \phi_1(x) = \phi_1(y)$, and T_4 maps isometrically $B_1(\rho^{-1}e_{k_n} + \sum_{l=n+1}^{\infty} \alpha_l \rho^{-1}e_{k_l})$ onto $B_1(\rho^{-1}e_{k_{n+1}} + \sum_{l=n+1}^{\infty} \alpha_l \rho^{-1}e_{k_{l+1}})$, thus $||T_4(x) - T_4(y)|| = ||x - y||$. In particular, $||x - y|| = 1$ if and only if $||T_4(x) - T_4(y)|| = 1$. If $n = m$ and $\phi_1(x) \neq \phi_1(y)$ let

$$q := \min\left\{n \in \mathbb{N} : \alpha_{n+1} \neq \alpha'_{n+1}\right\},$$

where $(\alpha_{n+1}, \alpha_{n+2}, ...) = \phi_1(x)$ and $(\alpha'_{n+1}, \alpha'_{n+2}, ...) = \phi_1(y)$. Then, $||x - y|| = |\rho^{-1}| \cdot s(k_{n+1}) > 1$ and $1 < ||T_4(x) - T_4(y)|| = |\rho^{-1}| \cdot s(k_{n+2}) < |\rho^{-1}| \cdot s(k_{n+1})$. Hence,

$$1 < ||T_4(x) - T_4(y)|| < ||x - y||.$$

ii). $x, y \in V_2$: We argue similarly as in the previous case. If

$$x \in B_{|\rho^{-n}|s(k_2)}(\rho^{-n}e_{k_1}),\ y \in B_{|\rho^{-m}|s(k_2)}(\rho^{-m}e_{k_1})$$

and $n > m$ then $||x - y|| = ||x|| = ||\rho^{-n}e_{k_1}|| > 1$ and

$$1 < ||T_4(x) - T_4(y)|| = ||T_4(x)|| = \left||\rho^{-n+1}e_{k_1}\right|| < ||x - y||.$$

If $n = m$ and $\phi_2(x) = \phi_2(y)$ then

$$x, y \in B_{|\rho^{-l+1}|}(\rho^{-l}e_{k_1} + \sum_{i=2}^{\infty} \delta_i \rho^{-l}e_{k_m} + \sum_{i=1}^{\infty} \beta_i \rho^{-l+1}e_{k_i})$$

where $(\delta_2, \beta_1, \delta_3, \beta_2, \delta_4, ...) = \phi_2(x) = \phi_2(y)$. However, this ball is isometrically transferred by T_4 onto

$$B_{|\rho^{-l+1}|}(\rho^{-l+1}e_{k_1} + \sum_{i=2}^{\infty} \rho^{-l+1}c_{i-1}e_{k_i})$$

where $(c_1, c_2, c_3, ...) = (\delta_2, \beta_1, \delta_3, \beta_2, \delta_4, ...)$. Thus, $||T_4(x) - T_4(y)|| = ||x - y||$. In particular, $||x - y|| = 1$ if and only if $||T_4(x) - T_4(y)|| = 1$.

Let $n = m$ and $\phi_2(x) \neq \phi_2(y)$. Denote $(c_1, c_2, c_3, ...) = (\delta_2, \beta_1, \delta_3, \beta_2, \delta_4, ...) = \phi_2(x)$ and $(c'_1, c'_2, c'_3, ...) = (\delta'_2, \beta'_1, \delta'_3, \beta'_2, \delta'_4, ...) = \phi_2(y)$. If $\delta_i \neq \delta'_i$ for some $i \in \mathbb{N}$ then, setting $j := \min\{i : \delta_i \neq \delta'_i\}$ we get $\left|\delta_j - \delta'_j\right| = 1$ and $||x - y|| = \left|\rho^{-l}\right| s(k_j)$. But $||T_4(x) - T_4(y)|| \leq \left|\rho^{-l+1}\right| s(k_2)$; thus, $||T_4(x) - T_4(y)|| < ||x - y||$.

Suppose now, that $\delta_i = \delta'_i$ for all $i \in \{2, 3, ...\}$; then, let $j := \min\{i : \beta_i \neq \beta'_i\}$. Consequently, $\left|\beta_j - \beta'_j\right| = 1$ and $||x - y|| = \left|\rho^{-l+1}\right| s(k_j)$. However,

$$||T_4(x) - T_4(y)|| = \left|\rho^{-l+1}\right| s(k_t),$$

where $t > j$. Hence,

$$||x - y|| > ||T_4(x) - T_4(y)||.$$

iii). $x \in V_1$, $y \notin V_1$ or $x \in V_2$, $y \notin V_2$: Then, $||x - y|| \geq ||x||$, hence $||x - y|| = \max\{||x||, ||y||\} > 1$. Since $1 < ||T_4(x)|| \leq ||x||$, $||T_4(y)|| \leq ||y||$ and similarly as above $||T_4(x) - T_4(y)|| = \max\{||T_4(x)||, ||T_4(y)||\}$, we get

$$1 < ||T_4(x) - T_4(y)|| \leq ||x - y||.$$

iv). The case while $x, y \notin V_1 \cup V_2$ is obvious.

This way, we proved that $T_4 : E_0 \to E_0$ is a non-expansive, surjective map with SDOPP which is not isometric. Observe that every $x \in E$ can be unequivocally written as $x = x_0 + x_1$ where $x_0 \in E_0$ and $x_1 \in E_1$. Define the map $T_4' : E \to E$ setting $T_4'(x) = T_4'(x_0 + x_1) := T_4(x_0) + x_1$. Then, for any $x, y \in E$ written as $x = x_0 + x_1$ and $y = y_0 + y_1$, where $x_0, y_0 \in E_0$ and $x_1, y_1 \in E_1$, we get

$$\begin{aligned} ||T_4'(x) - T_4'(y)|| &= ||T_4(x_0) + x_1 - T_4(y_0) - y_1|| \\ &= \max\{||x_1 - y_1||, ||T_4(x_0) - T_4(y_0)||\} \\ &\leq \max\{||x_1 - y_1||, ||x_0 - y_0||\} = ||x - y|| \end{aligned}$$

since $||T_4(x_0) - T_4(y_0)|| \leq ||x_0 - y_0||$ as we derived above. Thus, T_4' is non-expansive. Obviously, T_4' is not isometric, as T_4 is not isometric. If $||x - y|| = 1$, then $\max\{||x_1 - y_1||, ||x_0 - y_0||\} = 1$. If $||x_0 - y_0|| = 1$ then $||T_4(x_0) - T_4(y_0)|| = 1$ and conversely as T_4 has the SDOPP . Clearly $||x_1 - y_1|| = 1$ if and only if $||T_4'(x_1) - T_4'(y_1)|| = 1$. Hence, T_4' has the SDOPP. □

Now, we are ready to prove Theorem 1.

PROOF (OF THEOREM 1). ($\Longleftarrow$) : Let $T : E \to E$ be a surjective, non-expansive map with SDOPP. Assume that E is locally compact. Then $\mathbb{K}$ is locally compact. First, observe that we can assume $T(0) = 0$. Indeed, for surjective, non-expansive map $T_0 : E \to E$ with SDOPP, the map $T(x) := T_0(x) - T_0(0)$ is also surjective, non-expansive, with SDOPP, and additionally $T(0) = 0$. Since $||T_0(x) - T_0(y)|| = ||T(x) - T(y)||$ for all $x, y \in E$, proving that T is isometric, we get the same conclusion for T_0. Now, assumptions of [**3**, Theorem 5] are satisfied; hence, we conclude that T is isometric.

($\Longrightarrow$) : To complete this part of the proof we suppose that E is not locally compact. With this assumption we construct, partly using some ideas from [**3**, Examples 9 and 10], a surjective, non-expansive map $E \to E$ with SDOPP which is not isometric. First, assume that $\mathbb{K}$ is not locally compact. Then, by [**11**, 1.B], $card(\mathbb{k})$ is infinite or $\mathbb{K}$ is densely valued.

Suppose that $card(\mathbb{k})$ is infinite. Then, we can select an infinite sequence $(\lambda_n)_n \subset \mathbb{K}$ with $|\lambda_n| = 1$ $(n \in \mathbb{N})$ such that $|\lambda_i - \lambda_j| = 1$ if $i \neq j$. Take $\mu \in \mathbb{K}$ and $x_0 \in E$ such that $|\mu| < 1$ and $||x_0|| < 1$. Let $r := ||x_0||$. Now, form a sequence $(x_n)_n$ in E setting $x_n := \lambda_n x_0, n \in \mathbb{N}$. Then, the balls $B_r^-(x_1), B_r^-(x_2), ...$ are pairwise disjoint. Define the map $T_1 : E \to E$ as follows

$$T_1(x) = \begin{cases} \mu x & \text{if } x \in B_r^-(x_1) \\ x + x_n - x_{n+1} & \text{if } x \in B_r^-(x_{n+1}),\ n \in \mathbb{N} \\ x + \mu x_{n+1} - \mu x_n & \text{if } x \in B_{r\cdot|\mu|}^-(\mu x_n),\ n \in \mathbb{N} \\ x & \text{elsewhere.} \end{cases}$$

Then, $T_1(0) = 0$.

Let $U := \bigcup_{n \in \mathbb{N}} (B_{r\cdot|\mu|}^-(\mu x_n) \cup B_r^-(x_n))$. If $x \notin U$ then $T_1(x) = x$. Note also that

$$\begin{aligned} T_1(B_r^-(x_1)) &= B_{r\cdot|\mu|}^-(\mu x_1), \\ T_1(B_{r\cdot|\mu|}^-(\mu x_n)) &= B_{r\cdot|\mu|}^-(\mu x_{n+1}),\ n \in \mathbb{N} \end{aligned}$$

and

$$T_1(B_r^-(x_n)) = B_r^-(x_{n-1}) \quad (n = 2, 3, ...).$$

Hence, we imply that T_1 is surjective.

We can easily observe that $||T_1(x)|| \leq ||x||$ for every $x \in E$. Take $x, y \in E$, $x \neq y$. We see that $||x - y|| = 1$ or $||T_1(x) - T_1(y)|| = 1$ only if $\max\{||x||, ||y||\} \geq 1$, i.e. $x \notin U$ or $y \notin U$. If $x, y \notin U$ then $||x - y|| = ||T_1(x) - T_1(y)||$. Let $x \notin U$ and $y \in U$. Then, $||y|| < 1$ and $||x|| = 1$ if and only if $||T_1(x)|| = 1$. Hence, we deduce that T_1 has SDOPP.

To prove that T_1 is non-expansive, first assume that $x \notin B_r^-(x_1)$ or $y \notin B_r^-(x_1)$. If $x, y \in B_r^-(x_n)$ for some $n \in \mathbb{N}$ $(n > 1)$, then

$$||T_1(x) - T_1(y)|| = ||x + x_n - x_{n+1} - (y + x_n - x_{n+1})|| = ||x - y||.$$

Similarly, we get $||T_1(x) - T_1(y)|| = ||x - y||$ if $x, y \in B^-_{r\cdot|\mu|}(\mu x_n)$ for some $n \in \mathbb{N}$. If there exists $n \in \mathbb{N}$ such that $x \in B_r^-(x_n)$ $(x \in B^-_{r\cdot|\mu|}(\mu x_n))$ and $y \notin B_r^-(x_n)$ $(y \notin B^-_{r\cdot|\mu|}(\mu x_n))$, then, $||x - y|| = \max\{||x||, ||y||\}$; thus, $||T_1(x) - T_1(y)|| \leq ||x - y||$. Finally, suppose that $x, y \in B_r^-(x_1)$. Then

$$||T_1(x) - T_1(y)|| = |\mu| \cdot ||x - y|| < ||x - y||. \tag{3}$$

and we conclude that T_1 is non-expansive. From (3) we also imply that T_1 is not isometric.

Now, suppose that $\mathbb{K}$ is densely valued. Choose reals r_1, r_2 with $0 < r_1 < r_2 < 1$ and select two sequences $(a_n)_n, (b_n)_n \subset \mathbb{K}$ such that

$$\frac{r_1 + r_2}{2} < |a_n| < |a_{n+1}| < r_2$$

and

$$r_1 < |b_{n+1}| < |b_n| < \frac{r_1 + r_2}{2}.$$

for every $n \in \mathbb{N}$. Take $x_0 \in E$ with $||x_0|| = 1$. Let $A_n := B^-_{|a_n|}(a_n x_0)$ and $B_n := B^-_{|b_n|}(b_n x_0)$ $(n \in \mathbb{N})$. Define the map $T_2 : E \to E$ by

$$T_2(x) = \begin{cases} \frac{a_n}{a_{n+1}} x & \text{if } x \in A_{n+1},\ n \in \mathbb{N} \\ \frac{b_{n+1}}{b_n} x & \text{if } x \in B_n,\ n \in \mathbb{N} \\ \frac{b_1}{a_1} x & \text{if } x \in A_1 \\ x & \text{elsewhere.} \end{cases}$$

Then, $T_2(0) = 0$. Since $T_2(A_1) = B_1$, $T_2(B_n) = B_{n+1}$ $(n \in \mathbb{N})$, $T_2(A_n) = A_{n-1}$ $(n = 2, 3, ...)$ and if $x \notin \bigcup_{n\in\mathbb{N}}(A_n \cup B_n)$ then $T_2(x) = x$, we imply that T_2 is surjective.

Take $x, y \in E, x \neq y$. Then, $||x - y|| = 1$ or $||T_2(x) - T_2(y)|| = 1$ only if $\max\{||x||, ||y||\} \geq 1$. But then $||x - y|| = 1$ follows $||T_2(x) - T_2(y)|| = 1$ and conversely. Hence, T_2 has a SDOPP.

To prove that T_2 is non-expansive take first $x, y \notin \bigcup_{n\in\mathbb{N}}(A_n \cup B_n)$. Then $||T_2(x) - T_2(y)|| = ||x - y||$. If $x \in A_n$ $(x \in B_n)$ for some $n \in \mathbb{N}$ and $y \notin A_n$ $(y \notin B_n)$, then, $||x - y|| = \max\{||x||, ||y||\}$; thus, $||T_2(x) - T_2(y)|| \leq ||x - y||$ (note that $||T_2(x')|| \leq ||x'||$ for every $x' \in E$). If $x, y \in A_n$ or $x, y \in B_n$ for some $n \in \mathbb{N}$, then

$$||T_2(x) - T_2(y)|| < ||x - y|| < 1. \tag{4}$$

Hence, T_2 is non-expansive. (4) also shows us that T_2 is not isometric.

The case if $\mathbb{K}$ is locally compact and E is infinite-dimensional is provided by Proposition 2. □

Next result shows that the assumption about the SDOPP in Theorem 1 cannot be replaced by the DOPP.

PROPOSITION 3. *Let $\lambda \in \mathbb{K}$, $x_0 \in E$ be such that $|\lambda| > 1$ and $||x_0|| = 1$. Let $T : E \to E$ be defined as follows*

$$T(x) = \begin{cases} x - \lambda^k x_0 & x \in B^-_{|\lambda^k|}\left(\lambda^k x_0\right) \text{ for some } k \in \mathbb{N} \\ x & \text{elsewhere.} \end{cases}$$

Then, T is a surjective non-expansive map with the distance one preserving property, $T(0) = 0$ and T is not isometric.

PROOF. Note that the balls $B^-_{|\lambda^k|}\left(\lambda^k x_0\right)$ are disjoint, hence the map T is well defined. Denote $U := \bigcup_k B^-_{|\lambda^k|}\left(\lambda^k x_0\right)$. We will examine the desired properties of T.

(a): T has the DOPP: Let $x, y \in E$ be such that $||x - y|| = 1$. First, assume that $x \in U$. Then, there exists $k_0 \in \mathbb{N}$ for which $\left|\left|\lambda^{k_0} x_0 - x\right|\right| < \left|\lambda^{k_0}\right|$. Thus,

$$\left|\left|\lambda^{k_0} x_0 - y\right|\right| = \left|\left|\lambda^{k_0} x_0 - x + x - y\right|\right| \leq \max\left\{\left|\left|\lambda^{k_0} x_0 - x\right|\right|, ||x - y||\right\} < \left|\lambda^{k_0}\right|.$$

Then $T(x) = x - \lambda^{k_0} x_0$ and $T(y) = y - \lambda^{k_0} x_0$. Therefore we obtain

$$||T(x) - T(y)|| = \left|\left|x - \lambda^{k_0} x_0 - \left(y - \lambda^{k_0} x_0\right)\right|\right| = ||x - y|| = 1.$$

Now, suppose that $x \notin U$. Then, $\left|\left|\lambda^k x_0 - x\right|\right| \geq \left|\lambda^k\right| > 1$ for every $k \in \mathbb{N}$. Hence, we get

$$\left|\left|\lambda^k x_0 - y\right|\right| = \left|\left|\lambda^k x_0 - x + x - y\right|\right| = \left|\left|\lambda^k x_0 - x\right|\right| \geq \left|\lambda^k\right|$$

and conclude $y \notin U$. But then, $T(x) = x$, $T(y) = y$, thus $T(x) - T(y) = x - y$ and we are done.

(b): T is surjective: Take $x \in E$. If $x \notin U$, then $x = T(x)$. If $x \in U$, then we can find $k_0 \in \mathbb{N}$ such that $\left|\left|\lambda^{k_0} x_0 - x\right|\right| < \left|\lambda^{k_0}\right|$. Clearly, $||x|| = \left|\lambda^{k_0}\right|$ and

$$||\left(\lambda^{k_0+1} x_0 + x\right) - \lambda^{k_0+1} x_0|| < |\lambda^{k_0+1}|.$$

Therefore $T(\lambda^{k_0+1} x_0 + x) = \lambda^{k_0+1} x_0 + x - \lambda^{k_0+1} x_0 = x$.

(c): T is non-expansive: Take $x, y \in E$. If $x, y \notin U$, we get $||T(x) - T(y)|| = ||x - y||$. Assume that $||x - \lambda^{k_x}|| < |\lambda^{k_x}|$ and $||y - \lambda^{k_y}|| < |\lambda^{k_y}|$ for some $k_x, k_y \in \mathbb{N}$. Then

$$T(x) - T(y) = x - \lambda^{k_x} - (y - \lambda^{k_y}) = (x - y) - (\lambda^{k_x} - \lambda^{k_y}).$$

Hence, if $k_x = k_y$, we get $T(x) - T(y) = x - y$; otherwise, $||x - y|| = \max\left\{|\lambda^{k_x}|, |\lambda^{k_y}|\right\}$ since $||x|| = |\lambda^{k_x}|$ and $||y|| = |\lambda^{k_y}|$. Thus,

$$||T(x) - T(y)|| \leq \max\left\{||x - y||, ||\lambda^{k_x} - \lambda^{k_y}||\right\} = ||x - y||.$$

Take $z \notin U$. Then, $||x - z|| = \max\{||x||, ||z||\}$ since $||x - z|| \geq \left|\,|\lambda^{k_x}|\,\right| = ||x||$ and

$$||T(x) - T(z)|| = ||x - \lambda^{k_x} - z|| \leq \max\left\{||x - \lambda^{k_x}||, ||z||\right\} \leq ||x - z||.$$

(d): T is not isometric: Since $T\left(\lambda^k x_0\right) = 0$ for every $k \in \mathbb{N}$, $\left|\left|T\left(\lambda^k x_0\right) - T(0)\right|\right| = 0$. But $\left|\left|\lambda^k x_0\right|\right| \neq 0$.

□

For finite-dimensional non-Archimedean spaces over locally compact $\mathbb{K}$ we can give the following characterization for the map $T : E \to E$ to be isometric.

PROPOSITION 4. *Let E a finite-dimensional normed space over locally compact $\mathbb{K}$ and let $T : E \to E$ be a surjective non-expansive map with $T(0) = 0$. Then, $T|_{B_r}$ is isometric if and only if $T^{-1}(B_r) \subset B_r$ for some $r > 0$.*

PROOF. $(\Longrightarrow)$: Obvious. $(\Longleftarrow)$: Assume that $T : E \to E$ is a surjective non-expansive map with $T(0) = 0$ and $T^{-1}(B_r) \subset B_r$ for some $r > 0$. Suppose for a contradiction that there are $x, y \in B_r$ such that $||T(x) - T(y)|| < ||x - y||$. By assumption $\mathbb{K}$ is locally compact, thus, B_r is compact and we can find a finite partition $B^-_{||x-y||}(z_j)$, $j = 1, ..., m$ $(m > 1)$ of B_r. Without loss of generality, we can assume that $x \in B^-_{||x-y||}(z_1)$ and $y \in B^-_{||x-y||}(z_2)$. Since T is non-expansive and $T(0) = 0$, then for every $i \in \{1, ..., m\}$ there exists $j \in \{1, ..., m\}$ such that $T(B^-_{||x-y||}(z_i)) \subset B^-_{||x-y||}(z_j)$. In addition $T(x), T(y) \in B^-_{||x-y||}(z_k)$ for some $k \in \{1, ..., m\}$. Hence,

$$T\left(B^-_{||x-y||}(z_1) \cup B^-_{||x-y||}(z_2)\right) \subset B^-_{||x-y||}(z_k)$$

and we imply that there exists $l \in \{1, ..., m\}$ such that

$$T\left(\bigcup_{j=1}^{m} B^-_{||x-y||}(z_j)\right) \cap B^-_{||x-y||}(z_l) = \varnothing. \tag{5}$$

But T is surjective, therefore there is $z_0 \in E$ such that $T(z_0) = z_l \in B^-_{||x-y||}(z_l)$. It follows from (5) that $z_0 \notin B_r$, a contradiction. □

References

[1] F. S. Beckman and D. A. Quarles Jr., *On isometries of Euclidean spaces*, Proc. Amer. Math. Soc. **4** (1953), 810–815, DOI 10.2307/2032415. MR0058193

[2] S.-M. Jung and T. M. Rassias, *On distance-preserving mappings*, J. Korean Math. Soc. **41** (2004), no. 4, 667–680, DOI 10.4134/JKMS.2004.41.4.667. MR2068146

[3] A. Kubzdela, *Isometries, Mazur-Ulam theorem and Aleksandrov problem for non-Archimedean normed spaces*, Nonlinear Anal. **75** (2012), no. 4, 2060–2068, DOI 10.1016/j.na.2011.10.006. MR2870899

[4] B. Mielnik and T. M. Rassias, *On the Aleksandrov problem of conservative distances*, Proc. Amer. Math. Soc. **116** (1992), no. 4, 1115–1118, DOI 10.2307/2159497. MR1101989

[5] M. S. Moslehian and G. Sadeghi, *A Mazur-Ulam theorem in non-Archimedean normed spaces*, Nonlinear Anal. **69** (2008), no. 10, 3405–3408, DOI 10.1016/j.na.2007.09.023. MR2450548

[6] C. Perez-Garcia and W. H. Schikhof, *Locally convex spaces over non-Archimedean valued fields*, Cambridge Studies in Advanced Mathematics, vol. 119, Cambridge University Press, Cambridge, 2010. MR2598517

[7] T. M. Rassias and P. Šemrl, *On the Mazur-Ulam theorem and the Aleksandrov problem for unit distance preserving mappings*, Proc. Amer. Math. Soc. **118** (1993), no. 3, 919–925, DOI 10.2307/2160142. MR1111437

[8] T. M. Rassias, *Properties of isometric mappings*, J. Math. Anal. Appl. **235** (1999), no. 1, 108–121, DOI 10.1006/jmaa.1999.6363. MR1758670

[9] T. M. Rassias, *On the A. D. Aleksandrov problem of conservative distances and the Mazur-Ulam theorem*, Proceedings of the Third World Congress of Nonlinear Analysts, Part 4 (Catania, 2000), Nonlinear Anal. **47** (2001), no. 4, 2597–2608, DOI 10.1016/S0362-546X(01)00381-9. MR1972385

[10] T. M. Rassias, *On the Aleksandrov problem for isometric mappings*, Appl. Anal. Discrete Math. **1** (2007), no. 1, 18–28, DOI 10.2298/AADM0701018R. MR2316585

[11] A. C. M. van Rooij, *Non-Archimedean functional analysis*, Monographs and Textbooks in Pure and Applied Math., vol. 51, Marcel Dekker, Inc., New York, 1978. MR512894
[12] S. Xiang, *Mappings of conservative distances and the Mazur-Ulam theorem*, J. Math. Anal. Appl. **254** (2001), no. 1, 262–274, DOI 10.1006/jmaa.2000.7276. MR1807900

INSTITUTE OF CIVIL ENGINEERING, POZNAŃ UNIVERSITY OF TECHNOLOGY, UL. PIOTROWO 5, 61 – 138 POZNAŃ, POLAND
Email address: `albert.kubzdela@put.poznan.pl`

Contemporary Mathematics
Volume **704**, 2018
http://dx.doi.org/10.1090/conm/704/14168

Rational map $ax + 1/x$ on the projective line over $\mathbb{Q}_p$

Shilei Fan and Lingmin Liao

ABSTRACT. The dynamical structure of the rational map $ax + 1/x$ on the projective line $\mathbb{P}^1(\mathbb{Q}_p)$ over the field $\mathbb{Q}_p$ of p-adic numbers is described for $p \geq 3$.

1. Introduction

For a prime number p, let $\mathbb{Q}_p$ be the field of p-adic numbers and $\mathbb{P}^1(\mathbb{Q}_p)$ its projective line. Recently, polynomials and rational maps of $\mathbb{Q}_p$ have been studied as dynamical systems on $\mathbb{Q}_p$ or $\mathbb{P}^1(\mathbb{Q}_p)$. It turns out that these p-adic dynamical systems are quite different to the dynamical systems in Archimedean spaces. See for example, [**2**, **5**, **26**] and their bibliographies therein.

For polynomials and rational maps over $\mathbb{Q}_p$, we can find two different kinds of subsystems exhibiting totally different dynamical behavior. One is 1-Lipschitz dynamical systems and the other is p-adic repellers.

A 1-Lipschitz p-adic dynamical system can usually be fully described by showing all its minimal subsystems. Polynomials with coefficients in the ring $\mathbb{Z}_p$ of p-adic integers and rational maps with good reduction are two important families of 1-Lipschitz dynamical systems. In [**14**], the authors proved the following structure theorem for polynomials in $\mathbb{Z}_p[x]$. The same structure theorem for good reduction maps with degree at least 2 was proved in [**12**].

THEOREM 1 ([**14**], Theorem 1). *Let $f \in \mathbb{Z}_p[x]$ be a polynomial of integral coefficients with degree ≥ 2. We have the following decomposition*

$$\mathbb{Z}_p = \mathcal{P} \bigsqcup \mathcal{M} \bigsqcup \mathcal{B}$$

where $\mathcal{P}$ is the finite set consisting of all periodic points of f, $\mathcal{M} = \bigsqcup_i \mathcal{M}_i$ is the union of all (at most countably many) clopen invariant sets such that each $\mathcal{M}_i$ is a finite union of disks and each subsystem $f : \mathcal{M}_i \to \mathcal{M}_i$ is minimal, and each point in $\mathcal{B}$ lies in the attracting basin of a periodic orbit or of a minimal subsystem.

2010 *Mathematics Subject Classification.* Primary 37P05; Secondary 11S82, 37B05.

Key words and phrases. p-adic dynamical system, rational maps, minimal decomposition, subshift of finite type.

The research of the first author on this project has received funding from the European Research Council(ERC) under the European Union's Horizon 2020 research and innovation programme under grant agreement No 647133(ICHAOS). The first author has also been supported by NSF of China (Grant Nos. 11401236 and 11471132).

The decomposition in Theorem 1 is usually referred to as a *minimal decomposition* and the invariant subsets $\mathcal{M}_i$ are called *minimal components*. In the literature, the minimality of the polynomial (or 1-Lipschitz) dynamical systems on the whole space $\mathbb{Z}_p$ was widely studied [**1, 3, 7, 10, 13, 17, 18, 20, 24**].

The p-adic weak repellers (see Section 2) are expanding dynamical systems which have positive topological entropy and thus exhibit chaotic behaviors. In [**15**], it is proved that a transitive p-adic weak repeller is topologically conjugate to a subshift of finite type (see Theorem 5 in Section 2). A general method was also proposed in [**15**] to find subshifts of finite type subsystems in a p-adic polynomial dynamical system. We remark that Thiran, Verstegen and Weyers [**28**] and Dremov, Shabat and Vytnova [**9**] studied the chaotic behavior of p-adic quadratic polynomial dynamical systems. Woodcock and Smart [**29**] proved that the so-called p-adic logistic map $\frac{x^p-x}{p}$ is topologically conjugate to the full shift on p symbols.

On the other side, the dynamical properties of the fixed points of the rational maps have been studied in the space $\mathbb{C}_p$ of p-adic complex numbers [**4, 21, 23, 27**] and in the adelic space [**8**]. The Fatou and Julia theory of the rational maps on $\mathbb{C}_p$, and on the Berkovich space over $\mathbb{C}_p$, are also developed [**5, 6, 19, 25, 26**]. However, the global dynamical structure of rational maps on $\mathbb{Q}_p$ remains unclear, though the rational maps of degree one are totally characterized in [**11**] .

In the present article, we suppose $p \geq 3$ and investigate the following special class of rational maps of degree 2:

$$\phi(x) = ax + \frac{1}{x}, \quad a \in \mathbb{Q}_p \setminus \{0\}. \tag{1.1}$$

We distinguish three cases: (1) $|a|_p = 1$, (2) $|a|_p > 1$, (3) $|a|_p < 1$.

Observe that

$$\phi(x) = ax + \frac{1}{x} = \frac{ax^2 - 1}{x}.$$

When $|a|_p = 1$, the map ϕ has good reduction (see definition in page 58 of [**26**]). By Theorem 1.2 of [**12**], we immediately have the following structure theorem.

THEOREM 2. *Let $\phi(x) = ax + \frac{1}{x}$ with $a \in \mathbb{Q}_p \setminus \{0\}$. If $|a|_p = 1$, the dynamical system $(\mathbb{P}^1(\mathbb{Q}_p), \phi)$ can be decomposed into*

$$\mathbb{P}^1(\mathbb{Q}_p) = \mathcal{P} \bigsqcup \mathcal{M} \bigsqcup \mathcal{B}$$

where $\mathcal{P}$ is the finite set consisting of all periodic points of ϕ, $\mathcal{M} = \bigsqcup_i \mathcal{M}_i$ is the union of all (at most countably many) clopen invariant sets such that each $\mathcal{M}_i$ is a finite union of disks and each subsystem $\phi : \mathcal{M}_i \to \mathcal{M}_i$ is minimal, and each point in $\mathcal{B}$ lies in the attracting basin of a periodic orbit or of a minimal subsystem.

We are thus left to study the rest two cases. Denote by $\phi^{\circ n}$ the n-th iteration of ϕ. The following are our main theorems. We remark that in both cases, $\infty \in \mathbb{P}^1(\mathbb{Q}_p)$ is a fixed point of ϕ with multiplier $1/a$.

THEOREM 3. *Let $\phi(x) = ax + \frac{1}{x}$ with $a \in \mathbb{Q}_p \setminus \{0\}$. If $|a|_p > 1$, the dynamical structure of the system $\phi : \mathbb{P}^1(\mathbb{Q}_p) \to \mathbb{P}^1(\mathbb{Q}_p)$ is described as follows.*

(1) *If $\sqrt{1-a} \notin \mathbb{Q}_p$, then*

$$\forall x \in \mathbb{Q}_p, \quad \lim_{n\to\infty} \phi^{\circ n}(x) = \infty.$$

(2) *If $\sqrt{1-a} \in \mathbb{Q}_p$, then there exists an invariant subset $\mathcal{J} \subset \mathbb{Q}_p$ such that the subsystem $\phi : \mathcal{J} \to \mathcal{J}$ is topologically conjugate to the full shift on two symbols. Further,*

$$\forall x \in \mathbb{Q}_p \setminus \mathcal{J}, \quad \lim_{n\to\infty} \phi^{\circ n}(x) = \infty.$$

THEOREM 4. *Let $\phi(x) = ax + \frac{1}{x}$ with $a \in \mathbb{Q}_p \setminus \{0\}$. If $|a|_p < 1$, we distinguish two cases.*

(1) *If $\sqrt{-a} \notin \mathbb{Q}_p$, then $\phi(0) = \phi(\infty) = \infty$ is a repelling fixed point and the subsystem $\phi : \mathbb{Q}_p \setminus \{0\} \to \mathbb{Q}_p \setminus \{0\}$ has a minimal decomposition as stated in Theorem 2.*

(2) *If $\sqrt{-a} \in \mathbb{Q}_p$, then the dynamical system $\phi : \mathbb{P}^1(\mathbb{Q}_p) \to \mathbb{P}^1(\mathbb{Q}_p)$ has a subsystem which is conjugate to a subshift of finite type with positive entropy.*

2. Preliminaries

Let $p \geq 2$ be a prime number and $\mathbb{Q}_p$ be the field of p-adic numbers. Any $x \in \mathbb{Q}_p$ has a p-adic expansion

$$x = \sum_{n=v_p(x)}^{\infty} a_n p^n \quad (v_p(x) \in \mathbb{Z}, a_n \in \{0, 1, 2, \cdots, p-1\} \text{ and } a_{v_p(x)} \neq 0),$$

where the integer $v_p(x)$ is the p-valuation of x. Any point in the projective line $\mathbb{P}^1(\mathbb{Q}_p)$ of $\mathbb{Q}_p$ can be given in homogeneous coordinates by a pair $[x_1 : x_2]$ of points in $\mathbb{Q}_p$ which are not both zero. Two such pairs are equal if they differ by an overall (nonzero) factor $\lambda \in \mathbb{Q}_p^*$:

$$[x_1 : x_2] = [\lambda x_1 : \lambda x_2].$$

The field $\mathbb{Q}_p$ is identified with the subset of $\mathbb{P}^1(\mathbb{Q}_p)$ given by

$$\left\{[x : 1] \in \mathbb{P}^1(\mathbb{Q}_p) \mid x \in \mathbb{Q}_p\right\}.$$

This subset contains all points in $\mathbb{P}^1(\mathbb{Q}_p)$ except one: the point of infinity, which may be given as $\infty = [1 : 0]$.

The spherical metric defined on $\mathbb{P}^1(\mathbb{Q}_p)$ is analogous to the standard spherical metric on the Riemann sphere. If $P = [x_1, y_1]$ and $Q = [x_2, y_2]$ are two points in $\mathbb{P}^1(\mathbb{Q}_p)$, we define

$$\rho(P, Q) = \frac{|x_1 y_2 - x_2 y_1|_p}{\max\{|x_1|_p, |y_1|_p\} \max\{|x_2|_p, |y_2|_p\}}$$

or, viewing $\mathbb{P}^1(\mathbb{Q}_p)$ as $\mathbb{Q}_p \cup \{\infty\}$, for $z_1, z_2 \in \mathbb{Q}_p \cup \{\infty\}$ we define

$$\rho(z_1, z_2) = \frac{|z_1 - z_2|_p}{\max\{|z_1|_p, 1\} \max\{|z_2|_p, 1\}} \qquad \text{if } z_1, z_2 \in \mathbb{Q}_p,$$

and

$$\rho(z, \infty) = \begin{cases} 1, & \text{if } |z|_p \leq 1, \\ 1/|z|_p, & \text{if } |z|_p > 1. \end{cases}$$

Remark that the restriction of the spherical metric on the ring $\mathbb{Z}_p := \{x \in \mathbb{Q}_p, |x| \leq 1\}$ of p-adic integers is the same as the metric induced by the absolute value $|\cdot|_p$.

A rational map $\phi \in \mathbb{Q}_p(z)$ induces a transformation on $\mathbb{P}^1(\mathbb{Q}_p)$. Rational maps are always Lipschitz continuous on $\mathbb{P}^1(\mathbb{Q}_p)$ with respect to the spherical metric (see [**26**, Theorem 2.14.]).

In $\mathbb{Q}_p$, we denote by $D(a,r) := \{x \in \mathbb{Q}_p : |x|_p \leq r\}$ the closed disk centered at a with radius r and by $S(a,r) := \{x \in \mathbb{Q}_p : |x|_p = r\}$ its corresponding sphere. A closed disk in $\mathbb{P}^1(\mathbb{Q}_p)$ is either a closed disk in $\mathbb{Q}_p$ or the complement of an open disk in $\mathbb{Q}_p$.

We also recall the conditions under which a number in $\mathbb{Q}_p$ has a square root in $\mathbb{Q}_p$. An integer $a \in \mathbb{Z}$ is called a *quadratic residue modulo* p if the equation $x^2 \equiv a \pmod{p}$ has a solution $x \in \mathbb{Z}$. The following lemma characterizes those p-adic integers which admit a square root in $\mathbb{Q}_p$.

LEMMA 1 ([**22**]). *Let a be a nonzero p-adic number with its p-adic expansion*

$$a = p^{v_p(a)}(a_0 + a_1 p + a_2 p^2 + \cdots)$$

where $1 \leq a_0 \leq p-1$ and $0 \leq a_j \leq p-1$ $(j \geq 1)$. The equation $x^2 = a$ has a solution $x \in \mathbb{Q}_p$ if and only if the following conditions are satisfied

(i) *$v_p(a)$ is even;*

(ii) *a_0 is a quadratic residue modulo p if $p \neq 2$; or $a_1 = a_2 = 0$ if $p = 2$.*

Now we recall some standard terminology of the theory of dynamical systems. A *dynamical system* is a couple (X, ϕ), where X is a topological space and ϕ is a continuous transformation from X to itself. A subset $Y \subset X$ is *invariant* if $\phi(Y) \subset Y$. For such an invariant subset, let $\phi|_Y$ be the restriction of ϕ on Y. Then $(Y, \phi|_Y)$ becomes a *subsystem*. For simplicity, sometimes we still denote by ϕ the restricted transformation $\phi|_Y$.

A main interest in the theory of dynamical systems is to study the topological properties of the orbit $\{\phi^{\circ n}(x)\}_{n\geq 1}$ of a point $x \in X$ under the iterations $\phi^{\circ n}$. A point $x_0 \in X$ is called a *fixed point* of ϕ if $\phi(x_0) = x_0$. The set of all fixed points of ϕ is denoted by $\mathrm{Fix}(\phi)$. By definition, x_0 is called a *periodic point* of ϕ if $\phi^{\circ n}(x_0) = x_0$ for some $n \geq 1$. In this case, n is called a *period* of x_0, and the smallest n with this property is called the *exact period* of x_0.

For a periodic point $x_0 \in \mathbb{Q}_p$ of exact period n, the derivative $(\phi^{\circ n})'(x_0)$ (if exists) is called the multiplier of x_0. Remark that the multiplier is invariant by changing of coordinate. If ∞ is a periodic point of period n, then the multiplier of ∞ is $\psi'(0)$, where $\psi(x) = \frac{1}{\phi^{\circ n}(1/x)}$. A periodic point is called *attracting, indifferent,* or *repelling* accordingly as the absolute value of its multiplier is less than, equal to, or greater than 1. Periodic points of multiplier 0 are called *super attracting.*

A subsystem of a dynamical system is *minimal* if the orbit of any point in the subspace is dense in the subspace.

There is an important class of dynamical systems: symbolic dynamical systems. For $m \geq 1$, call the finite set $\mathcal{A} := \{0, 1, \cdots, m-1\}$ an alphabet. The infinite product space $\Sigma := \mathcal{A}^{\mathbb{N}}$ is said to be the symbolic space of m symbols. The (left)-shift σ on Σ is defined as

$$\forall x = x_0 x_1 x_2 \cdots \in \Sigma, \quad (\sigma(x))_i = x_{i+1}.$$

The couple (Σ, σ) is called a *full shift* on m symbols. Let A be a $k \times k$ matrix with entries in $\{0, 1\}$ and set

$$\Sigma_A = \left\{(x_j)_{j\geq 0} \in \Sigma : A_{x_i x_{i+1}} = 1, \ \forall i \geq 0\right\}.$$

Then Σ_A is an invariant subset of Σ. We thus obtain a subsystem (Σ_A, σ) called a *subshift of finite type*.

A large family of p-adic dynamical systems are conjugate to full shift or subshift of finite type. Let $\phi : X \to \mathbb{Q}_p$ be a map from a compact open set X of $\mathbb{Q}_p$ into $\mathbb{Q}_p$. Assume that (i) $\phi^{-1}(X) \subset X$; (ii) $X = \bigsqcup_{i\in I} D(c_i, p^{-\gamma_i})$ can be written as a finite disjoint union of disks of centers c_i and of the radius $p^{-\gamma_i}$ (with $-\gamma_i \in \mathbb{Z}$) such that for each $i \in I$ there is an integer $\tau_i \in \mathbb{Z}$ such that

$$|\phi(x) - \phi(y)|_p = p^{\tau_i}|x - y|_p \qquad (\forall x, y \in D(c_i, p^{-\gamma_i})). \tag{2.1}$$

For such a map ϕ, define its Julia set by

$$J_\phi = \bigcap_{n=0}^{\infty} (\phi^{-1})^{\circ n}(X). \tag{2.2}$$

It is clear that $\phi^{-1}(J_\phi) = J_\phi$ and then $\phi(J_\phi) \subset J_\phi$. We thus have a subsystem (J_ϕ, ϕ). Such a subsystem is called a *p-adic weak repeller* if all τ_i in (2.1) are nonnegative, but at least one is positive.

The dynamics of a p-adic weak repeller is determined by a matrix with entries in $\{0, 1\}$. Denote simply the disk $D(c_i, p^{-\gamma_i})$ by D_i. For any $i \in I$, let

$$I_i := \{j \in I : D_j \cap \phi(D_i) \neq \emptyset\} = \{j \in I : D_j \subset \phi(D_i)\}.$$

Then define the *incidence matrix* $A = (A_{i,j})_{I\times I}$ of (J_ϕ, ϕ), by

$$A_{ij} = 1 \;\text{ if } j \in I_i; \quad A_{ij} = 0 \;\text{ otherwise.}$$

The weak repeller (J_ϕ, ϕ) is called *transitive* if the matrix A is irreducible. In [**15**], the authors proved the following theorem.

THEOREM 5 ([**15**], Theorem 1.1). *Let (J_ϕ, ϕ) be a transitive p-adic weak repeller in $\mathbb{Q}_p$ with incidence matrix A. Then (J_ϕ, ϕ) is topologically conjugate to (Σ_A, σ), i.e., there exists a continuous bijection $h : J_\phi \to \Sigma_A$ such that $h \circ \phi = \sigma \circ h$.*

3. Dynamical structures

Now, let $p \geq 3$. We will study the dynamical structures of the rational maps $\phi(x) = ax + \frac{1}{x}$ with $a \in \mathbb{Q}_p \setminus \{0\}$ on the projective line $\mathbb{P}^1(\mathbb{Q}_p)$. In general, we can also consider the rational maps

$$ax + \frac{b}{x}, \quad \text{with } a, b \in \mathbb{Q}_p.$$

Remark that if $\sqrt{b}$ exists in $\mathbb{Q}_p$, then $ax + \frac{b}{x}$ is conjugate to $ax + \frac{1}{x}$ through the conjugacy $x \mapsto \frac{1}{\sqrt{b}}x$. The case that $\sqrt{b}$ does not exist in $\mathbb{Q}_p$ which is not included in the present paper could be a subject for future study.

The dynamical system $(\mathbb{P}^1(\mathbb{Q}_p), \phi)$ exhibits different dynamical structures according to different absolute values of a. When $|a|_p = 1$, the transformation ϕ has good reduction. The dynamical structure of $(\mathbb{P}^1(\mathbb{Q}_p), \phi)$, as shown in Theorem 2, can be deduced directly from Theorem 1.2 of [**12**].

We are thus concerned only with the cases: $|a|_p > 1$ and $|a|_p < 1$. We remark that for both cases, ∞ is a fixed point of ϕ.

3.1. Case $|a|_p > 1$.

PROPOSITION 1. *Suppose* $|a|_p > 1$. *If* $\sqrt{1-a} \notin \mathbb{Q}_p$, *then*

$$\forall x \in \mathbb{Q}_p, \quad \lim_{n\to\infty} \phi^{\circ n}(x) = \infty.$$

PROOF. By the assumption $|a|_p > 1$, for all $x \in \mathbb{Q}_p$ such that $|x|_p \geq 1$, we have

$$|\phi(x)|_p = \left|ax + \frac{1}{x}\right|_p = |ax|_p > |x|_p.$$

Thus the absolute values of the iterations $\phi^{\circ n}(x)$ are strictly increasing. Hence

$$\lim_{n\to\infty} \phi^{\circ n}(x) = \infty, \quad \text{for all } x \in \mathbb{Q}_p, |x|_p \geq 1. \tag{3.1}$$

That is to say, $\{x \in \mathbb{Q}_p : |x|_p \geq 1\}$ is included in the attracting basin of ∞.

Now we investigate the points in the open disk $\{x \in \mathbb{Q}_p : |x|_p < 1\}$. We make a partition of this disk:

$$A_1 := \left\{x \in \mathbb{Q}_p : |x|_p < 1, |ax|_p \neq \frac{1}{|x|_p}\right\}, \ A_2 := \left\{x \in \mathbb{Q}_p : |x|_p < 1, |ax|_p = \frac{1}{|x|_p}\right\}.$$

If $x \in A_1$, then

$$|\phi(x)|_p = \max\left\{|ax|_p, \frac{1}{|x|_p}\right\} > 1.$$

Thus by (3.1), $\phi(x)$ falls into the attracting basin of ∞, and $\lim_{n\to\infty} \phi^{\circ n}(x) = \infty$.

If $|ax|_p = \frac{1}{|x|_p}$, then $|a|_p = \frac{1}{|x|_p^2}$, which means that $v_p(a)$ is an even number. Since $|a|_p > 1$, by Lemma 1, the condition $\sqrt{1-a} \notin \mathbb{Q}_p$ means that $\sqrt{-a}$ does not exist in $\mathbb{Q}_p$. Hence the equation $\phi(x) = ax + \frac{1}{x} = 0$ has no solution in $\mathbb{Q}_p$. Since $p \geq 3$, by Lemma 1 again, the sum of the first digits of ax and $\frac{1}{x}$ in their p-adic expansions is not 0 modulo p. Thus $|\phi(x)|_p = |ax|_p$ or $|1/x|_p$. Hence, $|\phi(x)|_p > 1$. Therefore, by (3.1), $\lim_{n\to\infty} \phi^{\circ n}(x) = \infty$. □

LEMMA 2. *Suppose* $|a|_p > 1$. *If* $\sqrt{1-a} \in \mathbb{Q}_p$, *then* ϕ *has two repelling fixed points*

$$x_{1,2} = \pm\frac{1}{\sqrt{1-a}}.$$

PROOF. It is easy to check that $x_{1,2} = \pm\frac{1}{\sqrt{1-a}}$ are the two fixed points of ϕ. Note that

$$\phi'(x_1) = \phi'(x_2) = 2a - 1.$$

Since $|a|_p > 1$, we have

$$|\phi'(x_1)|_p = |\phi'(x_2)|_p = |2a-1|_p > 1.$$

□

LEMMA 3. *Suppose* $|a|_p > 1$. *If* $\sqrt{1-a} \in \mathbb{Q}_p$, *then*

$$\phi(D(x_1, p^{\frac{v_p(a)}{2}-1})) = \phi(D(x_2, p^{\frac{v_p(a)}{2}-1})) = D(0, p^{-\frac{v_p(a)}{2}-1}).$$

PROOF. Since $|a|_p > 1$ and $\sqrt{1-a} \in \mathbb{Q}_p$, the valuation $v_p(a)$ is a negative even number. Note that

$$|\phi(x_1)|_p = |\phi(x_2)|_p = |x_1|_p = |x_2|_p = p^{\frac{v_p(a)}{2}}.$$

We need only to show that for all x, y in the same disk $D(x_1, p^{\frac{v_p(a)}{2}-1})$ or $D(x_2, p^{\frac{v_p(a)}{2}-1})$,

$$|\phi(x) - \phi(y)|_p = p^{-v_p(a)}|x - y|_p.$$

Without loss of generality, we assume that $x, y \in D(x_1, p^{\frac{v_p(a)}{2}-1})$. Observe that

$$D(x_1, p^{\frac{v_p(a)}{2}-1}) = \left\{\frac{1}{x} : |x - \sqrt{1-a}|_p \le p^{-\frac{v_p(a)}{2}-1}\right\}.$$

Hence, there are $x', y' \in D(0, p^{-\frac{v_p(a)}{2}-1})$ such that

$$x = \frac{1}{\sqrt{1-a} + x'} \quad \text{and} \quad y = \frac{1}{\sqrt{1-a} + y'}.$$

So we have

$$\left|a - \frac{1}{xy}\right|_p = |2a - 1 - (x' + y')\sqrt{1-a} - x'y'|_p.$$

Observing that $|a|_p \ge 1$ and $|(x' + y')\sqrt{1-a}|_p \le |a|_p/p$, we have

$$\left|a - \frac{1}{xy}\right|_p = |a|_p.$$

Hence,

$$|\phi(x) - \phi(y)|_p = \left|(a - \frac{1}{xy})(x - y)\right|_p = |a|_p|x - y|_p = p^{-v_p(a)}|x - y|_p.$$

□

LEMMA 4. *Suppose* $|a|_p > 1$. *If* $\sqrt{1-a} \in \mathbb{Q}_p$, *then for all*

$$x \notin D(x_1, p^{\frac{v_p(a)}{2}-1}) \cup D(x_2, p^{\frac{v_p(a)}{2}-1}),$$

we have

$$\lim_{n\to\infty} \phi^{\circ n}(x) = \infty.$$

PROOF. Note that for all $x \notin D(0, 1/p)$, $|1/x|_p \le 1$. Thus

$$|\phi(x)|_p = |a|_p|x|_p > 1.$$

Hence

$$|\phi^{\circ n}(x)|_p = |a|_p^n|x|_p,$$

which implies

$$\forall\ x \notin D(0,1), \quad \lim_{n\to\infty} \phi^{\circ n}(x) = \infty.$$

To finish the proof, we will show that for all

$$0 \ne x \in D(0,1) \setminus (D(x_1, p^{\frac{v_p(a)}{2}-1}) \cup D(x_2, p^{\frac{v_p(a)}{2}-1})),$$

we have $|\phi(x)|_p > 1$.

We distinguish three cases.

Case (1) $|x|_p < 1/\sqrt{|a|_p}$. Since $|1/x|_p > |ax|_p$, we have

$$|\phi(x)|_p = |1/x|_p > 1.$$

Case (2) $|x|_p = 1/\sqrt{|a|_p}$. Observe that $|ax|_p = |1/x|_p = \sqrt{|a|_p}$. The assumption $x \notin D(x_1, p^{\frac{v_p(a)}{2}-1}) \cup D(x_2, p^{\frac{v_p(a)}{2}-1})$ implies that the sum of the first digits of the p-adic expansion of ax and $1/x$ is not 0 modulo p, which leads to

$$|\phi(x)|_p = |ax + 1/x|_p = |ax|_p = \sqrt{|a|_p} > 1.$$

Case (3) $1/\sqrt{|a|_p} < |x|_p < 1$. Since $|ax|_p > |1/x|_p > 1$, we have $|\phi(x)|_p = |ax|_p > 1$. □

Let $\Sigma_2 := \{0,1\}^{\mathbb{N}}$ and (Σ_2, σ) be the full shift on two symbols.

PROPOSITION 2. *Suppose $|a|_p > 1$ and $\sqrt{1-a} \in \mathbb{Q}_p$. Then there exists an invariant set $\mathcal{J}$ such that $(\mathcal{J}, \phi)$ is topologically conjugate to (Σ_2, σ), and*

$$\lim_{n\to\infty} \phi^{\circ n}(x) = \infty, \quad \forall x \in \mathbb{Q}_p \setminus \mathcal{J}.$$

PROOF. By the proof of Lemma 3, we obtain that both of the restricted maps

$$\phi : D(x_i, p^{\frac{v_p(a)}{2}-1}) \to D(0, p^{-\frac{v_p(a)}{2}-1}), \quad i = 1, 2$$

are expanding and bijective. Note that $D(x_i, p^{\frac{v_p(a)}{2}-1}) \subset D(0, p^{-\frac{v_p(a)}{2}-1})$ for $i = 1, 2$. Let

$$\Omega = D(x_1, p^{\frac{v_p(a)}{2}-1}) \cup D(x_2, p^{\frac{v_p(a)}{2}-1})$$

and

$$\mathcal{J} = \bigcap_{i=0}^{\infty} (\phi^{-1})^{\circ n}(\Omega).$$

By Theorem 5, $\mathcal{J}$ is ϕ-invariant and $(\mathcal{J}, \phi)$ is topologically conjugate to (Σ_2, σ). Note that all $x \in \Omega \setminus J$ will eventually fall into $\mathbb{Q}_p \setminus \Omega$ by iteration of ϕ. Thus by Lemma 4, we immediately get

$$\lim_{n\to\infty} \phi^{\circ n}(x) = \infty, \quad \forall x \in \mathbb{Q}_p \setminus \mathcal{J}.$$

□

PROOF OF THEOREM 3. It follows directly from Propositions 1 and 2. □

3.2. Case $|a|_p < 1$. We distinguish two sub cases: $\sqrt{-a} \notin \mathbb{Q}_p$ and $\sqrt{-a} \in \mathbb{Q}_p$.

3.2.1. $\sqrt{-a} \notin \mathbb{Q}_p$. We will use the notation $\lfloor \cdot \rfloor$ for the integer part of a real numbers.

LEMMA 5. *If $\sqrt{-a} \notin \mathbb{Q}_p$, then for all $-\lfloor v_p(a)/2 \rfloor \le i \le \lfloor v_p(a)/2 \rfloor$,*

$$\phi(S(0, p^i)) \subset S(0, p^{-i})$$

and $\phi^{\circ 2}$ is 1-Lipschitz continuous on $S(0, p^i) \cup S(0, p^{-i})$.

PROOF. If $x \in S(0, p^i)$ for some $-\lfloor v_p(a)/2 \rfloor \le i \le 0$, then by the assumption $|a|_p < 1$, we have

$$|\phi(x)|_p = |ax + 1/x|_p = p^{-i}.$$

Now let $x \in S(0, p^i)$ for some $0 \le i \le \lfloor v_p(a)/2 \rfloor$. When $i < \lfloor v_p(a)/2 \rfloor$, we have

$$|\phi(x)|_p = |ax + 1/x|_p = p^{-i}.$$

When $v_p(a)/2$ is even and $i = v_p(a)/2$, the condition $\sqrt{-a} \notin \mathbb{Q}_p$ implies that

$$|\phi(x)|_p = |ax + 1/x|_p = p^{-i}.$$

Hence, the first assertion of the lemma holds.

Let us show that $\phi^{\circ 2}$ is 1-Lipschitz continuous on $S(0,p^i)\cup S(0,p^{-i})$ for $-\lfloor v_p(a)/2\rfloor \le i \le \lfloor v_p(a)/2\rfloor$. Let $x, y \in S(0,p^i)\cap S(0,p^{-i})$. If $x\in S(0,p^i)$ and $y \in S(0,p^{-i})$, then $|xy|_p = 1$ and thus

$$|\phi(x)-\phi(y)|_p = \left|a - \frac{1}{xy}\right|_p |x-y|_p = |x-y|_p.$$

Hence, it suffices to show that for each $-\lfloor v_p(a)/2\rfloor \le i \le \lfloor v_p(a)/2\rfloor$,

$$\forall x,y\in S(0,p^i), \quad |\phi^{\circ 2}(x)-\phi^{\circ 2}(y)|_p \le |x-y|_p.$$

By observing that $|a|_p \le 1/|xy|_p$ and $|a|_p \le 1/|\phi(x)\phi(y)|_p$, we have

$$\begin{aligned}
|\phi^{\circ 2}(x)-\phi^{\circ 2}(y)|_p &= \left|a - \frac{1}{\phi(x)\phi(y)}\right|_p |\phi(x)-\phi(y)|_p \\
&= \left|a - \frac{1}{\phi(x)\phi(y)}\right|_p \left|a-\frac{1}{xy}\right|_p |x-y|_p \\
&\le \left|\frac{1}{\phi(x)\phi(y)}\right|_p \left|\frac{1}{xy}\right|_p |x-y|_p = |x-y|_p.
\end{aligned}$$

□

Denote by $\mathbb{C}_p$ the field of p-adic complex numbers. and by $B(a,r) = \{x\in\mathbb{C}_p : |x|_p \le r\}$ the closed ball in $\mathbb{C}_p$ centered at a with radius $r>0$.

LEMMA 6. *Assume $\sqrt{-a}\notin\mathbb{Q}_p$. Let $x_0\in\mathbb{Q}_p$ with $p^{-\lfloor v_p(a)/2\rfloor} \le |x_0|_p \le p^{\lfloor v_p(a)/2\rfloor}$. Then $\phi^{\circ 2}(B(x_0,|x_0|_p/p)) \subset B(\phi^{\circ 2}(x_0), |x_0|_p/p)$.*

PROOF. For any $x\in B(x_0,|x_0|_p/p)$, the condition $\sqrt{-a}\notin\mathbb{Q}_p$ implies

$$|x\phi(x)|_p = 1.$$

Hence

$$\begin{aligned}
|\phi^{\circ 2}(x)-\phi^{\circ 2}(x_0)|_p &= \left|a - \frac{1}{\phi(x)\phi(x_0)}\right|_p |\phi(x)-\phi(x_0)|_p \\
&= \left|a - \frac{1}{\phi(x)\phi(x_0)}\right|_p \left|a-\frac{1}{xx_0}\right|_p |x-x_0|_p \\
&\le \left|\frac{1}{\phi(x)\phi(x_0)}\right|_p \left|\frac{1}{xx_0}\right|_p |x-x_0|_p = |x-x_0|_p.
\end{aligned}$$

□

LEMMA 7. *Assume $\sqrt{-a}\notin\mathbb{Q}_p$. Let $x_0\in\mathbb{Q}_p$ with $p^{-\lfloor v_p(a)/2\rfloor} \le |x_0|_p \le p^{\lfloor v_p(a)/2\rfloor}$. Then the Taylor expansion of the map*

$$\phi^{\circ 2} : D(x_0,|x_0|_p/p) \to D(\phi^{\circ 2}(x_0), |x_0|_p/p)$$

can be written as

$$\phi^{\circ 2}(x+x_0) = \phi^{\circ 2}(x_0) + \sum_{i=1}^{\infty}\alpha_i(x+x_0)^i,$$

with $\alpha_i\in\mathbb{Q}_p$ such that

$$|\alpha_i|_p \le (p/|x_0|_p)^{i-1}.$$

PROOF. The assumption $\sqrt{-a} \notin \mathbb{Q}_p$ implies that there is no ϕ-preimage of 0 in $\mathbb{Q}_p$. However, there are two preimages $\pm\frac{1}{\sqrt{-a}}$ of 0 in the quadratic extension $\mathbb{Q}_p(\sqrt{-a})$. For each $x_0 \in \mathbb{Q}_p$ with $p^{-\lfloor v_p(a)/2 \rfloor} \leq |x_0|_p \leq p^{\lfloor v_p(a)/2 \rfloor}$, the condition $\sqrt{-a} \notin \mathbb{Q}_p$ implies that

$$\left| x_0 - \frac{1}{\sqrt{-a}} \right|_p = \left| x_0 + \frac{1}{\sqrt{-a}} \right|_p > \frac{|x_0|_p}{p}.$$

By Lemma 5, we also have

$$\left| \phi(x_0) - \frac{1}{\sqrt{-a}} \right|_p = \left| \phi(x_0) + \frac{1}{\sqrt{-a}} \right|_p > \frac{|\phi(x_0)|_p}{p}.$$

Hence, the ball

$$B(x_0, |x_0|_p/p) := \left\{ x \in \mathbb{C}_p, |x|_p \leq \frac{|x_0|_p}{p} \right\}$$

is disjoint from the set $\{\pm\frac{1}{\sqrt{-a}}, 0, \infty\}$ of polars of $\phi^{\circ 2}$. This implies that the Taylor expansion of $\phi^{\circ 2}$ at x_0

$$\phi^{\circ 2}(x + x_0) = \phi^{\circ 2}(x_0) + \sum_{i=1}^{\infty} \alpha_i (x + x_0)^i,$$

is convergent on the ball $B(x_0, |x_0|_p/p)$. By Lemma 6, we have $\phi^{\circ 2}(B(x_0, |x_0|_p/p)) \subset B(\phi^{\circ 2}(x_0), |x_0|_p/p)$. The Newton polygon [**26**, p. 249] gives

$$|\alpha_i|_p \left(\frac{|x_0|_p}{p} \right)^i \leq \frac{|x_0|_p}{p}.$$

That is $|\alpha_i|_p \leq (p/|x_0|_p)^{i-1}$. □

LEMMA 8. *If $\sqrt{-a} \notin \mathbb{Q}_p$, then for each $x \in \mathbb{Q}_p \setminus \{0, \infty\}$, there exists a positive integer N such that*

$$\phi^{\circ n}(x) \in S(0, p^i) \cup S(0, p^{-i}), \quad \forall n \geq N$$

for some $0 \leq i \leq \lfloor v_p(a)/2 \rfloor$.

PROOF. Note that $|\phi(x)|_p = |1/x|_p$, if $|x|_p < p^{-\lfloor v_p(a)/2 \rfloor}$. Thus we have

$$\phi(D(0, p^{-\lfloor v_p(a)/2 \rfloor - 1})) = \mathbb{P}^1(\mathbb{Q}_p) \setminus D(0, p^{\lfloor v_p(a)/2 \rfloor}).$$

It suffices to show that the statement holds for $x \in \mathbb{P}^1(\mathbb{Q}_p) \setminus D(0, p^{\lfloor v_p(a)/2 \rfloor})$. In fact, if $x \in \mathbb{P}^1(\mathbb{Q}_p) \setminus D(0, p^{\lfloor v_p(a)/2 \rfloor} + 1)$, one can check that

$$|\phi(x)|_p = |a|_p |x|_p.$$

So there exists an integer N such that $p^{-\lfloor v_p(a)/2 \rfloor} \leq |\phi^{\circ N}(x)|_p \leq p^{\lfloor v_p(a)/2 \rfloor}$. If $x \in S(0, p^{\lfloor v_p(a)/2 \rfloor})$, the conclusion is followed by Lemma 5. □

PROPOSITION 3. *Assume that $\sqrt{-a} \notin \mathbb{Q}_p$. Then $\mathbb{Q}_p \setminus \{0\}$ is ϕ-invariant and the subsystem $(\mathbb{Q}_p \setminus \{0\}, \phi)$ can be decomposed into*

$$\mathbb{P}^1(\mathbb{Q}_p) = \mathcal{P} \bigsqcup \mathcal{M} \bigsqcup \mathcal{B}$$

where $\mathcal{P}$ is the finite set consisting of all periodic points of ϕ, $\mathcal{M} = \bigsqcup_i \mathcal{M}_i$ is the union of all (at most countably many) clopen invariant sets such that each $\mathcal{M}_i$ is a finite union of disks and each subsystem $\phi : \mathcal{M}_i \to \mathcal{M}_i$ is minimal, and each point in $\mathcal{B}$ lies in the attracting basin of a periodic orbit or of a minimal subsystem.

PROOF. By Lemmas 5 and 8, it suffices to show that the subsystem

$$(S(0, p^i), \phi^{\circ 2})$$

has a minimal decomposition as stated in the proposition for all $-\lfloor v_p(a)/2 \rfloor \leq i \leq \lfloor v_p(a)/2 \rfloor$. Note that $\phi^{\circ 2}$ is equicontinuous on $S(0, p^i)$ and

$$S(0, p^i) = \bigcup_{j=1}^{p-1} D(jp^{-i}, p^{i-1}).$$

Thus for each disk $D(jp^{-i}, p^{i-1})$ in $S(0, p^i)$, its image $\phi^{\circ 2}(D(jp^{-i}, p^{i-1}))$ is still a disk in $S(0, p^i)$. Hence $\phi^{\circ 2}$ induces a map $\overline{\phi^{\circ 2}}$ from the set $\{D(jp^{-i}, p^{i-1}) : j = \{1, 2, \cdots, p-1\}\}$ of disks to itself:

$$\overline{\phi^{\circ 2}}(D(jp^{-i}, p^{i-1})) := \phi^{\circ 2}(D(jp^{-i}, p^{i-1})).$$

Assume that $(D(x_1 p^{-i}, p^{i-1}), D(x_2 p^{-i}, p^{i-1}), \cdots, D(x_k p^{-i}, p^{i-1}))$ is a k-cycle of $\overline{\phi^{\circ 2}}$, i.e.

$$\begin{aligned}
\overline{\phi^{\circ 2}}(D(x_1 p^{-i}, p^{i-1})) &= D(x_2 p^{-i}, p^{i-1}) \\
\overline{\phi^{\circ 2}}(D(x_2 p^{-i}, p^{i-1})) &= D(x_3 p^{-i}, p^{i-1}), \\
&\cdots \\
\overline{\phi^{\circ 2}}(D(x_k p^{-i}, p^{i-1})) &= D(x_1 p^{-i}, p^{i-1}).
\end{aligned}$$

Then $\phi^{\circ 2k}$ is a transformation on $D(x_1 p^{-i}, p^{i-1})$. We thus study the dynamical system $(D(x_1 p^{-i}, p^{i-1}), \phi^{\circ 2k})$. By Lemma 8, the Taylor expansion of the map

$$\phi^{\circ 2k} : D(x_1 p^{-i}, p^{i-1}) \to D(x_1 p^{-i}, p^{i-1})$$

can be written as

$$\phi^{\circ 2k}(x + x_1 p^{-i}) = \phi^{\circ 2k}(x_1 p^{-i}) + \sum_{j=1}^{\infty} \alpha_j (x + x_1 p^{-i})^j,$$

with $\alpha_j \in \mathbb{Q}_p$ such that

$$|\alpha_j|_p \leq 1/p^{(i-1)(j-1)}. \tag{3.2}$$

Thus one can check that the dynamical system $(D(x_1 p^{-i}, p^{i-1}), \phi^{\circ 2k})$ is conjugate to a dynamical system on $\mathbb{Z}_p$, denoted by $(\mathbb{Z}_p, \psi)$, through the conjugacy

$$h : D(x_1 p^{-i}, p^{i-1}) \to D(0, 1), \text{ with } h(x) = p^{i-1}(x - x_1 p^{-i}).$$

In fact, the inequality (3.2) implies that ψ is a convergent series with integer coefficients. By Theorem 1.1 of [**16**], the system $(\mathbb{Z}_p, \psi)$ has a minimal decomposition. Hence, the conjugated dynamical system $(D(x_1 p^{-i}, p^{i-1}), \phi^{\circ 2k})$ has a corresponding minimal decomposition which implies that

$$(S(0, p^i), \phi^{\circ 2})$$

has a minimal decomposition as stated in the proposition for all $-\lfloor v_p(a)/2 \rfloor \leq i \leq \lfloor v_p(a)/2 \rfloor$.

□

3.2.2. *Case* $\sqrt{-a} \in \mathbb{Q}_p$.

LEMMA 9. *If* $|a| < 1$ *and* $\sqrt{-a} \in \mathbb{Q}_p$, *then*

$$\phi\left(D\Big(\pm\frac{1}{\sqrt{-a}}, \frac{1}{p\sqrt{|a|_p}}\Big)\right) = D\Big(0, \frac{\sqrt{|a|_p}}{p}\Big),$$

and for all $x, y \in D(\pm\frac{1}{\sqrt{-a}}, \frac{1}{p\sqrt{|a|_p}})$,

$$|\phi(x) - \phi(y)|_p = |a|_p|x-y|_p.$$

PROOF. Note that $\phi(\pm 1/\sqrt{-a}) = 0$. It suffices to show that

$$\forall x, y \in D(\pm\frac{1}{\sqrt{-a}}, \frac{1}{p\sqrt{|a|_p}}), \quad |\phi(x) - \phi(y)|_p = |a|_p|x-y|_p.$$

Without loss of generality, we assume that $x, y \in D(\frac{1}{\sqrt{-a}}, \frac{1}{p\sqrt{|a|_p}})$. By the same arguments in the proof of Lemma 3, there exist $x', y' \in D(0, \frac{\sqrt{|a|_p}}{p})$ such that

$$x = \frac{1}{\sqrt{-a} + x'}, \quad y = \frac{1}{\sqrt{-a} + y'},$$

and

$$\left|a - \frac{1}{xy}\right|_p = \left|2a - (x'+y')\sqrt{-a} - x'y'\right|_p.$$

Since $x', y' \in D(0, \frac{\sqrt{|a|_p}}{p})$, we immediately get

$$\left|a - \frac{1}{xy}\right|_p = |a|_p.$$

So

$$|\phi(x) - \phi(y)|_p = \left|a - \frac{1}{xy}\right|_p |x-y|_p = |a|_p|x-y|_p.$$

□

PROPOSITION 4. *If* $|a|_p < 1$ *and* $\sqrt{-a} \in \mathbb{Q}_p$, *then there exists a* ϕ*-invariant subset* $\mathcal{J}$ *such that* $(\mathcal{J}, \phi)$ *is topologically conjugate to a subshift of finite type* (Σ_A, σ), *where the transition matrix is*

$$A = \begin{pmatrix} 0 & 0 & 1 & 0 \\ 0 & 0 & 1 & 0 \\ 0 & 0 & 0 & 1 \\ 1 & 1 & 0 & 1 \end{pmatrix}.$$

The topological entropy of the system $(\mathbb{P}^1(\mathbb{Q}_p), \phi)$ *is larger than* $\log 1.69562...$.

PROOF. Let $D_1 = D(1/\sqrt{-a}, 1/p)$, $D_2 = D(-1/\sqrt{-a}, 1/p)$, $D_3 = D(0, |a|_p/p)$ and $D_4 = \mathbb{P}^1(\mathbb{Q}_p) \setminus D(0, |a|_p)$. By Lemma 9, the restricted maps $\phi : D_1 \to D_3$ and $\phi : D_2 \to D_3$ are both bijective. One can also check directly that $\phi : D_3 \to D_4$ and $\phi : D_4 \to \phi(D_4) = \mathbb{P}^1(\mathbb{Q}_p) \setminus D(0,1)$ are bijective.

Set $\Omega = \bigcup_{i=1}^{4} D_i$ and consider the restricted map $\phi : \Omega \to \mathbb{P}^1(\mathbb{Q}_p)$. Let $g : x \mapsto 1/(x-1)$ and $\psi := g \circ \phi \circ g^{-1}$. We study the map $\psi : g(\Omega) \to \mathbb{P}^1(\mathbb{Q}_p)$. Observe that $g(\Omega) \subset \mathbb{Z}_p$. Take

$$\mathcal{J}' = \bigcap_{i=0}^{\infty} (\psi^{-1})^{\circ i}(\Omega).$$

One can check $(\mathcal{J}', \psi)$ is a transitive weak repeller with incidence matrix

$$A = \begin{pmatrix} 0 & 0 & 1 & 0 \\ 0 & 0 & 1 & 0 \\ 0 & 0 & 0 & 1 \\ 1 & 1 & 0 & 1 \end{pmatrix}.$$

Thus, by Theorem 5, $(\mathcal{J}', \psi)$ is topologically conjugate to the subshift of finite type (Σ_A, σ). Take $\mathcal{J} = g^{-1}(\mathcal{J}')$. Since $\psi = g \circ \phi \circ g^{-1}$, we deduce that $(\mathcal{J}, \phi)$ is topologically conjugate to (Σ_A, σ).

The topological entropy of (Σ_A, σ) is $\log(1.69562...)$ where $1.69562...$ is the maximal eigenvalue of the matrix A. Since (Σ_A, σ) is topologically conjugate to a subsystem of $(\mathbb{P}^1(\mathbb{Q}_p), \phi)$, we confirm the last assertion of the proposition. □

We remark that even though a chaotic subsystem is well described, the detailed dynamical structure of ϕ on the whole space for the case $|a|_p < 1$ and $\sqrt{-a} \in \mathbb{Q}_p$ is far from clear. There may exist more complicated subsystems.

PROOF OF THEOREM 4. It follows directly from Propositions 3 and 4. □

References

[1] V. S. Anashin, *Uniformly distributed sequences of p-adic integers* (Russian, with Russian summary), Mat. Zametki **55** (1994), no. 2, 3–46, 188, DOI 10.1007/BF02113290; English transl., Math. Notes **55** (1994), no. 1-2, 109–133. MR1275316

[2] V. Anashin and A. Khrennikov, *Applied algebraic dynamics*, De Gruyter Expositions in Mathematics, vol. 49, Walter de Gruyter & Co., Berlin, 2009. MR2533085

[3] V. S. Anashin, A. Yu. Khrennikov, and E. I. Yurova, *Characterization of ergodic p-adic dynamical systems in terms of the van der Put basis* (Russian), Dokl. Akad. Nauk **438** (2011), no. 2, 151–153, DOI 10.1134/S1064562411030100; English transl., Dokl. Math. **83** (2011), no. 3, 306–308. MR2857398

[4] S. Albeverio, U. A. Rozikov, and I. A. Sattarov, *p-adic (2, 1)-rational dynamical systems*, J. Math. Anal. Appl. **398** (2013), no. 2, 553–566, DOI 10.1016/j.jmaa.2012.09.009. MR2990080

[5] M. Baker and R. Rumely, *Potential theory and dynamics on the Berkovich projective line*, Mathematical Surveys and Monographs, vol. 159, American Mathematical Society, Providence, RI, 2010. MR2599526

[6] R. L. Benedetto, *Hyperbolic maps in p-adic dynamics*, Ergodic Theory Dynam. Systems **21** (2001), no. 1, 1–11, DOI 10.1017/S0143385701001043. MR1826658

[7] Z. Coelho and W. Parry, *Ergodicity of p-adic multiplications and the distribution of Fibonacci numbers*, Topology, ergodic theory, real algebraic geometry, Amer. Math. Soc. Transl. Ser. 2, vol. 202, Amer. Math. Soc., Providence, RI, 2001, pp. 51–70, DOI 10.1090/trans2/202/06. MR1819181

[8] B. Dragovich, A. Khrennikov, and D. Mihajlović, *Linear fractional p-adic and adèlic dynamical systems*, Rep. Math. Phys. **60** (2007), no. 1, 55–68, DOI 10.1016/S0034-4877(07)80098-X. MR2355465

[9] V. Dremov, G. Shabat, and P. Vytnova, *On the chaotic properties of quadratic maps over non-Archimedean fields*, p-adic mathematical physics, AIP Conf. Proc., vol. 826, Amer. Inst. Phys., Melville, NY, 2006, pp. 43–54, DOI 10.1063/1.2193109. MR2258672

[10] F. Durand and F. Paccaut, *Minimal polynomial dynamics on the set of 3-adic integers*, Bull. Lond. Math. Soc. **41** (2009), no. 2, 302–314, DOI 10.1112/blms/bdp003. MR2496506

[11] A. Fan, S. Fan, L. Liao, and Y. Wang, *On minimal decomposition of p-adic homographic dynamical systems*, Adv. Math. **257** (2014), 92–135, DOI 10.1016/j.aim.2014.02.007. MR3187646

[12] A. Fan, S. Fan, L. Liao, and Y. Wang, *Minimality of p-adic rational maps with good reduction*, Discrete Contin. Dyn. Syst. **37** (2017), no. 6, 3161–3182, DOI 10.3934/dcds.2017135. MR3622078

[13] A.-H. Fan, M.-T. Li, J.-Y. Yao, and D. Zhou, *Strict ergodicity of affine p-adic dynamical systems on* $\mathbb{Z}_p$, Adv. Math. **214** (2007), no. 2, 666–700, DOI 10.1016/j.aim.2007.03.003. MR2349716

[14] A. Fan and L. Liao, *On minimal decomposition of p-adic polynomial dynamical systems*, Adv. Math. **228** (2011), no. 4, 2116–2144, DOI 10.1016/j.aim.2011.06.032. MR2836116

[15] A. Fan, L. Liao, Y. F. Wang, and D. Zhou, *p-adic repellers in* $\mathbb{Q}_p$ *are subshifts of finite type* (English, with English and French summaries), C. R. Math. Acad. Sci. Paris **344** (2007), no. 4, 219–224, DOI 10.1016/j.crma.2006.12.007. MR2292990

[16] S. Fan and L. Liao, *Dynamics of convergent power series on the integral ring of a finite extension of* $\mathbb{Q}_p$, J. Differential Equations **259** (2015), no. 4, 1628–1648, DOI 10.1016/j.jde.2015.03.042. MR3345863

[17] S. Fan and L. Liao, *Dynamics of the square mapping on the ring of p-adic integers*, Proc. Amer. Math. Soc. **144** (2016), no. 3, 1183–1196, DOI 10.1090/proc12777. MR3447671

[18] S. Fan and L. Liao, *Dynamical structures of Chebyshev polynomials on* $\mathbb{Z}_2$, J. Number Theory **169** (2016), 174–182, DOI 10.1016/j.jnt.2016.05.014. MR3531236

[19] L.-C. Hsia, *Closure of periodic points over a non-Archimedean field*, J. London Math. Soc. (2) **62** (2000), no. 3, 685–700, DOI 10.1112/S0024610700001447. MR1794277

[20] S. Jeong, *Toward the ergodicity of p-adic 1-Lipschitz functions represented by the van der Put series*, J. Number Theory **133** (2013), no. 9, 2874–2891, DOI 10.1016/j.jnt.2013.02.006. MR3057052

[21] M. Khamraev and F. Mukhamedov, *On a class of rational p-adic dynamical systems*, J. Math. Anal. Appl. **315** (2006), no. 1, 76–89, DOI 10.1016/j.jmaa.2005.08.041. MR2196531

[22] K. Mahler, *p-adic numbers and their functions*, 2nd ed., Cambridge Tracts in Mathematics, vol. 76, Cambridge University Press, Cambridge-New York, 1981. MR644483

[23] F. Mukhamedov and U. Rozikov, *On rational p-adic dynamical systems*, Methods Funct. Anal. Topology **10** (2004), no. 2, 21–31. MR2067949

[24] R. Oselies and H. Zieschang, *Ergodische Eigenschaften der Automorphismen p-adischer Zahlen* (German), Arch. Math. (Basel) **26** (1975), 144–153, DOI 10.1007/BF01229718. MR0369301

[25] J. Rivera-Letelier, *Dynamique des fonctions rationnelles sur des corps locaux* (French, with English and French summaries), Astérisque **287** (2003), xv, 147–230. Geometric methods in dynamics. II. MR2040006

[26] J. H. Silverman, *The arithmetic of dynamical systems*, Graduate Texts in Mathematics, vol. 241, Springer, New York, 2007. MR2316407

[27] I. A. Sattarov, *p-adic* $(3,2)$*-rational dynamical systems*, p-Adic Numbers Ultrametric Anal. Appl. **7** (2015), no. 1, 39–55, DOI 10.1134/S2070046615010045. MR3305243

[28] E. Thiran, D. Verstegen, and J. Weyers, *p-adic dynamics*, J. Statist. Phys. **54** (1989), no. 3-4, 893–913, DOI 10.1007/BF01019780. MR988564

[29] C. F. Woodcock and N. P. Smart, *p-adic chaos and random number generation*, Experiment. Math. **7** (1998), no. 4, 333–342. MR1678087

School of Mathematics and Statistics, Hubei Key Laboratory of Mathematical Sciences, Central China Normal University, Wuhan, 430079, China && Aix-Marseille Université, Centrale Marseille, CNRS, Institut de Mathématiques de Marseille, UMR7373, 39 Rue F. Joliot Curie 13453, Marseille, France

Email address: `slfan@mail.ccnu.edu.cn`

LAMA, UMR 8050, CNRS, Université Paris-Est Créteil Val de Marne, 61 Avenue du Général de Gaulle, 94010 Créteil Cedex, France

Email address: `lingmin.liao@u-pec.fr`

Contemporary Mathematics
Volume **704**, 2018
http://dx.doi.org/10.1090/conm/704/14169

Non-Archimedean Hahn-Banach theorems and injective Banach spaces

M. González and C. Perez-Garcia

Abstract. We consider several non-Archimedean Hahn-Banach theorems and study their associated injective Banach spaces. We pay special attention to the relation between them and their hereditary properties. We also discuss certain problems that arise naturally. The results obtained show that the situation differs substantially from the real or complex case.

Introduction

Injective objects play an important role in many mathematical theories. Recall that in a given category an object E is injective if for any pair of objects D, F such that D embeds into F, every morphism $D \longrightarrow E$ admits an extension $F \longrightarrow E$.

For the category in which the objects E are classical Banach spaces (i.e. Banach spaces over $\mathbb{R}$ or $\mathbb{C}$), the injectivity properties for such E are formulated through the validity of certain Hahn-Banach extensions theorems for continuous linear maps with values in E. Several structural properties and examples of these classes of injective Banach spaces, as well as the relations between them, were studied in [**1**]. For more on real or complex injective Banach spaces, see [**2**] and the references therein.

This paper deals with non-Archimedean injective Banach spaces, when $\mathbb{R}$ or $\mathbb{C}$ is replaced by a non-Archimedean valued field. The non-Archimedean couples of the classical injectivity properties were introduced in [**13**]. Section 2 contains the most relevant achievements of that paper (without including their proofs) as well as some new related results. Section 3 is devoted to studying the hereditary properties of those non-Archimedean injectivity properties.

After the study carried out in Section 3 several interesting problems arise in a natural way. A discussion about them, including some partial answers as well as their relations with other problems already existing in the non-Archimedean literature, can be found in Section 4. In particular, we prove that Problem 4.1 is equivalent to the problem posed in Section 5 of [**13**]; as a by-product we obtain some partial affirmative answers to this last problem.

2010 *Mathematics Subject Classification.* Primary 46S10; Secondary 46A22.

Key words and phrases. non-Archimedean injective Banach spaces, classical injective Banach spaces, Hahn-Banach Theorem.

The first and second authors were supported in part by Ministerio de Economía y Competitividad, Grant MTM2013-45643-C2-2-P.

Finally, in Section 5 we compare the results and problems presented in this paper with their classical versions. This comparison reveals sharp an interesting contrasts between the classical injectivity properties and their non-Archimedean counterparts.

1. Preliminaries

Throughout this paper $K := (K, |\,.\,|)$ will be a non-Archimedean non-trivially valued field that is complete with respect to the metric induced by the valuation $|\,.\,| : K \longrightarrow [0, \infty)$. Also,

IN SECTIONS 1–4 E, F WILL BE NON-ARCHIMEDEAN BANACH SPACES OVER K. THE NORMS ON E AND F ARE BOTH DENOTED BY $\|\cdot\|$.

For fundamentals on non-Archimedean Banach spaces we refer to [**14**] and [**16**]. Here we only fix some notations and recall some basic concepts that will be involved in the paper.

E is called *spherically complete* if every family of mutually intersecting balls in E has a common point. *A spherical completion* of E is a pair $< F, T >$ consisting of a spherically complete space F and a linear isometry $T : E \longrightarrow F$ such that F has no spherically complete proper subspace containing $T(E)$. Every E has a spherical completion and any two spherical completions of E are linearly isometric. For this reason it is customary to use the name *the spherical completion* of E.

E is said to be *of countable type* if it contains a countable set whose linear hull is dense in E. Banach spaces of countable type are the adequate non-Archimedean substitutes of the real or complex separable ones.

An *operator* from E to F is a continuous linear map $T : E \longrightarrow F$. The set $L(E, F)$ of all operators $T : E \longrightarrow F$ is a non-Archimedean Banach space with the norm $\|T\| := \inf\{M > 0 : \|T(x)\| \leq M\,\|x\| \text{ for all } x \in E\}$.

E is *isomorphic* to F if there exists a bijective operator $T : E \longrightarrow F$ that is a homeomorphism. If, in addition, T is an isometry we say that E is *isometric* to F (sometimes denoted by $E \equiv F$).

An operator $P : E \longrightarrow E$ is called a *projection* if $P^2 = P$. A closed subspace M of E is said to be *complemented* in E if there exists a projection $P : E \longrightarrow E$ with range M. If, in addition, $\|P\| \leq 1$, we say that M is *orthocomplemented* in E.

$E' := L(E, K)$ is the *dual* of E. The elements of E' are called *functionals*. F is called a *dual space* if there exists an E such that $E' = F$.

Let $E'' := L(E', K)$ be the *bidual* of E. The formula $j_E(x)(f) := f(x)$ ($x \in E$, $f \in E'$) defines the canonical operator $j_E : E \longrightarrow E''$. E is called *reflexive* if j_E is a bijective isometry.

The most popular non-Archimedean Banach spaces are the polar ones, see [**14**, Section 2.5]. The norm on E is *polar* if for every $x \in E$,

$$\|x\| = \sup\{|f(x)| : f \in E',\ \|f\| \leq 1\}.$$

In this case we say that E is *norm polar*. If $\{\|x\| : x \in E\}$ is contained in the closure in K of $\{|\lambda| : \lambda \in K\}$ (e.g. when the valuation of K is dense), then E is norm polar if and only if $j_E : E \longrightarrow j_E(E)$ is an isometry. E is called *polar* if it has an equivalent polar norm, that is, if $j_E : E \longrightarrow j_E(E)$ is an homeomorphism.

Every E over a spherically complete K and every E of countable type over any K is polar. However, when K is not spherically complete there exist infinite-dimensional Banach spaces over K with trivial duals (e.g. ℓ^∞/c_0).

For a zero-dimensional compact Hausdorff space X, $C(X) := \{f : X \longrightarrow K : f \text{ is continuous}\}$, is a norm polar Banach space, equipped with the norm $\|f\| := \max_{x\in X} |f(x)|$.

Let I be an index set and let $(E_i)_{i\in I}$ be a family of non-Archimedean Banach spaces. Its *Banach product*, $\prod_i E_i$, is the vector space of all $(x_i)_{i\in I}$ in the cartesian product of $(E_i)_{i\in I}$ such that $\sup_i \|x_i\| < \infty$, normed by $(x_i)_{i\in I} \longmapsto \sup_i \|x_i\|$. The closed subspace of all $(x_i)_{i\in I} \in \prod_i E_i$ with $\lim_i \|x_i\| = 0$ (i.e. for every $\varepsilon > 0$, the set $\{i \in I : \|x_i\| \geq \varepsilon\}$ is finite), is called the *orthogonal direct sum* of $(E_i)_{i\in I}$ and is denoted by $\bigoplus_i E_i$.

Now, let $s : I \longrightarrow (0,\infty)$ and for each $i \in I$ let K_i be the vector space K equipped with the norm $\lambda \longmapsto |\lambda|\, s_i$, where $s_i := s(i)$. We write $\ell^\infty(I,s) := \prod_i K_i$, $c_0(I,s) := \bigoplus_i K_i$, $\ell^\infty(I) := \ell^\infty(I,\mathbf{1})$ and $c_0(I) := c_0(I,\mathbf{1})$, where $\mathbf{1}$ is the map $I \longrightarrow (0,\infty)$ with $\mathbf{1}_i = 1$ for all $i \in I$. All of these spaces are polar; they are norm polar if and only if $\{s_i : i \in I\}$ is contained in the closure in K of $\{|\lambda| : \lambda \in K\}$. When $I = \mathbb{N}$ then $\ell^\infty(\mathbb{N})$ and $c_0(\mathbb{N})$ are the well-known spaces ℓ^∞ and c_0 of all bounded sequences in K and of all sequences in K tending to 0, respectively. Recall that $\ell^\infty(I) = c_0(I)'$ for all I, in particular, $\ell^\infty = c_0'$.

Let $Y := \{e_i : i \in I\}$ be a subset of non-zero vectors of E. Y is called:

- A *base* of E if every $x \in E$ can be written uniquely as $x = \sum_{i\in I} \lambda_i\, e_i$, $\lambda_i \in K$.
- An *orthogonal system* in E if whenever $i_1, \ldots, i_n$ are distinct elements of I, one verifies

  $$\|\lambda_{i_1} e_{i_1} + \ldots + \lambda_{i_n} e_{i_n}\| = \max_{1\leq k\leq n} \|\lambda_{i_k} e_{i_k}\| \quad \text{for all } n \in \mathbb{N},\ \lambda_{i_1}, \ldots, \lambda_{i_n} \in K.$$

  If, in addition, $\|e_i\| = 1$ for all $i \in I$, we say that Y is an *orthonormal system* in E.
- An *orthogonal* (*orthonormal*) *base* of E if Y is an orthogonal (orthonormal) system that is also a base of E.

$C(X)$ has an orthonormal base consisting of characteristic functions on clopen (i.e. closed and open) subsets of X. Also, the unit vectors of K^I form an orthonormal base of $c_0(I)$. However, if I is an infinite set, $\ell^\infty(I)$ has a(n) (orthonormal) base if and only if the valuation of K is discrete.

The Banach spaces $c_0(I)$ and $C(X)$ are "the models" of non-Archimedean Banach spaces having a(n) (orthonormal) base, as we show in the next result.

Proposition 1.1.

(i) *The following are equivalent.*

 (i.α) *E has a base (resp. an orthonormal base).*

 (i.β) *E is isomorphic (resp. isometric) to a $c_0(I)$ space.*

 (i.γ) *E is isomorphic (resp. isometric) to a $C(X)$ space.*

(ii) *If E is of countable type then it has a base. E is isomorphic to K^n when E is n-dimensional ($n \in \mathbb{N}$), and isomorphic to c_0 when E is infinite-dimensional.*

(iii) *When the valuation of K is discrete E has a base; E has an orthonormal base if and only if $\{\|x\| : x \in E\} \subset \{|\lambda| : \lambda \in K\}$.*

PROOF. (i). By [**16**, Corollary 3.8] and its proof, if E has a base $\{e_i : i \in I\}$ then it is isomorphic to $c_0(I)$. The rest of (i.α) $\iff$ (i.β) is obvious. The equivalence (i.α) $\iff$ (i.γ) was proved in [**13**, Propositions 4.5, 4.6].

For (ii) use [**14**, Corollary 2.3.9].

The first part of (iii) follows from [**14**, Theorem 2.5.4]. The "only if" of the second part is trivial; for the "if" apply [**16**, Lemma 5.5 and Theorem 5.16] for the finite-dimensional and infinite-dimensional case, respectively. □

If E has a base $\{e_i : i \in I\}$, then all the bases of E have the same cardinality as I and every closed subspace of E has a base, whose cardinality is $\leq$ the cardinality of I.

2. Non-Archimedean injective Banach spaces

The following classes of non-Archimedean injective Banach spaces were considered in [**13**]. Here we use the same terminology as in [**1**] and [**2**], devoted to study injective Banach spaces over the real or complex field.

DEFINITION 2.1.
(i) E is said to be *injective* (resp. *universally countably injective*) if for every Banach space F and each subspace (resp. each subspace of countable type) D of F, every operator $t : D \longrightarrow E$ extends to an operator $T : F \longrightarrow E$. If some extension T exists with $\|T\| = \|t\|$, we say that E is 1-*injective* (resp. 1-*universally countably injective*).
(ii) E is said to be 1-*countably injective* if it satisfies the extension property in the above definition of 1-injective space, under the assumption that F is of countable type.

The non-Archimedean injective and 1-injective spaces of Definition 2.1(i) coincide with the weakly injective and injective ones defined in Chapter 4 of [**16**]. We refer to that chapter for the basic facts on non-Archimedean injectivity contained in the next proposition which will be used along the paper.

PROPOSITION 2.2.
(i) *E is 1-injective if and only if E is spherically complete.*
(ii) *For every sequence $E_1, E_2, \ldots$ of Banach spaces, $\prod_n E_n / \bigoplus_n E_n$ is 1-injective. In particular, ℓ^∞ / c_0 is 1-injective.*
(iii) *The map*

$$i_E : E \longrightarrow \prod_n E / \bigoplus_n E, \quad x \longmapsto \pi(x, x, \ldots, x, \ldots)$$

is a linear isometry of E into the 1-injective space $\prod_n E / \bigoplus_n E$ (by π we mean the canonical quotient map $\prod_n E \longrightarrow \prod_n E / \bigoplus_n E$).
(iv) *E is (1-)injective if and only if E is (ortho)complemented in any Banach space containing E.*
(v) *E is injective if and only if E is isomorphic to a 1-injective space.*

REMARKS 2.3.
1. Looking at Definition 2.1 it seems natural to think about the following injectivity property for E, which is the non-Archimedean counterpart of the classical separable injectivity, see Definition 5.1.

"For every Banach space of countable type F and each subspace D of F, every operator $D \longrightarrow E$ extends to an operator $F \longrightarrow E$".

But this property holds for every E, as closed subspaces of Banach spaces of countable type are complemented, [**14**, Theorem 2.3.13].

2. For any set I, $\ell^\infty(I)$ is 1-injective if and only if K is spherically complete, [**16**, Exercise 4.A].

3. Let I be an infinite set. Then $c_0(I)$ is 1-injective if and only if the valuation of K is discrete, [**16**, Exercise 4.B].

Observe that $c_0(I)$ cannot contain an isomorphic copy of $\ell^\infty(I)$. This is clear when the valuation of K is dense because $\ell^\infty(I)$ does not have a base; when the valuation of K is discrete this follows from the fact that $\ell^\infty(I)$ has a base with cardinality > cardinality of I, see [**11**].

More examples of the different classes of injective spaces considered in Definition 2.1 will be given along the paper.

The main results of [**13**] related to the injective spaces considered in Definition 2.1 are summarized in the next theorem.

THEOREM 2.4.

(i) ([**13**, Theorem 3.3]) *E is 1-injective $\Longleftrightarrow$ E is 1-(universally) countably injective.*

(ii) ([**13**, Theorem 4.3]) *Let $E := C(X)$ be an infinite-dimensional space. Then,*

$$\begin{aligned} E \text{ is 1-injective} &\Longleftrightarrow E \text{ is (universally countably) injective} \\ &\Longleftrightarrow \text{the valuation of } K \text{ is discrete.} \end{aligned}$$

(iii) ([**13**, Theorem 4.4]) *Let E be finite-dimensional. Then,*

$$\begin{aligned} E \text{ is 1-injective} &\Longleftrightarrow E \text{ is (universally countably) injective} \\ &\Longleftrightarrow K \text{ is spherically complete.} \end{aligned}$$

Therefore, for any non-Archimedean Banach space, the three 1-injectivity notions of Definition 2.1 coincide. Thus, we have the following diagram.

$$\begin{array}{ccccc} \text{1-injective} & \Longleftrightarrow & \text{1-universally countably injective} & \Longleftrightarrow & \text{1-countably injective} \\ \Downarrow & & & & \\ \text{injective} & \Longrightarrow & \text{universally countably injective} & & \end{array} \qquad \text{(I)}$$

Hence,

FROM NOW ON WE ONLY DEAL WITH 1-INJECTIVE AND (UNIVERSALLY COUNTABLY) INJECTIVE SPACES.

REMARKS 2.5.

1. If E is infinite–dimensional (resp. finite-dimensional) then E is 1-injective and isometric to a $C(X)$ space if and only if K is discretely valued (resp. spherically complete) and $\{\|x\| : x \in E\} \subset \{|\lambda| : \lambda \in K\}$, see Corollary 4.9 (resp. Remark 4.11.1) of [**13**].

2. From Proposition 1.1(i) and Theorem 2.4(ii) we obtain that for an infinite-dimensional Banach space with an orthonormal base i.e. for a $c_0(I)$ space with I an

infinite set, to have any of the injectivity properties of Definition 2.1 is equivalent to be K discretely valued.

3. There exist infinite-dimensional spaces E for which the converse of implication $\Downarrow$ of diagram (I) fails.

Indeed, when the valuation of K is discrete every Banach space over K is injective, [**16**, Corollary 4.14]. However, $c_0(\mathbb{N}, s)$, with $s_1, s_2, \ldots \in (0, \infty)$, $s_1 > s_2 > \ldots$ and $\lim_n s_n = 1$ is not 1-injective, see [**13**, Remark 3.5].

Examples of those E when the valuation of K is dense are the following. For each $n > 1$, let ℓ_n^∞ be the Banach space whose underlying vector space is ℓ^∞ and whose norm is

$$\|x\|_n := \max(\|x\|, n \operatorname{dist}(x, c_0)) \quad (x \in \ell^\infty),$$

where $\|\,.\,\|$ is the canonical supremum norm on ℓ^∞ and $\operatorname{dist}(x, c_0) := \inf\{\|x - y\| : y \in c_0\}$. Clearly $\|x\| \leq \|x\|_n \leq n\,\|x\|$ for all $x \in \ell^\infty$, so that ℓ_n^∞ is isomorphic to ℓ^∞. By Remark 2.3.2, ℓ_n^∞ is injective when K is spherically complete. But ℓ_n^∞ is not 1-injective unless K is discretely valued, see [**16**], p. 107.

4. The converse of implication $\Longrightarrow$ of diagram (I) will be discussed in Section 4.

The following characterizations of 1-injective spaces were crucial in [**13**] to get Theorem 2.4.

THEOREM 2.6. ([**13**, Theorem 3.4]) *The following are equivalent.*

(α) *E is 1-injective.*
(β) *The operator $\bigoplus_n E \longrightarrow E$, $(x_1, x_2, \ldots) \longmapsto \sum_n x_n$, extends to an operator $\prod_n E \longrightarrow E$ of norm ≤ 1.*
(γ) *For every subspace of countable type G of E, the operator $\bigoplus_n G \longrightarrow E$, $(x_1, x_2, \ldots) \longmapsto \sum_n x_n$, extends to an operator $\prod_n G \longrightarrow E$ of norm ≤ 1.*
(δ) *For every $s : \mathbb{N} \longrightarrow (0, \infty)$ and each subspace D of $c_0(\mathbb{N}, s)$, every operator $D \longrightarrow E$ of norm ≤ 1 extends to an operator $c_0(\mathbb{N}, s) \longrightarrow E$ of norm ≤ 1.*

In Theorem 2.8 we will analyze what happens when the condition "of norm ≤ 1" in (β) – (δ) is removed. We will show that the resulting properties in (β) and (γ) characterize the injective and universally countably injective spaces, respectively. On the other hand, although by Remark 2.3.1 this elimination in (δ) leads to a property that holds for every E, we will prove that by replacing there $c_0(\mathbb{N}, s)$ by a more adequate space, we get a new characterization of (universal countable) injectivity.

First, a basic lemma, which extends Corollary 4.3 of [**16**].

LEMMA 2.7. *Let E be a Banach space over a non-spherically complete K and such that every operator $c_0 \longrightarrow E$ extends to an operator $\ell^\infty \longrightarrow E$ (e.g. when E is universally countably injective). Then $E' = \{0\}$.*

PROOF. If $E' \neq \{0\}$ then there exist $x \in E$ and $g \in E'$ with $g(x) = 1$. Thus, the map $E \longrightarrow E$, $y \longmapsto g(y)\,x$ is a projection on E with range the one-dimensional subspace Kx, which is isomorphic to K via the map $\lambda\,x \longmapsto \lambda$ ($\lambda \in K$). It follows from the extension assumption on E that every continuous functional $c_0 (= \bigoplus_n K) \longrightarrow K$ extends to a continuous functional $\ell^\infty (= \prod_n K) \longrightarrow K$. By Theorem 2.6 K must be 1-injective i.e. spherically complete (Proposition 2.2(i)), a contradiction. $\square$

THEOREM 2.8.

(i) *The following are equivalent.*

(i.α) *E is injective.*

(i.β) *The operator $\bigoplus_n E \longrightarrow E$, $(x_1, x_2, \ldots) \longmapsto \sum_n x_n$, extends to an operator $\prod_n E \longrightarrow E$.*

If, in addition, $E' \neq \{0\}$, the above properties are equivalent to:

(i.γ) *For every set I, $s : I \longrightarrow (0, \infty)$ and each subspace D of $\ell^\infty(I, s)$, every operator $D \longrightarrow E$ extends to an operator $\ell^\infty(I, s) \longrightarrow E$.*

(ii) *The following are equivalent.*

(ii.α) *E is universally countably injective.*

(ii.β) *For every subspace of countable type G of E, the operator $\bigoplus_n G \longrightarrow E$, $(x_1, x_2, \ldots) \longmapsto \sum_n x_n$, extends to an operator $\prod_n G \longrightarrow E$.*

(ii.γ) *For every set I, $s : I \longrightarrow (0, \infty)$ and each subspace of countable type D of $\ell^\infty(I, s)$, every operator $D \longrightarrow E$ extends to an operator $\ell^\infty(I, s) \longrightarrow E$.*

PROOF. (i). Clearly (i.α) $\Longrightarrow$ (i.β). Conversely, suppose that (i.β) is true and let us see that (i.α) holds. Let F be a Banach space, D a subspace of F, $t : D \longrightarrow E$ an operator and $G := \overline{t(D)}$ the closure of $t(D)$ in E. Let $i_G : G \longrightarrow \prod_n G/\bigoplus_n G$ be the isometrical embedding of Proposition 2.2(iii) for the Banach space G. Since $\prod_n G/\bigoplus_n G$ is 1-injective, the operator $i_G \circ t : D \longrightarrow \prod_n G/\bigoplus_n G$ extends to an operator $\overline{i_G \circ t} : F \longrightarrow \prod_n G/\bigoplus_n G$.

Also, according to (i.β), there exists an operator $S : \prod_n G \longrightarrow E$ such that $S(x_1, x_2, \ldots) = \sum_n x_n$ for all $(x_1, x_2, \ldots) \in \bigoplus_n G$. Then, the map

$$Q : \prod_n G/\bigoplus_n G \longrightarrow E, \quad \pi(x_1, x_2, \ldots) \longmapsto x_1 + S(x_2 - x_1, x_3 - x_2, \ldots)$$

(where π is the canonical quotient map $\prod_n G \longrightarrow \prod_n G/\bigoplus_n G$) is a well-defined operator with $Q(i_G(x)) = x$ for all $x \in G$.

Finally, the operator $Q \circ (\overline{i_G \circ t}) : F \longrightarrow E$ extends t. Hence (i.α) holds.

Now, assume that $E' \neq \{0\}$. Clearly (i.α) $\Longrightarrow$ (i.γ). Let us prove (i.γ) $\Longrightarrow$ (i.β). From (i.γ) and Lemma 2.7 we have that K is spherically complete. Hence, every Banach space over K, in particular $\prod_n E$, is polar. Thus, $\prod_n E$ is linearly and homeomorphically embedded into some $\ell^\infty(I)$ ([**14**, Theorem 4.4.9]). We identify each element of $\prod_n E$ with its image by this embedding.

To get (i.β), let $t : \bigoplus_n E \longrightarrow E$, $(x_1, x_2, \ldots) \longmapsto \sum_n x_n$. Since $\bigoplus_n E \subset \prod_n E \subset \ell^\infty(I)$, it follows from (i.$\gamma$) that t extends to an operator $\ell^\infty(I) \longrightarrow E$. Therefore, the restriction of this operator to $\prod_n E$ gives the desired extension of t to $\prod_n E$.

(ii). Clearly (ii.α) $\Longrightarrow$ (ii.β), (ii.γ) (for the first, note that if G is of countable type then so is $\bigoplus_n G$).

For (ii.β) $\Longrightarrow$ (ii.α), proceed as in the proof of (i.β) $\Longrightarrow$ (i.α), observing that the subspaces D and G considered there are now of countable type.

Finally, to prove (ii.γ) $\Longrightarrow$ (ii.β), let G be a subspace of countable type of E. Then $\prod_n G$ is polar. Now, proceed as in the proof of (i.γ) $\Longrightarrow$ (i.β), just replacing $\prod_n E$ and $\bigoplus_n E$ by $\prod_n G$ and $\bigoplus_n G$, respectively. □

REMARKS 2.9.

1. From the proof of Theorem 2.8 we deduce that in (i.γ) and (ii.γ) of that theorem we can replace $\ell^\infty(I,s)$ by $\ell^\infty(I)$. However, in (δ) of Theorem 2.6, $c_0(\mathbb{N},s)$ can be replaced by c_0 only when the valuation of K is dense, see [**13**, Theorem 3.4, Remark 3.5].

2. Looking at Theorems 2.6 and 2.8, it is natural to ask whether properties (i) and (ii) of Theorem 2.8 are equivalent, that is, whether the converse of implication $\Longrightarrow$ of diagram (I) holds. This problem will be discussed in Section 4.

3. We also could ask whether the characterization of injective spaces given in Theorem 2.8(i.γ) is true when $E' = \{0\}$. Observe that a negative answer to this question would imply a negative answer for those E to the problem mentioned in the previous remark.

3. Hereditary properties

The main result of this section, Theorem 3.3, provides some hereditary properties for the non-Archimedean injectivity properties considered in Definition 2.1. The following lemma will be useful when dealing with Banach products.

LEMMA 3.1. *If E is injective (resp. universally countably injective) then there exists a $C > 1$ such that the following holds.*

$(*)^i_C$ *(resp. $(*)^u_C$): For every Banach space F and each subspace (resp. each subspace of countable type) D of F, every operator $t : D \longrightarrow E$ extends to an operator $T : F \longrightarrow E$ with $\|T\| \leq C\,\|t\|$.*

PROOF. Suppose first that E is injective and the conclusion is false; we will arrive at a contradiction. Let $\varrho \in K$ with $0 < |\varrho| < 1$. Let $C_1, C_2, \ldots$ be a sequence in $(1,\infty)$ with $\lim_n C_n = \infty$. For each $n \in \mathbb{N}$, $(*)^i_{C_n}$ fails, so that there exist a Banach space F_n, a subspace D_n of F_n and an operator $t_n : D_n \longrightarrow E$ with $|\varrho| \leq \|t_n\| \leq 1$ such that,

$$\text{if } T_n : F_n \longrightarrow E \text{ is an operator extending } t_n,\ \lim_n \|T_n\| = \infty. \tag{3.1}$$

Now,

$$t : \bigoplus_n D_n \longrightarrow E, \quad (d_1, d_2, \ldots) \longmapsto \sum_n t_n(d_n),$$

is a well-defined operator with $\|t\| \leq 1$. By injectivity of E, t extends to an operator $T : \bigoplus_n F_n \longrightarrow E$. Then, for each $n \in \mathbb{N}$, $T_n := T \circ i_n : F_n \longrightarrow E$ (where $i_n : F_n \longrightarrow \bigoplus_n F_n$ is the canonical embedding) is an extension of t_n. Also, $\sup_n \|T_n\| \leq \|T\|$, which contradicts (3.1).

When E is universally countably injective proceed as above, just noting that now the subspaces D_n of F_n ($n \in \mathbb{N}$) and $\bigoplus_n D_n$ of $\bigoplus_n F_n$ are of countable type. □

REMARKS 3.2.

1. The Banach spaces satisfying $(*)^i_C$ coincide with the $\frac{1}{C}$-injective ones defined in Chapter 4 of [**16**].

2. An alternative (more involved) proof of Lemma 3.1 for injective spaces can be found in [**16**, Theorem 4.12].

3. It follows from [**16**, Corollary 4.14] that if the valuation of K is discrete then every Banach space E over K satisfies $(*)^i_C$ for all $C > 1$; in particular E is injective.

THEOREM 3.3.

(i) (Isometries and Isomorphisms)

(i.1) *If E is isometric to F, then*

$$E \text{ is } 1\text{-injective} \iff F \text{ is } 1\text{-injective.}$$

(i.2) *If E is isomorphic to F, then*

$$E \text{ is (universally countably) injective} \iff$$
$$F \text{ is (universally countably) injective.}$$

(ii) (Subspaces) *Let M be a closed subspace of E.*

(ii.1) *If E is (1-)injective, then*

$$M \text{ is (1-)injective} \iff M \text{ is (ortho)complemented.}$$

(ii.2) *If E is universally countably injective, then*

$$M \text{ is universally countably injective} \Longleftarrow M \text{ is complemented.}$$

(iii) (Quotients) *If E is (1-)injective (resp. universally countably injective) and M is a closed subspace of E, then E/M is (1-)injective (resp. universally countably injective).*

(iv) (Banach Products) *Let $(E_i)_{i\in I}$ be a family of non-Archimedean Banach spaces. Then we have the following.*

(iv.1) *$\prod_i E_i$ is 1-injective $\iff$ E_i is 1-injective for all i.*

(iv.2) *$\prod_i E_i$ is injective (resp. universally countably injective) $\iff$ there exists a $C > 1$ such that each E_i satisfies $(*)^i_C$ (resp. $(*)^u_C$).*

(v) (Orthogonal Direct Sums) *If $(E_i)_{i\in I}$ is a family of non-Archimedean Banach spaces, then*

$$\bigoplus_i E_i \text{ is (1-)injective (resp. universally countably injective)} \Longrightarrow$$
$$E_i \text{ is (1-)injective (resp. universally countably injective) for all } i.$$

PROOF. It is straightforward to prove (i) as well as the "$\Longleftarrow$" of (ii). The converses "$\Longrightarrow$" of (ii.1) are direct consequences of Proposition 2.2(iv).

(iii). For 1-injective spaces the conclusion follows from Proposition 2.2(i) and the preservation of spherical completeness by taking quotients ([**16**, Theorem 4.2]). To get the conclusion for injective spaces apply the above and Proposition 2.2(v).

Next, we prove that if E is universally countably injective then so is E/M. For that, let F be a Banach space, D a subspace of countable type of F and $t : D \longrightarrow E/M$ an operator. By [**14**, Theorem 2.3.7], D has a base $\{d_1, d_2, \ldots\}$ such that

$$\|\sum_n \lambda_n d_n\| \geq \frac{1}{2} \max_n \|\lambda_n d_n\| \quad \text{for all } \textstyle\sum_n \lambda_n\, d_n \in D.$$

If $\pi : E \longrightarrow E/M$ is the canonical quotient map, then for each $n \in \mathbb{N}$ there exists $x_n \in E$ with $\pi(x_n) = t(d_n)$ and $\|x_n\| \leq 2\,\|t(d_n)\|$. The formula

$$h(\sum_n \lambda_n\, d_n) := \sum_n \lambda_n\, x_n \quad (\lambda_1, \lambda_2, \ldots \in K),$$

defines an operator $h : D \longrightarrow E$ with $\|h\| \leq 4\,\|t\|$ and such that $\pi \circ h = t$. Since E is universally countably injective, h can be extended to an operator $\overline{h} : F \longrightarrow E$. Then, $\pi \circ \overline{h} : F \longrightarrow E/M$ extends t.

(iv). From Proposition 2.2(i) and the behaviour of spherical completeness by taking Banach products ([**16**, Exercise 4.A]) we obtain (iv.1).

Next, let us prove (iv.2). By Lemma 3.1, $\prod_i E_i$ is injective (resp. universally countably injective) if and only if there exists a $C > 1$ such that $\prod_i E_i$ satisfies $(*)^i_C$ (resp. $(*)^u_C$). Also, it is easily verified that

$$\prod_i E_i \text{ satisfies } (*)^i_C \text{ (resp. } (*)^u_C) \iff \text{each } E_i \text{ satisfies } (*)^i_C \text{ (resp. } (*)^u_C).$$

Then, the conclusion of (iv.2) follows immediately.

(v). It is a direct consequence of (i) and (ii) as each E_i is isometric to an orthocomplemented subspace of $\bigoplus_i E_i$. □

REMARKS 3.4.

1. From Theorem 2.4(iii) we have that, for any finite-dimensional 1-injective space, all its isomorphic spaces and all its subspaces are again 1-injective.

2. For infinite-dimensional spaces the above facts are false. Indeed, let E be an infinite-dimensional injective space that is not 1-injective (see Remark 2.5.3). By Proposition 2.2 such an E is isomorphic to a 1-injective space and is an, even complemented, subspace of its (1-injective) spherical completion.

The question whether the converse of "$\Longleftarrow$" of Theorem 3.3(ii.2) holds will be discussed in the next section.

3. Let K be discretely valued. From Remark 3.2.3 we have that every Banach space over K is (universally countably) injective. But 1-injectivity may fail. For an example, take $E := c_0(\mathbb{N}, s)$ as in Remark 2.5.3. Recall that $E = \bigoplus_n E_n$, where each E_n is the vector space K equipped with the norm $\lambda \longmapsto |\lambda|\, s_n$, which is 1-injective by Theorem 2.4(iii). Hence, this example also shows that 1-injective spaces are not preserved by taking orthogonal direct sums.

4. Let K be densely valued. There exist infinite-dimensional 1-injective spaces having (non-complemented) closed subspaces that are not even universally countably injective. For an example, let $E := \ell^\infty/c_0$ which is 1-injective (Proposition 2.2(ii)) and let M be an infinite-dimensional closed subspace of countable type of E. By Proposition 1.1 M is isomorphic to a $C(X)$ space and, since by Theorem 2.4(ii) $C(X)$ is not universally countably injective, neither is M.

For orthogonal direct sums the situation is very restrictive:

Let $(E_i)_{i\in I}$ be an infinite family of Banach spaces over K such that one of the E_i has non-trivial dual. Then $\bigoplus_i E_i$ is not universally countably injective.

PROOF. Suppose $\bigoplus_i E_i$ is universally countably injective; we will arrive at a contradiction. Since this orthogonal direct sum has non-trivial dual, from Lemma 2.7 we deduce that K is spherically complete. Let $\varrho \in K$ with $0 < |\varrho| < 1$. For each $i \in I$ select $x_i \in E_i$ with $\varrho \leq \|x_i\| \leq 1$. The one-dimensional subspace Kx_i of each E_i is 1-injective (Theorem 2.4(iii)) and by Theorem 3.3(ii.1) orthocomplemented in E_i i.e. there exists a projection $P_i : E_i \longrightarrow E_i$ with $\|P_i\| \leq 1$ and range Kx_i. Thus, $\bigoplus_i Kx_i$ is also orthocomplemented in $\bigoplus_i E_i$ via the projection

$$\bigoplus_i E_i \longrightarrow \bigoplus_i Kx_i, \quad (y_i)_{i\in I} \longmapsto (P_i(y_i))_{i\in I}.$$

Hence $\bigoplus_i K x_i$ is universally countably injective, and then also $c_0(I)$, as it is isomorphic to $\bigoplus_i K x_i$. It follows from Proposition 1.1(i) and Theorem 2.4(ii) that the valuation of K is discrete, a contradiction. $\square$

5. If I is a finite set, it is easily seen that $\bigoplus_i E_i$ $(= \prod_i E_i)$ is (universally countably) injective $\Longleftrightarrow$ E_i is (universally countably) injective for all $i \in I$.

The question whether infinite Banach products of (universally countably) injective spaces are again (universally countably) injective, will be discussed in the next section.

4. Some open problems and some partial answers

In this section we deal with some open problems resulting from Theorem 3.3 and their relations with other problems already existing in the non-Archimedean literature.

We will denote by $\mathcal{C}_{na}$ the class of all non-Archimedean Banach spaces over K and by $\mathcal{C}$ a subclass of $\mathcal{C}_{na}$ i.e. a family of non-Archimedean Banach spaces satisfying a certain property (e.g. to have a base, to be a dual space, ...).

4.1. Subspaces of universally countably injective spaces. Looking at Theorem 3.3(ii) the following arises naturally.

PROBLEM 4.1. ***Is every universally countably injective closed subspace M of a universally countably injective space complemented in it?***

We will see in Corollary 4.4 that the above problem is equivalent to:

PROBLEM 4.2. ([**13**, Section 5]) ***Is every universally countably injective space E injective?***

First a basic result.

PROPOSITION 4.3. *Let $\mathcal{C} \subset \mathcal{C}_{na}$. Then the following are equivalent.*

(α) *Problem 4.1 has an affirmative answer if $M \in \mathcal{C}$.*

(β) *Problem 4.2 has an affirmative answer if $E \in \mathcal{C}$.*

PROOF. $(\alpha) \Longrightarrow (\beta)$. Let $E \in \mathcal{C}$ be universally countably injective. By assumption E is complemented in its spherical completion, which is 1-injective (Proposition 2.2(i)). Thus, E is injective, by Theorem 3.3(ii.1).

$(\beta) \Longrightarrow (\alpha)$. Let E be universally countably injective and $M \in \mathcal{C}$ an universally countably injective closed subspace of E. By assumption M is injective, and by Theorem 3.3(ii.1), M is complemented in E. $\square$

By taking $\mathcal{C} = \mathcal{C}_{na}$ in Proposition 4.3 we deduce the announced result.

COROLLARY 4.4. *Problem 4.1 is equivalent to Problem 4.2.*

Our next purpose is to find some classes $\mathcal{C} \subset \mathcal{C}_{na}$ such that Problem 4.2 has an affirmative answer when $E \in \mathcal{C}$. As a by-product, thanks to Proposition 4.3, we will obtain an affirmative answer to Problem 4.1 when $M \in \mathcal{C}$, see Theorem 4.12 for a summary.

It follows from Theorem 2.4 that Problem 4.2 has an affirmative answer when either E is finite-dimensional or has an orthonormal base (in this last case, use also Proposition 1.1(i)). This can be extended to Banach spaces with a base, as we show in the next result.

PROPOSITION 4.5. *Let E be infinite-dimensional. Then,*

$$E \text{ has a base and is (universally countably) injective} \iff \text{the valuation of } K \text{ is discrete.}$$

PROOF. The "$\Longleftarrow$" follows from Proposition 1.1(iii) and Remark 3.2.3. The "$\Longrightarrow$" from Theorem 2.4(ii) and the fact that every Banach space with a base is isomorphic to a $C(X)$ space (Proposition 1.1(i)). □

REMARKS 4.6.

1. Since every Banach space E of countable type has a base, from the above result and comments we deduce that Problem 4.2 has an affirmative answer for E.

2. There exist infinite-dimensional Banach spaces of countable type over any discretely valued K having an orthogonal base which are not 1-injective, see the first part of Remark 2.5.3. Observe that by Remark 2.5.2 any such space cannot have an orthonormal base.

The infinite-dimensional dual spaces $\ell^\infty(I)$ ($\equiv c_0(I)'$) are some of the most popular examples of Banach spaces without a base when K is not discretely valued. The next result provides an affirmative answer for them to Problem 4.2; even more:

PROPOSITION 4.7. *Let E be a non-trivial dual space (e.g. $E = \ell^\infty(I)$). Then the following are equivalent.*

(α) *E is 1-injective.*
(β) *E is (universally countably) injective.*
(γ) *K is spherically complete.*

PROOF. (α) $\Longrightarrow$ (β) is obvious and (β) $\Longrightarrow$ (γ) is a consequence of Lemma 2.7, as $E' \neq \{0\}$. For (γ) $\Longrightarrow$ (α), use [**16**, Corollary 4.5] to get that E is spherically complete; then apply Proposition 2.2(i). □

REMARK 4.8. If, in addition, E is reflexive then (α) – (γ) of Proposition 4.7 are equivalent to

(δ) *K is spherically complete and E is finite-dimensional.*

Indeed, (γ) $\Longrightarrow$ (δ) follows from the fact that a reflexive E over a spherically complete K is finite-dimensional, [**14**, Corollary 7.4.20]; (δ) $\Longrightarrow$ (α) follows from Theorem 2.4(iii).

We also have the following extension of Proposition 4.7.

THEOREM 4.9. *Let E be a Banach space with non-trivial dual and such that every subspace of countable type of E is contained in an isometric copy of a dual space inside E. Then properties (α) – (γ) of Proposition 4.7 are equivalent for E.*

PROOF. For (α) $\Longrightarrow$ (β) $\Longrightarrow$ (γ) proceed as in the proof of Proposition 4.7.

To prove (γ) $\Longrightarrow$ (α), suppose that K is spherically complete and let us see that E is 1-injective. By Theorem 2.4(i) it is equivalent to see that E is 1-universally countably injective.

Let F be a Banach space, D a subspace of countable type of F and $t : D \longrightarrow E$ an operator. By assumption $t(D)$, which is a subspace of countable type of E, is contained in an isometric copy of a dual space $M \subset E$. Let $t_M : D \longrightarrow M$, $d \in D \longmapsto t(d) \in M$. Since by Proposition 4.7 M is 1-injective, t_M extends to an operator $\bar{t}_M : F \longrightarrow M$ with $\|\bar{t}_M\| = \|t_M\| = \|t\|$. Let $i : M \longrightarrow E$ be the canonical inclusion. Then $i \circ \bar{t}_M : F \longrightarrow E$ is an extension of t preserving its norm. □

Now, we give some examples of Banach spaces for which Theorem 4.9 applies. They are the non-Archimedean couple of the classical Pełczyński-Sudakov spaces, which were introduced in [**12**]. For an uncountable set I, let $\ell_c^\infty(I)$ be the vector space of all bounded functions $f : I \longrightarrow K$ that are countably supported (i.e. $\{i \in I : f(i) \neq 0\}$ is countable), equipped with the inherited supremum norm of $\ell^\infty(I)$. $\ell_c^\infty(I)$ is a closed subspace of $\ell^\infty(I)$. Also, it is straightforward to check that every subspace of countable type of $\ell_c^\infty(I)$ is contained in an isometric copy of ℓ^∞ $(\equiv c_0')$ inside $\ell_c^\infty(I)$. Therefore, from Theorem 4.9 we derive the following.

COROLLARY 4.10. *Properties* $(\alpha) - (\gamma)$ *of Proposition* 4.7 *are equivalent for* $E := \ell_c^\infty(I)$.

REMARKS 4.11.

1. Let E be a Banach space with non-trivial dual and such that every subspace of countable type of E is contained in an isomorphic (non-necessarily isometric) copy of a dual space inside E.

With the same procedure as in the proof of Theorem 4.9 one can see that

$$E \text{ is universally countably injective} \Longleftrightarrow K \text{ is spherically complete.}$$

However, spherical completeness of K does imply that E is 1-injective. In fact, there exist non-1-injective spaces over such K which are isomorphic to the dual space ℓ^∞, see the second part of Remark 2.5.3.

On the other hand, it is unknown whether spherical completeness of K implies that E is injective i.e. whether Problem 4.2 has an affirmative answer for these E.

2. When $E' = \{0\}$ the conclusions of Theorem 4.9 and Remark 4.11.1 are false. Indeed, by Proposition 2.2(ii), $E := \ell^\infty/c_0$ is always 1-injective. Then E has trivial dual when K is not spherically complete, by Lemma 2.7. Also, for those K, every closed subspace of countable type of E is reflexive, [**14**, Corollary 7.4.30], hence isometric to a dual space.

3. The Banach spaces $\ell_c^\infty(I)$ of Corollary 4.10 are "new" examples of 1-injective spaces when K is spherically complete. By "new" we mean that it is the first time we find them in the non-Archimedean context.

The next theorem summarizes the partial affirmative answers to Problems 4.1 and 4.2 that are deduced from previous results.

THEOREM 4.12. *Let $\mathcal{C}_1$ be the class of all $E \in \mathcal{C}_{na}$ that are isomorphic to $C(X)$ spaces (i.e. of all $E \in \mathcal{C}_{na}$ having a base, Proposition* 1.1(*i*)). *Let $\mathcal{C}_2$ be the class of all $E \in \mathcal{C}_{na}$ with non-trivial dual such that every subspace of countable type of E is contained in an isometric copy of a dual space inside E. Then we have the following.*

(i) *Problem* 4.1 *has an affirmative answer if $M \in \mathcal{C}_1 \cup \mathcal{C}_2$.*

(ii) *Problem* 4.2 *has an affirmative answer if $E \in \mathcal{C}_1 \cup \mathcal{C}_2$.*

PROOF. First of all note that (i) follows from (ii) and Proposition 4.3.

To get the conclusion of (ii) if $E \in \mathcal{C}_1$ and is finite-dimensional apply Theorem 2.4(iii); for infinite-dimensional $E \in \mathcal{C}_1$ apply Proposition 4.5. Finally, if $E \in \mathcal{C}_2$ the conclusion follows from Theorem 4.9. □

4.2. Banach products of injective spaces. For a family $(E_i)_{i\in I}$ of non-Archimedean Banach spaces Theorem 3.3(iv) says that

$$\prod_i E_i \text{ is 1-injective} \iff \text{each } E_i \text{ is 1-injective,} \tag{4.1}$$

and

$$\prod_i E_i \text{ is (universally countably) injective} \implies$$
$$\text{each } E_i \text{ is (universally countably) injective.}$$

If I is a finite set it is easily seen that the previous implication becomes an equivalence. But if I is infinite it is unknown whether the converse of that implication holds, which leads to:

PROBLEM 4.13. ***Let $(E_i)_{i\in I}$ be an infinite family of non-Archimedean Banach spaces. If each E_i is (universally countably) injective then so is $\prod_i E_i$?***

We will see in Corollary 4.16 that if Problem 4.13 has a negative answer then also has the following "old problem" in the theory of non-Archimedean Banach spaces, which was posed in 1974 by L. Gruson and M. van der Put, [**6**] (see also [**16**], p. 107).

PROBLEM 4.14. ([**6**, Problem 2.21(i)]) ***Does every injective space E satisfy $(*)^i_C$ for all $C > 1$?***

For that we need a preliminary result.

PROPOSITION 4.15. *Let $\mathcal{C} \subset \mathcal{C}_{na}$. Then the following are equivalent.*

(α) *Problem 4.13 has an affirmative answer if $E_i \in \mathcal{C}$ for all $i \in I$.*
(β) *There exists a $C > 1$ such that every injective (resp. universally countably injective) space $E \in \mathcal{C}$ satisfies $(*)^i_C$ (resp. $(*)^u_C$).*

PROOF. $(\beta) \implies (\alpha)$ follows from Theorem 3.3(iv.2). We will prove $(\alpha) \implies (\beta)$ for injective spaces; the same reasoning works for universally countably injective. Suppose (α) is true and (β) is false; we will arrive at a contradiction. For each $n \in \mathbb{N}$ and for $C_n := n+1$ there exists an injective space $E_n \in \mathcal{C}$ for which $(*)^i_{C_n}$ fails. From (α) we have that $\prod_n E_n$ is injective, and by Theorem 3.3(iv.2) we can find a $C_0 > 1$ such that each E_n satisfies $(*)^i_{C_0}$. Choose $m \in \mathbb{N}$ with $C_0 < C_m$. Then E_m satisfies $(*)^i_{C_m}$, a contradiction. □

By taking $\mathcal{C} = \mathcal{C}_{na}$ in Proposition 4.15 we deduce the announced result.

COROLLARY 4.16. *If Problem 4.13 has a negative answer then also has Problem 4.14.*

Now, the couple of Theorem 4.12 reads as follows.

THEOREM 4.17. *Let $\mathcal{C}_1$ and $\mathcal{C}_2$ be the subclasses of $\mathcal{C}_{na}$ considered in Theorem 4.12. Then Problem 4.13 (resp. Problem 4.14) has an affirmative answer when either*

(i) *there is an $i \in I$ such that E_i is infinite-dimensional and $E_i \in \mathcal{C}_1$ (resp. E is infinite-dimensional and $E \in \mathcal{C}_1$), or*
(ii) *each $E_i \in \mathcal{C}_2$ (resp. $E \in \mathcal{C}_2$).*

PROOF. (i) follows from Remark 3.2.3 and Proposition 4.5. To get the conclusion of (ii) related to Problem 4.13 use Theorem 4.9 and (4.1); for Problem 4.14 we only need Theorem 4.9. □

REMARKS 4.18.

1. Here we give some examples of non-Archimedean Banach spaces that belong to $\mathcal{C}_1 \cup \mathcal{C}_2$ which, according to Theorems 4.12 and 4.17, lead to affirmative answers to all the problems posed in Section 4.

From Proposition 1.1(i) we have: - $E \in \mathcal{C}_1$ if and only if E is isomorphic to a $c_0(I)$ space; - every space of countable type is in $\mathcal{C}_1$; - when the valuation of K is discrete $\mathcal{C}_1 = \mathcal{C}_{na}$.

Also, if E is either a dual space or a Pełczyński-Sudakov space $\ell_c^\infty(I)$ then $E \in \mathcal{C}_2$.

2. Clearly every finite-dimensional E is in $\mathcal{C}_1 \cap \mathcal{C}_2$ as it is a reflexive space with a base. Even more, if K is densely valued and spherically complete, then

$$\mathcal{C}_1 \cap \mathcal{C}_2 = \{E \in \mathcal{C}_{na} : E \text{ is finite-dimensional}\}.$$

To see that, suppose that there exists an infinite-dimensional $E \in \mathcal{C}_1 \cap \mathcal{C}_2$. Then E has a base and also an infinite-dimensional closed subspace M isometric to a dual space. Thus, M is 1-injective (Proposition 4.7) and has a base. By Proposition 4.5 the valuation of K is discrete, a contradiction.

However, if K is not spherically complete, every Banach space E of countable type is in $\mathcal{C}_1 \cap \mathcal{C}_2$. Indeed, $E \in \mathcal{C}_1$ (see the previous remark) and, as E is reflexive ([**14**, Corollary 7.4.30]), $E \in \mathcal{C}_2$.

3. Let K be densely valued and spherically complete. In the same spirit as in [**16**], p. 107 an interesting case of Problem 4.14 is whether the injective space ℓ_n^∞ $(n > 1)$ considered in Remark 2.5.3 satisfies $(*)_C^i$ for all $C > 1$ (trivially it satisfies $(*)_C^i$ for all $C > n$). This fact leads naturally to the following interesting case of Problem 4.13:

Is $\prod_n \ell_n^\infty$ an (universally countably) injective space?

5. Comparison with the classical theory

By "classical theory" we mean Functional Analysis over $\mathbb{R}$ or $\mathbb{C}$. For fundamentals on classical Banach spaces we refer to [**10**]. All the Banach spaces considered in this section are over the real field $\mathbb{R}$. For injective-like complex Banach spaces, see e.g. Section 2.7.2 of [**2**]. Hence,

IN THE SEQUEL E, F WILL BE BANACH SPACES OVER $\mathbb{R}$.

We equip $\mathbb{R}^n$ $(n \in \mathbb{N})$ with its maximum norm. Also, for a compact Hausdorff space X, $C(X) := \{f : X \longrightarrow \mathbb{R} : f \text{ is continuous}\}$, equipped with the usual maximum norm. For a set I, by βI we mean the Stone-Čech compactification of I when it is equipped with the discrete topology.

The following notions of injective spaces are the classical versions of the non-Archimedean ones considered in Definition 2.1 and Remark 2.3.1.

DEFINITION 5.1.

(i) E is said to be *injective* (resp. *universally separably injective*) if for every Banach space F and each subspace (resp. each separable subspace) D of F, every operator $t : D \longrightarrow E$ extends to an operator $T : F \longrightarrow E$. If some extension T exists with $\|T\| = \|t\|$, we say that E is 1-*injective* (resp. 1-*universally separably injective*).

(ii) E is said to be (1-)*separably injective* if it satisfies the extension property in the above definition of (1-)injective space, under the assumption that F is separable.

Clearly we have

$$\begin{array}{ccccc} \text{1-injective} & \Longrightarrow & \text{1-universally separably injective} & \Longrightarrow & \text{1-separably injective} \\ \Downarrow & & \Downarrow & & \Downarrow \\ \text{injective} & \Longrightarrow & \text{universally separably injective} & \Longrightarrow & \text{separably injective} \end{array} \qquad \text{(II)}$$

An analysis of diagram (II) was carried out in [**1**]. For more on classical injective Banach spaces see [**2**] and the references therein.

We are going to compare the results of previous sections with their classical counterparts. This comparison reveals sharp and interesting contrasts between the Archimedean and non-Archimedean injectivity properties.

5.1. Basic facts. Firstly we say a few words about finite-dimensional injective spaces (compare Theorem 2.4(iii) and Remark 3.4.1). It is well-known that, for an n-dimensional space E ($n \in \mathbb{N}$), one has that E is 1-injective (1-(universally) separably injective) if and only if E is isometric to $\mathbb{R}^n$. Observe that this implies that every finite-dimensional space E is injective.

Now, let $E := \mathbb{R}^n$ and let $F := \ell^2_n$ i.e. the real vector space $\mathbb{R}^n$ with the norm induced by ℓ^2. Then, E is 1-injective, F is isomorphic to E but, by the above, F is not 1-separably injective when $n > 1$.

FROM NOW ON ALL THE REAL BANACH SPACES ARE ASSUMED TO BE INFINITE-DIMENSIONAL AND ALL THE SETS I ARE ASSUMED TO BE INFINITE.

• (Compare Remarks 2.3.1 and 4.6.1) According to a deep result of M. Zippin, [**18**], c_0 is the only separable Banach space that is separably injective. In particular, ℓ^2 is not separably injective.

Also, by [**1**, Proposition 3.5], separable Banach spaces cannot be universally separably injective, as they do not contain an isomorphic copy of ℓ^∞.

• (Compare Remark 2.3.2 and Proposition 4.7) A classical result of R.S. Philips, [**15**], says that the dual space $\ell^\infty(I)$ ($\equiv \ell^1(I)'$) is 1-injective. Among them, ℓ^∞ is the best known classical example of 1-injective space.

On the other hand, there exist injective dual Banach spaces that are not 1-separably injective. For an example, let $n \in \mathbb{N}$ and $E := \ell^2_n \oplus_\infty \ell^\infty$. Then E is isomorphic to ℓ^∞, hence injective. Also, E is isometric to the dual $(\ell^2_n \oplus_1 \ell^1)'$. But, if $n > 1$, E is not 1-separably injective; otherwise it would also be ℓ^2_n and we have seen that it is not possible (by $\oplus_\infty$ and $\oplus_1$ we mean a direct sum equipped with the $\|\,.\,\|_\infty$ and $\|\,.\,\|_1$ norm, respectively).

However, every separably injective dual space E is injective. Indeed, separable injectivity implies that E is an $\mathcal{L}_\infty$-space ([**1**, Proposition 3.5]) and then injective, by [**2**, Proposition 1.9].

• (Compare Remarks 2.3.3 and 2.5.2) By Sobczyk's theorem, $c_0(I)$ is separably injective (see [**3**] for a rather detailed survey). But $c_0(I)$ is not universally separably injective, [**1**], p. 201.

Also, a well-known classical result, proved by H.P. Rosenthal in [**17**], states that an injective space containing an isomorphic copy of $c_0(I)$ must also contain an isomorphic copy of $\ell^\infty(I)$.

• (Compare Proposition 2.2(i) and Remark 2.5.1) It was proved in [**19**, Theorem 2.1] that the following are equivalent.

(α) E is 1-injective.

(β) Every family of mutually intersecting closed balls in E has a common point.

(γ) E is isometric to a $C(X)$ space with X extremely disconnected (i.e. the closure of any open set in X is open).

Observe that βI is an extremely disconnected compact space and $C(\beta I)$ is isometric to $\ell^\infty(I)$, [**2**], p. 8.

Although $c_0(I)$ is not universally separably injective, it is isomorphic to a space of continuous functions, more concretely to $C(I_\infty)$, where I_∞ is the one-point compactification of I when this set is equipped with the discrete topology. However, $c_0(I)$ cannot be isometric to a $C(X)$ space (compare Proposition 1.1(i)). Indeed, the closed unit ball of $c_0(I)$ does not have extreme points whereas, for any compact X, the characteristic function on X and its opposite are extreme points of the closed unit ball of $C(X)$.

ℓ^2 is not even isomorphic to a $C(X)$ space as this last contains an isomorphic copy of c_0 and ℓ^2 does not.

• (Compare Proposition 2.2(ii),(iii)) In the classical case $\prod_n E/\bigoplus_n E$ may not be even injective, e.g. $\prod_n \mathbb{R}/\bigoplus_n \mathbb{R}$ $(= \ell^\infty/c_0)$, see [**1**, Section 4.3].

On the other hand, it is well-known that every Banach space is isometrically embedded in an 1-injective space, $\ell^\infty(I)$ for a certain I. The non-Archimedean counterpart of this classical result holds for norm polar Banach spaces over spherically complete K (apply Proposition 4.7 and [**14**, Theorem 4.4.9]).

• (Compare Theorem 3.3(iii)) Quotients of 1-injective spaces may even fail to be separably injective. In fact, as with all separable classical Banach spaces, ℓ^1 is isometrically embedded in the 1-injective space $E := \ell^\infty$. Now, by [**4**, Corollary 4.16], E has a quotient isomorphic to ℓ^2, which is not separably injective.

5.2. Open problems. Firstly we pay attention to some open problems in the classical theory, with answers in the non-Archimedean case. For more open problems on classical injective spaces, see Chapter 6 of [**2**].

• (Compare Remark 2.3.1) It is unknown whether some classical Banach spaces are separably injective or not. This is for example the case for $\ell^\infty/C[0,1]$ and $C(\beta\mathbb{N} \times \beta\mathbb{N})$, see Problems 8 and 10 of [**2**, Chapter 6], respectively.

• (Compare Proposition 2.2(v)) A basic classical problem is whether each injective space is isomorphic to a 1-injective space; for an overview see [**2**, Section 1.6.1] and the references therein.

• (Compare Theorem 2.4(i),(ii)) In 1964 J. Lindenstrauss, [**8**], [**9**], obtained that, under the continuum hypothesis, 1-separably injective spaces are 1-universally separably injective. He left open the question in the usual setting of set theory with the Axiom of Choice. This problem was solved negatively in 2013, [**1**], where it was even proved that, for any of the implications of diagram (II), there exists a $C(X)$ space for which its converse fails (see Theorems 3.1 and 4.1 of [**13**] for an account).

On the other hand, it was asked in Section 7.2 of [**1**] if, without the continuum hypothesis, 1-separably injective spaces must be universally separably injective, even for $C(X)$ spaces. As the authors say in [**1**], it would be interesting to characterize (universally) separably injective $C(X)$ spaces in terms of topological properties of X, see also [**2**, Chapter 6, Problem 2].

Secondly we show that the partial affirmative answers given in Section 4 to the non-Archimedean open problems posed in that section, are false in the classical case. According to Corollary 4.4 we center our attention in Problems 4.2, 4.13 and 4.14.

• (Compare Theorem 4.12) It was proved in Section 4 of [**1**] that the Pełczyński-Sudakov Banach spaces $\ell^\infty_c(I)$, I an uncountable set (provided in [**12**]) and the Banach space ℓ^∞/c_0 are 1-universally separably injective but not injective. Recall that ℓ^∞/c_0 is isometric to $C(\beta\mathbb{N}\setminus\mathbb{N})$, see e.g. [**7**, (2.11)]. Also, it was proved in [**1**] that $\ell^\infty_c(I)$ is isomorphic to a (but not isometric to any) $C(X)$ space.

• (Compare Theorem 4.17) There exists a sequence $(E_n)_n$ of injective Banach spaces for which: **(i)** $\prod_n E_n$ is not even separably injective; **(ii)** there is no $C > 1$ such that E_n satisfies $(*)^u_C$ for all n.

Once we have an example satisfying (i), it is straightforward to check (ii) for it. In order to get such an example, take for each $n \in \mathbb{N}$, $E_n := \ell^2_n \bigoplus_\infty \ell^\infty$ (already considered in Section 5.1), which is an injective dual space. Suppose $\prod_n E_n$ is separably injective. Since $\prod_n \ell^2_n$ is isometric to a complemented subspace of $\prod_n E_n$, we obtain that $\prod_n \ell^2_n$ is also separable injective. But this last Banach product has a complemented subspace isomorphic to ℓ^2 (see [**5**]), from which we derive that ℓ^2 is separably injective, a contradiction.

Observe that each E_n is isomorphic to ℓ^∞ and hence to $C(\beta\mathbb{N})$. However, if $n > 1$, E_n cannot be isometric to a $C(X)$ space; if it were then every n-dimensional complemented subspace of E_n with projection of norm $= 1$, in particular ℓ^2_n, would be isometric to $\mathbb{R}^n$, an impossibility.

References

[1] A. Avilés, F. Cabello Sánchez, J. M. F. Castillo, M. González, and Y. Moreno, *On separably injective Banach spaces*, Adv. Math. **234** (2013), 192–216, DOI 10.1016/j.aim.2012.10.013. MR3003929

[2] A. Avilés, F. Cabello Sánchez, J. M. F. Castillo, M. González, and Y. Moreno, *Separably injective Banach spaces*, Lecture Notes in Mathematics, vol. 2132, Springer, [Cham], 2016. MR3469461

[3] F. Cabello Sánchez, J. M. F. Castillo, and D. Yost, *Sobczyk's theorems from A to B*, Extracta Math. **15** (2000), no. 2, 391–420. III Congress on Banach Spaces (Jarandilla de la Vera, 1998). MR1823898

[4] J. Diestel, H. Jarchow, and A. Tonge, *Absolutely summing operators*, Cambridge Studies in Advanced Mathematics, vol. 43, Cambridge University Press, Cambridge, 1995. MR1342297

[5] M. González, *On the duality problem for weakly compact, strictly singular operators*, Quaest. Math. **28** (2005), no. 1, 37–38, DOI 10.2989/16073600509486113. MR2136180

[6] L. Gruson and M. van der Put, *Banach spaces*, Table Ronde d'Analyse non archimédienne (Paris, 1972), Soc. Math. France, Paris, 1974, pp. 55–100 Bull. Soc. Math. France, Mém. No. 39-40. MR0365173

[7] I. E. Leonard and J. H. M. Whitfield, *A classical Banach space: l_∞/c_0*, Rocky Mountain J. Math. **13** (1983), no. 3, 531–539, DOI 10.1216/RMJ-1983-13-3-531. MR715776

[8] J. Lindenstrauss, *Extension of compact operators*, Mem. Amer. Math. Soc. No. **48** (1964), 112 pp. MR0179580

[9] J. Lindenstrauss, *On the extension of operators with range in a $C(K)$ space*, Proc. Amer. Math. Soc. **15** (1964), 218–225, DOI 10.2307/2034038. MR0167822

[10] J. Lindenstrauss and L. Tzafriri, *Classical Banach spaces. I. Sequence spaces*, Springer-Verlag, Berlin-New York, 1977. Ergebnisse der Mathematik und ihrer Grenzgebiete, Vol. 92. MR0500056

[11] J. Martínez-Maurica and C. Perez-Garcia, *Orthogonal bases in $l^\infty(X)$*, Collect. Math. **35** (1984), no. 2, 183–188. MR814358

[12] A. Pełczyński and V. N. Sudakov, *Remark on non-complemented subspaces of the space $m(S)$*, Colloq. Math. **9** (1962), 85–88. MR0139001

[13] C. Perez-Garcia, *Non-Archimedean countably injective Banach spaces*, J. Convex Anal. **22** (2015), no. 3, 733–746. MR3400152

[14] C. Perez-Garcia and W. H. Schikhof, *Locally convex spaces over non-Archimedean valued fields*, Cambridge Studies in Advanced Mathematics, vol. 119, Cambridge University Press, Cambridge, 2010. MR2598517

[15] R. S. Phillips, *On linear transformations*, Trans. Amer. Math. Soc. **48** (1940), 516–541, DOI 10.2307/1990096. MR0004094

[16] A. C. M. van Rooij, *Non-Archimedean functional analysis*, Monographs and Textbooks in Pure and Applied Math., vol. 51, Marcel Dekker, Inc., New York, 1978. MR512894

[17] H. P. Rosenthal, *On relatively disjoint families of measures, with some applications to Banach space theory*, Studia Math. **37** (1970), 13–36. MR0270122

[18] M. Zippin, *The separable extension problem*, Israel J. Math. **26** (1977), no. 3-4, 372–387, DOI 10.1007/BF03007653. MR0442649

[19] M. Zippin, *Extension of bounded linear operators*, Handbook of the geometry of Banach spaces, Vol. 2, North-Holland, Amsterdam, 2003, pp. 1703–1741, DOI 10.1016/S1874-5849(03)80047-5. MR1999607

Departamento de Matemáticas, Facultad de Ciencias, Universidad de Cantabria, Avda. de los Castros s/n, 39071 Santander, Spain

Email address: manuel.gonzalez@unican.es

Departamento de Matemáticas, Facultad de Ciencias, Universidad de Cantabria, Avda. de los Castros s/n, 39071 Santander, Spain

Email address: perezmc@unican.es

Contemporary Mathematics
Volume **704**, 2018
http://dx.doi.org/10.1090/conm/704/14170

G-modules with a convex base

H. Ochsenius and E. Olivos

ABSTRACT. In the theory of Banach spaces over valued fields developed, for example, in the classical book of van Rooij, **Non-archimedean Functional Analysis**, the group G of a valuation on a field K is a subgroup of the multiplicative subgroup $(\mathbb{R}^+, \cdot)$. Also, if E is a K-vector space, the set of norms of E is a subset of $[0, \infty)$. Yet this theory can be subsumed in a larger one, by allowing G to be any linearly ordered multiplicative group, and considering the norms of the space as elements of a new structure, a G-module. Such a setting includes infinite dimensional spaces, where a bilinear form defines a (non-archimedean) norm, the space is complete in the induced topology and the Projection Theorem holds.

Time and time again it has turned out that the structure of G-modules has a strong say in the features of non-archimedean spaces over fields with valuations of arbitrary rank. This paper, that continues a line which started with the work of W. Schikhof and E. Olivos, is the starting point in the study of a particular class of G-modules, those in which an independent generating set is a convex interval in the order of the G-module.

The authors could not forget to thank Wim Schikhof. It was during his stays in Temuco that this line of research started.

1. Preliminaries

We state first the necessary definitions and facts.
Let $(G, \cdot)$ be an abelian multiplicatively written totally ordered group. If G is isomorphic to a subgroup of $(\mathbb{R}^+, \cdot)$ we say that it has **rank one**. If that is not the case, we shall define the rank of G in the following way. Let $\mathcal{H}$ be the set of **convex subgroups** of G, that is those subgroups H such that for all $h \in H$ and $g \in G$, $h^{-1} < g < h$ implies $g \in H$. Call such a subgroup H **principal** if there exists an $h \in H$ such that $H = \bigcup_{n \in \mathbb{N}} (h^{-n}, h^n)$. Now, the set of principal convex subgroups different from G is totally ordered by inclusion. The order type of this set is the **rank** of G.
Another simple yet powerful clasification of totally ordered groups is obtained by considering the set of the elements greater than 1_G. If this set has a minimal element, usually denoted by g_0, we shall say that it is **quasidiscrete**; otherwise it is **quasidense**.

2010 *Mathematics Subject Classification.* Primary 06F99.
Key words and phrases. G-module, tight submodule.
The authors were partially supported by DIUFRO DI15-0043.

Let $(X, \leq)$ be a linearly ordered set without ending points. We say X is a G-**module** if there is an action of G on X, $(g, z) \mapsto gz$, which is increasing in both variables and such that for any $z \in X$ the orbit Gz is coinitial (and cofinal) on X. In what follows we will adjoin to every G-module, a zero element, 0, such that $0 < z$, for each $z \in X$ and $g0 = 0$ for each $g \in G$.
Important examples of G-modules are the group G itself, the action of G on G is the multiplication of G; the Dedekind completion of G, denoted by $G^{\#}$, with the action $g * s = \inf_{G^{\#}}\{gh : h \in G \text{ and } h \geq s\}$; and the group $\sqrt{G}$, whose elements are the square roots of the elements of G where the action is the multiplication in $\sqrt{G}$.
For another example, we recall that if H is a convex subgroup of G, then G/H is totally ordered by the relation $g_1H \leq g_2H$ if and only if $(g_1H \neq g_2H$ and $g_1 < g_2)$ or $(g_1g_2^{-1} \in H)$ (see [**3**]). With the action $(g_1, g_2H) \mapsto g_1g_2H$, G/H is also a G-module.
As usual the **Stabilizer** of an element $x \in X$ is the convex subgroup of G, $\text{Stab}(x) = \{g \in G : gx = x\}$. A G-module (submodule) X is called **faithful** if for every $x \in X$, $\text{Stab}(x) = \{1_G\}$.
A G-module X is **continuous** if for every $W \subset G$ for which $\inf_G W$ exists, the following holds: for every $z \in X$, $(\inf_G W)z = \inf_X Wz$. If G is quasidiscrete, every G-module X is continuous, on the other hand, if G is quasidense then X is continuous if for every $z \in X$, $z = \inf_X\{gz : g \in G, g > 1\}$. In any case, as a G-module both G and $G^{\#}$ (the Dedekind completion of G) are continuous. For details, see [**2**].

2. Basic concepts

We collect here the basic definitions and facts concerning two closely intertwined structures, the value group G and the G-module X. Throughout §2, §3 and §4 G is a linearly ordered abelian multiplicative group with unit 1. We assume $G \neq \{1\}$. The next definition borrows the classical notion in linear space theory.

DEFINITION 2.1. *Let X be a G-module. A subset B of X is called* (i) **generating** *if $GB = X$ in which case we say that X is generated by B;* (ii) **independent** *if $b_1, b_2 \in B, b_1 \neq b_2 \Rightarrow Gb_1 \cap Gb_2 = \varnothing$,* (iii) **a base** *(of X) if it is generating and independent.*

If we look at the orbits defined by the action of G on X, it is clear that for a set $B \subseteq X$,
(i) B is generating if and only if $X = \bigcup_{b \in B} Gb$.
(ii) B is independent if for any $b_1 \neq b_2$ in B the orbits Gb_1 and Gb_2 are disjoint.
(iii) B is a base if it contains one and only one element of each of the orbits Gb, $b \in B$.
It follows that

COROLLARY 2.2. *(i) Two bases of a G-module have the same cardinality.*
(ii) $B \subseteq X$ is a base of X if and only if it is a minimal generating set, if and only if it is a maximal independent set.
(iii) $B \subseteq X$ is a base of X if and only if for every $x \in X$ there is a unique $b \in B$ such that $x \in Gb$.

Now we arrive at the heart of our subject of study.

DEFINITION 2.3. *A* **convex base** *of a G-module X is nothing more than a base that is convex as a subset of X. A G-module X with convex base will be called for simplicity a* **tight G-module**.

Obviously we have that if B is a convex base of X, $b_1 < b_2 < b_3$ in B and for some $g \in G$, $g \neq 1_G$, $gb_2 < b_3$, then $gb_2 < b_1$.

LEMMA 2.4. *If B is a convex base of X, then for each $g \in G$, gB is also a convex base of X.*

PROOF. Only convexity of gB merits a proof. But $gb_1 < x < gb_2$ implies $b_1 < g^{-1}x < b_2$. Therefore, by convexity of B, $g^{-1}x \in B$, that is $x \in gB$. □

We start the study of tight G-modules with a basic property.

THEOREM 2.5. *Let X be a tight G-module. Then Stab is constant on X.*

PROOF. It suffices to prove that Stab is constant on a convex base B. So, let $b_1, b_2 \in B$, $b_1 < b_2$, we prove $\mathrm{Stab}(b_1) = \mathrm{Stab}(b_2)$.
(a) $\mathrm{Stab}(b_1) \subseteq \mathrm{Stab}(b_2)$. We may suppose Stab $(b_1) \neq \{1\}$, so there is a $g \in \mathrm{Stab}(b_1)$, $g < 1$. We have $b_1 = gb_1 \leq gb_2 \leq b_2$, so, by convexity, $gb_2 \in B$. If $gb_2 \neq b_2$ B would not be independent. Hence $gb_2 = b_2$ i.e. $g \in \mathrm{Stab}(b_2)$.
(b) $\mathrm{Stab}(b_2) \subseteq \mathrm{Stab}(b_1)$. We may suppose Stab $(b_2) \neq \{1\}$, so there is a $g \in \mathrm{Stab}(b_2)$, $g > 1$. Now $b_1 \leq gb_1 \leq gb_2 = b_2$ and by convexity $gb_1 \in B$. Hence, $gb_1 = b_1$ so $g \in \mathrm{Stab}(b_1)$. □

It is time for some examples.

EXAMPLE 2.6. *Let $X := \mathbb{R}^+$, $G = (\langle 2 \rangle, \cdot)$, the action is the usual multiplication in $\mathbb{R}^+$. Since any positive real number x can be written as $x = 2^m y$ where $m \in \mathbb{Z}$ and $y \in [1, 2)$ we see that X is a tight G-module with convex base $B = [1, 2)$.*

EXAMPLE 2.7. *Two-orbits G-modules are not necessarily tight. We shall present two examples. In the first place let $(G, \cdot)$ any group. To every $g \in G$ we adjoin an immediate predecessor g^- and define $G^- = \{g^- : g \in G\}$. Then $X := G \cup G^-$ is a G-module under the action $g_1 g_2^- := (g_1 g_2)^-$ which extends the multiplication on G (see* [**1**] *Section 1.5). It is easy to check that X is tight with convex base $\{g^-, g\}$ for any $g \in G$.*
On the other hand, if $G = (\mathbb{Q}^+, \cdot)$, then $X := \mathbb{Q}^+ \cup \mathbb{Q}^+\sqrt{2}$ with the usual multiplication in $\mathbb{Q}^+$ is not tight.

EXAMPLE 2.8. *If G has a non trivial convex subgroup H, then the G-module $G^{\#}$, the Dedekind completion of G under the action $gx := \sup_{G^{\#}}\{gg_1 : g_1 \leq x\}$, can never have a convex base, since for all $h \in H$, Stab$(h) = 1_G$ but Stab$(\sup H) = H$ and thus the Stabilizer is not constant on $G^{\#}$.*

It will turn out that we can fully describe the class of tight G-modules.
Let B a non-empty chain. The G-module $X := B \times G$ with anti-lexicographic ordering and action $g'(b, g) := (b, g'g)$, $(g, g' \in G, b \in B)$, is indeed a tight G-module. Notice that its convex base is $B \times \{1_G\}$.
One can get a "picture" of X as the ordered group G, but now where each $g \in G$ is replaced by a copy of B (i.e. (B, g)), and where the ordering on (B, g) is inherited from B.

Varying on this last example, let H be a proper convex subgroup of G and consider, with B as above, the G/H-module $B \times G/H$. Then the formula

$$g'(b, gH) = (b, gg'H)$$

defines an action of G on $B \times G/H$ making it into a G-module (this is a general way to view a G/H-module as a G-module with convex base $B \times H$). Notice that $\mathrm{Stab}(z) = H$ for all $z \in B \times G/H$.
We will see now that these are the only G-modules with a convex base:

THEOREM 2.9. *For a G-module X the following are equivalent.*
(α) X is isomorphic to $B \times G/H$ (as a G-module) for some non-empty chain B and some proper convex subgroup H.
(β) X has a convex base.

PROOF. $(\alpha) \Rightarrow (\beta)$ has been explained above, we proceed to prove $(\beta) \Rightarrow (\alpha)$. Let B be a convex base of X and let (Theorem 2.5) $H := \mathrm{Stab}(x)$ for all $x \in X$, let $\pi : G \to G/H$ be the canonical quotient map. The formula $\varphi(b, \pi(g)) = gb$ defines a map $\varphi : B \times G/H \to X$ (as $\pi(g_1) = \pi(g_2)$ implies $g_1 b = g_2 b$). We prove that φ is an isomorphism. Clearly φ is surjective. To prove injectivity, let $b_1, b_2 \in B$, $g_1, g_2 \in G$ be such that $\varphi(b_1, \pi(g_1)) = \varphi(b_2, \pi(g_2))$. Then $g_1 b_1 = g_2 b_2$, and since B is a base, we have $b_1 = b_2 =: b$ and $g_1 g_2^{-1} \in \mathrm{Stab}(b) = H$. So $\pi\left(g_1 g_2^{-1}\right) = 1$, i.e. $\pi(g_1) = \pi(g_2)$. We conclude that $(b_1, \pi(g_1)) = (b_2, \pi(g_2))$ proving injectivity of φ. That φ respects the action is simple to prove.
Finally we show that φ is increasing. Let $(b_1, \pi(g_1)) < (b_2, \pi(g_2))$. We must show that $g_1 b_1 \le g_2 b_2$. We consider two possibilities.
(a) $\pi(g_1) < \pi(g_2)$. Then $g_1 < g_2$, and if $b_1 \le b_2$ there is nothing to prove, so we may assume $b_1 > b_2$. Now $\pi(g_1 g_2^{-1}) < \pi(1)$, so $\pi(g_1 g_2^{-1}) < H = \mathrm{Stab}(b_1)$. Thus $g_1 g_2^{-1} b_1 < b_1$ and since $[b_2, b_1]$ is a subset of the convex base B, then $g_1 g_2^{-1} b_1 < b_2$.
(b) $\pi(g_1) = \pi(g_2)$. Then $b_1 < b_2$ so that $g_1 b_1 = \varphi(b_1, \pi(g_1)) = \varphi(b_1, \pi(g_2)) = g_2 b_1 \le g_2 b_2$ and we are done. □

3. Continuity of G-modules with convex base

In this little Section we will describe which tight G-modules are continuous and which are not. We remark that our proofs rely heavily on the facts for continuous G-modules stated in the Preliminaries.

LEMMA 3.1. *Let X have a convex base B, let* $H := \mathrm{Stab}(x)$ *for all* $x \in X$. *The following are equivalent.*
(α) X is continuous.
(β) G is quasidiscrete, or $H \neq \{1\}$, or $\#B$(the cardinality of B)$= 1$.

PROOF. $(\beta) \Rightarrow (\alpha)$. If G is quasidiscrete every G-module is continuous. If $H \neq \{1\}$ and G is quasidense we have for all $x \in X$,

$$\inf_X\{gx : g \in G,\, g > 1\} = \inf_X\{gx : g \in H,\, g > 1\} = x$$

so X is continuous.
Finally, if $\#B = 1$ and $H = \{1\}$ we have that $X = Gb$ where b is faithful. Then $X \cong G$, therefore it is continuous.
$(\alpha) \Rightarrow (\beta)$. Suppose (β) is not true; we show that X is not continuous. By assumption G is quasidense, X is faithful, $\#B \ge 2$. So, there are $b_1, b_2 \in B$, $b_1 < b_2$. Let $g \in G$, $g > 1$. Then $b_1 < b_2 < gb_1$ (as $gb_1 \notin B$). We see that b_2 is a

lower bound of $\{gb_1 : g > 1\}$, and b_2 is larger than b_1; we conclude that X is not continuous at b_1. $\square$

One-orbit G-modules are always tight and isomorphic to G/H for some proper convex subgroup H, so they are continuous and have a convex base (viz. $\{1_{G/H}\}$). Now 2-orbit modules play a key role in a new criterion for continuity.

THEOREM 3.2. *For a G-module X the following are equivalent.*
(α) X is not continuous.
(β) G is quasidense, X contains a faithful 2-orbit submodule Y with convex base.

PROOF. (β) $\Rightarrow$ (α). By hypothesis Y does not satisfy (β), therefore (Lemma 3.1) it cannot be continuous. But then neither is X.
(α) $\Rightarrow$ (β). Clearly G is quasidense. As X is not continuous there is an $a \in X$ for which $\inf_X\{ga : g \in G,\ g > 1\}$ is different from a.
Notice that a is faithful. Clearly a is a lower bound of $\{ga : g > 1\}$, so there is a lower bound $b > a$ i.e. $a < b \leq ga$ for all $g \in G$, $g > 1$.
By quasidenseness of G and faithfulness of a we even have $a < b < ga$ for all $g \in G$, $g > 1$.
Put $Y := Ga \cup Gb$. Y is a 2-orbit G-module since $b \notin Ga$. Thus $\{a, b\}$ is a base of Y. To show it is convex (in Y), let $a < b$, $z \in Y$, $a < z < b$. Then clearly $z \notin Ga$. From $g^{-1}b < a < b$ for all $g \in G$, $g > 1$ we deduce that $z \notin Gb$, a contradiction as $z \in Y$. By Theorem 2.5, b must be faithful, so Y is faithful and we are done. $\square$

COROLLARY 3.3. *For a 2-orbit G-module X the following are equivalent.*
(α) X is not continuous.
(β) G is quasidense. X is faithful and has a convex base.

4. Maximal tight G-submodules

In this study we shall use a different approach than the classical one of Section §2. We searched there for the base of a given G-module X, but here we will consider a subset B of X and decide whether it is a convex base of the submodule it generates. Such a B is called a c-subbase. For example, pick an element $a \in X$, clearly $\{a\}$ is a c-subbase. We will describe explicitly the construction of maximal c-subbases that contain a. Our main Theorem (Theorem 4.16) states the relationships between these c-subbases and maximal tight G-submodules of X.
IN WHAT FOLLOWS G IS AN ARBITRARY LINEARLY ORDERED GROUP $\neq \{1\}$ AND, FOR SIMPLICITY, ALL G-MODULES ARE ASSUMED TO BE FAITHFUL.

DEFINITION 4.1. *A subset B of a G module X is called a* **c-subbase** *if it is a base of the submodule GB and convex as a subset of it. In that case we shall say that GB is a* **tight submodule** *of X.*

Notice that it can happen that such a B may not be convex as a subset of X, see Lemma 4.6 below.

LEMMA 4.2. *Every submodule Y of a tight module X is also tight. More precisely, if B is a convex base of X then $B \cap Y$ is a convex base of Y.*

PROOF. Clearly $B \cap Y$ is independent (being a subset of B). Let $y \in Y$. Then $y = gb$ for some $g \in G$, $b \in B$. Then $b = g^{-1}y \in Y$, so $b \in B \cap Y$. We see that $B \cap Y$ generates Y; it follows that $B \cap Y$ is a base of Y. As B is convex in X we have $B \cap Y$ is convex in Y, and Lemma 4.2 is proved. □

We have the following useful characterization.

LEMMA 4.3. *The following statements are equivalent:*
(i) B is a c-subbase.
(ii) $gB > B$ for all $g \in G$, $g > 1$.

PROOF. *(i)* ⇒ *(ii)* Let B be a convex base of GB, $g \in G$, $g > 1$, $b \in B$. By faithfulness $gb > b$, by independence $gb \notin B$. Since B is convex and $g > 1$, we must have $gB > B$.
(ii) ⇒ *(i)* Obviously B generates GB. For independence suppose $b_1 \neq b_2 \in B$ and that for some $g_1 < g_2 \in G$ we have $g_1b_1 = g_2b_2$. Then $b_1 = g_1^{-1}g_2b_2$ which is not possible by *(ii)*. Thus we only have to prove convexity. If not, there are $g \in G, g \neq 1$ and b, b_1, $b_2 \in B$ such that $b_1 < gb < b_2$. This conflicts with $gB > B$ ($gb < b_2$ for $g > 1$ and $g^{-1}b_1 < b$, for $g < 1$). □

We can conclude:

COROLLARY 4.4. *Nonempty subsets of c-subbases are c-subbases.*

The interesting question is the converse one. If B_1 is a c-subbase in a G-module X, does there exist another c-subbase B_2 in X that properly contains B_1? If no such B_2 exists, we shall say that B_1 is a **maximal c-subbase**. Their existence and description will be our next task.
First we have, by invoking Zorn's Lemma, the following result.

LEMMA 4.5. *Each c-subbase in a G-module X can be extended to a maximal c-subbase.*

PROOF. Let B be any c-subbase, let $\mathfrak{B}$ be the collection of all c-subbases containing B. Then $\mathfrak{B} \neq \varnothing$ and by Zorn it is enough to show that each non-empty linearly ordered $\mathfrak{C} \subseteq \mathfrak{B}$ has an upper bound. Now take $\tilde{B} := \bigcup \mathfrak{C}$. Clearly $\tilde{B}$ contains every element of $\mathfrak{C}$, so it remains to be shown that $\tilde{B}$ is a c-subbase i.e that b_1, $b_2 \in \tilde{B}$, $g > 1$ implies $gb_1 > b_2$. Now there is a $\Gamma \in \mathfrak{C}$ for which b_1, $b_2 \in \Gamma$. Since Γ is a c-subbase we have $gb_1 > b_2$ and we are done. □

An interesting fact is the following.

LEMMA 4.6. *Let B be a maximal c-subbase. Then B is convex in X.*

PROOF. Suppose that for b_1, $b_2 \in B$, $x \in X$ we have $b_1 < x < b_2$. Consider $\tilde{B} := B \cup \{x\}$. We shall prove that, for $g > 1$, $g\tilde{B} > \tilde{B}$. This will mean that $\tilde{B}$ is a c-subbase and by maximality $\tilde{B} = B$ and $x \in B$.
Therefore, pick $g > 1$, thus $gx > x$ and since $gB > B$, we have $gB > \tilde{B}$, and $gx > gb_1 > B$, hence $gx > g\tilde{B}$ and we are done. □

We aim next to give an explicit description of maximal c-subbases. For that we need to discuss separately the cases G quasidense and G quasidiscrete. It will turn out that both of them can be obtained as a kind of "neighbourhood" of some element $a \in X$.
CASE I. G IS QUASIDENSE.

For each $a \in X$ we define directly a set B_a which will turn out to be the unique maximal c-subbase containing a.

DEFINITION 4.7. *Let X be a G-module and $a \in X$. Define*

$$B_a := \bigcap\{(g_1a,\, g_2a) : g_1,\, g_2 \in G,\; g_1 < 1 < g_2\}$$

Clearly B_a contains a and it is convex in X.

We collect properties of B_a in the following.

LEMMA 4.8. *For each $a, b \in X$ and $g \in G$ we have*
(i) $gB_a = B_{ga}$.
(ii) $a \in B_b \Leftrightarrow b \in B_a \Leftrightarrow B_a = B_b$.
(iii) $GB_a = GB_b$ if and only if $B_a = g\,B_b$ for some $g \in G$.

PROOF. (*i*) $x \in gB_a \Leftrightarrow g^{-1}x \in B_a \Leftrightarrow \forall g_1, g_2 \in G(g_1 < 1 < g_2 \Rightarrow g_1a < g^{-1}x < g_2a) \Leftrightarrow \forall g_1, g_2 \in G(g_1 < 1 < g_2 \Rightarrow g_1ga < x < g_2ga) \Leftrightarrow x \in B_{ga}$.
(*ii*) Let $a \in B_b$. For $g_1, g_2 \in G$ with $g_1 < 1 < g_2$ we have $g_2^{-1} < 1 < g_1^{-1}$, so $g_2^{-1}b < a < g_1^{-1}b$ i.e. $g_1a < b < g_2a$. Hence, $b \in B_a$. By symmetry we therefore have $a \in B_b \Leftrightarrow b \in B_a$.
To complete the proof it suffices to establish: $a \in B_b \Rightarrow B_a \subseteq B_b$. To do this, let $g_1, g_2 \in G, g_1 < 1 < g_2$. Then $x \in (g_1a,\, g_2a)$ and $a \in (g_1b,\, g_2b)$ so that $x \in (g_1^2b,\, g_2^2b)$. By quasidenseness of G it follows that $x \in B_b$.
(*iii*) ($\Rightarrow$) Since $GB_a = GB_b$, there are $g \in G$ and $b' \in B_b$ such that $a = gb'$. By (i) and (ii), $B_a = B_{gb'} = gB_{b'} = gB_b$. The other implication is trivial. $\square$

Now we are ready to prove the following theorem.

THEOREM 4.9. *For each $a \in X$ the set B_a is the unique maximal c-subbase containing a.*

PROOF. To show that B_a is a c-subbase, let $g \in G, g > 1$, and $x, y \in B_a$; we shall prove that $gx > y$. By quasidenseness of G there are $g_1, g_2 \in G$, $g_1 < 1 < g_2$ such that $g > g_2g_1^{-1} > 1$. Since $x, y \in B_a$ it follows that $x > g_1a$ and $y < g_2a$. Then $gx > gg_1a > g_2a > y$ and we are done.
Next we turn to maximality. Let B be a c-subbase, $B \supseteq B_a$; we prove $B \subseteq B_a$. Let $x \in B, g_1, g_2 \in G$, $g_1 < 1 < g_2$. B is c-subbase, so $g_2B > B > g_1B$; in particular $g_1a < x < g_2a$. So $x \in B_a$. Uniqueness follows from the same proof. $\square$

CASE II. G IS QUASIDISCRETE. As usual $g_0 := \min\{g \in G : g > 1\}$. We start with a lemma.

LEMMA 4.10. *For each $a \in X$ the sets $[a,\, g_0a)$ and $(a,\, g_0a]$ are maximal c-subbases.*

PROOF. We only consider $B := [a,\, g_0a)$, for the other interval the proof runs similarly.
First notice that $g_0B > B$ so B is a c-subbase. To prove maximality, let $z \in X \setminus B$ and suppose $B' = B \cup \{z\}$ is a c-subbase, we derive a contradiction.
From $g_0B' > B'$ it follows that $g_0a > z$. By assumption, $z \notin [a,\, g_0a)$ implying that $z < a$. Since $g_0B' > B'$ we have $g_0z > a$. But then $a < g_0z < g_0a$, so $g_0z \in B$ contradicting $g_0z > B'$. $\square$

But there are maximal c-subbases that cannot be described in that way. For example, consider the group $G = \langle 2 \rangle$, the G-module $X = \mathbb{Q}^+$ and the maximal c-subbase $B = \{q \in \mathbb{Q} : q \in (\sqrt{3}, 2\sqrt{3})\}$.
Since we aim to find a description of the collection of all maximal c-subbases of a G-module X, we shall consider its completion $X^\#$.

THEOREM 4.11. *For each $a \in X^\#$ the sets*

$$B_a := [a, g_0 a)_{X^\#} \cap X \qquad \text{and} \qquad B_a^+ := (a, g_0 a]_{X^\#} \cap X$$

are maximal c-subbases.

PROOF. If $a \in X$ then B_a, B_a^+ are just the intervals $[a, g_0 a)$ and $(a, g_0 a]$ of Lemma 4.10, so we may suppose that $a \in X^\# \setminus X$. Then $B_a = B_a^+$, so we only have to prove that $(a, g_0 a)_{X^\#} \cap X := B$ is a maximal c-subbase. Obviously B is a c-subbase.
For the maximality proof let $z \in X \setminus B$ and suppose $B' := B \cup \{z\}$ is a c-subbase, we derive a contradiction. From $g_0 B' > B'$ it follows that $g_0 b > z$ for all $b \in B'$. Now $a = \inf_{X^\#} B$ so that $a \geq g_0^{-1} z$. We see that $z \leq g_0 a$. But they cannot be equal as $z \in X$, $g_0 a \notin X$. Hence, $z < g_0 a$.
By assumption $z \notin [a, g_0 a)$, so $z < a$. But then $a < g_0 z < g_0 a$, so $g_0 z \in B$ contradicting $g_0 z > B'$.

□

As could be expected we prove now:

THEOREM 4.12. *Each maximal c-subbase of X has the form B_a or B_a^+ of Theorem 4.11.*

PROOF. Let B be a maximal c-subbase in X. From $g_0 B > B > g_0^{-1} B$ we infer that B is bounded. Put $a := \inf_{X^\#} B$. We consider two cases.
1) $a \in B$. If there is some $b \in B$, $b > g_0 a$ then $a \leq g_0 a \leq b$, so by convexity $g_0 a \in B$ contradicting $g_0 a > B$. It follows that $B \subseteq [a, g_0 a)_{X^\#} \cap X = B_a$. By maximality, $B = B_a$.
2) $a \notin B$. Since $a = \inf_{X^\#} B$ we have for all $b \in B$, $g_0 a = \inf_{X^\#} g_0 B \geq b$ (as $g_0 B > B$). So we have $b \leq g_0 a$, implying $B \subseteq (a, g_0 a]_{X^\#} \cap X = B_a^+$. By maximality $B = B_a^+$, which proves our theorem. □

However, we have the following result.

PROPOSITION 4.13. *For all $a \in X^\#$ there exists $b \in X$ such that $GB_b = GB_a = GB_a^+$.*

PROOF. If $a \in X$, $GB_a = GB_a^+$ trivially. So we assume that $a \notin X$. Let us call $A := (a, g_0 a) \cap X$, thus $A = B_a = B_a^+$. For $b \in A$, put $B = [b, g_0 b)$. Suppose $z \in GA$, then for some $g \in G$ and $x \in A$, $z = gx$. Since both x and b belong to A either $a < b < x < g_0 a < g_0 b$ or $g_0^{-1} b < a < x < b < g_0 a$. In any case $z = gx \in GB$. The proof of the other inclusion runs in a similar way. □

Remark. In what follows we shall study the tight G submodule generated by a maximal c-subbase. By the preceding proposition we shall frequently name this c-subbase as B_a, $a \in X$ (as in the quasidense case).
The following example shows that GB_a can be different from X.

EXAMPLE 4.14. *Let* $G := \{(2^k, q) : k \in \mathbb{Z},\ q \in \mathbb{Q}^+\}$ *with the antilexicographic ordering, and* $X := \{(2^{\frac{m}{2}}, r) : m \in \mathbb{Z},\ r \in \mathbb{R}^+\}$. *Thus* X *is a* G*-module under the natural multiplication* $(2^k, q) \cdot (2^{\frac{m}{2}}, r) = \left(2^{\frac{2k+m}{2}}, qr\right)$. *Set* $a := (2, 1) \in X$, *then* B_a, *the interval with ending points* $(2, 1)$ *and* $(2^2, 1)$, *is the set* $\left\{(2, 1), \left(2^{\frac{3}{2}}, 1\right)\right\}$. *But it is clear that any element of* GB_a *has the form* $(2^{\frac{m}{2}}, q)$, *with* $q \in \mathbb{Q}$. *Hence* $GB_a \neq X$.

The next lemma is the quasidiscrete version of Lemma 4.8.

LEMMA 4.15. *Let* G *be a quasidiscrete group,* X *a* G*-module. For any* $a, b \in X, g \in G$ *we have that*
i) $gB_a = B_{ga}$.
ii) If $a \neq b$, *then* $a \in B_b \Leftrightarrow g_0 b \in B_a$.
iii) $GB_a = GB_b \Leftrightarrow B_b \cap Ga \neq \varnothing$.

PROOF.
i) $a \in X$, $B_a = [a, g_0 a)$ and $gB_a = B_{ga}$.
ii) $a \in B_b \Leftrightarrow b < a < g_0 b < g_0 a \Leftrightarrow g_0 b \in B_a$.
iii) We shall prove firstly $GB_a = GB_b \Rightarrow B_b \cap Ga \neq \varnothing \Rightarrow GB_a \subseteq GB_b$. Then by symmetry $GB_a = GB_b \Rightarrow B_a \cap Gb \neq \varnothing \Rightarrow GB_b \subseteq GB_a$. This shall complete the proof.
If $GB_a = GB_b$, then $a \in gB_b$ for some $g \in G$ hence $g^{-1}a \in B_b \cap Ga$. Now suppose $B_b \cap Ga \neq \varnothing$ thus for some $b' \in B_b$ and some $g \in G$ we have $b' = ga$. Pick $x \in B_a$, then $gx \in gB_a = B_{ga} = B_{b'}$, hence $b' \leq gx < g_0 b'$. Since $b \leq b' < g_0 b$, there are two possible cases, either $gx < g_0 b$ or $g_0 b \leq gx$. In the first, $b \leq b' \leq gx < g_0 b$; in the second, $g_0 b \leq gx < g_0 b' < g_0^2 b$. Either way $gx \in GB_b$ as we claimed. □

Since we have now identified the maximal c-subbases of X for G quasidense and G quasidiscrete, we can now state our main Theorem.

THEOREM 4.16. *Let* X *be a* G*-module.*
(i) For any $a \in X$ *the set* GB_a *is the unique maximal tight submodule of* X *containing* a.
(ii) Each maximal tight submodule Y *of* X *is of the form* GB_a *for some* $a \in X$. *Even more, if* G *is quasidense, each convex base of such a* Y *is of the form* B_a *for some* $a \in X$.

PROOF. *(i)* Uniqueness and maximality of GB_a. This will be done separately according to whether G is quasidense or quasidiscrete.
a) G quasidense. Clearly GB_a is a tight submodule with convex base B_a. For maximality let Y be a tight submodule containing GB_a, we show that $Y \subseteq GB_a$. In fact, let C be a convex base of Y. By Lemma 4.2, $C \cap GB_a$ is a convex base of GB_a, in particular it is non empty. Let $z \in C \cap GB_a$. Then there is a $g \in G$ such that $g^{-1}z \in B_a$ implying that $B_{g^{-1}z} = B_a$, by Lemma 4.8. On the other hand, $g^{-1}C$ is also a convex base of Y containing $g^{-1}z$ therefore $g^{-1}C \subseteq B_{g^{-1}z} = B_a$. We see that $Y \subseteq GB_a$.
b) G quasidiscrete. Let Y be a maximal tight submodule containing GB_a and let C be a c-subbase of Y. Put $u := \inf_{X^\#} C$. Since C is a convex subset of X, either $C = B_u$ or B_u^+. By Proposition 4.13 we can assume that $u \in X$ and $C = B_u$. So,

$Y = GB_u$ for some $u \in X$. Therefore $B_a \subseteq GB_u$ so $a \in gB_u$ for some $g \in G$. We conclude that $Ga \cap B_u \neq \emptyset$, then by Lemma 4.15$(ii)$, $GB_a = GB_u = Y$. This shows both, maximality and uniqueness.
(ii) Let Y be a maximal tight submodule of X, we contend that $Y = GB_a$ for some $a \in X$. Let B be a convex base of Y. By Lemma 4.5 it extends to a maximal c-subbase which is of the form B_a for some $a \in Y$ (see Remark above). Thus the maximal tight submodule $Y = GB$ lies in GB_a and $Y = GB_a$. Even more, by Theorem 4.9, when G is quasidense, B must be equal to B_a for some $a \in X$. □

COROLLARY 4.17. *Let X be a tight G-module. For any $a \in X$ the c-subbase B_a is a convex base of X.*

PROOF. Let $a \in X$, and consider the c-subbase B_a . By (i), GB_a is the unique maximal tight submodule containing a. Thus $GB_a = X$. □

Remark. It is now easy to decide if a G-module X is tight. Just consider GB_a for any $a \in X$.

References

[1] H. Ochsenius and W. H. Schikhof, *Banach spaces over fields with an infinite rank valuation*, p-adic functional analysis (Poznań, 1998), Lecture Notes in Pure and Appl. Math., vol. 207, Dekker, New York, 1999, pp. 233–293. MR1703500

[2] H. Ochsenius and W. H. Schikhof, *Lipschitz operators on Banach spaces over Krull valued fields*, Ultrametric functional analysis, Contemp. Math., vol. 384, Amer. Math. Soc., Providence, RI, 2005, pp. 203–233, DOI 10.1090/conm/384/07137. MR2174786

[3] P. Ribenboim, *Théorie des valuations* (French), Deuxième édition multigraphiée. Séminaire de Mathématiques Supérieures, No. 9 (Été, vol. 1964, Les Presses de l'Université de Montréal, Montreal, Quebec, 1968. MR0249425

[4] A. C. M. van Rooij, *Non-Archimedean functional analysis*, Monographs and Textbooks in Pure and Applied Math., vol. 51, Marcel Dekker, Inc., New York, 1978. MR512894

DEPARTAMENTO DE MATEMÁTICA Y ESTADÍSTICA. FACULTAD DE INGENIERÍA Y CIENCIAS. UNIVERSIDAD DE LA FRONTERA. TEMUCO. CHILE
Email address: herminia.ochsenius.a@gmail.com
Email address: elena.olivos@ufrontera.cl

Contemporary Mathematics
Volume **704**, 2018
http://dx.doi.org/10.1090/conm/704/14171

A journey throughout the history of p-adic numbers

Yvette Perrin

ABSTRACT. We describe the context where p-adic numbers were created at the end of the nineteenth century. Next, algebraic and topologic structures were constructed on that set of numbers by K. Hensel. They are the only complete valued extensions of $\mathbb{Q}$, different from $\mathbb{R}$. All non-Archimedean mathematics grew up during the twentieth century where the fields $\mathbb{Q}_p$ play the role of $\mathbb{R}$. Now, p-adic analysis finds applications in various scientific domains: quantum physic, dynamical systems, computers. The paper insists on the analogies and overall the differences between methods and results obtained in classical and non-Archimedean mathematics.

1. Introduction

The history of p-adic numbers, which appeared in the late nineteenth century, gives a very good illustration of how the mathematical theory was born, grew autonomously and is now reinvested in areas other than mathematics. Thus, p-adic numbers were introduced by the German mathematician Kurt Hensel in a 1897 article designed to solve divisibility problems in algebraic numbers fields [**12**]. Around 1900, topology concepts spread and then in 1907, Hensel [**15**] generated the concept of an absolute value on a field of p-adic numbers, for any prime number p. Equipped with algebraic and topological structures, any field of p-adic numbers is a valued extension of the field $\mathbb{Q}$ of rational numbers, as the field $\mathbb{R}$ of real numbers is. So, it could now be the basis for new, non Archimedean mathematics that were developed throughout the 20-th century. Around 1990, p-adic numbers have mainly entered in physics and quantum physics. Until recently, physical theories used the usual absolute value on $\mathbb{Q}$ and its extension to the field of real numbers. But some mathematical physicists thought that the real continuum is not suitable for studying quantum phenomena and have attempted to develop mathematical models of quantum physics based on extensions of $\mathbb{Q}$ other than $\mathbb{R}$, particularly the p-adic fields $\mathbb{Q}_p$.

It is out of question to display here the sprawling development of p-adic mathematics or even to highlight the main steps. Our purpose is just to describe two or three moments of the story and show how a mathematical tool born from the need to solve a specific problem in a very limited domain, is then taken as an object of study itself, grows richer with unexpected properties and gives birth to a whole new mathematical universe that is sometimes more appropriate to represent the

2010 *Mathematics Subject Classification.* Primary 12J25, 46S10.
Key words and phrases. p-adic analysis, ultrametric fields.

reality of the world. Here we highlight the important role played by analogy in the construction of these new entities and also by the Local-Global principle to which Hensel used to refer for the construction of p-adic numbers and the establishment of some of their properties. Expanded to number theory, this principle has become an important investigative tool, especially for solving Diophantine equations.

In the second part we will show how Hensel was led to the idea of p-adic numbers and then we will define these numbers and their algebraic calculations. We will then give some simple examples of their use in algebra and number theory.

In the third part we will examine properties of p-adic numbers and the fourth part will focus on the definition and properties of p-adic absolute values. We know from the work of Hensel (1905) and Ostrowski (1918) that the field of rational numbers can be provided with two types of absolute values and only two ones: the usual absolute value that is Archimedean and the p-adic absolute values that are not Archimedean. Each one of the absolute values gives a completion of $\mathbb{Q}$: the field $\mathbb{R}$ for the usual absolute value and the field $\mathbb{Q}_p$ for each p-adic absolute value. Following Cauchy's work, Hensel [**16**] sketched out an analysis on p-adic fields, especially a theory of analytic functions, based on expansions in Taylor series. This work has known a very important development until now. It is on this domain of p-adic analysis we will insist in the fourth part of this paper, focusing on algebraic and topological properties which are shared by $\mathbb{R}$ and the filed $\mathbb{Q}_p$ but also those which are different.

In the fifth part, we will highlight the properties of p-adic numbers which have led some physicists and mathematicians to favour p-adic extensions of the field of rational numbers as a base for a new mathematical model of quatum mechanics that would solve some of the paradoxes of the classical theory.

2. How Hensel was led to create the p-adic numbers

The algebraic numbers theory has played a major role in mathematical research in the second half of the 19th century mainly for the greatest German mathematicians: Gauss, Dirichlet, Kummer, Weber. It is in this domain that Hensel's works led to the construction of p-adic numbers. Hensel was trying to solve the problem of divisibility of a prime integer in a field of algebraic numbers. Specifically, consider a number field $\mathbb{Q}(\theta)$ i.e. a finite extension of $\mathbb{Q}$. This field may be generated by a single element θ algebraic over $\mathbb{Q}$. Let p be a prime integer that we want to put as a product of integers of $\mathbb{Q}(\theta)$. Consider the minimal polynomial $P(X)$ of θ and its reductions modulo the powers of p. There exists an integer r such that the factorisation of $P(X)$ modulo the powers of p higher than r leads to the factorisation of p in $\mathbb{Q}(\theta)$, whence the idea of Hensel to write the coefficients of $P(X)$ and more generally all the integers, in base p. Then, to get to the notion of p-adic numbers, it seems that Hensel's main motivation was the analogy between the ring of integers $\mathbb{Z}$ and the field of fractions $\mathbb{C}(X)$ [**11**]. To be specific:

A rational number is the quotient of two integers $f = \frac{a}{b}$, $a,\ b \in \mathbb{Z},\ b \neq 0$.

Similarly an element of $\mathbb{C}(X)$ is the quotient of two polynomials with complex coefficients:

$$f(X) = \frac{P(X)}{Q(X)},\ P(X),\ Q(X) \in \mathbb{C}[X], Q(X) \neq 0$$

The two rings $\mathbb{Z}$ and $\mathbb{C}[X]$ have similar properties: any element of $\mathbb{Z}$ has a unique factorisation into a product of prime numbers, any polynomial can be uniquely factorized into a product of irreducible polynomials $P(X) = a(X - \alpha_1)...(X - \alpha_n)$, $a \in \mathbb{C}$ and $\alpha_i \in \mathbb{C}$ for $i = 1, ..., n$.

Hensel noted that the analogy goes farther: Given a prime number p, any positive integer m can be written in base p in the form:

$$a_0 + a_1 p + a_2 p^2 + ... + a_n p^n \tag{1}$$

where $n \in \mathbb{N}$ and the coefficients a_i are integers which satisfy the inequality $0 \leq a_i \leq p - 1$

Similarly, given an irreducible polynomial $X - \alpha$ of $\mathbb{C}[X]$, any polynomial $P(X)$ can be written: $P(X) = a_0 + a_1(X - \alpha) + ... + a_n(X - \alpha)^n$ where the coefficients α and a_i are complex numbers. Such expansions are interesting because they give local information. For example the expansion of $P(X)$ indicates if it vanishes at α and to what order, as the expansion of m indicates whether it is divisible by p and to what order. We can extend this method of expansion for positive rational numbers of $\mathbb{Q}$ as one can do for rational functions in $\mathbb{C}(X)$. Any rational function $f(X) = \dfrac{P(X)}{Q(X)}$ admits an expansion in a Laurent series in the neighbourhood of any point $\alpha \in \mathbb{C}$ by dividing, according to increasing powers of $x - \alpha$, the expansion of the polynomial $P(X)$ by that of $Q(X)$:

$$f(X) = a_{n_0}(X - \alpha)^{n_0} + a_{n_0+1}(X - \alpha)^{n_0+1} + ... = \sum_{i \geq n_0} a_i (X - \alpha)^i \ , \ n_0 \in \mathbb{Z}$$

(n_0 can be negative, this is the case when the smallest power of $X - \alpha$ in $Q(X)$ is greater than that of $(X - \alpha)$ in $P(X)$. We then say that α is *a pole* of $f(x)$). Usually, expansions of rational functions are infinite. The fundamental point is that any rational function can be represented by a Laurent series in terms of each of the primes $(X - \alpha)$ of the ring $\mathbb{C}[X]$. Note, however, that all these series do not always represent a rational function: for example exponential or logarithmic functions are series that do not represent rational functions. Hensel uses the analogy between $\mathbb{Z}$ and $\mathbb{C}[X]$ to generalize the expansion in base p of the integers to the positive rational numbers. To pass from positive integers to positive rational numbers we proceed as for rational functions, by dividing numerator expansion by denominator expansion. So any positive rational number admits an expansion of the form:

$$\frac{a}{b} = \sum_{i \geq n_0} a_i p^i \text{ with } 0 \leq a_i \leq p - 1, \ n_0 \in \mathbb{Z}. \tag{2}$$

Noting that -1 can be formally represented by:

$$-1 = (p - 1) + (p - 1)p + (p - 1)p^2 + ... + (p - 1)p^n + ...$$

and defining the product of two expansions usually, we show that any positive and negative element of $\mathbb{Q}$ admits expansion of type (2). Such an expansion is called its Hensel's expansion. The analogy with Laurent series goes on. But, any Hensel's expansion does not represent a rational number. Hensel's original idea consisted of considering all these expansions, even those that do not correspond to a rational number. They constitute *the set $\mathbb{Q}_p$ of p-adic numbers.*

3. Definition and properties of p-adic numbers

Let p be a prime integer and let $\mathbb{Q}_p$ be the set of p-adic numbers $\sum_{i \geq n_0} a_i p^i$ where $0 \leq a_i \leq p - 1$, $n_0 \in \mathbb{Z}$. We can define an addition and a multiplication of two p-adic numbers as we define an addition and a multiplication of two decimal expansions without forgetting what to carry. Armed with these operations, $\mathbb{Q}_p$ is a field that contains the field of rational numbers. All the expansions $\sum_{i \geq n_0} a_i p^i$ where n_0 is positive, form a ring denoted by $\mathbb{Z}_p$ which is called *the ring of integers of* $\mathbb{Q}_p$. It is an integral domain and its field of fractions is $\mathbb{Q}_p$. Its ideals are the subsets $p^n\mathbb{Z}_p$, $n \geq 1$, its unique maximal ideal is the ideal $p\mathbb{Z}_p$ and its invertible elements called *units* are the elements of $\mathbb{Z}_p \setminus p\mathbb{Z}_p$. One of the most important algebraic properties of p-adic numbers was obtained by Hensel. Exposed in lemmas that bear his name [**13**] it made the p-adic numbers an important tool for solving algebra and number theory problems. We still state the simplest form of this lemma and give some applications that illustrate the local-global principle.

HENSEL'S LEMMA. *Let $F(X) = a_0 + a_1X + ... + a_nX^n$ be a polynomial with coefficients in $\mathbb{Z}_p$. Suppose there exists a p-adic integer α_1 such that $F(\alpha_1) \equiv 0$ mod$(p\mathbb{Z}_p)$ and $F'(\alpha_1) \neq 0$ mod$(p\mathbb{Z}_p)$. Then, there exists $\alpha \in \mathbb{Z}_p$ such that $\alpha \equiv \alpha_1$ mod$(p\mathbb{Z}_p)$ and $F(\alpha) = 0$.*

One consequence of Hensel's lemma is that it is easy to know whether a polynomial with coefficients in $\mathbb{Z}$ has roots in $\mathbb{Z}_p$ since it is sufficient, in most cases, to find roots modulo p. For example the polynomial $F(X) = X^2 + 1$ has roots in $\mathbb{Q}_5$ because 2 and 3 are roots modulo 5. Of course if a polynomial $F(X)$ has roots in $\mathbb{Q}$, it has roots in each field $\mathbb{Q}_p$ since $\mathbb{Q}_p$ contains $\mathbb{Q}$. But it is more interesting to ask whether a polynomial that has roots in each field $\mathbb{Q}_p$ has roots in $\mathbb{Q}$. In some cases the answer is positive. Let us give two examples:

THEOREM 1. *A positive rational m is a square of a rational number, (that is to say the polynomial $X^2 - m$ has roots in $\mathbb{Q}$) if and only if it is a square in each p-adic field $\mathbb{Q}_p$.*

2. Minkowski's Theorem *Let $P(X_1, X_2, ..., X_n) \in \mathbb{Q}[X_1, ..., X_n]$ be a quadratic form (that is to say a homogeneous polynomial of degree 2 for n variables). The equation $P(X_1, ..., X_n) = 0$ has a non-trivial solution if and only if it has a non-trivial solution in each set $(\mathbb{Q}_p)^n$.*

Unfortunately, the existence of solutions in each field $\mathbb{Q}_p$ does not always imply the existence of solutions in $\mathbb{Q}$. We give an example and its proof which nicely illustrate Hensel's lemma.

EXAMPLE. Let $P(X) = (X^2 - 2)(X^2 - 17)(X^2 - 34)$. The equation $P(X) = 0$ has roots in each field $\mathbb{Q}_p$ but does not have any in $\mathbb{Q}$.

PROOF. It is clear that the equation has no roots in $\mathbb{Q}$. Let us prove it has roots in each $\mathbb{Q}_p$. Suppose $p \neq 2$ and $p \neq 17$. The only non-trivial case is when 2 and 17 are not squares in $\mathbb{Q}_p$. Let us stay in this hypothesis, that is to say the polynomials $P(X) = X^2 - 2$ and $Q(X) = X^2 - 17$ have no roots in $\mathbb{Q}_p$. By Hensel's lemma, $P(X)$ and $Q(X)$ have no roots mod p i.e. 2 and 17 are not quadratic residues mod p. Let us use Legendre's symbols $\left(\frac{n}{p}\right)$ defined for any integer n and for a prime p by

$\left(\frac{n}{p}\right) = 0$ if n is a multiple of p,

$\left(\frac{n}{p}\right) = 1$ if n is not a multiple of p but is a quadratic residue mod p,

$\left(\frac{n}{p}\right) = -1$ otherwise.

Then, for any two integers n and m and by Gauss theory, we have

$$\left(\frac{m}{p}\right)\left(\frac{n}{p}\right) = \left(\frac{mn}{p}\right) \quad \forall m,\ n \in \mathbb{Z}.$$

Now, here we have, $\left(\frac{2}{p}\right) = -1$, $\left(\frac{17}{p}\right) = -1$ and so $\left(\frac{2}{p}\right)\left(\frac{17}{p}\right) = 1$. But on the other hand, $\left(\frac{2}{p}\right)\left(\frac{17}{p}\right) = \left(\frac{34}{p}\right)$, hence $\left(\frac{34}{p}\right) = 1$. It follows that 34 is a quadratic residue modulo p and hence, by Hensel's lemma, the polynomial $X^2 - 34$ has a root in $\mathbb{Q}_p$

Suppose now $p = 2$. It is proven that an integer m is a square in $\mathbb{Z}_2$ if and only if $m \equiv 1 \pmod 8$. So, $17 = 1 + 2 \times 8$ is a square in $\mathbb{Z}_2$. Finally, suppose $p = 17$. Then $6^2 \equiv 2 \pmod{17}$, hence 6 is a root of $P(X)$ modulo 17, hence by Hensel's Lemma $P(X)$ has a root in $\mathbb{Q}_{17}$.

In many case, however, the existence or non-existence of roots in $\mathbb{Q}$ for a Diophantine equation can be detected by studying the solution in $\mathbb{Q}_p$ for each prime p. Continuing that analogy principle, we can say that what happens locally at each point p, for such an equation, provides information on its global behaviour. □

4. p-adic absolute values

p-adic numbers were first considered as formal expansions in powers of p. The idea of introducing topological notions in p-adic fields does not really appear to Hensel before 1905 [**14**] and it was only in 1907, after writing the book *Theorie der algebraischen Zahlen* [**16**] where he completely re-exposed the theory of algebraic numbers, that he got the definition of the p-adic absolute values and its main properties [**2**].

Fix a prime number p and let x be a non-zero p adic number

$$x = a_{n_0}p^{n_0} + \ldots + a_n p^n + \ldots = \sum_{i \geq n_0} a_i p^i, \quad 0 \leq a_i \leq p-1,\ n_0 \in \mathbb{Z}, \quad a_{n_0} \neq 0.$$

By definition, the valuation $v_p(x)$ of x is the integer n_0 and its p-adic absolute value is defined by $|x| = \dfrac{1}{p^{v_p(x)}}$. We convent that $v_p(0) = +\infty$ and this implies $|0|_p = 0$. The p-adic absolute value satisfies the relations:

$|x|_p = 0$ if and only if $x = 0$,

$|xy|_p = |x|_p |y|_p$

and the strong triangle inequality:

$|x + y|_p \leq \max(|x|_p, |y|_p)$

which implies the usual triangle inequality:

$|x + y|_p \leq |x|_p + |y|_p$.

A distance is associated with this absolute value, defined as usual for any two elements x and y of $\mathbb{Q}_p$ by : $d(x, y) = |x - y|_p$.

With the definition of this distance, it is now possible to view any p-adic number $x = \sum_{i=n_0}^{\infty} a_i p^i$ as the limite of the sequence of rational numbers $x_k = \sum_{i=n_0}^{k} a_i p^i$. Then $\mathbb{Q}$ is obviously dense in $\mathbb{Q}_p$ for the p-adic absolute value just as $\mathbb{Q}$ is dense in $\mathbb{R}$ for the usual absolute value. By analogy we will note $\mathbb{R} = \mathbb{Q}_\infty$.

In 1913, a student of Hensel showed that $\mathbb{Q}_p$ is the completion of $\mathbb{Q}$ for the p-adic distance. In 1918, Ostrovski proved that any absolute value on $\mathbb{Q}$ is equivalent either to one of a p-adic absolute value $| \, . \, |_p$ for some p, or to the usual absolute value. Consequently there are only two ways to complete the field $\mathbb{Q}$ to a valued field on which it is easier to do analysis: one leads to Archimedean field $\mathbb{R}$ and the other leads to the non-Archimedean fields. (Note that if p and q are two distinct primes, the fields $\mathbb{Q}_p$ and $\mathbb{Q}_q$ that we obtain are not isomorphic). The sets $\mathbb{R}$ and $\mathbb{Q}_p$ share many properties both algebraic and topological; we will recall the main ones:

Algebraic structure:

$\mathbb{Q}$ is a subfield of both $\mathbb{R}$ and $\mathbb{Q}_p$.

Any real number admits an expansion in base p : $\sum_{-\infty}^{k} a_n p^n$, where $k \in \mathbb{Z}$ and $0 \leq a_n \leq p-1$. The periodic expansions characterize the rational numbers.

Any p-adic number admits a Hensel's expansion $\sum_{n=k}^{+\infty} a_n p^n$, where $k \in \mathbb{Z}$ and $0 \leq a_n \leq p-1$. Similarly the periodic expansions characterize the rational numbers.

Neither $\mathbb{R}$ nor the fields $\mathbb{Q}_p$ are algebraically closed.

Topological structure:

$\mathbb{Q}$ is dense in both $\mathbb{R}$ and $\mathbb{Q}_p$. Both $\mathbb{R}$ and $\mathbb{Q}_p$ are locally compact.

All of these similarities incite us to do on each such field $\mathbb{Q}_p$ what one did on $\mathbb{R}$. This leads to develop the foundation of a theory of analytic functions, functional analysis, theory of integration, probability, where $\mathbb{Q}_p$ takes place the of $\mathbb{R}$.

However, we must note that, even though there are many similarities between $\mathbb{R}$ and $\mathbb{Q}_p$, there are also big differences. The algebraic closure of $\mathbb{R}$ is a finite extension of $\mathbb{R}$: $\mathbb{C} = \mathbb{R}[i]$. Armed with the absolute value that extends that of $\mathbb{R}$, the field $\mathbb{C}$ is complete. For the p-adic fields the situation is less simple. Whatever p, the algebraic closure of the field $\mathbb{Q}_p$ has an infinite dimension. Provided with the p-adic absolute value extending that of $\mathbb{Q}_p$, this algebraic closure is not complete. One has to take its completion, which fortunately, is also algebraically closed. Thus, we can get an algebraically closed, complete valued field. This field, which is denoted by $\mathbb{C}_p$, then plays for $\mathbb{Q}_p$ a role similar to that of $\mathbb{C}$ for $\mathbb{R}$.

Next, $\mathbb{R}$ is an ordered field while $\mathbb{Q}_p$ is not. The valued field $\mathbb{R}$ is Archimedean, while the p-adic valued fields are non-Archimedean. It follows that the geometry of any field $\mathbb{Q}_p$ is very different from that of $\mathbb{R}$. For example, all triangles are isosceles, the two equal sides being the largest (in absolute value), any point of a disk is the center of the disk and consequently, if two disks have a common point, one of these disks is contained in the other.

$\mathbb{R}$ is connected, while $\mathbb{Q}_p$ is totally disconnected.

These substantive differences lead to most of the differences between the real analysis and the p-adic analysis. As we said in the introduction, Hensel was the first who exploited new ideas in topology to develop an analysis on $\mathbb{Q}_p$, referring to the complex analysis developed by Cauchy. This one is based primarily on Taylor series that allow access to transcendental functions. Thus Hensel defined logarithm and exponential functions, as in the classical case, by the series:

$$Log(1+x) = \sum_{n=1}^{\infty}(-1)^{n+1}\frac{x^n}{n}, \quad Exp(x) = \sum_{n=0}^{\infty}\frac{x^n}{n!}.$$

The fact that p-adic fields are complete valued fields as $\mathbb{R}$ or $\mathbb{C}$ implies that most of the properties of power series in $\mathbb{C}_p$ are the same as in $\mathbb{C}$ and have similar proofs. For example:

The region of convergence of a power series $\sum_{n=0}^{\infty} a_n x^n$ is a disk of radius $\liminf_{n\to+\infty} \frac{1}{\sqrt[n]{|a_n|_p}}$ (It follows that, for example, the series defining the logarithm has a radius of convergence equal to one, whereas the one that defines the exponential function has the finite radius of convergence $\rho = p^{-\frac{1}{p-1}}$).

The radius of convergence has the usual properties: if r is the radius of convergence, the series converges on the disk $|x| < r$ and diverges on the set $|x| > r$. But, contrary to complex power series, what happens on the circle $|x| = r$ is rather simple: either the series converges everywhere or it converges nowhere.

The sum and product of two convergent series inside a disk is a convergent series in this disk.

The sum of a convergent power series is a differentiable function and its derivative is expanded in a power series of the same radius of convergence.

The same properties of course apply to power series $\sum_{n=0}^{\infty} a_n(x-\alpha)^n$, with $\alpha \in \mathbb{C}_p$.

Now, let D be an open set of $\mathbb{C}_p$. A function f from D to $\mathbb{C}_p$ is said to be *analytic* if it can be expanded in a power series converging in a neighborhood of α whenever $\alpha \in D$. The sum of a convergent power series is an analytic function on its region of convergence. Weierstrass gave a process to expand analytically the sum of a convergent power series out of its convergence disk in $\mathbb{C}$. But by Schöbe's thesis (1936), we know that this method does not work in an ultrametric valued field. Indeed, let $f(x) = \sum_{n=0}^{\infty} a_n(x-\alpha)^n$ be a power series in $\mathbb{C}_p$ and let D be its convergence disk. If we expand $f(x)$ with respect to a point $\beta \in D$, we obtain a power series which has the same region of convergence. Thus through that method, it is never possible to get out of the convergence disk.

On the other hand, a uniform analytic function on a connected region of $\mathbb{C}$ is characterized by its values on an open subset however small it is. Indeed, if two functions defined on the same connected set take the same values on a disk of radius as small as we want, they coincide on the whole set. We say that *they satisfy the principle of uniqueness.* If we want to define on a p-adic field a method of analytic extension that satisfies the principle of uniqueness, it is not possible to define the analytic functions by local properties as, for example, having a convergent expansion in power series around any point, because the fields $\mathbb{C}_p$ are totally disconnected. Under these conditions, for realizing the analytic continuation it is necessary to seek other methods where analytic elements (that is to say the bricks of extension) are not necessarily power series with their convergence disk. The emergence of a theory of analytic functions on p-adic fields is due to Marc Krasner in the late forties and the fifties [**17**]. It was then developed by the author until 1974 [**18**] and generalized by a French school: M. Lazard [**21**], E. Motzkin [**23**], Ph. Robba [**24**], A. Escassut [**5**], [**6**], [**7**], A. Boutabaa [**3**], Y. Feneyrol-Perrin [**8**], [**9**] (let us also mention Yvette Amice's work [**1**]). Let Krasner expose his idea. *My attention was drawn in 1946 by Runge's method (equivalent to what of Weierstrass in the complex field), where the analytic elements in question are analytic functions defined on some connected sets characterized by Runge (which bear his name). They are uniform limits of rational fractions without pole on these sets. Of course, the area I have identified for the same purpose in ultrametric valued fields are not connected because the field itself is not, but they have some properties of the connected sets in the complex field. That is why I called them quasi-connected sets.*

Krasner showed that analytic elements satisfy the principle of uniqueness and then defined analytic functions, as in the complex case, by a chain of analytic elements [**18**]. This notion of analytic elements was then widespread and their properties analyzed in depth as well as those of their analytic extensions, guided by the properties of complex analytic functions. We do not go farther into details but refer to the work of the authors mentioned above.

5. The p-adic numbers at the foundation of quantum mechanics

The use of p-adic numbers was not confined to the domain of analytic functions opened by Hensel, but has invested throughout the 20-th century all branches of analysis: theory of integration, normed vector spaces, Hilbert spaces, operators theory, probability, dynamical systems. All conceptual tools, necessary to give a mathematical framework to quantum physics, were now available to physicists and mathematicians. In the 1990's, a team of Russian mathematicians: V. Vladimirov, I. Volovich, E. Zelenov [**25**], [**26**], [**27**] and A. Khrennikov [**20**] took possession of these tools to found a new theory of quantum mechanics. We will not describe this theory but we just refer to the work of those authors. We want to only emphasize two strong ideas developed especially by Khrennikov [**20**] which, in our point of view, justify the attempt to abandon the real continuum and substitute totally disconnected fields to found a mathematical model of the quantum world. The first idea is the necessity to take into account the limit of exactness of physical measurement in the mathematical model itself and not only in the result of experiments. The second idea is to justify negative probabilities involved in some formulas of the theory of quantum mechanics.

The measures of physical phenomena are rational numbers. With an absolute value on $\mathbb{Q}$ it is possible to have a notion of size of the phenomena, to evaluate their distance and to make calculations on the data. Mathematical models of the physical world used the usual absolute value and its extension to the field of real numbers. Since $\mathbb{R}$ is an Archimedean ordered field, this model assumes that the measurements of physical quantities (called observables in the language of quantum mechanics) can be obtained with infinite exactness. The quantization of energy, the discrete nature of the spectra of physical observables require that the mathematical formalism which models these phenomena should be based on a system of numbers which limits the accuracy of physical measurement. This is the case of p-adic numbers, as we are going to see.

Measurement of physical quantities.
The measure L of a physical quantity is a rational number which is in the form

$$L = \frac{a_{-k}}{m^k} + ... + \frac{a_{-1}}{m} + n_0 \tag{3}$$

where $n_0 \in \mathbb{N}$, $m \in \mathbb{N}$, $0 \leq a_{-j} \leq m-1$, $j = 1, ..., k$. (For simplicity and to stay within our topic, we assume that m is a prime number. But what follows is valid for any integer, except that the ring $\mathbb{Q}_m$ is not a field when m is not prime.) The exactness of the measurement is given by the term $\frac{1}{m^k}$. This means that the measurement process is continued in $k+1$ steps, the unit of measurement being divided by m at each step. All physical measurement are made with finite exactness. In formula (3) let us develop the entire part n_0 in base m and denote by B_m the set of numbers of the form :

$\frac{a_{-k}}{m^k} + ... + \frac{a_{-1}}{m} + a_0 + ... + a_q m^q, \quad 0 \leq a_i \leq m-1, \ -k \leq i \leq q, \ (k \in \mathbb{N}, \ q \in \mathbb{N})$

and let us set $-B_m = \{-x \mid x \in B_m\}$. If we complete the set $B_m \cup (-B_m)$ around to the usual absolute value of $\mathbb{Q}$ we obtain all the real numbers. Some of these numbers have infinite negative powers of m in their expansion. In any case these numbers may represent quantified physical objects, such as energy. If, instead, we consider a complete extension of B_m relative to the m-adic absolute value we obtain the field of m-adic numbers. These numbers can be represented by their Hensel's expansion $\sum_{n=-k}^{\infty} a_n m^n$ whose fractional part $\sum_{n=-k}^{-1} a_n m^n$ is finite. These numbers can be interpreted as physical quantities measurements made with limited exactness. Interpreted in terms of measures of magnitude, the elements of $\mathbb{Q}_m$ which have an infinite number of positive powers of m, can be described as ideal elements corresponding to infinitely large physical quantities, the measurement process of these quantities having infinite steps where at each step the measurement unit is multiplied by m. But an important restriction is to do. Indeed, Hensel's expansion of any negative rational number has an infinite number of positive powers of m. Suppose $a = \sum_{n=j}^{q} a_n m^n$. Then

$$\forall a \in \mathbb{Q}^+, \ -a = (-1)a = \Big(\sum_{n=0}^{\infty}(m-1)m^n\Big)\Big(\sum_{n=j}^{q} a_n m^n\Big) = \sum_{n=j}^{\infty} \alpha_n m^n.$$

and therefore it cannot be considered as a measure of physical quantity. Consequently, using a m-adic numeration for measurement of physical quantities imposes not to use two sides scales where results of physical quantities can have different sign: The sign minus is reserved for infinitely large quantities [**20**].

Probabilities.

Probability theory plays an essential role both in the classical theory of quantum mechanics and in the p-adic one. The notion of measurable quantity is replaced by the distribution of values, each of them assigned with a probability. According to Kolmogorov's axioms, adopted unanimously by scientists, a probability is a measure with values in the range $[0, 1]$ of the real line. But in standard quantum mechanics and quantum field theory, negative probabilities appear very naturally in certain formulas, so that some physicists have used them in their calculations without giving a precise meaning. For example, this is the case of P. Dirac [**4**] and R.

Feynman [10]. The presence of negative probabilities led to the hidden variable theory. Suppose that in Bayes formula:

$$P(i) = \sum_{\alpha} P(\frac{i}{\alpha})P(\alpha)$$

with $\sum_{\alpha} P(\alpha) = 1$, each $P(i)$ is positive and some $P(\frac{i}{\alpha})$ or $P(\alpha)$ are negative. If it is impossible to measure $P(\frac{i}{\alpha})$ or $P(\alpha)$ one considers that α is a hidden variable. We know that this theory of hidden variables has raised much controversy, especially from A. Einstein. The theory of p-adic probabilities proposed in the 1990s by A. Khrennikov [19] gives a very natural meaning at these negative probabilities and no longer requires to use hidden variables for their justification. The idea is to return to the notion of frequential probabilities proposed by R. von Mises [22]. (Note that Kolmogorov's axioms are the theorems of von Mises probabilities, except the axiom of sigma-additivity). Schematically the probability of an event is the limit of a sequence of relative frequencies $f_n = \frac{c_n}{N}$ where c_n is the number of realisations of the event after N experiments. The f_n' are rational numbers between 0 and 1. The limit of the sequence $(f_n)_{n\in\mathbb{N}}$ in $\mathbb{R}$ belongs to the interval $[0,1]$ of $\mathbb{R}$, but taken in a p-adic completion of $\mathbb{Q}$, this limit can take any value of $\mathbb{Q}_p$. Indeed, it is easy to prove the following proposition:

PROPOSITION. *The closure of* $\mathbb{Q}\cap[0,1]$ *in* $\mathbb{Q}_p$ *for the topology associated to the* p*-adic absolute value is* $\mathbb{Q}_p$.

Let x be a p-adic number and let $\sum_{n=n_0}^{+\infty} a_n p^n$ be its Hensel's expansion. Then x is the limit of the sequence $x_k = \sum_{n=n_0}^{k} a_n p^n$ of rational numbers. There exists a sequence $(u_k)_{k\in\mathbb{N}}$ of rational numbers such that $x_k u_k \in [0,1]\cap\mathbb{Q}$ and $\lim_{k\to+\infty} |u_k - 1|_p = 0$, for example $u_k = 1 + p^{n_0+k+1}$. Then $x = \lim_{k\to+\infty} u_k x_k$ in $\mathbb{Q}_p$.

It follows that any negative rational numbers can be, a priori, a value of probability. We know that the theory due to von Mises was strongly criticized. It is based on the notion of random sequence, called collective in von Mises terminology. The definition of collective raised great difficulties which contributed to giving up the theory. Today some mathematicians have rehabilitated this theory and solve the problem of collectives on the bases of Kolmogorov's notion of algorithms complexity. We refer for example to the work of Khrennikov [20], to have the development of a comprehensive theory of probability with values in a p-adic field.

6. Conclusion

While two famous German mathematicians G. Cantor and R. Dedekind presented a rigorous construction of the real numbers, on which would be based all mathematics and physics theories in the 20th century, another German mathematician, K. Hensel, opened another mathematical universe, not Archimedean, totally disconnected, that seemed exotic to most of mathematicians, but now finds echoes, not only in mathematics but also in the world of physical and human sciences. It is not yet known what promising future the use of p-adic fields in modelling the quantum world may have, but it reveals how our conception of physical reality depends on mathematical constructions that man has created over the centuries.

References

[1] Y. Amice, *Les nombres p-adiques* (French), Presses Universitaires de France, Paris, 1975. Préface de Ch. Pisot; Collection SUP: Le Mathématicien, No. 14. MR0447195

[2] N. Bourbaki, *Éléments d'histoire des mathématiques* (French), Hermann, Paris, 1974. Nouvelle édition revue, corrigée et augmentée; Histoire de la Pensée, No. IV. MR0349314

[3] A. Boutabaa, *Théorie de Nevanlinna p-adique* (French, with English summary), Manuscripta Math. **67** (1990), no. 3, 251–269, DOI 10.1007/BF02568432. MR1046988

[4] P. A. M. Dirac, *The physical interpretation of quantum mechanics*, Proc. Roy. Soc. London. Ser. A. **180** (1942), 1–40, DOI 10.1098/rspa.1942.0023. MR0010295

[5] A. Escassut, *Analytic elements in p-adic analysis*, World Scientific Publishing Co., Inc., River Edge, NJ, 1995. MR1370442

[6] J.-P. Bézivin, K. Boussaf, and A. Escassut, *Zeros of the derivative of a p-adic meromorphic function*, Bull. Sci. Math. **136** (2012), no. 8, 839–847, DOI 10.1016/j.bulsci.2012.07.003. MR2995004

[7] K. Boussaf, A. Boutabaa, and A. Escassut, *Growth of p-adic entire functions and applications*, Houston J. Math. **40** (2014), no. 3, 715–736. MR3275619

[8] Y. Feneyrol-Perrin, *Fonctions analytiques sur un corps algébriquement clos, muni d'une valuation au sens de Krull, et complet* (French, with English summary), C. R. Acad. Sci. Paris Sér. A-B **287** (1978), no. 2, A74–A73. MR0491618

[9] Y. Feneyrol-Perrin and L. Haddad, *Un théorème de Mittag-Leffler pour les fonctions méromorphes sur un corps valué au sens de Krull* (French), Compositio Math. **57** (1986), no. 2, 249–269. MR827354

[10] R. P. Feynman, *Negative probability*, Quantum implications, Routledge & Kegan Paul, London, 1987, pp. 235–248. MR944983

[11] F. Q. Gouvêa, *p-adic numbers. An introduction*, Universitext, Springer-Verlag, Berlin, 1993. MR1251959

[12] Hensel, K. *über die Fundamentalgleichnung und die ausserwesentlichen Diskrimantenheiler eines algebraischen Karpers.* Gott. Nach. (1897).

[13] K. Hensel, *Neue Grundlagen der Arithmetik* (German), J. Reine Angew. Math. **127** (1904), 51–84, DOI 10.1515/crll.1904.127.51. MR1580632

[14] Hensel, K. *über die Arithmetische Eigenschaften der algebraischen und transzendenten Zahlen.* Jahresber.der D.M.V., t. XIV (1905).

[15] Hensel, K. *über die Arithmetischen Eigenschaften der Zahlen*, Jahresber.der D.M.V., t. XVI (1907).

[16] Hensel, K. *Theorie der algebraischen Zahlen.* Leipzig (Teubner) (1908).

[17] M. Krasner, *Prolongement analytique dans les corps valués complets: éléments analytiques, préliminaires du théorème d'unicité* (French), C. R. Acad. Sci. Paris **239** (1954), 468–470. MR0067925

[18] M. Krasner, *Rapport sur le prolongement analytique dans les corps valués complets par la méthode des éléments analytiques quasi-connexes* (French), Table Ronde d'Analyse non archimédienne (Paris, 1972), Soc. Math. France, Paris, 1974, pp. 131–254. Bull. Soc. Math. France, Mém. No. 39–40. MR0385168

[19] A. Yu. Khrennikov, *On the extension of the von Mises frequency approach and the A. N. Kolmogorov axiomatic approach to p-adic probability theory* (Russian, with Russian summary), Teor. Veroyatnost. i Primenen. **40** (1995), no. 2, 458–464, DOI 10.1137/1140040; English transl., Theory Probab. Appl. **40** (1995), no. 2, 371–376 (1996). MR1346483

[20] A. Khrennikov, *Non-Archimedean analysis: quantum paradoxes, dynamical systems and biological models*, Mathematics and its Applications, vol. 427, Kluwer Academic Publishers, Dordrecht, 1997. MR1746953

[21] M. Lazard, *Les zéros des fonctions analytiques d'une variable sur un corps valué complet* (French), Inst. Hautes Études Sci. Publ. Math. **14** (1962), 47–75. MR0152519

[22] R. von Mises, *Probability, statistics and truth*, The Macmillan Company, New York, 1957. Second revised English edition, prepared by Hilda Geiringer. MR0090158

[23] E. Motzkin, *Prolongement analytique en analyse p-adique* (French), Colloque de Théorie des Nombres (Univ. Bordeaux, Bordeaux, 1969), Soc. Math. France, Paris, 1971, pp. 151–158. Bull. Soc. Math. France, Mém. No. 25. MR0367278

[24] P. Robba, *Fonctions analytiques sur les corps valués ultramétriques complets* (French, with English summary), Prolongement analytique et algèbres de Banach ultramétriques, Société Mathématique de France, Paris, 1973, pp. 109–218, 219–220. Astérisque, No. 10. MR0357841

[25] V. S. Vladimirov and I. V. Volovich, *p-adic quantum mechanics*, Comm. Math. Phys. **123** (1989), no. 4, 659–676. MR1006299

[26] V. S. Vladimirov, I. V. Volovich, and E. I. Zelenov, *p-adic analysis and mathematical physics*, Series on Soviet and East European Mathematics, vol. 1, World Scientific Publishing Co., Inc., River Edge, NJ, 1994. MR1288093

[27] E. I. Zelenov, *p-adic Heisenberg group and Maslov index*, Comm. Math. Phys. **155** (1993), no. 3, 489–502. MR1231639

Université Clermont Auvergne, Laboratoire de Mathématiques Blaise Pascal, UMR-CNRS, BP 10448, F-63000 Clermont-Ferrand, France

Email address: Yvette.Perrin@math.univ-bpclermont.fr

Contemporary Mathematics
Volume **704**, 2018
http://dx.doi.org/10.1090/conm/704/14172

On some classes of non-Archimedean Köthe spaces

Wiesław Śliwa and Agnieszka Ziemkowska

Abstract. We introduce some classes of non-Archimedean Köthe spaces and study relations between them.

1. Introduction

In this paper all linear spaces are over a non-Archimedean non-trivially valued field $\mathbb{K}$ which is complete under the metric induced by the valuation $|\cdot| : \mathbb{K} \to [0, \infty)$. For fundamentals on locally convex Hausdorff spaces (lcs) and normed spaces we refer to [3, 4, 5].

By a *Köthe space* we mean an infinite-dimensional Fréchet space with a Schauder basis and with a continuous norm. The power series spaces $A_p(a)$ and their generalizations $D_f(a, r)$ are the most known and important examples of nuclear Köthe spaces (see [1, 7, 8, 9, 10]).

An infinite matrix $A = (a_n^k)$ of positive real numbers is a *Köthe matrix* if $a_n^k \leq a_n^{k+1}$ for all $k, n \in \mathbb{N}$.

The space $K(A) = \{(\alpha_n) \subset \mathbb{K} : \lim_{n\to\infty} |\alpha_n| a_n^k = 0 \text{ for every } k \in \mathbb{N}\}$ with the base (p_k) of norms, where

$$p_k((\alpha_n)) = \max_n |\alpha_n| a_n^k, k \in \mathbb{N},$$

is a Köthe space. The sequence (e_j), where $e_j = (\delta_{j,n})$, is an unconditional Schauder basis in $K(A)$. Any Köthe space is isomorphic to the space $K(A)$ for some Köthe matrix A (see [**1**, Proposition 2.4]).

A Köthe space $K(A)$ is nuclear if and only if for any $i \in \mathbb{N}$ there exists $j \in \mathbb{N}$ with $j > i$ such that $\lim_n a_n^i / a_n^j = 0$ ([**1**, Proposition 3.5]).

Let Γ be the family of all non-decreasing sequences $a = (a_n)$ of positive real numbers with $\lim a_n = \infty$. Let $a = (a_n) \in \Gamma$. Then the following Köthe spaces are nuclear (see [**1**]):

(1) $A_1(a) = K(A)$, where $A = (a_n^k), a_n^k = \exp(-a_n/k)$;
(2) $A_\infty(a) = K(A)$, where $A = (a_n^k), a_n^k = \exp(k a_n)$.

$A_1(a)$ and $A_\infty(a)$ are the *power series spaces* (*of finite type* and *of infinite type*, respectively).

Denote by Φ_c the family of all strictly increasing odd functions $f : \mathbb{R} \to \mathbb{R}$ such that f is convex in $[0, \infty)$. For $r \in (-\infty, \infty]$ we denote by Λ_r the family of all

2010 *Mathematics Subject Classification.* Primary 46S10, 46A35.
Key words and phrases. Non-Archimedean Köthe space, Schauder basis, bounded operator.

strictly increasing sequences $(r_k) \subset \mathbb{R}$ with $\lim_k r_k = r$ such that $r_k r_j > 0$ for all $k, j \in \mathbb{N}$.

Let $f \in \Phi_c, a \in \Gamma, r \in (-\infty, \infty]$ and $(r_k) \in \Lambda_r$. Denote by $D_f(a,r)$ the Köthe space $K(A)$, where $A = (a_n^k) = (\exp f(r_k a_n))$; it is clear that the space $D_f(a,r)$ does not depend on the choice of the sequence $(r_k) \in \Lambda_r$. It is said to be a *generalized power series space* (see [9, 10]). It is nuclear ([9, Proposition 2]).

If $f : \mathbb{R} \to \mathbb{R}, f(x) = x$ and $r = 0$ (respectively, $r = \infty$), then $D_f(a,r)$ is isomorphic to $\Lambda_1(a)$ (respectively, $\Lambda_\infty(a)$).

We will say that a Köthe matrix $A = (a_n^k)$ is of type

$$(d_0) \text{ if } \frac{a_{n+1}^k}{a_{n+1}^{k+1}} \leq \frac{a_n^k}{a_n^{k+1}} \text{ for all } k, n;$$

$$(d_1) \text{ if } \exists k : \forall j\ \exists l : \sup_n \frac{(a_n^j)^2}{a_n^k a_n^l} < \infty;$$

$$(d_2) \text{ if } \forall k\ \exists j : \sup_n \frac{a_n^k a_n^l}{(a_n^j)^2} < \infty \text{ for any } l;$$

$$(d_3) \text{ if } \frac{a_n^{k+1}}{a_n^k} \leq \frac{a_n^{k+2}}{a_n^{k+1}} \text{ for all } k, n;$$

$$(d_4) \text{ if } \frac{a_n^{k+1}}{a_n^k} \geq \frac{a_n^{k+2}}{a_n^{k+1}} \text{ for all } k, n;$$

$$(d_5) \text{ if } \exists M \geq 1 : \frac{a_n^{k+1}}{a_n^k} \leq \left(\frac{a_n^{k+2}}{a_n^{k+1}}\right)^M \quad \text{for all } k, n.$$

Let $0 \leq i \leq 5$. We say that a Schauder basis (x_n) in a Köthe space E is of type (d_i) if there exists a Köthe matrix $A = (a_n^k)$ of type (d_i) such that the map

$$T : E \to K(A), x = \sum_{n=1}^{\infty} \alpha_n x_n \to Tx = (\alpha_n)$$

is an isomorphism. A Schauder basis of type (d_0) in a Köthe space is called regular. The Schauder basis (e_n) in $D_f(a,r)$ is regular (see [9, Proposition 2] and its proof).

In [6, Proposition 3] we have shown that any two regular Schauder bases (x_n) and (y_n) in a Köthe space E are semi-equivalent i.e. for some sequence (β_n) of non-zero scalars the map

$$T : E \to E, x = \sum_{n=1}^{\infty} \alpha_n x_n \to Tx = \sum_{n=1}^{\infty} \beta_n \alpha_n y_n$$

is an isomorphism.

Any Schauder basis in a Köthe space E with a regular Schauder basis is a permutation of a regular Schauder basis in E ([6, Proposition 4]). Thus any Köthe space with a regular Schauder basis has the quasi-equivalence property, that is any two Schauder bases (x_n) and (y_n) in E are quasi-equivalent, i.e. (x_n) is semi-equivalent to some permutation $(y_{\pi(n)})$ of (y_n) ([6, Theorem 5]).

The generalized power series spaces $D_f(a,r)$ have the quasi-equivalence property ([9, Proposition 2]).

A Köthe space E is called of type (d_i) if it is isomorphic to the Köthe space $K(A)$ for some Köthe matrix A of type (d_i) $(0 \leq i \leq 5)$.

Clearly, $D_f(a,r)$ is of type (d_0).

$D_f(a,r)$ is of type: (d_1) if $r=\infty$ or f is rapidly increasing (i.e.

$$\lim_{t\to\infty}[f(ct)/f(t)]=\infty$$

for any $c>0$) and $r>0$; (d_2) if $r\le 0$; (d_3) if $r=\infty$; (d_4) if $r\le 0$ or $f(t)=t, t\in\mathbb{R}$; (d_5) if $r>0$ or $f(t)=t, t\in\mathbb{R}$ (Proposition 3.8). In particular: $A_1(a)$ is of type $(d_0),(d_2),(d_4)$ and (d_5); $A_\infty(a)$ is of type $(d_0),(d_1),(d_3),(d_4)$ and (d_5).

N. De Grande-De Kimpe has proved in [1, Proposition 4.3] that every continuous linear map from $A_1(a)$ to $A_\infty(b)$ is bounded. We generalize this result by proving that every continuous linear map from a Köthe space of type (d_2) to a Köthe space of type (d_1) is bounded (Proposition 3.10).

The main results of this paper are the following two theorems.

A nuclear Köthe space E is isomorphic to a power series space of infinite type if and only if E has a Schauder basis (x_n) of type (d_3) and (d_4) (Theorem 3.12). *A nuclear Köthe space E is isomorphic to a power series space of finite type if and only if E has a Schauder basis (x_n) of type (d_2) and (d_5)* (Theorem 3.14).

In our paper we use and develop some ideas of [2, Chapter II] and [11].

2. Preliminaries

By a *seminorm* on a linear space E we mean a function $p: E\to[0,\infty)$ such that $p(\alpha x)=|\alpha|p(x)$ for all $\alpha\in\mathbb{K}, x\in E$ and $p(x+y)\le\max\{p(x),p(y)\}$ for all $x,y\in E$. A seminorm p on E is a *norm* if $\ker p:=\{x\in E: p(x)=0\}=\{0\}$. The set of all continuous norms on a locally convex space E is denoted by $\mathcal{P}_c(E)$.

A sequence (x_n) in a lcs E is a *Schauder basis* in E if each $x\in E$ can be written uniquely as $x=\sum_{n=1}^\infty \alpha_n x_n$ with $\alpha_n\in\mathbb{K}, n\in\mathbb{N}$, and the coefficient functionals $f_n: E\to\mathbb{K}, x\to\alpha_n$ $(n\in\mathbb{N})$ are continuous.

A metrizable complete lcs is called a *Fréchet space.* A series $\sum_{n=1}^\infty x_n$ in a Fréchet space E is convergent if and only if $\lim_n x_n=0$.

An increasing sequence (p_k) of continuous norms on a Fréchet space E is a *base* of continuous norms on E if for any $p\in\mathcal{P}_c(E)$ there is $k\in\mathbb{N}$ such that $p\le kp_k$.

Let E,F be locally convex spaces. A map $T: E\to F$ is called a linear homeomorphism if it is linear, bijective and the maps T,T^{-1} are continuous. If there exists a linear homeomorphism $T: E\to F$, then we say that E is isomorphic to F. A linear map $T: E\to F$ is *bounded* if for some neighbourhood U of 0 in E the set $T(U)$ is bounded in F.

If E and F are nuclear Köthe spaces and every linear continuous map from E to F is bounded, then E has no quotient isomorphic to F and F has no subspace isomorphic to E.

Two Köthe matrices $A=(a_n^k)$ and $B=(b_n^k)$ are *equivalent* if

$$(1)\ \forall k\ \exists l: \sup_n a_n^k/b_n^l<\infty;\ (2)\ \forall l\ \exists t: \sup_n b_n^l/a_n^t<\infty;$$

clearly, $A=(a_n^k)$ and $B=(b_n^k)$ are *equivalent* if and only if $K(A)$ and $K(B)$ are equal as sets and the identity map $I: K(A)\to K(B)$ is an isomorphism.

A Köthe matrix $A=(a_n^k)$ is a *representation of a Schauder basis* (x_n) in a Köthe space E if the map

$$T: E\to K(A), x=\sum_{n=1}^\infty \alpha_n x_n\to Tx=(\alpha_n)$$

is an isomorphism.

For a sequence (y_n) and a set Y, by $(y_n) \subset Y$ we mean that $y_n \in Y, n \in \mathbb{N}$.

Let $\varphi \in \mathbb{K}$ with $|\varphi| > 1$. Then for any $a > 0$ there exists $k \in \mathbb{Z}$ such that $|\varphi^k| \in [a, |\varphi|a)$. Let $(c_n) \subset (0, \infty)$. If $A = (a_n^k)$ and $D = (d_n^k)$ are Köthe matrices with $a_n^k = c_n d_n^k$ for $n, k \in \mathbb{N}$, then the Köthe spaces $K(A)$ and $K(D)$ are isomorphic; if $(\varphi_n) \subset \mathbb{K}$ with $|\varphi_n| \in [c_n, |\varphi| c_n), n \in \mathbb{N}$, then the map

$$T : K(A) \to K(D), (\alpha_n) \to (\alpha_n \varphi_n)$$

is an isomorphism.

3. Results

We start with the following two simple propositions.

PROPOSITION 3.1. A Schauder basis (x_n) in a Köthe space E is of type (d_1) if and only if it is of type (d_3).

Proof. Assume that (x_n) has a representation $A = (a_n^k)$ of type (d_1). Then there exists a strictly increasing sequence $(j_k) \subset \mathbb{N}$ such that $\sup_n [(a_n^{j_k})^2 / a_n^{j_1} a_n^{j_{k+1}})] < \infty$ for every $k \in \mathbb{N}$. Let $B = (b_n^k)$, where $b_n^k = a_n^{j_k}$ for all $k, n \in \mathbb{N}$. Clearly B and A are equivalent Köthe matrices. We have

$$\forall k \, \exists M_k > 0 : \frac{(b_n^k)^2}{b_n^1 b_n^{k+1}} < M_k \text{ for any } n,$$

so for all $k, n \in \mathbb{N}$ we have

$$\frac{b_n^{k+1}}{b_n^k} \leq \frac{b_n^{k+1}}{b_n^1} \leq M_{k+1} \frac{b_n^{k+2}}{b_n^{k+1}}.$$

Let $(t_k) \subset (0, \infty)$ be an increasing sequence with $t_k t_{k+2} > M_{k+1} t_{k+1}^2$ for any $k \in \mathbb{N}$. Let $c_n^k = t_k b_n^k$ for all $k, n \in \mathbb{N}$. Then $C = (c_n^k)$ is a representation of (x_n) and

$$\frac{(c_n^{k+1})^2}{c_n^k c_n^{k+2}} = \frac{(t_{k+1})^2}{t_k t_{k+2}} \frac{(b_n^{k+1})^2}{b_n^k b_n^{k+2}} < M_{k+1}^{-1} M_{k+1} = 1$$

for all $k, n \in \mathbb{N}$, so C is of type (d_3). Thus (x_n) is of type (d_3).

Assume now that (x_n) has a representation $A = (a_n^k)$ of type (d_3). Then for all $k, n, m \in \mathbb{N}$ we have

$$\frac{a_n^{k+1}}{a_n^k} \leq \frac{a_n^{k+m+1}}{a_n^{k+m}}.$$

Hence for all $k, n \in \mathbb{N}$ we get

$$\frac{a_n^k}{a_n^1} = \frac{a_n^2}{a_n^1} \frac{a_n^3}{a_n^2} \cdots \frac{a_n^k}{a_n^{k-1}} \leq \frac{a_n^{k+1}}{a_n^k} \frac{a_n^{k+2}}{a_n^{k+1}} \cdots \frac{a_n^{2k-1}}{a_n^{2k-2}} = \frac{a_n^{2k-1}}{a_n^k}.$$

Thus $\sup_n [(a_n^k)^2 / a_n^1 a_n^{2k-1}] \leq 1$ for every $k \in \mathbb{N}$, so A is of type (d_1). □

COROLLARY 3.2. A Köthe space is of type (d_1) if and only if it is of type (d_3).

PROPOSITION 3.3. A Köthe space E of type (d_0) is nuclear if and only if it is non-normable.

Proof. Without loss of generality we can assume that $E = K(A)$, where $A = (a_n^k)$ is a Köthe matrix of type (d_0). Then the sequence $(a_n^l / a_n^k)_{n=1}^{\infty}$ is monotone for all $l, k \in \mathbb{N}$, so

$$\forall l \exists k : \lim_n a_n^l / a_n^k = 0 \Leftrightarrow \forall l \exists k : \inf_n a_n^l / a_n^k = 0.$$

Thus E is nuclear if and only if E is non-normable. □

Corollary 3.4. Any non-normable Köthe space with a regular Schauder basis is nuclear.

The next proposition is clear.

Proposition 3.5. Let $1 \leq i \leq 5$. (1) Let (x_n) and (y_n) be quasi-equivalent Schauder bases in a Köthe space E. Then (x_n) is of type (d_i) if and only if (y_n) is of type (d_i). (2) Let E be a Köthe space with the quasi-equivalence property. Then the following conditions are equivalent: (a) E is of type (d_i); (b) E has a Schauder basis of type (d_i); (c) Any Schauder basis in E is of type (d_i).

Corollary 3.6. Let E be a Köthe space with a regular Schauder basis. Then for every $1 \leq i \leq 5$ the following conditions are equivalent: (a) E is of type (d_i); (b) E has a Schauder basis of type (d_i); (c) Any Schauder basis in E is of type (d_i).

We shall need the following.

Lemma 3.7. *Let $f \in \Phi_c$ and $z > y > x \geq 0$. Then*

$$\frac{f(z) - f(y)}{f(y) - f(x)} \geq \frac{z - y}{y - x}.$$

Proof. Let $t = [(z - y)/(z - x)]$. Then $t \in (0, 1), 1 - t = [(y - x)/(z - x)]$ and $y = tx + (1 - t)z$. Hence $f(y) \leq tf(x) + (1 - t)f(z)$, so

$$\frac{f(z) - f(y)}{f(y) - f(x)} \geq \frac{f(z) - tf(x) - (1 - t)f(z)}{tf(x) + (1 - t)f(z) - f(x)} = \frac{t(f(z) - f(x))}{(1 - t)(f(z) - f(x))} = \frac{t}{1 - t} = \frac{z - y}{y - x}.$$

□

Using Lemma 3.7 we get the following.

Proposition 3.8. Let $f \in \Phi_c, a \in \Gamma$ and $r \in (-\infty, \infty]$. Then $D_f(a, r)$ is
(1) of type (d_1) if $r = \infty$ or f is rapidly increasing and $r > 0$;
(2) of type (d_2) if $r \leq 0$;
(3) of type (d_3) if $r = \infty$ or f is rapidly increasing and $r > 0$;
(4) of type (d_4) if $r \leq 0$ or $f(t) = t, t \in \mathbb{R}$;
(5) of type (d_5) if $r > 0$ or $f(t) = t, t \in \mathbb{R}$.

Proof. Let $A = (a_n^k) = (\exp f(r_k a_n))$ where $(r_k) \in \Lambda_r$.

(1) Let $r = \infty$. Let $p \in \mathbb{N}$ and $k \in \mathbb{N}$. Then there exists $l > k$ with $r_l > 2r_k$. If $k > p$ then by Lemma 3.7 we get

$$\frac{f(r_l a_i) - f(r_k a_i)}{f(r_k a_i) - f(r_p a_i)} \geq \frac{r_l - r_k}{r_k - r_p} > 1$$

for all $i \in \mathbb{N}$; so

$$f(r_l a_i) - f(r_k a_i) \geq f(r_k a_i) - f(r_p a_i)$$

for all $i, k, p \in \mathbb{N}$. Thus $2f(r_k a_i) - f(r_p a_i) - f(r_l a_i) \leq 0$ for all $i, k, p \in \mathbb{N}$. It follows that for all $p \in \mathbb{N}, k \in \mathbb{N}$ there exists $l \in \mathbb{N}$ such that $(a_i^k)^2/(a_i^p a_i^l) \leq 1$ for all $i \in \mathbb{N}$. Thus A is of type (d_1).

Let $r \in (0, +\infty)$ and f be rapidly increasing. Let $p \in \mathbb{N}$ and $k \in \mathbb{N}$. Let $l > k$. Then $[f(r_l a_i)/f(r_k a_i)] \to \infty$, as $i \to \infty$, so there exists $i_0 \in \mathbb{N}$ such that $f(r_l a_i) \geq 2f(r_k a_i)$ for all $i \geq i_0$. Hence $2f(r_k a_i) - f(r_p a_i) - f(r_l a_i) \leq 0$ for all $i \geq i_0$. It follows that for all $p \in \mathbb{N}, k \in \mathbb{N}$ there exists $l \in \mathbb{N}$ such that $\sup_i[(a_i^k)^2/(a_i^p a_i^l)] < \infty$. Thus A is of type (d_1).

(2) Let $r \le 0$. Let $k \in \mathbb{N}$. Then there exists $q \in \mathbb{N}$ with $2r_q > r + r_k$. Hence $-r_k + r_q \ge -r_q + r_v > 0$ for all $v > q$. By Lemma 3.7 we get

$$\frac{f(-r_k a_j) - f(-r_q a_j)}{f(-r_q a_j) - f(-r_v a_j)} \ge \frac{-r_k + r_q}{-r_q + r_v} \ge 1$$

for all $j \in \mathbb{N}$, so

$$f(r_q a_j) - f(r_k a_j) \ge f(r_v a_j) - f(r_q a_j)$$

for $v > q$ and $j \in \mathbb{N}$. Thus $f(r_k a_j) + f(r_v a_j) - 2f(r_q a_j) \le 0$ for all $v, j \in \mathbb{N}$. It follows that for any $k \in \mathbb{N}$ there exists $q \in \mathbb{N}$ such that $[a_j^k a_j^v/(a_j^q)^2] \le 1$ for all $v, j \in \mathbb{N}$. Thus A is of type (d_2).

(3) It is enough to use (1) and Proposition 3.1.

(4) Let $r \le 0$. Let $(r_k) = (-k^{-1})$ if $r = 0$ and $(r_k) = ((k+1)k^{-1}r)$ if $r < 0$. Then $[(-r_k + r_{k+1})/(-r_{k+1} + r_{k+2})] = (k+2)k^{-1} > 1$ for all $k \in \mathbb{N}$. By Lemma 3.7 we get

$$\frac{f(-r_k a_n) - f(-r_{k+1} a_n)}{f(-r_{k+1} a_n) - f(-r_{k+2} a_n)} \ge \frac{-r_k + r_{k+1}}{-r_{k+1} + r_{k+2}} > 1$$

for all $k, n \in \mathbb{N}$. Hence

$$f(r_{k+1} a_n) - f(r_k a_n) \ge f(r_{k+2} a_n) - f(r_{k+1} a_n)$$

for all $k, n \in \mathbb{N}$. It follows that $(a_n^{k+1})^2 \ge a_n^k a_n^{k+2}$ for all $k, n \in \mathbb{N}$, so A is of type (d_4).

Let $r > 0$ and $f(t) = t, t \in \mathbb{R}$. Let $(r_k) = (kr(k+1)^{-1})$ if $r < \infty$ and $(r_k) = (k)$ if $r = \infty$. Then $[(a_n^{k+1})^2/(a_n^k a_n^{k+2})] = \exp[(2r_{k+1} - r_k - r_{k+2})a_n] \ge 1$ for all $k, n \in \mathbb{N}$, so A is of type (d_4).

(5) Let $r \in (0, \infty)$. Let $(r_k) = (kr(k+1)^{-1})$. Then $[(r_{k+2} - r_{k+1})/(r_{k+1} - r_k)] = (k+1)(k+3)^{-1} \ge 2^{-1}$ for $k \in \mathbb{N}$. By Lemma 3.7 we get

$$\frac{f(r_{k+2} a_n) - f(r_{k+1} a_n)}{f(r_{k+1} a_n) - f(r_k a_n)} \ge \frac{r_{k+2} - r_{k+1}}{r_{k+1} - r_k} \ge \frac{1}{2}$$

for all $k, n \in \mathbb{N}$. Hence

$$f(r_{k+1} a_n) - f(r_k a_n) \le 2[f(r_{k+2} a_n) - f(r_{k+1} a_n)]$$

for all $k, n \in \mathbb{N}$. It follows that

$$\frac{a_n^{k+1}}{a_n^k} \le \left(\frac{a_n^{k+2}}{a_n^{k+1}}\right)^2$$

for all $k, n \in \mathbb{N}$, so A is of type (d_5).

If $r = \infty$, then A is of type (d_3) so it is of type (d_5). Let $r \le 0$ and $f(t) = t, t \in \mathbb{R}$. Let $(r_k) = (-k^{-1})$ if $r = 0$ and $(r_k) = ((k+1)k^{-1}r)$ if $r < 0$. Then

$$\left(\frac{a_n^{k+2}}{a_n^{k+1}}\right)^3 \frac{a_n^k}{a_n^{k+1}} = \exp([3(r_{k+2} - r_{k+1}) + (r_k - r_{k+1})]a_n) \ge 1$$

for all $k, n \in \mathbb{N}$, so A is of type (d_5). □

COROLLARY 3.9. Let $a \in \Gamma$. Then (1) $A_1(a)$ is of type $(d_0), (d_2), (d_4)$ and (d_5); (2) $A_\infty(a)$ is of type $(d_0), (d_1), (d_3), (d_4)$ and (d_5).

Our next proposition is a generalization of some result of N. De Grande - De Kimpe ([1, Proposition 4.3]) on operators between power series spaces. It follows that $A_1(a)$ is not of type (d_1) (since it is of type (d_2)) and $A_\infty(a)$ is not of type (d_2) (since it is of type (d_1)) for any $a \in \Gamma$.

PROPOSITION 3.10. Let $A = (a_n^k)$ and $B = (b_n^k)$ be Köthe matrices of type (d_2) and (d_1), respectively. Then every continuous linear map $T : K(A) \to K(B)$ is bounded.

Proof. Let (e_j) and (h_i) be the coordinate Schauder bases in $K(A)$ and $K(B)$, respectively. For some $(t_{i,j}) \subset \mathbb{K}$ we have $Te_j = \sum_{i=1}^\infty t_{i,j}h_i$ for all $j \in \mathbb{N}$. By the continuity of T we have

$$(*_1)\ \forall k\, \exists \varphi(k), C(k) : |t_{i,j}|b_i^k \le C(k)a_j^{\varphi(k)} \text{ for all } i, j.$$

We shall prove that

$$(*_2)\ \exists q : \forall k\, \exists D(k) : |t_{i,j}|b_i^k \le D(k)a_j^q \text{ for all } i, j.$$

Since B is of type (d_1) we have

$$\exists l_0 : \forall k\, \exists l_k, S(k) : (b_i^k)^2 \le S(k)b_i^{l_0}b_i^{l_k} \text{ for any } i.$$

Since A is of type (d_2) we have

$$\exists q : \forall v\, \exists P_v : a_j^{\varphi(l_0)}a_j^v \le P_v(a_j^q)^2 \text{ for any } j.$$

Hence, for $k \in \mathbb{N}$, $v = \varphi(l_k)$ and $i, j \in \mathbb{N}$ we have

$$\left(|t_{i,j}|\frac{b_i^k}{a_j^q}\right)^2 \le \left(|t_{i,j}|\frac{b_i^{l_0}}{a_j^{\varphi(l_0)}}\right)\left(|t_{i,j}|\frac{b_i^{l_k}}{a_j^{\varphi(l_k)}}\right) S(k)P_{\varphi(l_k)} \le C(l_0)C(l_k)S(k)P_{\varphi(l_k)}.$$

Thus we have shown $(*_2)$ with $D(k) = [C(l_0)C(l_k)S(k)P_{\varphi(l_k)}]^{1/2}$. It follows that T is bounded. Indeed, for $k \in \mathbb{N}$ and $x = (x_j) \in K(A)$ we have

$$p'_k(Tx) = p'_k(T(\sum_{j=1}^\infty x_je_j)) = p'_k(\sum_{j=1}^\infty x_jTe_j) \le \max_j |x_j|p'_k(Te_j) \le$$

$$\max_j(|x_j| \max_i |t_{i,j}|p'_k(h_i)) = \max_j(|x_j| \max_i |t_{i,j}|b_i^k) \le \max_j |x_j|D(k)a_j^q =$$

$$D(k)\max_j |x_j|p_q(e_j) = D(k)p_q(\sum_{j=1}^\infty x_je_j) = D(k)p_q(x).$$

Thus

$$\exists q : \forall k\, \exists D(k) : p'_k(Tx) \le D(k)p_q(x) \text{ for any } x \in K(A),$$

so T is bounded. □

By Proposition 3.10 we get the following (see [9, Theorem 8]).

COROLLARY 3.11. Let $f, g \in \Phi_c, a, b \in \Gamma$ and $r \le 0$. Then any continuous linear map $T : D_f(a, r) \to D_g(b, \infty)$ is bounded.

Now we prove that a nuclear Köthe space is isomorphic to a power series space of infinite type [respectively, of finite type] if and only if it has a Schauder basis of type (d_3) and (d_4) [respectively, (d_2) and (d_5)].

THEOREM 3.12. *A nuclear Köthe space E is isomorphic to a power series space of infinite type if and only if E has a Schauder basis (x_n) of type (d_3) and (d_4).*

Proof. By Proposition 3.5 and Corollary 3.9 we get the implication ($\Rightarrow$).

($\Leftarrow$) By assumptions there exist Köthe matrices $A = (a_n^k)$ and $B = (b_n^k)$ of type (d_3) and (d_4), respectively, such that the maps $E \to K(A), x = \sum_{n=1}^{\infty} \alpha_n x_n \to (\alpha_n)$ and $E \to K(B), x = \sum_{n=1}^{\infty} \alpha_n x_n \to (\alpha_n)$ are isomorphisms.

A is of type (d_3), so $(a_n^{k+1}/a_n^k) \leq (a_n^{k+s}/a_n^{k+s-1})$ for $1 \leq k < s$ and $s, n \in \mathbb{N}$; hence

$$(*_1)\ (a_n^s/a_n^1) \leq (a_n^{2s-1}/a_n^s) \text{ for } n, s \in \mathbb{N}.$$

Let $p_k : E \to [0, \infty), x = \sum_{n=1}^{\infty} \alpha_n x_n \to p_k(x) = \max_n |\alpha_n| a_n^k$ and
$q_k : E \to [0, \infty), x = \sum_{n=1}^{\infty} \alpha_n x_n \to q_k(x) = \max_n |\alpha_n| b_n^k$ for $k \in \mathbb{N}$.

Then (p_k) and (q_k) are bases in $\mathcal{P}_c(E)$.

First we shall construct a strictly increasing sequence $(j_k) \subset \mathbb{N}$ and a sequence $(\delta_k) \subset (0, \infty)$ such that

$$(*_2)\ \ b_n^{j_k} b_n^{j_{k+2}} \geq \delta_k (b_n^{j_{k+1}})^2$$

for all $n, k \in \mathbb{N}$.

Let $j_1 \in \mathbb{N}$ and $C > 0$ with $p_1 \leq C q_{j_1}$. Let $j_2 > j_1$. Assume that for some $k \in \mathbb{N}$ we have $j_t \in \mathbb{N}$ for $1 \leq t \leq k+1$ and $\delta_t > 0$ for $1 \leq t < k$. Let $s \in \mathbb{N}$ and $M_k > 0$ with $q_{j_{k+1}} \leq M_k p_s$. Let $j_{k+2} > j_{k+1}$ and $W_k > 0$ with $p_{2s-1} \leq W_k q_{j_{k+2}}$. Then

$$p_1 \leq C q_{j_k} \leq C q_{j_{k+1}} \leq C M_k p_s \leq C M_k p_{2s-1} \leq C M_k W_k q_{j_{k+2}}.$$

Hence $a_n^1 \leq C b_n^{j_k} \leq C b_n^{j_{k+1}} \leq C M_k a_n^s \leq C M_k a_n^{2s-1} \leq C M_k W_k b_n^{j_{k+2}}$ for all $n \in \mathbb{N}$, so

$$\frac{b_n^{j_{k+1}}}{b_n^{j_k}} \leq \frac{C M_k a_n^s}{a_n^1} \leq C M_k \frac{a_n^{2s-1}}{a_n^s} \leq C M_k^2 W_k \frac{b_n^{j_{k+2}}}{b_n^{j_{k+1}}}.$$

Let $\delta_k = (C M_k^2 W_k)^{-1}$. Then $b_n^{j_k} b_n^{j_{k+2}} \geq \delta_k (b_n^{j_{k+1}})^2$ for $n \in \mathbb{N}$.

Put $\varepsilon_k = \prod_{i=1}^{k} \delta_i$ and $t_k = \prod_{i=1}^{k} \varepsilon_i$ for $k \in \mathbb{N}$. Let $\beta_n = b_n^{j_1+1}/b_n^{j_1}$ for $n \in \mathbb{N}$. By $(*_2)$ we get

$$\frac{b_n^{j_{k+2}}}{b_n^{j_{k+1}}} \geq \varepsilon_k \frac{b_n^{j_2}}{b_n^{j_1}} \geq \varepsilon_k \beta_n$$

for all $n, k \in \mathbb{N}$. Hence we have for $n, k \in \mathbb{N}$:

$$b_n^{j_{k+2}} \geq t_k (\beta_n)^k b_n^{j_2} \geq t_k (\beta_n)^{k+1} b_n^{j_1} = t_k c_n^k,$$

where $c_n^k = b_n^{j_1} (\beta_n)^{k+1}$. Clearly, $C = (c_n^k)$ is a Köthe matrix. B is of type (d_4), so $\beta_n \geq b_n^{k+1}/b_n^k$ for all $n, k \in \mathbb{N}$ with $k \geq j_1$. Hence we get

$$(*_3)\ \ b_n^k \leq b_n^{k+1} \leq (\beta_n)^{k+1-j_1} b_n^{j_1} \leq c_n^k$$

for $k, n \in \mathbb{N}$ with $k \geq j_1$. Thus $b_n^k \leq c_n^k \leq (t_k)^{-1} b_n^{j_{k+2}}$ for all $k, n \in \mathbb{N}$, so the Köthe matrices B and C are equivalent. Thus E is isomorphic to $K(C)$.

By the nuclearity of E there exists $k > j_1$ with $\lim_n (b_n^k / b_n^{j_1}) = \infty$. Hence, by $(*_3)$ we infer that $\lim_n (\beta_n)^{k+1-j_1} = \infty$, so $\lim \beta_n = \infty$. Let $(\beta_{\pi(n)})$ be a non-decreasing permutation of (β_n). Let $\gamma_n = \ln(2\beta_{\pi(n)}), n \in \mathbb{N}$. Then $\gamma = (\gamma_n) \in \Gamma$.

Let $\varphi \in \mathbb{K}$ with $|\varphi| > 1$. Let $(\varphi_n) \subset \mathbb{K}$ with $|\varphi_n| \in [b_n^{j_1}, |\varphi| b_n^{j_1}), n \in \mathbb{N}$. The linear map $T : K(C) \to A_\infty(\gamma), (\alpha_n) \to (\varphi_{\pi(n)} \alpha_{\pi(n)})$, is a bijection, since for any $k \in \mathbb{N}$ the following conditions are equivalent: $|\varphi_{\pi(n)}| |\alpha_{\pi(n)}| \exp[(k+1)\gamma_n] \to 0$; $|\varphi_n| |\alpha_n| (2\beta_n)^{k+1} \to 0$; $|\alpha_n| c_n^k = |\alpha_n| b_n^{j_1} \beta_n^{k+1} \to 0$.

T is an isomorphism since

$$\begin{aligned}\max_n |\varphi_{\pi(n)}|\,|\alpha_{\pi(n)}|\exp[(k+1)\gamma_n] &= \max_n |\varphi_n|\,|\alpha_n|(2\beta_n)^{k+1}\\ &\le \max_n 2^{k+1}|\varphi|b_n^{j_1}(\beta_n)^{k+1}|\alpha_n|\\ &= 2^{k+1}|\varphi|\max_n|\alpha_n|c_n^k \le |\varphi|\max_n|\varphi_n|\,|\alpha_n|(2\beta_n)^{k+1}\\ &= |\varphi|\max_n|\varphi_{\pi(n)}|\,|\alpha_{\pi(n)}|\exp[(k+1)\gamma_n].\end{aligned}$$

Thus E is isomorphic to $A_\infty(\gamma)$. □

Using Corollary 3.6 we get

COROLLARY 3.13. A non-normable Köthe space E is isomorphic to a power series space of infinite type if and only if E is of type $(d_0), (d_3)$ and (d_4).

THEOREM 3.14. *A nuclear Köthe space E is isomorphic to a power series space of finite type if and only if E has a Schauder basis (x_n) of type (d_2) and (d_5).*

Proof. By Proposition 3.5 and Corollary 3.9 we get the implication $(\Rightarrow)$.

$(\Leftarrow)$ By assumptions there exist Köthe matrices $B=(b_n^k)$ and $C=(c_n^k)$ of type (d_2) and (d_5), respectively, such that the maps $E\to K(B), x=\sum_{n=1}^\infty \alpha_n x_n \to (\alpha_n)$ and $E\to K(C), x=\sum_{n=1}^\infty \alpha_n x_n\to(\alpha_n)$ are isomorphisms.

C is of type (d_5), so

$$\exists M\ge 1: \frac{c_n^{k+1}}{c_n^k}\le\left(\frac{c_n^{k+2}}{c_n^{k+1}}\right)^M \quad \text{for all}\, n,k.$$

Hence

$$\frac{c_n^{k+m}}{c_n^k}\le\left(\frac{c_n^{k+m+1}}{c_n^{k+1}}\right)^M \quad\text{for all}\, n,k,m,$$

so

$$(*_0)\ \frac{c_n^{k+m}}{c_n^k}\le\left(\frac{c_n^{k+m+t}}{c_n^{k+t}}\right)^{M^t}\quad\text{for all}\, n,k,m,t.$$

B is of type (d_2), so $\forall k\,\exists j_0>k:\forall l\,\exists C_0>0: b_n^k b_n^l\le C_0(b_n^{j_0})^2$ for any n. By nuclearity of E for any $j_0\in\mathbb{N}$ there exists $k_0>j_0$ with $\lim_n b_n^{k_0}/b_n^{j_0}=\infty$, then

$$\forall C_0\exists n_0: C_0(b_n^{j_0})^2\le(b_n^{k_0})^2 \text{ for any } n\ge n_0.$$

It follows that there exists a strictly increasing sequence $(j_k)\subset\mathbb{N}$ such that

$$(*_1)\ \forall k,l\,\exists n_0: b_n^{j_k}b_n^{j_l}\le(b_n^{j_{k+1}})^2 \text{ for any}\, n\ge n_0.$$

Clearly the Köthe matrix $\hat{B}=(\hat{b}_n^k)=(b_n^{j_k})$ is equivalent to $B=(b_n^k)$. Let $k\in\mathbb{N}$; of course $\hat{b}_n^k\hat{b}_n^1\le(\hat{b}_n^{k+1})^2$ for any $n\in\mathbb{N}$. Put $s_n^k=\max\{1\le l\le n:\hat{b}_n^k\hat{b}_n^l\le(\hat{b}_n^{k+1})^2\}$ for $n\in\mathbb{N}$. Then $\lim_n s_n^k=\infty$. Indeed, in the opposite case for some $m\in\mathbb{N}$ the set $V=\{n>m: s_n^k<m\}$ is infinite and $\hat{b}_n^k\hat{b}_n^m>(\hat{b}_n^{k+1})^2$ for $n\in V$ - a contradiction by $(*_1)$.

Let $(i_k)\subset\mathbb{N}$ be a strictly increasing sequence with $i_1=1$ such that $s_n^v\ge k$ for all k,v,n with $1\le v\le k$ and $n\ge i_k$. Put $s_n=k$ when $i_k\le n<i_{k+1}, k\in\mathbb{N}$. Clearly $(s_n)\subset\mathbb{N}$ is a non-decreasing sequence with $\lim_n s_n=\infty$. Then

$$(*_2)\ \hat{b}_n^k\hat{b}_n^{s_n}\le(\hat{b}_n^{k+1})^2 \text{ for all}\, n,k \,\text{with}\, n\ge i_k$$

since $s_n\le s_n^k$ for all $k,n\in\mathbb{N}$ with $n\ge i_k$. For $k,n\in\mathbb{N}$ we put $\hat{a}_n^k=\begin{cases}\hat{b}_n^k/\hat{b}_n^{s_n} & \text{if } n\ge i_k\\ 1 & \text{if } n<i_k.\end{cases}$

Then $0 < \hat{a}_n^k \le \hat{a}_n^{k+1} \le 1$ and $\hat{a}_n^k \le (\hat{a}_n^{k+1})^2$ for all $k,n \in \mathbb{N}$; hence $\hat{a}_n^k \hat{a}_n^l \le (\hat{a}_n^{k+1})^2$ for all $k,l,n \in \mathbb{N}$.

Indeed, let $k \in \mathbb{N}$. If $n \ge i_k$, then $s_n \ge k$, so $\hat{a}_n^k \le 1$. If $n < i_k$, then $\hat{a}_n^k = \hat{a}_n^{k+1} = 1$. If $i_k \le n < i_{k+1}$, then $\hat{a}_n^k \le \hat{a}_n^{k+1} = 1$. If $n \ge i_{k+1}$, then $\hat{b}_n^k/\hat{b}_n^{s_n} \le (\hat{b}_n^{k+1}/\hat{b}_n^{s_n})^2$, so $\hat{a}_n^k \le (\hat{a}_n^{k+1})^2 \le \hat{a}_n^{k+1}$.

Let $\varphi \in \mathbb{K}$ with $|\varphi| > 1$. Let $(\varphi_n) \subset \mathbb{K}$ with $|\varphi_n| \in [1/\hat{b}_n^{s_n}, |\varphi|/\hat{b}_n^{s_n}), n \in \mathbb{N}$.

The Köthe matrix $\hat{A} = (\hat{a}_n^k)$ is a representation of the Schauder basis $(y_n) = (\varphi_n x_n)$. Indeed, the linear map $T : E \to K(\hat{A}), \sum_{n=1}^\infty \alpha_n y_n \to (\alpha_n)$ is a bijection, since for every $(\alpha_n) \subset \mathbb{K}$ we have the following equivalences: $\alpha_n y_n \to 0$ in E $\Leftrightarrow |\alpha_n|\,|\varphi_n| b_n^k \to 0$ for any $k \Leftrightarrow |\alpha_n|\,|\varphi_n| \hat{b}_n^k \to 0$ for any $k \Leftrightarrow |\alpha_n|(\hat{b}_n^k/\hat{b}_n^{s_n}) \to 0$ for any $k \Leftrightarrow |\alpha_n| \hat{a}_n^k \to 0$ for any $k \Leftrightarrow (\alpha_n) \in K(\hat{A})$.

The map T is continuous, since

$$\max_n |\alpha_n| \hat{a}_n^k \le \max_n |\alpha_n| (\hat{b}_n^k/\hat{b}_n^{s_n}) \le \max_n |\alpha_n|\,|\varphi_n| \hat{b}_n^k = \max_n |\alpha_n|\,|\varphi_n| b_n^{j_k}$$

for every $k \in \mathbb{N}$. By the open mapping theorem the map T is an isomorphism.

Put $\tilde{a}_n^k = \exp(-2^{-k})\hat{a}_n^k$ for $k,n \in \mathbb{N}$. Then $\tilde{A} = (\tilde{a}_n^k)$ is a Köthe matrix equivalent to $\hat{A}$ and

$$(*_3)\ \ 0 < \tilde{a}_n^k < \tilde{a}_n^{k+1} < 1 \wedge \tilde{a}_n^k \le (\tilde{a}_n^{k+1})^2 \text{ for } k,n \in \mathbb{N};$$

$$(*_4)\ \ \tilde{a}_n^k \tilde{a}_n^l < (\tilde{a}_n^{k+1})^2 \text{ for all } k,l,n \in \mathbb{N}.$$

Let $p_k : E \to [0,\infty), \sum_{n=1}^\infty \alpha_n x_n \to \max_n |\alpha_n| c_n^k$ and $q_k : E \to [0,\infty), \sum_{n=1}^\infty \beta_n y_n \to \max_n |\beta_n| \tilde{a}_n^k$ for $k \in \mathbb{N}$. Then (p_k) and (q_k) are bases in $\mathcal{P}_c(E)$. Thus

$$\forall k\, \exists l, C_0 > 0 : p_k \le C_0 q_l \wedge q_k \le C_0 p_l.$$

By nuclearity of E we have:
$\forall k\, \exists l > k : [p_k(y_n)/p_l(y_n)] \to_n 0$ and $[q_k(y_n)/q_l(y_n)] \to_n 0$. Hence we get

$$\forall k\, \exists l, n_0 : [p_k(y_n) \le q_l(y_n) \wedge q_k(y_n) \le p_l(y_n)] \text{ for any } n \ge n_0.$$

Thus there exist a strictly increasing sequences $(m_k), (l_k), (n_k) \subset \mathbb{N}$ such that $p_1(y_n) \le q_{m_1}(y_n)$ for any $n \ge n_1$ and

$$q_{m_{k+1}}(y_n) \le p_{l_{k+1}}(y_n) \le p_{2l_{k+1}-l_{k-1}}(y_n) \le q_{m_{k+2}}(y_n) \text{ for all } n,k \text{ with } n \ge n_{k+1},$$

where $l_0 = 1$. Hence we get

$$\tilde{a}_n^{m_{k+1}} \le |\varphi_n| c_n^{l_{k+1}} \le |\varphi_n| c_n^{2l_{k+1}-l_{k-1}} \le \tilde{a}_n^{m_{k+2}} \text{ for all } n,k \text{ with } n \ge n_{k+1}.$$

Let $k,n \in \mathbb{N}$ with $n \ge n_{k+1}$. Using $(*_0)$ we get

$$\frac{\tilde{a}_n^{m_{k+1}}}{\tilde{a}_n^{m_k}} \le \frac{c_n^{l_{k+1}}}{c_n^{l_{k-1}}} \le \left[\frac{c_n^{2l_{k+1}-l_{k-1}}}{c_n^{l_{k+1}}}\right]^{M^{l_{k+1}-l_{k-1}}} \le \left(\frac{\tilde{a}_n^{m_{k+2}}}{\tilde{a}_n^{m_{k+1}}}\right)^{M^{l_{k+1}-l_{k-1}}}.$$

Thus for $A = (a_n^k) = (\tilde{a}_n^{m_k})$ and some $(M_k) \subset \mathbb{N}$ with $M_k > M^{l_{k+1}-l_{k-1}} \ge 1$ for $k \in \mathbb{N}$, we have

$$(*_5)\ \frac{a_n^{k+1}}{a_n^k} \le \left(\frac{a_n^{k+2}}{a_n^{k+1}}\right)^{M_k} \text{ for all } n,k.$$

By $(*_3)$ and $(*_4)$ we have

$$(*_6)\ a_n^k a_n^l < (a_n^{k+1})^2 \text{ for all } k,l,n \text{ and } a_n^k < a_n^{k+1} < 1 \text{ for all } k,n.$$

For all $n, k \in \mathbb{N}$ there exists $M_n^k \in (0, M_k]$ with

$$(*_7)\ \frac{a_n^l}{a_n^{k+1}} < \frac{a_n^{k+1}}{a_n^k} = \left(\frac{a_n^{k+2}}{a_n^{k+1}}\right)^{M_n^k} \quad \text{for all } l \in \mathbb{N}.$$

Let $n \in \mathbb{N}$. Put $t_n = a_n^2/a_n^1, W_n^1 = W_n^2 = 0$ and

$$W_n^k = \frac{1}{M_n^1} + \frac{1}{M_n^1 M_n^2} + \ldots + \frac{1}{M_n^1 M_n^2 \ldots M_n^{k-2}}$$

for $k \geq 3$. Then $(t_n) \subset (1, \infty)$ and $W_n^{k+1} = W_n^k + \frac{1}{M_n^1 M_n^2 \ldots M_n^{k-1}} > W_n^k$ for $k \geq 3$.

By $(*_7)$ we have for $k \geq 2$

$$\frac{a_n^{k+1}}{a_n^k} = \left(\frac{a_n^2}{a_n^1}\right)^{\frac{1}{M_n^1} \cdot \frac{1}{M_n^2} \cdot \ldots \cdot \frac{1}{M_n^{k-1}}} = (t_n)^{W_n^{k+1} - W_n^k}.$$

Hence we get $a_n^l/a_n^k = (t_n)^{W_n^l - W_n^k}$ for $l \geq k \geq 2$. So $a_n^k = a_n^2 (t_n)^{W_n^k}$ for $k \geq 3$. Hence, by $(*_6)$, we have

$$(*_8)\ W_n^l + W_n^k \leq 2W_n^{k+1}$$

for $l, k \geq 3$, so $W_n := \lim_k W_n^k = \sup_k W_n^k < \infty$. Put $d_n^k = (t_n)^{W_n^k - W_n}$ for $k, n \in \mathbb{N}$. Then $D = (d_n^k)$ is a Köthe matrix and $K(A)$ is isomorphic to $K(D)$, since $a_n^k = a_n^2 t_n^{W_n} d_n^k$ for $n \in \mathbb{N}$ and $k \geq 2$. Let $\beta_n = (t_n)^{W_n - W_n^3}$ for $n \in \mathbb{N}$. Then $(\beta_n) \subset (1, \infty)$. Clearly, $H = (h_n^k) = (\beta_n^{-1/k})$ is a Köthe matrix. We shall prove that $K(D)$ is isomorphic to $K(H)$. Let $k \geq 3$. For $n \in \mathbb{N}$ we have

$$\frac{W_n - W_n^{k+1}}{W_n^{k+1} - W_n^k} \geq \frac{W_n^{k+2} - W_n^{k+1}}{W_n^{k+1} - W_n^k} = \frac{1}{M_n^k} \geq \frac{1}{M_k} \quad \text{and}$$

$$\frac{W_n - W_n^{k+1}}{W_n - W_n^k} = 1 - \frac{1}{1 + \frac{W_n - W_n^{k+1}}{W_n^{k+1} - W_n^k}} \geq 1 - \frac{1}{1 + \frac{1}{M_k}} = \frac{1}{1 + M_k},$$

so

$$\inf_n \frac{W_n - W_n^{k+1}}{W_n - W_n^k} \geq \frac{1}{1 + M_k}.$$

Hence

$$\inf_n \frac{W_n - W_n^{k+1}}{W_n - W_n^3} \geq \prod_{i=3}^{k} (1 + M_i)^{-1}$$

for $k > 3$. By $(*_8)$ we have $W_n + W_n^k \leq 2W_n^{k+1}$ for $n \in \mathbb{N}$, so

$$\sup_n \frac{W_n - W_n^{k+1}}{W_n - W_n^k} \leq \frac{1}{2}.$$

Hence

$$\sup_n \frac{W_n - W_n^k}{W_n - W_n^3} \leq \frac{1}{2^{k-3}}.$$

Let $k \geq 3$, $j > \prod_{i=3}^k (1 + M_i)$ and $l > 3 + \log_2 j$. Then $W_n^k - W_n \leq j^{-1}(W_n^3 - W_n) \leq W_n^l - W_n$ for $n \in \mathbb{N}$, so $d_n^k \leq h_n^j \leq d_n^l$ for all $n \in \mathbb{N}$. It follows that $K(D)$ is isomorphic to $K(H)$.

By nuclearity of E we have $\lim_n a_n^l/a_n^3 = \infty$ for some $l > 3$. Thus $\lim_n \beta_n = \infty$, since $\beta_n \geq (t_n)^{W_n^l - W_n^3} = a_n^l/a_n^3$ for $n \in \mathbb{N}$.

Let $(\beta_{\pi(n)})$ be a non-decreasing permutation of (β_n). Let $(\gamma_n) = (\ln \beta_{\pi(n)})$; clearly $\gamma = (\gamma_n) \in \Gamma$. Clearly, the map $K(H) \to A_1(\gamma), (\alpha_n) \to (\alpha_{\pi(n)})$ is an isomorphism. We have shown that E is isomorphic to $A_1(\gamma)$. □

Using Corollary 3.6 we get

COROLLARY 3.15. A non-normable Köthe space E is isomorphic to a power series space of finite type if and only if E is of type $(d_0), (d_2)$ and (d_5).

References

[1] N. De Grande-De Kimpe, *Non-Archimedean Fréchet spaces generalizing spaces of analytic functions*, Nederl. Akad. Wetensch. Indag. Math. **44** (1982), no. 4, 423–439. MR683530

[2] E. Dubinsky, *The structure of nuclear Fréchet spaces*, Lecture Notes in Mathematics, vol. 720, Springer, Berlin, 1979. MR537039

[3] C. Perez-Garcia and W. H. Schikhof, *Locally convex spaces over non-Archimedean valued fields*, Cambridge Studies in Advanced Mathematics, vol. 119, Cambridge University Press, Cambridge, 2010. MR2598517

[4] A. C. M. van Rooij, *Non-Archimedean functional analysis*, Monographs and Textbooks in Pure and Applied Math., vol. 51, Marcel Dekker, Inc., New York, 1978. MR512894

[5] P. Schneider, *Nonarchimedean functional analysis*, Springer Monographs in Mathematics, Springer-Verlag, Berlin, 2002. MR1869547

[6] W. Śliwa, *On the quasi-equivalence of orthogonal bases in non-Archimedean metrizable locally convex spaces*, Bull. Belg. Math. Soc. Simon Stevin **9** (2002), no. 3, 465–472. MR2016585

[7] W. Śliwa, *On relations between non-Archimedean power series spaces*, Indag. Math. (N.S.) **17** (2006), no. 4, 627–639, DOI 10.1016/S0019-3577(06)81038-4. MR2320118

[8] W. Śliwa and A. Ziemkowska, *On complemented subspaces of non-Archimedean power series spaces*, Canad. J. Math. **63** (2011), no. 5, 1188–1200, DOI 10.4153/CJM-2011-018-0. MR2866075

[9] W. Śliwa, *On non-Archimedean generalized power series spaces*, Advances in ultrametric analysis, Contemp. Math., vol. 596, Amer. Math. Soc., Providence, RI, 2013, pp. 281–291, DOI 10.1090/conm/596/11903. MR3156551

[10] W. Śliwa, *On relations between non-Archimedean generalized power series spaces*, J. Convex Anal. **22** (2015), no. 2, 569–590. MR3346204

[11] V. P. Zahariuta, *On the isomorphism of cartesian products of locally convex spaces*, Studia Math. **46** (1973), 201–221. MR0330991

FACULTY OF MATHEMATICS AND NATURAL SCIENCES, UNIVERSITY OF RZESZÓW, UL. PIGONIA 1, 35- 310 RZESZÓW, POLAND

Email address: wsliwa@ur.edu.pl; sliwa@amu.edu.pl

INSTITUTE OF MATHEMATICS, POZNAŃ UNIVERSITY OF TECHNOLOGY, UL. PIOTROWO 3A, 60-965 POZNAŃ, POLAND

Email address: agnieszka.ziemkowska@put.poznan.pl

Contemporary Mathematics
Volume **704**, 2018
http://dx.doi.org/10.1090/conm/704/14173

On the sub-coordinate representation of p-adic functions

Ekaterina Yurova Axelsson

ABSTRACT. In this paper we introduce a new way of representation of p-adic functions, namely, the sub-coordinate representation. The main feature of such representation is that the values of a function f are given in the canonical form of representation of p-adic number. In the sub-coordinate representation the function f is determined by a set of p-valued functions that map a set $\{0, 1, \dots, p-1\}$ into itself, and by the order of these functions. As one of the applications of the sub-coordinate representation, we study a problem of generalization of Hensel's lifting lemma.

1. Introduction

Any p-adic function can be written in various ways. For instance, p-adic functions can be set by means of power series, Mahler series, van der Put series, and so forth, see for example [**4**], [**8**], [**9**], [**10**]. In general, the choice of representation of p-adic functions is determined by the specifics of the problem and usability of the corresponding mathematical formalism. In this paper we introduce new sub-coordinate representation of p-adic functions.

In Theorem 2.1, we define the sub-coordinate representation for 1-Lipschitz functions. This Theorem establishes the existence and uniqueness of such a representation. In Remark 2.2 we give the relation whereby we obtain the sub-coordinate representation from table setting of 1-Lipschitz function. In Remark 2.3 we show a connection between sub-coordinate and well-known coordinate representations. Furthermore, the sub-coordinate representation exists for p^α-Lipschitz functions, $\alpha \geq 1$ due to the fact that any p^α-Lipschitz function is uniquely defined by the set of p^α elements which are the 1-Lipschitz functions, see Theorem 3.1 in [**14**]. In the section 3 we consider examples of the sub-coordinate representation.

In Example 3.3 we discuss an application of the sub-coordinate representation to generalization of p-adic version of Hensel's Lemma. Classical formulation of Hensel's Lemma (see, for instance, [**11**] and [**12**]) is a criterion for the existence of the roots of a polynomial f over the ring of p-adic integers $\mathbb{Z}_p$. Namely, the equation $f(x) = 0$ is solvable over $\mathbb{Z}_p$, once the congruence $f(x) \equiv 0 \pmod p$ is solvable (under certain restrictions on f). In the paper [**14**] the question on generalization of Hensel's Lemma was considered in the following formulation: There exists a natural number R such that the equation $f(x) = M$, $M \in \mathbb{Z}_p$ (and, in particular, for $M = 0$)

2010 *Mathematics Subject Classification.* Primary 11S82, 37P99.
Key words and phrases. p-adic, Hensel's lifting lemma, Lipschitz function.

can be solved as soon as the congruence $f(x) \equiv M \pmod{p^R}$ is solvable. Which p-adic 1-Lipschitz functions satisfy this proposition? Note that in general case (f is 1-Lipschitz function) one should check solvability of infinite number of congruences $f(x) \equiv M \pmod{p^k}$, $k \geq 1$ (for $M = 0$ see Theorem 2.1 from [**14**]) in order to solve the equation $f(x) = M$, $M \in \mathbb{Z}_p$. In Theorem 2.4 from [**14**], conditions by means of the van der Put series were presented for those 1-Lipschitz functions, that satisfy the proposition above. In this paper we introduce an analogous statement (see Theorem 3.4) in terms of the sub-coordinate representation. This representation has allowed to estimate the possible generalization boundaries of Hensel's Lemma. Theorem 3.5 tells us that there are two alternatives. The first one says that if the number of non-bijective sub-functions $\varphi_{k,m}$, $m \in \{0, 1, \dots, p^k - 1\}$, $k \geq 1$ from the sub-coordinate representation of the function f (see 2.1) is finite, then there exists $R \in \mathbb{N}$ such that solvability of the equation $f(x) = M$, $M \in \mathbb{Z}_p$ reduces to verifying the solvability of the congruence $f(x) \equiv M \pmod{p^R}$, that is Hensel's Lemma can be generalized. The second alternative says that if the number of such sub-functions is infinite, then one cannot construct generalization of Hensel's Lemma in the sense that for any $k \in \mathbb{N}$ there exists $C \in \mathbb{Z}_p$ such that $f(x) \equiv C \pmod{p^k}$, but $f(x) \neq C$.

2. Sub-coordinate representation

In this paper we consider functions $f : \mathbb{Z}_p \to \mathbb{Z}_p$ ($\mathbb{Z}_p$ is a set of p-adic integers), which satisfy the Lipschitz condition with constant p^α, $\alpha \geq 0$ (or p^α-Lipschitz functions). Recall that $f : \mathbb{Z}_p \to \mathbb{Z}_p$ is p^α-Lipschitz function if $|f(x) - f(y)|_p \leq p^\alpha |x - y|_p$, for all $x, y \in \mathbb{Z}_p$. This condition is equivalent to that from $x \equiv y \pmod{p^k}$ follows $f(x) \equiv f(y) \pmod{p^{k-\alpha}}$ for all $k \geq 1 + \alpha$.

A class of 1-Lipschitz functions takes a special place (i.e. $\alpha = 0$). This is due to the following reason. For all $k \geq 1$ a 1-Lipschitz transformation $f : \mathbb{Z}_p \to \mathbb{Z}_p$ the reduced mapping modulo p^k is $f_{mod\ p^k} : \mathbb{Z}/p^k\mathbb{Z} \to Z/p^k\mathbb{Z}$, $z \mapsto f(z) \pmod{p^k}$. Mapping $f_{mod\ p^k}$ is a well-defined (the $f_{mod\ p^k}$ does not depend on the choice of representative in the ball $z + p^k\mathbb{Z}_p$). By Theorem 3.1 from [**14**], any p^α-Lipschitz function f can be defined by the set of p^α suitable 1-Lipschitz functions, namely,

$$f(a + p^\alpha x) = \sum_{i=0}^{p^\alpha - 1} I_i(a) \cdot f_i(x), \ a \in \{0, \dots, p^\alpha - 1\},$$

where $f_i(x) : \mathbb{Z}_p \to \mathbb{Z}_p$, $i \in \{0, \dots, p^\alpha - 1\}$ are 1-Lipschitz functions. Here $I_i(a) = 1$ as $a = i$, and $I_i(a) = 0$ otherwise.

The main feature of the sub-coordinate representation is that the values of the function f are defined through the canonical representation of a p-adic number. In this case, the function f itself is defined by the set of p-valued functions that map a set $\{0, 1, \dots, p - 1\}$ into itself, and by the order of the usage of these functions.

In the next theorem, the sub-coordinate representation is provided for p-adic functions $f \colon \mathbb{Z}_p \to \mathbb{Z}_p$, which satisfy the Lipschitz condition with constant 1. However, one can generalize the theorem for the case of p^α-Lipschitz functions ($\alpha \geq 1$) because of Theorem 3.1 from [**14**].

THEOREM 2.1. *Let $f \colon \mathbb{Z}_p \to \mathbb{Z}_p$ be a 1-Lipschitz function. Then there exist p-valued functions $\varphi_0 \colon \{0, 1, \dots, p - 1\} \to \{0, 1, \dots, p - 1\}$ and*

$$\varphi_{k,a} \colon \{0, 1, \dots, p - 1\} \to \{0, 1, \dots, p - 1\} \ \textit{for } a \in \{0, 1, \dots, p^k - 1\}, \ k \geq 1$$

such that

$$(2.1)\quad f(x) = f(x_0 + p\cdot x_1 + \ldots + p^k\cdot x_k + \ldots) = \\ = \varphi_0(x_0) + \sum_{k=1}^{\infty} p^k \sum_{a=0}^{p^k-1} I_a([x]_k)\varphi_{k,a}(x_k),$$

where $I_a([x]_k) = 1$, *if* $[x]_k = a$, *and* $I_a([x]_k) = 0$ *otherwise*, $[x]_k \equiv x \pmod{p^k}$.

We call the series 2.1 *sub-coordinate* representation of the function f.

REMARK 2.2. Let us set residues of the ring $\mathbb{Z}/p\mathbb{Z}$ as minimum positive integers, then the functions φ_0 and $\varphi_{k,a}$, $a \in \{0, 1, \ldots, p^k - 1\}$, $k \geq 1$ clearly define p-valued functions on $\{0, 1, \ldots, p-1\}$. The functions φ_0 and $\varphi_{k,a}$, are defined as follows:

$$\varphi_0(x) \equiv f(x)(\mathrm{mod} p)$$
$$\varphi_{k,a}(x) \equiv \frac{f(a + p^k x) - f(a)(\mathrm{mod} p^k)}{p^k}(\mathrm{mod} p).$$

Thus the functions $\varphi_{k,a}$ could be considered as:

(1) a p-valued function that maps the set $\{0, 1, \ldots, p-1\}$ into itself;
(2) a transformation of the field $\mathbb{Z}/p\mathbb{Z}$ into itself. In other words, $\varphi_{k,a}$ could be a polynomial over $\mathbb{Z}/p\mathbb{Z}$, i.e. $\varphi_{k,a}(x) = \alpha_{p-1}x^{p-1} + \ldots + \alpha_1 x + \alpha_0$ for $\alpha_s \in \mathbb{Z}/p\mathbb{Z}$;
(3) a substitution on the set $\{0, 1, \ldots, p-1\}$ (or as an element of the symmetric group of substitutions S_p), if $\varphi_{k,a}$ defines a bijective map.

REMARK 2.3. There is a simple relationship between the sub-coordinate representation 2.1 and the coordinate representation of p-adic functions. Recall that any map $f : \mathbb{Z}_p \to \mathbb{Z}_p$ can be represented in the coordinate form:

$$f(x) = \delta_0(f(x)) + p\delta_1(f(x)) + \ldots + p^k\delta_k(f(x)) + \ldots,$$

where the functions $\delta_k : \mathbb{Z}_p \to \{0, 1, \ldots, p-1\}$ are defined in the following way.

Let any p-adic integer x be written in the canonical form $x = x_0 + px_1 + \ldots + p^k x_k + \ldots$, $x_k \in \{0, 1, \ldots, p-1\}$, then $\delta_k(x) = x_k$ for $k \geq 0$. By Proposition 3.33 from [**4**], the function f is 1-Lipschitz if and only if for each $k \geq 1$ the k-th coordinate function $\delta_k(f(x))$ does not depend on $\delta_{k+s}(x)$ for all $s \geq 1$, i.e. $\delta_k(f(x + p^{k+1}\mathbb{Z}_p)) = \delta_k(f(x))$ for all $x \in \{0, 1, \ldots, p^{k+1} - 1\}$ or

$$\delta_k(f(x)) = \delta_k(f(x_0 + px_1 + \ldots + p^k x_k + \ldots)) = \Delta_{f,k}(x_0, x_1, \ldots, x_k).$$

Then

$$f(x) = f(x_0 + px_1 + \ldots + p^k x_k + \ldots) = \sum_{k=0}^{\infty} p^k \Delta_{f,k}(x_0, x_1, \ldots, x_k),$$

where $\Delta_{f,k}$ is a p-valued function that maps the Cartesian product of $k+1$ sets $\{0, 1, \ldots, p-1\}$ into itself, and $k \geq 0$.

The functions $\Delta_{f,k}$ can be set by its sub-functions by fixing the variables $x_0, x_1, \ldots, x_{k-1}$. These sub-functions coincide with φ_0 and $\varphi_{k,a}(x)$, $a \in \{0, 1, \ldots, p^k - 1\}$, i.e.

$$\Delta_{f,0} = \varphi_0; \quad \Delta_{f,k} = \sum_{a=0}^{p^k-1} I_a([x]_k)\varphi_{k,a},\ k \geq 1,$$

where

$$I_a([x]_k) = \begin{cases} 1, & \text{if } [x]_k = a; \\ 0, & \text{otherwise.} \end{cases}$$

3. Examples of the sub-coordinate representation

In this section we will give some examples on the sub-coordinate representation of p-adic functions.

EXAMPLE 3.1. Here is the sub-coordinate representation of some function f. Note that a particular determination of the function occurs by means of the sub-functions φ_0 and $\varphi_{k,a}$. In this example these sub-functions are not defined.

$$p = 3 \qquad f(21) = f(0 + 3 \cdot 1 + 3^2 \cdot 2) =$$

$$= \varphi_0(0) + 3\varphi_{1,[0]}(1) + 3^2\varphi_{2,[3]}(2) + \textstyle\sum_{k=3}^{\infty} \varphi_{k,[21]}(0)$$

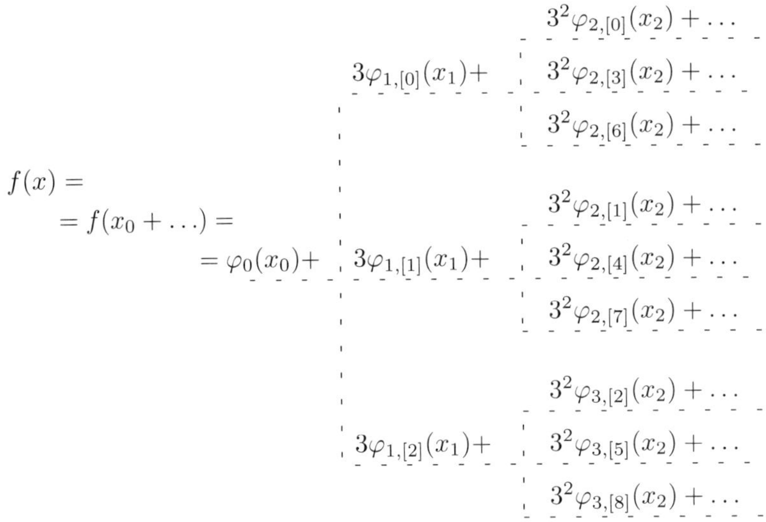

FIGURE 1. Sub-coordinate representation

EXAMPLE 3.2. Let us construct the sub-coordinate representation for the concrete function. Let $p = 2$ and $f(x_0 + 2x_1 + \ldots + 2^k x_k + \ldots) = f(x) = x^2 + 1$.

Let us find the sub-functions φ_0 and $\varphi_{k,a}$ from 2.1 for $a \in \{0, 1, \ldots, 2^k - 1\}$, $0 \le k \le 3$. We set these functions as polynomials over $\mathbb{Z}/2\mathbb{Z}$. Then

$$\begin{array}{lllll} \varphi_0 = x_0 \oplus_2 1; & \varphi_{1,0} = 0; & \varphi_{2,0} = 0; & \varphi_{3,0} = 0; & \varphi_{3,4} = 0; \\ & \varphi_{1,1} = 1; & \varphi_{2,1} = 0; & \varphi_{3,1} = 0; & \varphi_{3,5} = 1; \\ & & \varphi_{2,2} = 1; & \varphi_{3,2} = 0; & \varphi_{3,6} = 0; \\ & & \varphi_{2,3} = 0; & \varphi_{3,3} = 1; & \varphi_{3,7} = 0. \end{array}$$

Let us find several values of the function f modulo 2^4 :

$$f(7) = f(1 + 2 + 2^2) \equiv$$

$$\equiv \varphi_0(1) + 2\varphi_{1,1}(1) + 2^2\varphi_{2,3}(1) + 2^3\varphi_{3,7}(0) \equiv 2 \pmod{2^4};$$

$$f(11) \equiv f(1+2+2^3) \equiv$$
$$\equiv \varphi_0(1) + 2\varphi_{1,1}(1) + 2^2\varphi_{2,3}(0) + 2^3\varphi_{3,3}(1) \equiv 10 \pmod{2^4};$$

$$f(15) \equiv f(1+2+2^2+2^3) \equiv$$
$$\equiv \varphi_0(1) + 2\varphi_{1,1}(1) + 2^2\varphi_{2,3}(1) + 2^3\varphi_{3,7}(1) \equiv 2 \pmod{2^4}.$$

EXAMPLE 3.3. In this example, we will use the sub-coordinate representation of p-adic functions $f\colon \mathbb{Z}_p \to \mathbb{Z}_p$ to look into the possibility of generalization of Hensel's lemma. In this regard, we consider the problem of constructing the solvability criterion of the equation $f(x) = M$, $M \in \mathbb{Z}_p$ for 1-Lipschitz function f. The results of this section are based on the use of generalized Hensel's lemma, which has been obtained in [**14**]. A generalization of Hensel's lemma in terms of the sub-coordinate representation is presented in the following theorem.

THEOREM 3.4. *Let $f : \mathbb{Z}_p \to \mathbb{Z}_p$ be a 1-Lipschitz function defined in the sub-coordinate form* 2.1. *If*

(1) *for some natural number R there exist $\bar{a} \in \{0, 1, \ldots, p^R - 1\}$ such that $f(\bar{a}) \equiv 0 \pmod{p^R}$ and*
(2) *for any $k \geq R$ and $a \geq p^R$, where $a \equiv \bar{a} \pmod{p^R}$, the functions $\varphi_{k,a}$ are bijective,*

then there exist an unique p-adic integer $r \in \mathbb{Z}_p$ such that $f(r) = 0$ and $r \equiv \bar{a}$ $\pmod{p^R}$.

The next theorem discusses the possibility of further generalization of Hensel's Lemma. Here we are interested in the generalizations for which the problem of availability of the root of the function can be solved by finite number of steps. Let $\mathcal{B}$ be a set of 1-Lipschitz functions $f\colon \mathbb{Z}_p \to \mathbb{Z}_p$ such that the sub-coordinate representation of f consists of only finitely many non-bijective functions $\varphi_{k,a}$. In other words, there exists $R \in \mathbb{N}$ such that the sub-functions $\varphi_{k,a}$ are bijective for any $a \in \{0, 1, \ldots, p^k - 1\}$, $k \geq R$.

THEOREM 3.5. *Let $f : \mathbb{Z}_p \to \mathbb{Z}_p$ be a 1-Lipschitz function defined in the sub-coordinate form* 2.1. *Then*

(1) *If $f \in \mathcal{B}$, then there exists $R \in \mathbb{N}$ such that the equation $f(x) = C$, $C \in \mathbb{Z}_p$ is solvable iff the equation $f(x) \equiv C \pmod{p^R}$ is solvable;*
(2) *If $f \notin \mathcal{B}$ (i.e. there are infinitely many non-bijective $\varphi_{k,a}$ in* 2.1*), then for any $k \in \mathbb{N}$ there exists $C \in \mathbb{Z}_p$ such that $f(x) \equiv C \pmod{p^k}$, but $f(x) \neq C$;*
(3) *The function f is bijective on $\mathbb{Z}_p$ if and only if any function φ_0, $\varphi_{k,a}$, $a \in \{0, 1, \ldots, p^k - 1\}$, $k \geq 1$ in* 2.1 *is bijective.*

References

[1] V. S. Anashin, *Uniformly distributed sequences of p-adic integers* (Russian, with Russian summary), Mat. Zametki **55** (1994), no. 2, 3–46, 188, DOI 10.1007/BF02113290; English transl., Math. Notes **55** (1994), no. 1-2, 109–133. MR1275316

[2] V. S. Anashin, *Uniformly distributed sequences of p-adic integers* (Russian, with Russian summary), Diskret. Mat. **14** (2002), no. 4, 3–64; English transl., Discrete Math. Appl. **12** (2002), no. 6, 527–590. MR1964120

[3] V. Anashin, Ergodic Transformations in the Space of p-adic Integers, in: p-adic Mathematical Physics. 2-nd Int. Conference (Belgrade, Serbia and Montenegro, 21 September 2005), AIP Conference Proceedings. 826 (2006) 3–24.

[4] Vladimir Anashin and Andrei Khrennikov, *Applied algebraic dynamics*, De Gruyter Expositions in Mathematics, vol. 49, Walter de Gruyter & Co., Berlin, 2009. MR2533085

[5] V. S. Anashin, A. Yu. Khrennikov, and E. I. Yurova, *Characterization of ergodic p-adic dynamical systems in terms of the van der Put basis* (Russian), Dokl. Akad. Nauk **438** (2011), no. 2, 151–153, DOI 10.1134/S1064562411030100; English transl., Dokl. Math. **83** (2011), no. 3, 306–308. MR2857398

[6] Vladimir Anashin, Andrei Khrennikov, and Ekaterina Yurova, *T-functions revisited: new criteria for bijectivity/transitivity*, Des. Codes Cryptogr. **71** (2014), no. 3, 383–407, DOI 10.1007/s10623-012-9741-z. MR3189123

[7] Andrei Khrennikov and Ekaterina Yurova, *Criteria of measure-preserving for p-adic dynamical systems in terms of the van der Put basis*, J. Number Theory **133** (2013), no. 2, 484–491, DOI 10.1016/j.jnt.2012.08.013. MR2994369

[8] Kurt Mahler, *p-adic numbers and their functions*, 2nd ed., Cambridge Tracts in Mathematics, vol. 76, Cambridge University Press, Cambridge-New York, 1981. MR644483

[9] M. van der Put, Algèbres de fonctions continues p-adiques, Universiteit Utrecht, 1967.

[10] W. H. Schikhof, *Ultrametric calculus. An introduction to p-adic analysis*, Cambridge Studies in Advanced Mathematics, vol. 4, Cambridge University Press, Cambridge, 1984. MR791759

[11] Svetlana Katok, *p-adic analysis compared with real*, Student Mathematical Library, vol. 37, American Mathematical Society, Providence, RI; Mathematics Advanced Study Semesters, University Park, PA, 2007. MR2298943

[12] Neal Koblitz, *p-adic numbers, p-adic analysis, and zeta-functions*, Graduate Texts in Mathematics, vol. 58, Springer-Verlag, New York-Heidelberg, 1977. MR0466081

[13] Andrei Khrennikov and Ekaterina Yurova, *Criteria of ergodicity for p-adic dynamical systems in terms of coordinate functions*, Chaos Solitons Fractals **60** (2014), 11–30, DOI 10.1016/j.chaos.2014.01.001. MR3168575

[14] Ekaterina Yurova Axelsson and Andrei Khrennikov, *Generalization of Hensel's lemma: finding the roots of p-adic Lipschitz functions*, J. Number Theory **158** (2016), 217–233, DOI 10.1016/j.jnt.2015.06.004. MR3393548

Linnaeus University, Department of Mathematics, Vejdes Plats 6, 352 52, Växjö, Sweden

Email address: `Ekaterina.Yurova@lnu.se`